AF333966

AN INTRODUCTION TO DIGITAL AND ANALOG INTEGRATED CIRCUITS AND APPLICATIONS

AN INTRODUCTION TO DIGITAL AND ANALOG INTEGRATED CIRCUITS AND APPLICATIONS

SANJIT K. MITRA

University of California at Santa Barbara

HARPER & ROW, PUBLISHERS, New York

Cambridge, Philadelphia, San Francisco,
London, Mexico City, São Paulo, Sydney

1817

Sponsoring Editor: Charlie Dresser
Project Editor: Céline Keating
Designer: Robert Sugar
Production Manager: Marion A. Palen
Compositor: Composition House Limited
Printer and Binder: The Maple Press Company
Art Studio: J & R Technical Services Inc.

An Introduction to Digital and Analog Integrated Circuits and Applications

Copyright © 1980 by Sanjit K. Mitra

Library of Congress Cataloging in Publication Data

Mitra, Sanjit Kumar.
   An introduction to digital and analog integrated circuits and applications.

   Includes bibliographies and index
   1. Integrated circuits.   I. Title.
TK7874.M532   621.381'73   79-19936
ISBN 0-700-22521-8

To
Sai Baba

CONTENTS

3 Combinational digital circuits   65

4   Sequential digital circuits   153

5 Amplifier fundamentals   225

6 Operational amplifiers   269

7 Nonlinear function modules   341

PREFACE

Research and development efforts in electronic circuit and system design in recent years have been influenced primarily by the progress and breakthroughs in analog and digital integrated circuit (IC) technology. For design purposes, the engineer now has at his disposal a large variety of IC modules, available as off-the-shelf items, equaling and often excelling in performance and usefulness their discrete counterparts as well as being cheaper, smaller, and more reliable. For many applications circuit designers can either find a circuit already available in IC form or can alter the performance of an existing IC module to suit their requirements by connecting a few external components. It is now possible to design fairly complex systems using these modules without knowing too much about their internal construction and device physics.

This text provides a systems oriented introduction to the analog and digital integrated electronic circuits and systems. It discusses the fundamental concepts, principles, and characterization of the basic IC modules along with some useful applications. The internal construction and fabrication procedures of these modules are not discussed; rather the modules are treated as multiterminal building blocks having specified relations between their terminal variables. This approach makes possible a very wide coverage of the various aspects of analog and digital circuits even in a text of this size.

In the discussion of the various IC modules both the ideal and the nonideal characteristics have been treated. A large number of worked-out examples have been included. Although simple enough for the reader to follow, most examples were evolved from real life. Over 300 problems augment the text.

This book can be used as a text in a first course on electronic circuits and systems at the junior level following a sophomore level course on linear circuit analysis. Such a junior level course can then be followed by upper division courses on semiconductor device physics and discrete electronic circuit analysis and design. Alternately, this text can be used for an elective course on IC applications to follow courses on device physics and discrete electronic circuits. The book is entirely self-contained and is also suitable for the practicing engineer who is interested in learning the modern aspects of electronic circuits. The prerequisite for this text is a knowledge of linear circuit theory fundamentals and analysis techniques as taught in sophomore level courses.

The text is divided into eight chapters. Chapter 1 provides a short introduction to the main aspects of integrated electronic circuits and systems, describing the properties, characterization, and representation of signals and those of the electronic circuits and systems for processing these signals. Advantages and limitations of using integrated circuits are discussed along with a brief review of the steps involved in designing an electronic system.

The next three chapters deal with the analysis and design of digital systems. The mathematical foundations of digital signal and system characterization are provided in Chapter 2. Here the fundamental concepts and important results of switching algebra are reviewed. This chapter also includes a discussion of commonly used representations of numbers in digital systems and arithmetic operations involving these numbers. Chapter 3 deals with the memoryless digital circuits, commonly known as combinational circuits. In particular, implementation of switching functions using practical digital integrated circuits discussed and the design of some typical useful circuits and systems is included. Chapter 4, on the other hand, focuses on digital circuits with memory, the so-called sequential circuits. Here also the emphasis is on the implementation of such circuits using commercially available integrated circuits. Designs of several typical useful sequential circuits are included.

The next three chapters are devoted to the design of analog systems. An understanding of these and the last chapters in the book requires a knowledge of elementary linear circuit analysis techniques as taught on the sophomore level in most schools. Chapter 5 reviews the characterization and representation of ideal and practical amplifiers, and outlines several typical applications of these devices. Chapter 6 discusses operational amplifiers, the most widely used analog IC, and the characteristics of ideal and practical operational amplifiers. The implementations of a wide variety of useful linear circuits are also described. The design of nonlinear circuits using several commercially available nonlinear function modules is considered in Chapter 7. In particular, this chapter describes the characterization of analog multipliers, logarithmic amplifiers, sine–cosine modules, analog comparators, and analog switches, and outlines a number of interesting practical applications of these devices.

Finally, Chapter 8 deals with ICs involved in the conversion of analog signals to digital forms, and vice-versa. Four important units in this category are the sample-and-hold, analog-to-digital converter, digital-to-analog converter, and the analog multiplexer. The characterization and simple applications of these four units are discussed.

The book can be used for either a one-quarter or a one-semester course. It has evolved from teaching a junior level one-quarter course at the University of California, Davis. This course is given as a core course to all engineering majors following a required sophomore level course on linear circuit analysis. The book's present form is the result of four years of teaching.

I am indebted to many of my friends former and present colleagues—for their help, advice, and encouragement, without which the writing of this book would have been difficult. These colleagues include Professors V. R. Algazi, K. W. Current, L. Hatfield, J. A. Howard, H. H. Loomis, K. Mondal, E. W. Owen, and M. A. Soderstrand, who reviewed various parts of the manuscript and offered many constructive criticisms. The preliminary version of the manuscript was reviewed by Dr. A. Barna, and Professors C. S. Burrus, J. F. Delansky, S. W. Director, M. O'Flynn, and B. A. Shenoi. The manuscript was later reviewed by Professors J. D. Bargainer, A. J. Brodersen, and R. E. Lee. I acknowledge with gratitude their numerous helpful suggestions and comments. Several of my former and present students assisted me in the development of the course and the course notes. In particular, I thank P. Ananthakrishna and Drs. A. Bhumiratana, D. Bukofzer, S. Chakrabarti, E. Fields, and R. Gnanasekaran for their assistance.

Sanjit K. Mitra

AN INTRODUCTION TO DIGITAL AND ANALOG INTEGRATED CIRCUITS AND APPLICATIONS

CHAPTER 1

Introduction

The age of electronics began in the early twentieth century and is still young compared to many other developments. But one must agree that electronics probably has influenced our day-to-day life more than any other single development. Without it, the modern world would stand still. It is involved directly or indirectly in every facet of our daily activities, whether at work or at recreation. It has found widespread applications in education, health care, business, transportation, communication, entertainment, space exploration, defense, and innumerable other places. "Electronics" has become a household word as have many items that belong to its world like "vacuum tube," "transistor," "computer," and the like.

Unlike many other fields, the field of electronics has been experiencing rapid and phenomenal growth during the last two decades. In the early period (1900–1950), the main components of electronics were the vacuum tubes. Then came the revolutionary new components—semiconductor devices like transistors, which radically influenced the design of systems during the period 1950–1960. Probably the most important contribution in recent years has been the development of the integrated circuit technology and now we are in the age of integrated circuits. The integrated circuit (often abbreviated as IC) has made dramatic inroads into the design of electronic circuits and systems. This does not mean vacuum tubes and transistors have become obsolete; these components are still being used because of either the nonavailability of suitable ICs for certain specific purposes or because of practical limitations of present-day ICs. However, continuing breakthroughs in the IC technology coupled with increasing new applications indicate that in the foreseeable future ICs will constitute a major part of all electronic circuits and systems design.

This text is intended to provide an introduction to electronic integrated circuits and systems. We plan to concentrate on the tools and techniques necessary to design such systems, since a large variety of ICs are readily available as off-the-shelf items and it is possible to design useful practical circuits and systems using these ICs as building blocks. Details on the internal construction and fabrication procedures of ICs has not been included in this text; however, the ICs are treated as multiterminal building blocks having specified relations between their terminal variables.

1-1   Signals

The main purpose of electronic circuits is to process electrical signals whose amplitudes or variations with time contain useful *information*. By processing we mean the circuit accepts the signal at one point (called the *input*), performs certain operations on this signal, and then delivers it to another point (called the *output*). The characteristic of the output signal in some sense is more desirable to the user than that of the input signal.

Electrical signals are voltages and currents that vary with time. These signals can be one of the two basic types: *continuous-time* and *discrete-time* signals. The former type of signal, usually called an *analog* signal, is a continuous function of time; that is, it is defined for every instant of time. An analog signal is usually denoted with a lower case letter with the time dependence explicitly shown, for example, $e(t)$. If the analog signal is a constant and independent of time, it is denoted by an upper case letter, for example, E. A typical analog signal is depicted in Fig. 1-1(a). Another property of an analog signal in most cases is that the level of the signal waveform at any instant of time can take any real value (usually within a prescribed range). For example, consider a purely sinusoidal analog voltage $e(t)$ of frequency 100 Hz and amplitude 5 V:

$$e(t) = 5 \sin(200\pi t) \tag{1-1}$$

Note that $e(t)$ can assume any value between $+5$ and -5 V, depending on the value of time t. Amplitudes of $e(t)$ at various values of t are shown in Table 1-1.

The discrete-time signal, on the other hand, is defined only at prescribed discrete instants of time. An example of a discrete-time signal is shown in Fig. 1-1(b) which is seen to be defined at time instants labeled t_a, t_b, t_c, and so on, but not defined at any time between these specific instants. A discrete-time signal is thus a sequence of numbers and is usually denoted by a lower case letter inside brackets with the discrete-time dependence explicitly shown, for example, $\{x(t_n)\}$, where $x(t_n)$ is the value assumed by the discrete-time signal at the time instant t_n. In the general case, the amplitude of a discrete-time signal can take any real value within a prescribed range. In most applications, for a discrete-time signal, the time instants defining the signal are equally spaced.

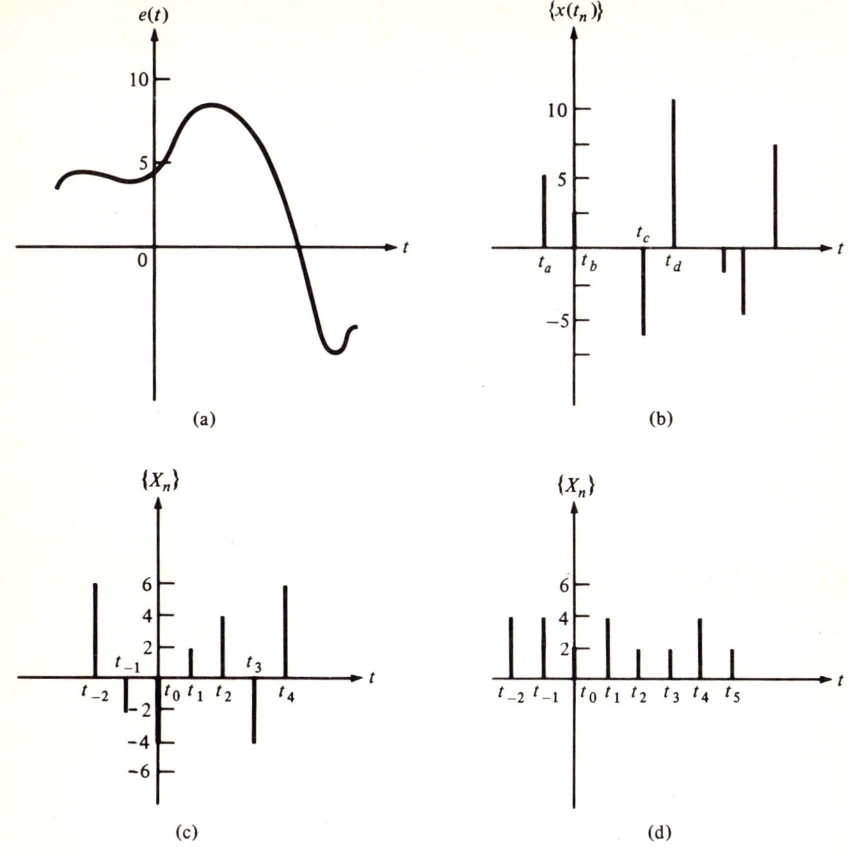

Figure 1-1   Typical electrical signals: (a) analog, (b) discrete-time, (c) five-valued digital, and (d) two-valued digital signals.

The discrete-time signal example of Fig. 1-1(b) has been shown as a train of spikes of negligible width. Such a signal may be generated by sampling a continuous-time signal by means of an ideal switch which closes instantaneously at the time instants $\{t_n\}$ and remains open at all other times. However, in practice, a discrete-time signal is a train of pulses of narrow width as may be obtained by sampling with a nonideal switch which closes for a finite time period.

A commonly found form of the discrete-time signal is the *digital* signal, the amplitude of which is restricted to one of a fixed number of possible "values."

TABLE 1-1
Variation of Amplitudes of an Analog Signal

t (msec)	0	1	2	2.5	6.3
$e(t)$ (V)	0	2.93893	4.75528	5.0	-3.644843

An example of such a signal is shown in Fig. 1-1(c). A more widely used version of the digital signal is depicted in Fig. 1-1(d) in which the amplitude can take one of the two possible values. In this text we consider only this type of two-valued digital signals, also known as *binary* signals. Note that as in the general case of discrete-time signals, a practical digital signal is again a series of pulses of finite width. Other types of digital signals use dc levels or modulated sine waves with the former representation being most commonly used in digital ICs.

A digital signal is usually denoted by an upper case letter inside brackets with a subscript showing the discrete-time dependence, for example, $\{X_n\}$. The two allowable values of a digital signal are usually labeled HIGH and LOW, with the former being used to indicate the more positive level. Frequently, these symbols are replaced by their first letters H and L, or by the symbols 1 and 0, respectively. Thus, an equivalent representation of the digital signal of Fig. 1-1(d) is as shown in Table 1-2, where 4 is HIGH or 1, and 2 is LOW or 0. Alternately, it can be represented as a sequence of 1's and 0's. An individual sample of such a digital signal is called a *bit*.

TABLE 1-2
A Tabular Representation of a Digital Signal $\{X_n\}$

t_n	t_1	t_2	t_3	t_4	t_5	$\cdots$
X_n	1	0	0	1	0	$\cdots$

It should be noted that in practice, actual high and low values of a digital signal are allowed a wide variation in prescribed ranges. Thus, any signal value above a certain value is identified as HIGH and any signal below a certain other value is identified as LOW provided there is no overlap between these two ranges.

1-2   Role of Electronic Circuits

An electronic circuit is an interconnection of components and devices. It has a set of accessible input terminals to which signals to be processed are fed and a set of accessible output terminals from which processed signals are extracted. An interconnection of electronic circuits and components constitute an electronic system. Based on the forms of signals at various points in the circuit, an electronic circuit can be roughly classified into three broad groups: analog, digital, and hybrid. As the name implies, in an analog circuit, all pertinent signals are analog in nature. Likewise, in a digital circuit, the signals are digital in form. Finally, in a hybrid circuit, certain signals are analog and the others are digital in nature. The same type of classification is also applied to electronic systems.

We mentioned earlier that the role of electronic circuits is to operate on signals to develop signals with more desirable properties. Let us now examine some typical signal-processing operations performed by electronic circuits.

Examples of operations done by analog circuits are amplification, attenuation, addition, subtraction, multiplication, division, filtering, modulation, demodulation, and so on. Of all these operations, *amplification* is probably the most widely used operation in analog system design. In many applications, electrical signals are derived from nonelectrical sources by means of devices called *transducers*. An example of a transducer is the microphone which generates an electrical signal by converting the sound pressure on its diaphragm. Invariably, such electrical signals are always very weak. Similarly, electrical signals transmitted over cables are attenuated by copper and dielectric losses in the cable. Weakening of signals also occurs in numerous other situations. In all of these cases, the weak signal needs to be strengthened or *amplified* before any other type of signal-processing operation can be performed on it. The amplification operation is executed by an electronic circuit called an *amplifier*. Mathematically, if $e(t)$ is the input signal to an amplifier, the output is the signal $K \cdot e(t)$, where $|K|$ is greater than 1 and is known as the *gain* of the amplifier. The opposite of amplification is *attenuation* which is performed by an *attenuator* network.

The analog circuit which *adds* two or more signals is usually called a *summing amplifier*. If $e_1(t)$ and $e_2(t)$ are the two input signals to a two-input summing amplifier, the output is given by $K_1 e_1(t) + K_2 e_2(t)$ with K_1 and K_2 being of the same sign. The *subtraction* operation is performed by a *difference amplifier*. Here if $e_1(t)$ and $e_2(t)$ are the two inputs, then the output signal is $K_1 e_1(t) - K_2 e_2(t)$ where K_1 and K_2 are again of the same sign. By interchanging the input connections, $e_1(t)$ can be subtracted from $e_2(t)$.

The *multiplication* operation is accomplished by the analog *multiplier*. It receives two analog signals $e_1(t)$ and $e_2(t)$ at its two input terminals and develops an output signal that is proportional to the product $e_1(t) e_2(t)$ at its output. Likewise, the *division* of two analog signals $e_1(t)$ and $e_2(t)$ to produce a signal proportional to $e_1(t)/e_2(t)$ is performed by the *divider* circuit. These operations are increasingly being used by designers in developing complex systems.

Another frequently used operation is *filtering* of signals composed of frequency components in a wide range. The corresponding circuit is known as a *filter*. It is essentially a frequency selective network passing frequency components in a certain range (*passband*) while stopping altogether frequency components in some other range (*stopband*). Depending on the locations of the passband and stopband, various types of filters can be defined. If the passband is from dc to f_c Hz, then it is called a *low-pass filter*. Thus, if the input to the low-pass filter is $e(t) = A \sin(2\pi f_1 t) + B \sin(2\pi f_2 t)$ where $f_1 < f_c < f_2$, then the output of the low-pass filter is $A \sin(2\pi f_1 t)$. On the other hand, if it is a *high-pass filter* with passband from f_c Hz to infinity, then the same input will lead to a signal $B \sin(2\pi f_2 t)$ appearing at the filter output.

The *modulation* and *demodulation* processes play key roles in almost all wireless transmission of signals such as radio and television. The two commonly used modulation techniques are the *amplitude modulation* (AM) and *frequency modulation* (FM). In both of these cases, the information-carrying low-frequency signal to be transmitted is used to modify (or *modulate*) a very high frequency sinusoidal *carrier* waveform. In the case of amplitude modulation, the low-frequency signal is made to vary the amplitude of the carrier, whereas, in the case of frequency modulation, it is employed to vary the frequency of the carrier. If $v_m(t)$ denotes the modulating low-frequency signal and $v_c(t) = E \cos(\omega_c t + \phi)$ denotes the carrier, then the result of the amplitude modulation process is to develop a modulated carrier of the form $[E + \alpha \cdot v_m(t)]\cos(\omega_c t + \phi)$. On the other hand, the modulated carrier obtained via frequency modulation is of the form $E \cos\{\omega_c t + \gamma \int v_m(t)dt\}$. Figure 1-2 illustrates the above two types of modulation. The process of recovering the original low-frequency signal from the modulated carrier is known as *demodulation*. The corresponding electronic circuits are called the *modulator* and *demodulator*.

In the case of digital signals, typical operations performed by digital circuits are logical operations, arithmetic operations, shifting, comparison, counting, and so on. Some of these operations are unique to the digital domain. The others have equivalent operations in the analog domain, but are implemented

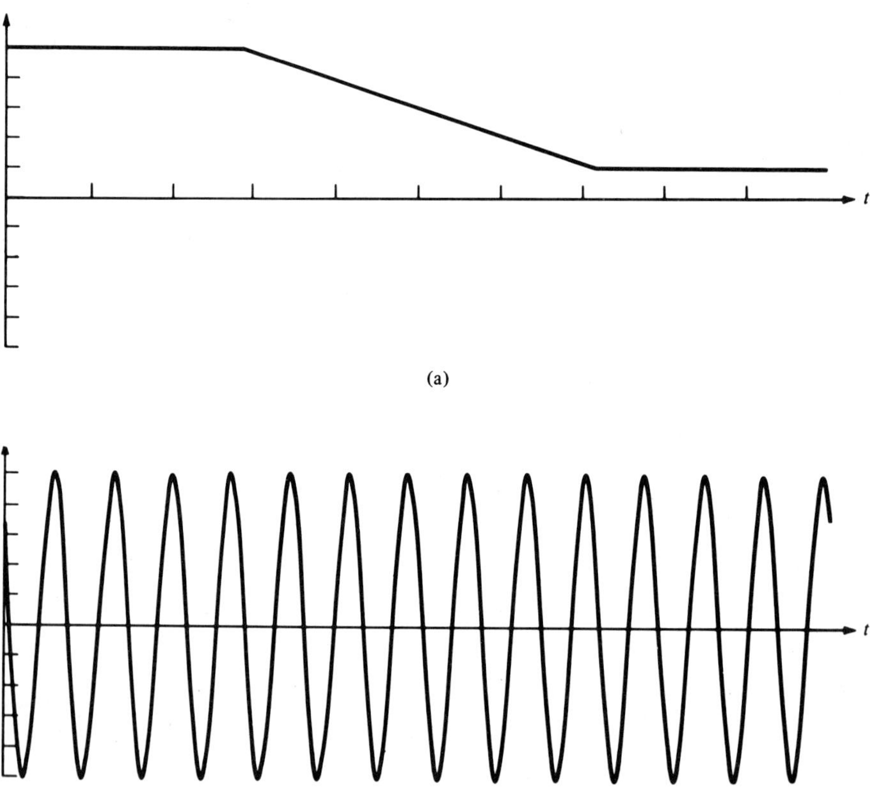

(a)

(b)

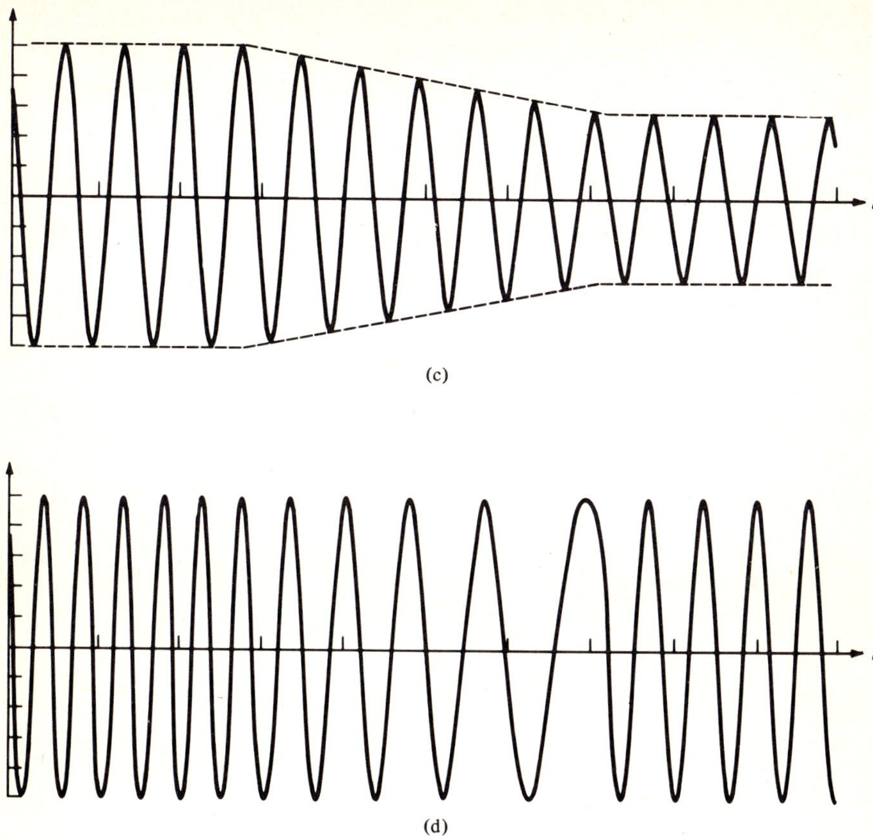

Figure 1-2 Illustration of modulation: (a) modulating signal, (b) carrier, (c) amplitude modulation, and (d) frequency modulation.

entirely differently. The logical operations are executed by circuits known as *gates*. There are quite a variety of logical operations; the simplest of these is the *negation* which is performed by a *NOT* gate. As the name implies, the NOT gate produces an output digital signal whose value is exactly the opposite (complement) of that of the input digital signal. That is, if the input variable is HIGH, the output is LOW, and vice versa. An example of a logic operation involving two or more digital signals is the *OR* operation. The output of an OR gate is LOW if and only if all the input signals are at LOW level; otherwise it is HIGH.

Arithmetic operations performed using digital circuits are the *addition, subtraction, multiplication,* and *division*. The corresponding circuits are the *adder, subtractor, multiplier,* and *divider*. The arithmetic operations are performed on numbers represented in digital form using a set of bits. An example of such a representation is shown in Table 1-3.

TABLE 1-3
A Digital Representation of Decimal Integers

Decimal number	0	1	2	3	4	5	6	7
Digital equivalent	000	001	010	011	100	101	110	111

Another useful digital operation is performed by a circuit called a *digital comparator* which compares two digital numbers and develops an output to indicate whether or not the two numbers are equal. It may also have additional output terminals to indicate which number is " greater " than the other.

There are only a few operations involving both analog and digital signals. The simplest of these is performed by the *analog comparator*. It compares the difference of two analog signals and develops a digital output which is HIGH if the difference is positive and is LOW if it is negative. The conversion of an arbitrary analog signal to a digital " equivalent " is implemented by the *analog-to-digital converter*. Likewise, the conversion of an arbitrary digital number to its analog equivalent is performed by the *digital-to-analog converter*. The analog comparator plays a central role in the design of these converters.

So far we have touched upon a few of the numerous signal-processing operations performed by analog, digital, and hybrid circuits. The remaining chapters of this text provide further details on these and many other types of operations along with descriptions of specific circuit implementations of these operations using ICs. These chapters also include some simple system design examples using these basic circuits.

1-3 General Characteristics of Electronic Circuits

An electronic system is designed by interconnecting various electronic circuits. Often it may be necessary to connect additional components and devices to modify the operation and properties of some commercially available electronic circuit modules.

Almost all electronic circuits, either available in modular form or designed specifically by a circuit designer, have certain properties in common. To understand these properties, let us consider a circuit which develops an output signal as a function of two input signals. Whether the signal is a voltage or a current, it is either applied or extracted across a pair of terminals called a *port*. Hence the circuit under consideration is a 3-port network. In most cases, one of the terminals of each port is a common terminal usually connected to the system ground as shown in Fig. 1-3. In addition to the input and output terminals, most electronic circuits require external dc supplies or batteries for biasing and supplying power; these are usually not shown in the circuit diagram.

A major feature of such circuits is that they are unidirectional, that is, the input terminals are " isolated " from the output terminals. Or, in other words,

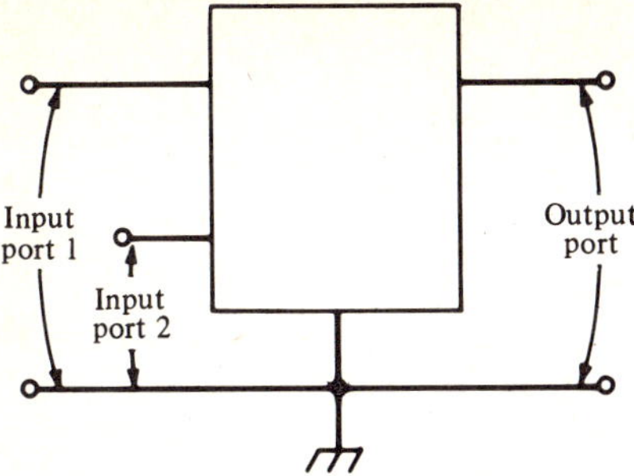

Figure 1-3   A 3-port electronic circuit.

electrical conditions at the input are independent of those at the output. This property in particular allows the output of one electronic circuit to act as an input to one or more circuits and enables the design of a complex system by means of an interconnection of simpler, noninteracting circuits.

The input and output impedances of an electronic circuit are, in practice, finite which may cause difficulty in the operation of a system if not taken into account in the design phase. To understand the effect of these impedances, consider a NOT gate, a 2-port electronic circuit, whose input and output signal variables are voltage waveforms. Let us assume that for this gate, any voltage in the range 4 V and above is considered to be in the HIGH state and a voltage level below 2.5 V is considered to be in the LOW state. The input and output impedances of this gate are 10 and 1 kΩ, respectively. A model of this gate showing these impedances is shown in Fig. 1-4. The operation of this gate with output open-circuited is assumed to be as follows:

$$E_o = 2 \text{ V} \qquad \text{for} \qquad E_i > 4 \text{ V}$$
$$E_o = 5 \text{ V} \qquad \text{for} \qquad E_i < 2.5 \text{ V}$$

$$(1\text{-}2)$$

This implies that the gate acts as an inverter with the output in open-circuit condition. If this gate is driving one or more other logic circuits, their input

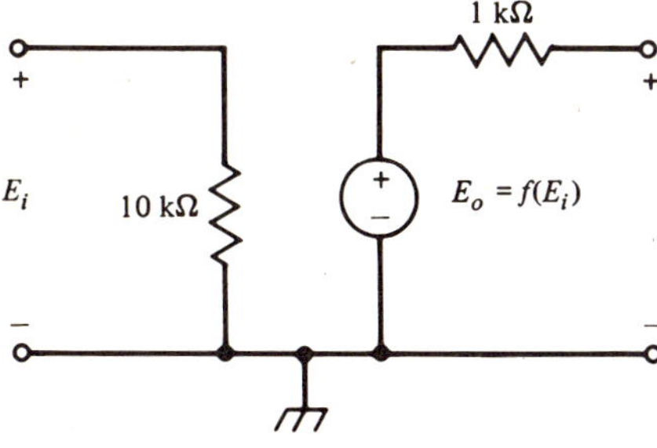

Figure 1-4   The model of a nonideal NOT gate.

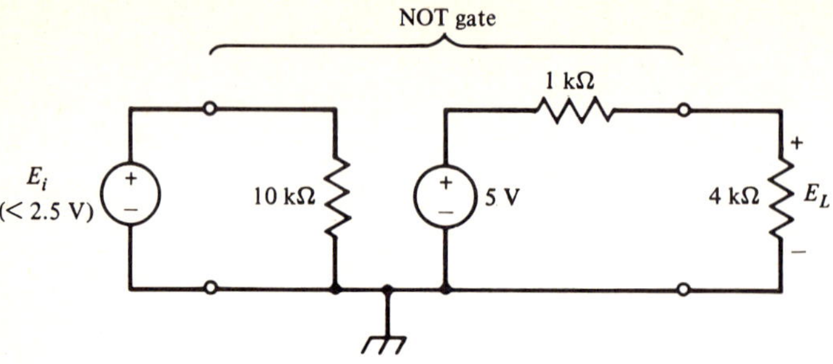

Figure 1-5 The NOT gate with a load at its output.

impedances will "load" the output of this gate and in practice the output voltage will be lower than those shown in Eq. (1-2) for both input conditions. For example, if the effective load resistance connected at the output is 4 kΩ (Fig. 1-5), the voltage E_L produced at the output with the input in the LOW state will be given as

$$E_L = \frac{4000}{1000 + 4000} \times 5 = 4 \text{ V} \tag{1-3}$$

If the load resistance is less than 4 kΩ, then the output E_L will be less than 4 V and will no longer be identified as HIGH. This type of situation may occur

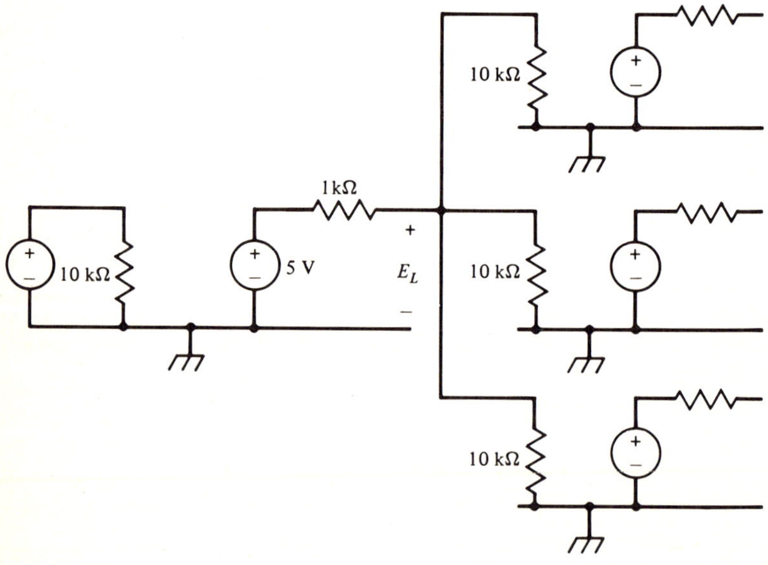

Figure 1-6 The NOT gate driving three other logic circuits.

if the NOT gate is driving three other similar circuits with each having an input impedance of 10 kΩ as shown in Fig. 1-6.

In addition to the above limitations, in general, the circuit design imposes constraints on the values the currents and voltages at the input and output are allowed to take. If, in actual operation of the circuit, any one of the pertinent variables exceed the allowable range, the performance of the circuit may no longer be acceptable. This may cause improper operation of the whole system.

1-4 Examples of Electronic Systems

At this point it would be instructive to examine more or less qualitatively several typical electronic systems. This will provide us with some ideas of how different types of electronic circuits are connected to form a complex system.

An Analog Electronic System

Most naturally occurring signals such as speech, pressure, and so on, are analog in form. As a result, a major portion of all electronic systems is still analog systems. Some easily recognizable analog electronic systems are the AM/FM radio receiver, the tape recording/reproducing unit, and television. We consider another system—a typical electronic music synthesis system.[1]

The process of combining electronically several individually produced parts of a musical score to form the entire composition is usually known as electronic music synthesis. It involves the modification and/or combination of natural and electronically generated sound sources and is used currently in the production of almost every form of recorded music.

An example of electronic music synthesis is in the generation of 2-channel stereophonic magnetic tape from an 8-channel magnetic tape in each channel of which the sound of individual instruments or performers has been recorded. A block diagram representation of such a system is shown in Fig. 1-7.

When the 8-channel tape is moved, it causes a voltage to be developed in the reproducing head in response to the flux density variation at the gap of the head. This voltage is fed first to an amplifier whose gain can be adjusted externally. It is followed by a *tone modifier* unit—the key part of the music synthesis system. This unit actually consists of several circuits as shown in Fig. 1-8.

The first circuit in the tone modifier is the *delayor* whose purpose is to delay the input signal by a prescribed amount. By controlling the amount of delay and the gain of the preceding amplifier, it is possible to "place" any one of the original eight sound sources at any "position" between the right or left loudspeaker. Consider, for example, the sound source of channel 8. If the gain of the amplifier in channel 8-R is set to zero, then the complete signal generated will be fed to the left loudspeaker and it would appear to the listener that the sound source is at the position of the left loudspeaker. Similarly, by setting the gain of the amplifier in channel 8-L to zero, the same sound source can be made to appear from the right loudspeaker. By feeding the same audio signal to both speakers, the sound source can be made to appear at the middle of the two

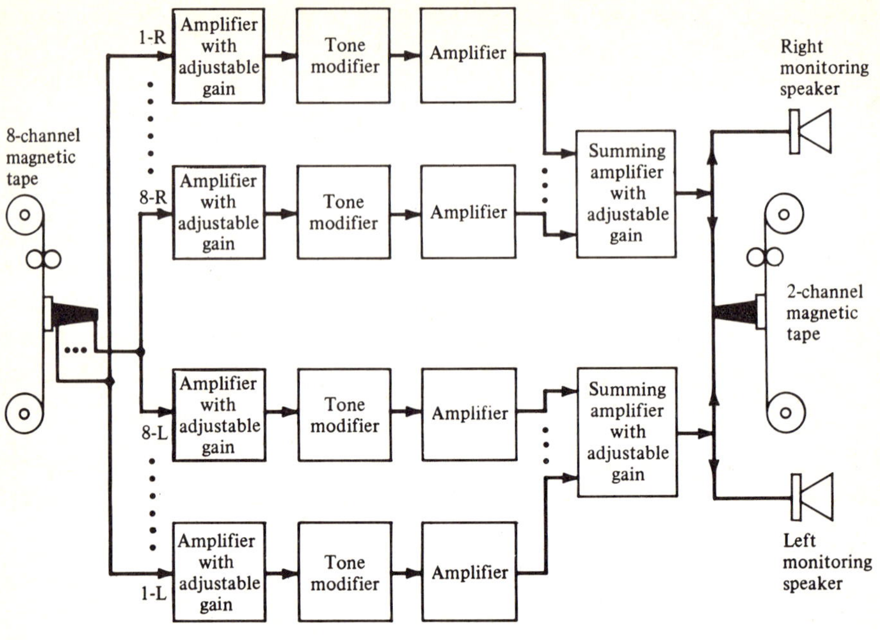

Figure 1-7   An electronic music synthesis system.

loudspeakers. If a delay is introduced in the left channel path with the audio signals fed to both loudspeakers kept at the same amplitude, then the sound source can be made to appear anywhere between the center and the right speaker. By introducing a delay only in the right channel path, the sound source position can be reversed.

The *frequency-response modifier* unit is essentially a frequency selective network and is used to modify the waveform of the audio signal by emphasizing or deemphasizing selected frequencies. The *timbre modifier* circuit performs a somewhat similar operation by allowing the introduction of new frequency components. The purpose of the *vibrato* or *tremolo generator* unit is to modulate

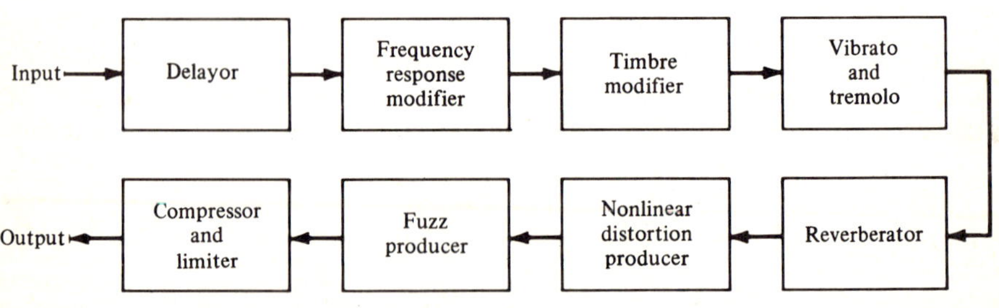

Figure 1-8   The tone modifier unit.

the audio signal by amplitude or frequency modulation. In the next unit, *reverberation* effect is introduced artificially into the original signal which is initially recorded without any reverberation. Often the subjective response of the listener can be made more attractive by providing an appropriate amount of *distortion* to the audio waveform which is introduced by the following circuit. The *fuzz producer* unit adds to the audio signal certain amounts of random noise and electronically generated high harmonics which in some cases may make the sound more desirable for listening.

The *compressor* circuit is essentially an amplifier unit with a nonlinear gain characteristic such that the gain automatically decreases in a gradual manner whenever the amplitude of the input signal level exceeds a preselected value, and its use provides a better discrimination of the signal against the unwarranted noises generated in the system. The *limiter* is used to protect against sudden overloads by ensuring that the output level does not increase beyond a certain value.

Additional details on this system and other electronic systems used in the recording of music are found in Reference 1.

A Digital Electronic System

Digital systems are finding increasing applications in many diverse areas due to the phenomenal growth in digital IC technology during the last decade. One commonly used digital system is the digital computer. The digital computer is literally becoming an integral part of our life. It is being used for making out paychecks and airline reservations, keeping inventories, guiding the airplanes and satellites, and more importantly, solving complex arithmetic problems which would otherwise take hundreds of man-years to solve. Some other easily recognizable examples are pocket calculators, digital clocks, and watches introduced barely a few years ago. The control circuitry in many modern-day washing machines, microwave ovens, toys, and so on, are again all digital circuits.

We consider here in some detail the operation of an electronic digital watch.[2] A block diagram representation of a typical electronic watch is shown in Fig. 1-9. The quartz crystal-controlled oscillator circuit generates a pulse train whose frequency is in the kilohertz range. This pulse train is fed into a series of counters. The first counter circuit develops one pulse per second (pps). If the oscillator produces f_o pulses per second (that is, the oscillator frequency is f_o Hz), then the counter circuit produces one output pulse for every f_o input pulses. Since the number of output pulses in a counter is less than the number of input pulses, it is also known as a *divide-by-f_o counter*. This counter is followed by a divide-by-60 counter whose output is now one pulse per minute. The output is then fed into another divide-by-60 counter, developing one pulse every hour at its output. Finally, the one-pulse-per-hour pulse train is counted by a divide-by-24 counter. If the last three counters are reset to zero at midnight, the contents of these counters at any instant provide an exact indication of the time. Appropriate connection of a *decoder* network to each of these counters and a display

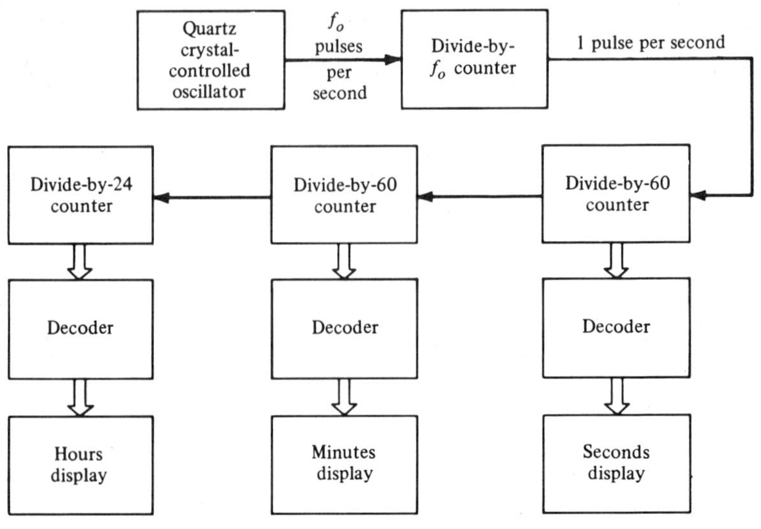

Figure 1-9 Block-diagram representation of an electronic watch.

device is used to provide visual read-out. A common type of display is a 7-segment light-emitting-diode (LED) unit providing a decimal read-out. This is a 24-hr watch. Thus 1 p.m. is displayed as 13 hr, and so on. For a calendar watch with date and day-of-the-week display, two additional counters are needed at the output of the divide-by-24 counter.

Most of the above circuitry (without the quartz crystal and the LED displays) is currently produced as a single IC by some manufacturers. In some electronic watches, the decoder network is replaced by a motor-driving logic circuit which generates appropriate pulses to drive the hands-driving motor to provide a more conventional mechanical display. Additional details on electronic watches are found in References 2–5.

A Hybrid Electronic System

Both analog and digital signals are involved in a hybrid system. Invariably as a result the system contains some type of converter circuits to convert one form of electrical signal to the other. A simple example of such a system is the digital voltmeter which is used to provide a digital read-out of an analog dc voltage. Essentially, it is an analog-to-digital converter with additional decoder circuitry at the output connecting a suitable display device. Other examples of the hybrid system are the Pulse-Code-Modulation (PCM) telephone communication system[6] and the music distribution system in commercial airplanes.[7]

The basic idea behind the last two examples is the conversion of an analog signal to a digital form, the transmission of the digital signal over wires, and then at the receiving end, the reconstruction of the original analog waveform

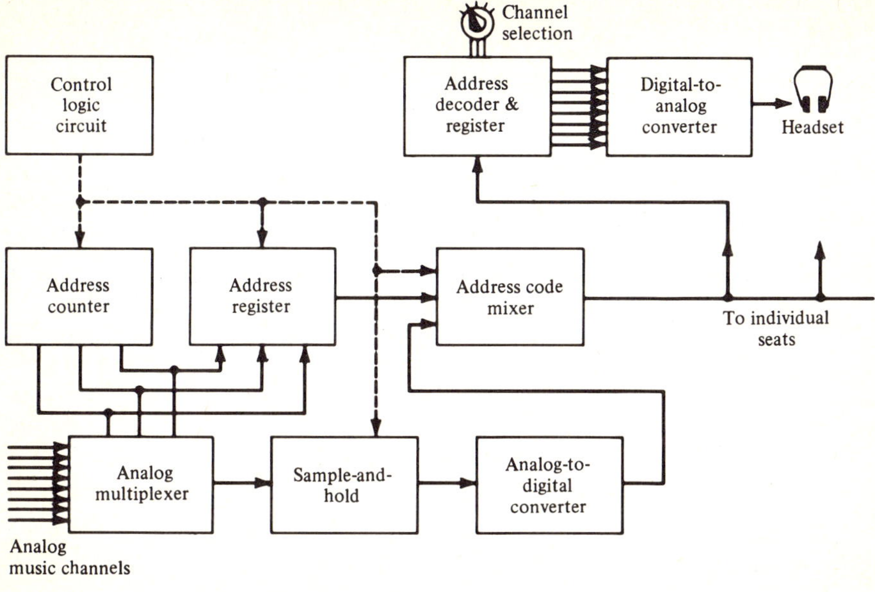

Figure 1-10   Aircraft music distribution system.

by another conversion process. We illustrate here the operation of the aircraft
music distribution system shown in block diagram form in Fig. 1-10.

The eight analog (music) input signals are first sequentially combined onto
a single channel by the *analog multiplexer*. This electronic circuit essentially
acts like a rotary switch connecting the output terminal to one of the input
terminals at a time with the input terminal selected by a digital address counter.
The output of the multiplexer is fed into a sample-and-hold circuit. It samples
the waveform by a periodically operated switch and holds the output of the
switch at a constant level just long enough to allow the analog-to-digital
converter to code each sample by a digital equivalent represented by 8 bits. To
identify the input channel associated with the sample being coded, a 3-bit
address code is added to the output of the analog-to-digital converter. The
digital signal appearing at the output of the code mixer is thus a binary bit
stream of 11-bit *words* which is wired to every seat in the plane. At each seat,
an address decoder selects only those words whose address code correspond
to the address selected by the listener by means of a mechanical channel selection switch. These words (without the address code) are then fed into a digital-
to-analog converter whose output after appropriate smoothing forms the
analog music and is fed finally into the headset. The digital conversion of multiplexed analog signals allows a single wire transmission and processing by
digital ICs which conserves wiring and provides a dramatic decrease in total
weight of electronic circuitry.

1-5 Types of Electronic Circuits[8-10]

Electronic circuits can be broadly classified into two groups: discrete and integrated circuits. *A discrete circuit* contains only individually manufactured, packaged, tested, and specified circuit elements such as transistors, resistors, capacitors, and so on, which have been interconnected by wires or plated conductors. Each interconnection here is accessible to the user if necessary. On the other hand, an *integrated circuit* is an interconnection of circuit elements inseparably associated on or within a continuous supporting material known as *substrate.* Here, except for those interconnections that are specifically designed for external connections, remaining interconnections are inaccessible to the user. In many ICs, individual measurements of constituent components are impossible. The IC must be specified and tested in accordance with its electronic function.

The integrated circuit can be further classified into several types. In a *monolithic integrated circuit*, the substrate is a thin block of semiconductor material, usually silicon, in which circuit elements are formed by the diffusion of specific impurities to yield the appropriate electrical characteristics. An insulating oxide layer is grown over the semiconductor surface. After a thin metal layer (usually alumina) is deposited on top of this oxide, it is selectively etched to leave the interconnection pattern (called *metallization*) connecting the circuit elements.

A very large number of identical but separate circuits are simultaneously processed and fabricated on one semiconductor block called a *wafer*, which is approximately about 25 mil thick.* Typically, a circular wafer of 3-in. diameter may contain nearly 1000 individual small circuits. Figure 1-11 shows the photograph of such a wafer. The wafer is cut into small sizes, called *chips*, with each chip containing only one circuit. The size of a typical chip is about 100×100 mil.

There are basically three levels of integration in the case of monolithic ICs, primarily for digital circuits: (i) small-scale integration (SSI), (ii) medium-scale integration (MSI), and (iii) large-scale integration (LSI). The circuit complexity per chip varies from several simple logic circuits containing a few transistors in the SSI case to a complete system with over 10,000 transistors in the LSI case. Figure 1-12 illustrates the evolution of the monolithic IC technology since 1960.

The *thin-film integrated circuit* makes use of an insulating ceramic or glass substrate. The size of the substrate is typically 5×5 in. The circuit is fabricated on top of the substrate using either a photo-etch process or a vacuum-deposition process. In the former case, thin layers of resistive, dielectric, and conductive tantalum-based materials are deposited on the substrate using a sputtering technique. Portions of the overlay, which are not wanted, are selectively removed by a photo-etching process. This results in a network of resistors and capacitors, interconnected by conducting paths.

---

* 1 mil = 0.001 in.

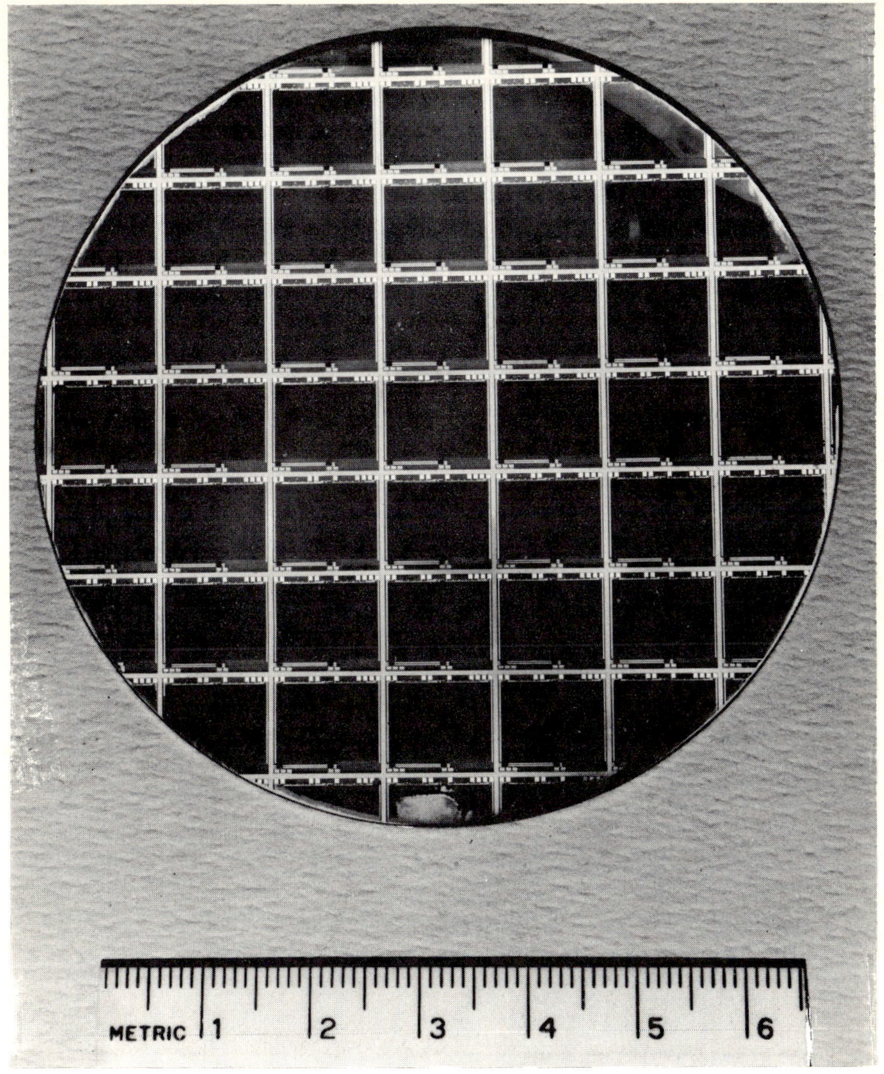

Figure 1-11   A typical IC wafer. (Courtesy Rockwell International.)

In the second approach to the development of thin-film ICs, resistive, dielectric, and conductive materials are first vaporized in a vacuum and then selectively deposited on the substrate through appropriate masks. Insulating layers are later deposited on top followed by additional film covering. As before, this approach also leads to an interconnected R–C network.

In both cases, usually a number of identical circuits (typically of the order of 20) are fabricated on a single substrate at one time. Individual circuits are then obtained by cutting. Active elements, such as transistors and monolithic ICs,

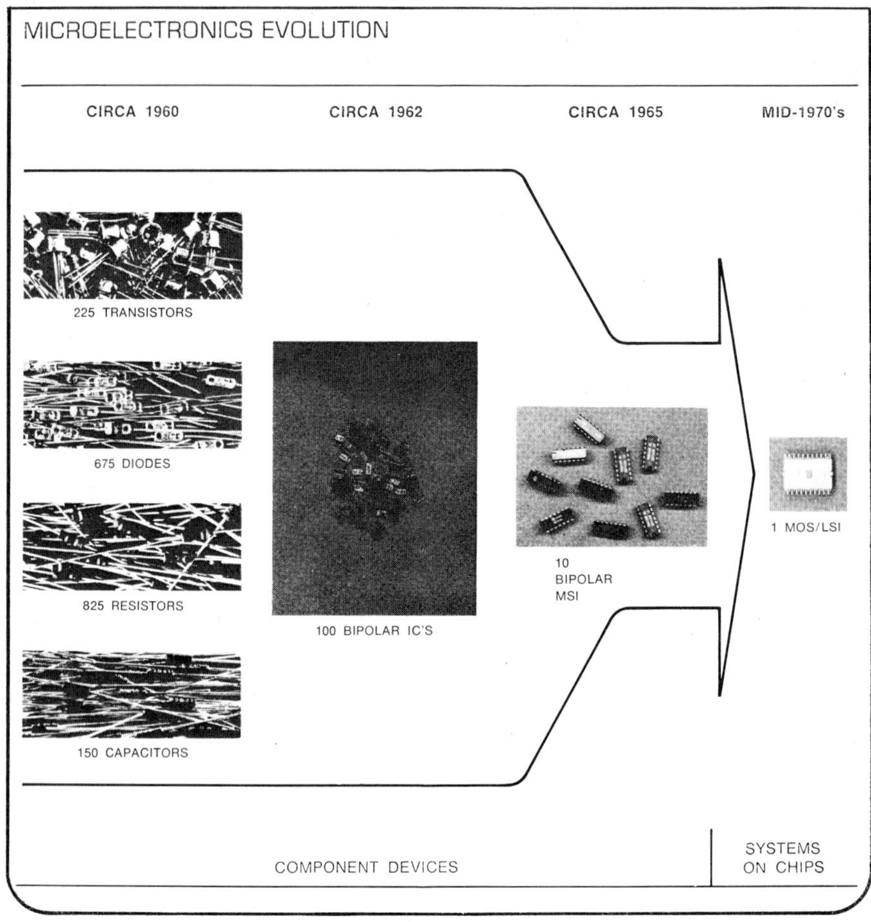

Figure 1-12   The evolution of monolithic IC technology (Reference 10). (Courtesy Rockwell International.)

are attached by bonding them to the thin-film R–C circuit, if necessary. Resistor values in the range of a few ohms to about a megohm and capacitor values from as low as 0.1 pF to about 0.5 μF can be obtained using thin-film techniques. These components can further be trimmed to within 0.1 percent of their nominal design values after fabrication.

The *thick-film integrated circuit* is fabricated using a ceramic printed circuit process. Here the resistor and conductor patterns are "printed" on a ceramic substrate by pressing appropriate materials individually through a screen. The wafer can be as small as 0.625 sq. in. in area. The pattern on the wafer is then dried by placing it in a low-temperature oven. In the next step, the resistor and conductor patterns are "fired" in place on the wafer in a high-temperature oven. Active elements and thin-film capacitors are attached later. Thick-film circuits are cheaper to fabricate than the thin-film circuits, but have higher component tolerances and temperature coefficients.

A number of commercially available electronic circuits are also produced in a *hybrid integrated circuit* form by combining one or more monolithic IC chips with discrete components and/or film-type ICs.

1-6 Packaging of Integrated Circuits

Most commercially available ICs are marketed in standard packages. The packaging process involves the mounting of the IC chip inside the package, the attachment of outside terminals, and the hermetic sealing of the package. Numerous styles of standard housing are used in IC packaging and it is not possible to list all of these here. Instead we describe here some commonly utilized cases.[11]

Some earlier packagings for SSI ICs made use of the metal-can-type cases originally developed for mounting transistors. These are the TO-3, TO-5, and TO-8 cans. Figure 1-13(a) shows one version of the 10-lead TO-5-type can. Its bottom view, shown in Fig. 1-13(b), illustrates the pigtail numbering scheme used to identify the leads. A more common type of package is the ceramic flat-pack (F/P) usually available with 14 leads, as sketched in Fig. 1-13(c). A notch on the package is used to identify the ordering of the terminal numbers [Fig. 1-13(d)]. The most commonly used package is the ceramic or plastic dual-in-line package (DIP). A 14-lead DIP plastic package is shown in Fig.

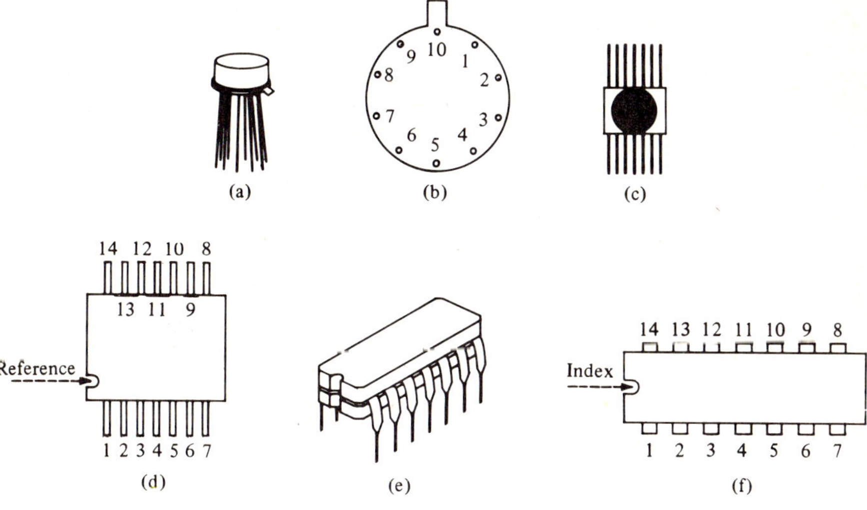

Figure 1-13 Typical IC packages: (a) 10-lead TO-5 can, (b) bottom view of TO-5 can, (c) 14-lead flat-pack, (d) top view of 14-lead flat-pack, (e) 14-lead dual-in-line package, and (f) top view of 14-lead dual-in-line package.

1-13(e). Its top view showing the lead-numbering scheme (identified by an index notch) is sketched in Fig. 1-13(f).

For MSI and LSI circuits again there are a large variety of packages. For digital MSI circuits, the 16-lead DIP is more common. Packaging is a difficult problem for LSI circuits, which typically require more than 40 leads. The DIP here is about 3 in. long and occupies a lot of space. As a result, new types of packages such as chip- and film-carriers are being developed particularly for LSI packaging.[12]

1-7   Integrated Circuits in System Design

We may be curious by now to ask ourselves the question—why is the IC becoming so popular? The main reason for the increasing popularity of the IC is economic; each IC costs less than its equivalent discrete counterpart because a very large number of circuits can be simultaneously processed and produced in a batch, cutting the cost of an individual circuit.

The IC fabrication leads to improved reliability due to a number of factors such as elimination of individual interconnections between components, resistance to vibration and shock, and lower thermal stress. To illustrate the improvement of reliability, consider the digital flip-flop circuit of Fig. 1-14.[9] A fabrication of this circuit in discrete-circuit form requires 20 resistors, 6 capacitors, 6 transistors, and 4 diodes—a total of 36 components. A monolithic IC fabrication of this circuit reduces the parts count by a factor of 36 (from 36 components to 1 IC). The total number of connections in the discrete-circuit version is 234 which reduces to 28 in the monolithic IC version—a reduction by a factor of 8.5. Case and lead seals are also reduced dramatically.

Improved reliability implies a decrease in down time of the electronic systems designed using ICs. The maintenance of these systems is easier. ICs have smaller

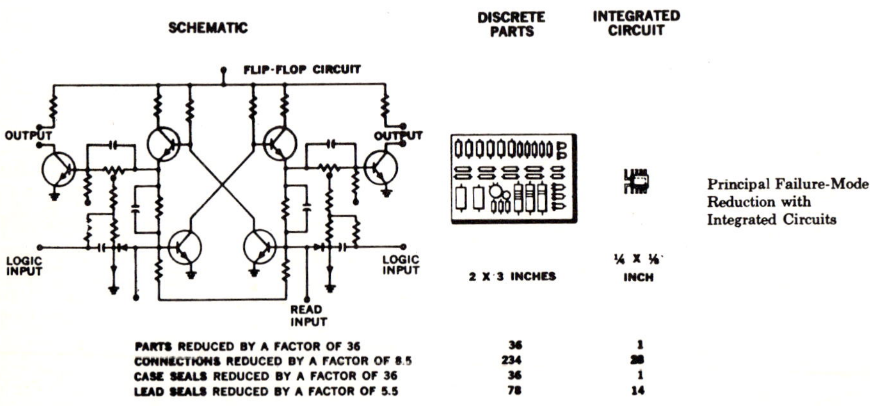

Figure 1-14 A comparison of discrete and integrated circuits (Reference 9). (Courtesy Rockwell International.)

size and weight and have a lower power requirement. All these factors lead to lower cost in the design of the system.

Note, however, for each new circuit the initial development cost to produce it in IC form is very high. Thus the cost saving per circuit can come only if the demand for that circuit is high enough so that a large number of these can be manufactured. Consequently, for most design problems, an engineer, planning to use ICs, will have to choose the circuits from the existing available ICs and often may have to alter the circuit function by connecting additional discrete components and/or ICs.

1-8   The Design Process

The design of an electronic system consists of several steps. In the first step, a circuit diagram of the complete system showing the interconnections of con-stituent functional blocks is developed from the system specifications. The specifications of the individual functional blocks are developed by the system designer so that when all the blocks are interconnected, the complete system meets the overall system specifications. These circuit blocks are initially assumed to be ideal noninteracting devices with precise input–output relationships described by mathematical equations. In many cases, some of the necessary functional blocks are commercially available as off-the-shelf IC modules. In other cases, these blocks may have to be individually designed using available ICs and other components. In the latter situation, circuit diagrams of the functional blocks need to be developed by the circuit designer. In general, the circuit diagrams of a system or its constituent functional blocks do not show the IC power supply connections for simplicity. However, all connections need to be shown in a detailed circuit diagram (schematic) to facilitate the final assembly of the circuit and the system.

In the second step, a mathematical analysis of the individual circuits is carried out using appropriate models of circuit components to ensure that each of the circuits built with practical components will function as designed and will meet the specifications. Invariably, this step is now carried out on a digital computer. If the analysis indicates failures in the circuit's performance, it may have to be redesigned using different configurations and/or components. If this redesign is still unsatisfactory, the specifications of each of the circuits in the system may have to be altered.

In the final step, a prototype of each of the circuits is built in the laboratory and tested individually. If each circuit operates successfully, the complete system is built using these prototypes and checked for its overall performance.

1-9   Summary

Electrical signals are either in analog or digital form (Section 1-1). In the former case, they are continuous functions of time whereas in the latter case, they are discrete functions of time. Moreover, the amplitude of a digital signal can take one of a fixed number (usually two) of values. The purpose of an electronic circuit is to operate on an input signal with certain characteristics. Several typical operations performed by analog, digital, and hybrid electronic circuits are described in Section 1-2. A complex electronic system is designed by interconnecting a number of individually designed and tested electronic circuits performing simple operations. To facilitate the design of the overall system, the constituent electronic circuits should have certain properties which are discussed somewhat qualitatively in Section 1-3. With the aid of block diagram representation, three typical electronic systems are described in Section 1-4 to illustrate how separate electronic circuits are interconnected to form the system. There are several types of integrated circuits (ICs) depending upon the fabrication procedures used and circuit density. These are described in Section 1-5. Some commonly used IC packages are illustrated in Section 1-6. Section 1-7 reviews the basic advantages associated with the use of ICs in system design. Finally, the basic steps in the design of an electronic system are outlined in Section 1-8.

References

1. H. F. Olson, "Electronic Music Synthesis for Recordings," *IEEE Spectrum*, Vol. 8, no. 1, April 1971, pp. 18–30.
2. M. Eleccion, "The Electronic Watch," *IEEE Spectrum*, Vol. 10, no. 4, April 1973, pp. 24–32.
3. G. W. Taylor and I. Lefkowitz, "Improved Electronic Watches," *Proc. IEEE* (Letters), Vol. 61, no. 4, April 1973, pp. 487–489.
4. V. H. Grinich and H. G. Jackson, *Introduction to Integrated Circuits*, McGraw-Hill Book Co., New York, 1975.
5. G. M. Walker, "Choosing Sides in Digital Watch Technology," *Electronics*, June 10, 1976, pp. 91–99.
6. E. F. Gallagher, "The Military Goes Digital," *IEEE Spectrum*, Vol. 14, no. 2, February 1977, pp. 42–45.
7. W. R. Spofford, Jr., "Putting D-A Converters to Work: 10 Examples Show Versatility," *Electronics*, October 26, 1970, pp. 91–97.
8. F.C. Fitchen, *Electronic Integrated Circuits and Systems*, Van Nostrand Reinhold Co., New York, 1970.
9. M. S. Parks, *The Story of Microelectronics—First, Second, and Future Generations*, North American Aviation-Autonetics Division, Anaheim, Calif., 1966.
10. M. S. Parks, *Microelectronics in the 1970's*, Rockwell International Corp., Anaheim, California, 1974.
11. R. P. Turner, *ABC's of Integrated Circuits*, Howard W. Sams & Co., Indianapolis, Indiana, 1971.
12. J. Lyman, "Growing Pin Count is Forcing LSI Package Changes," *Electronics*, March 17, 1977, pp. 81–91.

CHAPTER 2

Digital Systems Fundamentals

As indicated earlier, the principle of operation of digital systems is quite different from that of the analog systems. In addition, signal processing and storing concepts in the digital domain are also not similar to their counterparts in the analog domain. It is thus quite important to understand thoroughly the basic concepts and rules of digital signal processing and it is the purpose of this chapter to provide these fundamentals. To keep the mathematical rigor to a minimum, no attempt has been made to provide the information in a formal manner.

2-1   Switching Variable

Just as in the case of analog electronic systems, voltages and currents are also signal variables of interest in a digital electronic circuit. However, here they are discrete-time functions and are defined at specific time instants. Moreover, their amplitudes are restricted to have certain specific values. In the type of digital circuits used in most applications, the signal amplitudes are allowed two "values" and in this text we shall deal exclusively with such circuits. In practice, the signal amplitude is assigned one "value" if its amplitude is within one range, and is assigned the other "value" if it is in another range. For example, in the commonly used transistor-transistor-logic (TTL) circuits, one range is from 2.4 to 5.5 V, and the other range is from 0 to 0.4 V. In order to determine the "value" of the digital signal in such circuits we have to know in which range the actual value of the signal lies. If the amplitude is in the range 2.4 to 5.5 V, we say the digital signal is HIGH; if it is in the range 0 to

0.4 V, then it is LOW. For proper operation of a digital circuit it is not necessary to know the precise value of its amplitude as long as it is in one of the two allowable ranges. We assume, of course, that the circuits are properly designed so that signals of interest are not outside these two ranges. Note that similar high and low ranges exist for other types of digital circuits available in the market.

A signal variable that can take one of two possible values is called a *switching variable*. The concept of switching variables is the most basic concept in digital systems on which the complete theory of digital systems is based. It is a usual practice to denote a switching variable with a capital letter.

2-2   Logic Assignment

In a digital circuit the input and output variables are switching variables. In general, the output switching variables are functions of the input switching variables. There are basically two types of digital circuits as we shall discuss later. The digital circuit is also called a *switching* or *logic circuit*.

Truth Value

As defined in the previous section, a switching variable can have at any instant one of two possible values. In other words, we can say that the switching variable at specified instants of time can be in either one of two *states*. The value or state of the switching variable at a given instant is its *truth value*.

It is usual practice to label the two possible states of a switching variable as LOGICAL ZERO and LOGICAL ONE or simply as 0 and 1 (not to be confused with the numbers 0 and 1 in decimal number system). This type of labeling simplifies the characterization of digital systems.

Let us now illustrate this labeling procedure in more detail.

Positive and Negative Logic

In the TTL digital circuit the value of the switching variable can be either in the HIGH range or in the LOW range. In the *positive logic* assignment scheme we assign the truth value of 1 if the signal is in the HIGH range and assign the truth value of 0 if it is in the LOW range. In the *negative logic* assignment scheme exactly the opposite procedure is followed. Here the signal in the HIGH range is given the truth value of 0. If it is assigned a truth value of 1, it is in the LOW range. A similar convention is followed for other types of logic circuits. Unless otherwise stated, we shall follow the *positive logic* assignment in this text.

Region of Uncertainty

In all digital circuits the two allowable range of signal values are separated by a region called the *region of uncertainty*. In TTL circuits the region of uncertainty is from 0.4 to 2.4 V. If the actual value of the signal is in this range, then we are unable to assign a specific truth value irrespective of the logic assignment being followed. A signal may take a value in the region of uncertainty due to faulty design, change in circuit parameter, and so on. In this book we shall, for simplicity, assume that a switching variable never takes a value in the region of uncertainty.

2-3   Basic Operations

Switching variables that are functions of several independent switching variables can be generated by making use of a number of algebraic operations and their combinations. We now discuss the three most basic operations. A few other operations will be discussed later.

OR Operation

Two switching variables X and Y can be combined by an " OR " operation to form a new switching variable A as shown symbolically by

$$A = X + Y \tag{2-1}$$

In words, we say that "A is equal to X OR Y." The symbol "$+$" used in Eq. (2-1) indicates the OR operation and should not be confused with the operation of *addition* in the usual sense. The OR operation is defined as follows:

> "A has a truth value of 1 if either X or Y (or both) has a truth value of 1, otherwise A has a truth value of 0."

This basic operation can be best explained in terms of a table explicitly showing the truth values of A for every possible combination of truth values of X and Y. Such tables are commonly known as *truth tables*. The truth table for the OR operation is shown in Table 2-1. Since X and Y can each take one of two possible values, there are four possible combinations of their truth values which are shown in the first two columns of Table 2-1. We read the truth table row-wise. For example, in the first row, $X = 0$ and $Y = 0$, and by definition, $A = 0 + 0 = 0$.

TABLE 2-1
The Truth Table for
the OR Operation

X	Y	$A = X + Y$
0	0	0
0	1	1
1	0	1
1	1	1

The OR operation is commutative, which means

$$X + Y = Y + X \tag{2-2}$$

From the definition of the OR operation it follows that

$$0 + 0 = 0 \tag{2-3}$$

$$0 + 1 = 1 \tag{2-4}$$

$$1 + 1 = 1 \tag{2-5}$$

The following theorems also are a consequence of the definition and can be proved using truth tables.

$$X + 0 = X \tag{2-6}$$

$$X + 1 = 1 \tag{2-7}$$

$$X + X = X \tag{2-8}$$

Example 2-1.   Prove the theorem given by Eq. (2-8).

The pertinent truth table is as indicated in Table 2-2. The first column in this table lists the two possible truth values of X. The entries in the second column have been obtained using Eqs. (2-3) and (2-5). Thus, when $X = 0$, $X + X = 0 + 0$ which by Eq. (2-3) is equal to 0. Likewise, when $X = 1$, $X + X = 1 + 1$ which is equal to 1 by Eq. (2-5). The two columns are identical, which proves Eq. (2-8).

In a similar manner we can prove the theorems given by Eqs. (2-6) and (2-7).

The OR operation can be extended to more than two variables. For example, $A = X + Y + Z$ implies that A has a truth value of 0 if X, Y, and Z have each a truth value of 0; otherwise, A has a truth value of 1.

The OR operation is also associative, implying

$$X + Y + Z = (X + Y) + Z = X + (Y + Z) \tag{2-9}$$

The use of parentheses on the right-hand side of Eq. (2-9) indicates that the expression inside the parentheses has to be evaluated first. For example, to find the truth value of $(X + Y) + Z$, we first determine that of $X + Y$, and OR

TABLE 2-2
The Truth Table to
Prove $X + X = X$

X	$X + X$
0	0
1	1

$\uparrow$———— Identical ————$\uparrow$

the result with the truth value of Z. As can be seen from the two different pairings on the right-hand side of Eq. (2-9), $X + Y + Z$ can be evaluated several different ways, and the result is always the same.

AND Operation

An alternate way to combine two switching variables X and Y to form a new variable A is by means of the "AND" operation shown symbolically as

$$A = X \cdot Y \tag{2-10}$$

In words, we state that "A is equal to X AND Y." Note that the symbol "$\cdot$" indicates here the AND operation, not a multiplication. The AND operation is defined as follows:

"A has a truth value of 1 if both X and Y have a truth value of 1; otherwise A has a truth value of 0."

The AND operation can easily be explained by means of a truth table shown in Table 2-3. It is often convenient to indicate the AND operation without the symbol "$\cdot$". Thus $AB \equiv A \cdot B$.

This second type of basic operation can also be extended to more than two variables. For example, $A = XYZ$ has a truth value of 1 if and only if the three variables X, Y, and Z each have a truth value of 1; otherwise A has a truth value of 0.

Like the previous operation, the AND operation also is commutative, that is,

$$X \cdot Y = Y \cdot X \tag{2-11}$$

and is associative, that is,

$$X \cdot Y \cdot Z = (X \cdot Y) \cdot Z = X \cdot (Y \cdot Z) \tag{2-12}$$

From the definition of the AND operation it is evident that

$$0 \cdot 0 = 0 \tag{2-13}$$

$$0 \cdot 1 = 0 \tag{2-14}$$

$$1 \cdot 1 = 1 \tag{2-15}$$

TABLE 2-3
The Truth Table for
the AND Operation

X	Y	$A = XY$
0	0	0
0	1	0
1	0	0
1	1	1

The following three theorems are a consequence of the definitions and can be readily proved using truth tables:

$$X \cdot X = X \tag{2-16}$$

$$X \cdot 0 = 0 \tag{2-17}$$

$$X \cdot 1 = X \tag{2-18}$$

NEGATION Operation

Another basic operation is the "NEGATION" operation shown symbolically as

$$A = \bar{X} \tag{2-19}$$

In words, we say that the switching variable A is the *negative*, or *complement*, of X. The negation operation, also called a *NOT* operation, is symbolically shown by a " bar " on top of the variable and is defined as:

"A has a truth value of 1 if X has a truth value of 0, and A has a truth value of 0 if X has a truth value of 1."

The negation operation is illustrated in Table 2-4.

TABLE 2-4
The Truth Table for the
NEGATION Operation

X	$A = \bar{X}$
0	1
1	0

It follows from the definition that

$$\bar{\bar{X}} = X \tag{2-20}$$

The following two theorems are a consequence of the three basic operations as defined above:

$$X + \bar{X} = 1 \tag{2-21}$$

$$X \cdot \bar{X} = 0 \tag{2-22}$$

The validity of these two theorems can be easily verified. Consider, for example, Eq. (2-21). Note, when $X = 0$, $\bar{X} = 1$. Hence $X + \bar{X} = 0 + 1$ which by Eq. (2-4) is equal to 1. Similarly, when $X = 1$, $\bar{X} = 0$, and $X + \bar{X} = 1 + 0$ which is again equal to 1 by Eq. (2-4).

Switching Function

A switching variable generated by combining several switching variables using the three switching operations discussed above is more commonly known as a *switching function*. Thus the switching variable A defined by Eq. (2-1) is a switching function. Similarly $A = XY$ is also a switching function. Examples of more complex switching functions are as follows:

$$A = \bar{X} + XY \tag{2-23}$$

$$B = XY\bar{Z}W + XY(\bar{Z} + X\bar{Y}) \tag{2-24}$$

$$C = \overline{X + Y\bar{\bar{Z}}} + Z \tag{2-25}$$

In many cases, it may be necessary to explicitly show the switching variables generating the switching function. Thus the switching functions A, B, and C of Eqs. (2-23)–(2-25) may be written as $A(X, Y)$, $B(W, X, Y, Z)$, and $C(X, Y, Z)$, respectively. It should be noted that we have defined only two distinct operations (OR, AND) to combine two switching variables. It is possible to define 14 other such binary operations (Problem 2-11). Any one or more of these can be used to generate switching functions. We show later that any switching function can always be expressed in terms of the parent switching variables combined with the aid of the three basic operations defined earlier.

In evaluating the truth values of a switching function generated using the three basic operations, the following rules have to be followed:* First, the NEGATION operation, then the AND, and finally the OR operation are performed. For example, in evaluating A given by Eq. (2-23), first the truth value of $\bar{X}$ is determined, next XY is evaluated, and finally these two truth values are combined using the OR operation to determine the truth value of A. If parentheses are present, then the expression inside the parentheses is evaluated first. Thus, in the case of Eq. (2-24), $(\bar{Z} + X\bar{Y})$ is first evaluated, the result is then ANDed with the truth value of XY. The truth value of the combination is next ORed with that of $XY\bar{Z}W$. In the case of Eq. (2-25), the expression $\overline{X + Y\bar{Z}}$ is equivalent to $(\overline{X + Y\bar{Z}})$ and has to be evaluated first.

Example 2-2. Evaluate the switching function A of Eq. (2-23).

A simple way to evaluate this is by a truth table, shown in Table 2-5. The first two columns give all possible combinations of truth values of X and Y. The third column gives the truth values of $\bar{X}$ and the fourth column gives the truth values of XY. The fifth column is obtained by performing an OR operation of the truth values of $\bar{X}$ and XY, and thus describes the explicit dependence of the switching function A on the switching variables X and Y.

---

* Unless specified otherwise by parentheses or bars.

TABLE 2-5
The Truth Table for $A = \bar{X} + XY$

X	Y	$\bar{X}$	XY	A
0	0	1	0	1
0	1	1	0	1
1	0	0	0	0
1	1	0	1	1

In evaluating a switching function of two switching variables X and Y, all possible combinations of truth values of X and Y must be examined. Since both X and Y can have either one of two possible values, X and Y together can have $2^2 = 4$ possible combinations. Similarly, in evaluating a switching function of n switching variables $X_1, \ldots, X_n$, 2^n possible combinations of the truth values of these variables must be examined.

Example 2-3. Evaluate the switching function C defined in Eq. (2-25).

We observe that C is a function of three switching variables. Hence there are $2^3 = 8$ possible combinations of the truth values of these three variables, for each of which the truth value of C must be determined. As before, we form the corresponding truth table, shown in Table 2-6. In this table, the first three columns indicate the eight possible combinations of the truth values of X, Y, and Z. The next four columns list the truth values of $\bar{Z}$, $Y\bar{Z}$, $X + Y\bar{Z}$, and $\overline{X + Y\bar{Z}}$, respectively. Finally, by combining the entries in the third and the seventh columns row-wise by an **OR** operation, we obtain the truth values of the switching function C as given by the last column.

We next define two very special types of switching functions.

Minterms and Maxterms

A switching function formed by the **AND** combination of each and every switching variable or its complement is called a *minterm*. For example, if X, Y,

TABLE 2-6
Truth Table for $C = \overline{X + Y\bar{Z}} + Z$

X	Y	Z	$\bar{Z}$	$Y\bar{Z}$	$X + Y\bar{Z}$	$\overline{X + Y\bar{Z}}$	C
0	0	0	1	0	0	1	1
0	0	1	0	0	0	1	1
0	1	0	1	1	1	0	0
0	1	1	0	0	0	1	1
1	0	0	1	0	1	0	0
1	0	1	0	0	1	0	1
1	1	0	1	1	1	0	0
1	1	1	0	0	1	0	1

and Z are three switching variables, then $X\bar{Y}Z$ is a minterm. When n switching variables are involved, there are 2^n different minterms. Any one of these minterms has a truth value of 1 for only one unique combination of the truth values of the parent switching variables. For all of the remaining $2^n - 1$ combinations, this particular minterm will have a truth value of 0. Thus the minterm $X\bar{Y}Z$ is equal to 1 for $X = 1$, $Y = 0$, and $Z = 1$. For all other seven possible combinations of the truth values of X, Y, and Z, $X\bar{Y}Z$ will take the value 0.

A switching variable or its complement is often labeled as a *literal*. A minterm is formed by taking the AND combination of *each* literal. So, if X, Y, and Z are the three switching variables of interest, then $X\bar{Y}$ is not a minterm, because the variable Z or its complement is missing.

On the other hand, a *maxterm* is the switching function formed by taking the OR combination of each literal. For example, with X, Y, and Z as the switching variables, a typical maxterm will be $\bar{X} + Y + Z$. Once again, there are 2^n possible maxterms if n switching variables are involved. A maxterm takes the value 0 only for one particular unique combination of the truth values of the constituent switching variables. For all other $2^n - 1$ combinations of truth values, the particular maxterm will take the value 1.

Later in this chapter, we show that any arbitrary switching function can always be expressed in two canonic forms. The first canonic form is given as an OR combination of minterms and the second canonic form is given as an AND combination of maxterms. To prove these results, we need to use certain theorems in switching algebra, which is the algebra of two-valued switching variables.

2-4   Switching Algebra Theorems

The *switching algebra* in essence forms the basis of digital system design. With the aid of switching algebra we can analyze the operation of a given digital circuit and describe its operation in terms of a switching function. With the use of the switching algebra, we can simplify a complex expression defining a switching function which leads to a simpler (and often cheaper) digital circuit realization.

In this text, we do not intend to develop the theory of the switching algebra in a formal manner. Instead we summarize here the main theorems that are frequently used in the design and analysis of digital systems. For ease of future reference, all the pertinent theorems are summarized in two tables. Table 2-7 lists nine theorems involving a single switching variable. Twelve additional theorems involving two or more switching variables are included in Table 2-8. Some additional theorems have already been mentioned earlier in this chapter. All of these theorems can be proved in a number of ways; we outline the proofs of some of them. The proof of the remaining theorems is left as an exercise.

Theorems 2-6 and 2-8 have been proved earlier. All of the other 19 theorems can be proved in a similar manner with the aid of truth tables. For example,

TABLE 2-7
Switching Algebra Theorems Involving a Single Variable

Theorem Number	Theorem Statement	Remarks
2-1 2-2 2-3 2-4	$X + 0 = X$ $X + 1 = 1$ $X \cdot 0 = 0$ $X \cdot 1 = X$	Theorems on the aggressivity and neutrality of the elements 0 and 1
2-5	$\bar{\bar{X}} = X$	Theorem on double negation
2-6 2-7	$X + \bar{X} = 1$ $X\bar{X} = 0$	Theorems of the excluded middle
2-8 2-9	$X + X = X$ $X \cdot X = X$	Theorems of absorption

TABLE 2-8
Switching Algebra Theorems Involving More Than One Variable

Theorem Number	Theorem Statement	Remarks
2-10 2-11	$X + Y = Y + X$ $XY = YX$	Commutative theorems
2-12 2-13	$(X + Y) + Z = X + (Y + Z) = X + Y + Z$ $(XY)Z = X(YZ) = XYZ$	Associative theorems
2-14 2-15	$X(Y + Z) = XY + XZ$ $X + YZ = (X + Y)(X + Z)$	Distributive theorems
2-16 2-17	$X + XY = X$ $X(X + Y) = X$	Additional theorems of absorption
2-18 2-19	$X + \bar{X}Y = X + Y$ $X(\bar{X} + Y) = XY$	Theorems of absorption of negation
2-20 2-21	$\overline{X + Y} = \bar{X} \cdot \bar{Y}$ $\overline{X \cdot Y} = \bar{X} + \bar{Y}$	DeMorgan's theorems

TABLE 2-9
Truth Table to Prove Distributive Theorem 2-14

X	Y	Z	$Y + Z$	$X(Y + Z)$	XY	XZ	$XY + XZ$
0	0	0	0	0	0	0	0
0	0	1	1	0	0	0	0
0	1	0	1	0	0	0	0
0	1	1	1	0	0	0	0
1	0	0	0	0	0	0	0
1	0	1	1	1	0	1	1
1	1	0	1	1	1	0	1
1	1	1	1	1	1	1	1

$\llcorner$———— Identical columns ————$\lrcorner$

the validity of Theorems 2-14, 2-16, and 2-18 are readily established from their corresponding truth tables given in Tables 2-9, 2-10, and 2-11, respectively.

Instead of proving the other theorems in a similar way, we now use a different approach. We prove a theorem by making use of previously proven switching algebra theorems as illustrated by the following examples.

TABLE 2-10
Truth Table to Prove Theorem 2-16

X	Y	XY	$X + XY$
0	0	0	0
0	1	0	0
1	0	0	1
1	1	1	1

$\llcorner$———— Identical columns ————$\lrcorner$

TABLE 2-11
Truth Table to Prove Theorem 2-18

X	Y	$\bar{X}$	$\bar{X}Y$	$X + \bar{X}Y$	$X + Y$
0	0	1	0	0	0
0	1	1	1	1	1
1	0	0	0	1	1
1	1	0	0	1	1

$\llcorner$———— Identical columns ————$\lrcorner$

Example 2-4. Prove the second distributive Theorem 2-15.
The right-hand side can be expressed as

$$(X + Y)(X + Z) = (X + Y)X + (X + Y)Z \qquad \text{(by Theorem 2-14)}$$
$$= XX + YX + XZ + YZ \qquad \text{(by Theorem 2-14)}$$
$$= X + YX + XZ + YZ \qquad \text{(by Theorem 2-9)}$$
$$= X + XY + XZ + YZ \qquad \text{(by Theorem 2-11)}$$
$$= X \cdot 1 + XY + XZ + YZ \qquad \text{(by Theorem 2-4)}$$
$$= X(1 + Y + Z) + YZ \qquad \text{(by Theorem 2-14)}$$
$$= X \cdot 1 + YZ \qquad \text{(by Theorem 2-2)}$$
$$= X + YZ \qquad \text{(by Theorem 2-4)}$$

Note that the last five steps, in essence, also prove Theorem 2-16.

Example 2-5. Prove Theorem 2-19.
We express

$$X(\bar{X} + Y) = X\bar{X} + XY \qquad \text{(by Theorem 2-14)}$$
$$= 0 + XY \qquad \text{(by Theorem 2-7)}$$
$$= XY \qquad \text{(by Theorem 2-1)}$$

Example 2-6. Show $\overline{X + Y} = \bar{X} \cdot \bar{Y}$.
By Theorems 2-7 and 2-6, $(\overline{X + Y})(X + Y) = 0$ and $(\overline{X + Y}) + (X + Y) = 1$. So if we can show

$$(\bar{X} \cdot \bar{Y})(X + Y) = 0$$
$$(\bar{X} \cdot \bar{Y}) + (X + Y) = 1$$

then $\bar{X} \cdot \bar{Y}$ must be the complement of $(X + Y)$.* Now

$$(\bar{X} \cdot \bar{Y})(X + Y) = (\bar{X} \cdot \bar{Y}) \cdot X + (\bar{X} \cdot \bar{Y}) \cdot Y \qquad \text{(by Theorem 2-14)}$$
$$= \bar{X} \cdot \bar{Y} \cdot X + \bar{X} \cdot \bar{Y} \cdot Y \qquad \text{(by Theorem 2-13)}$$
$$= (\bar{X} \cdot X)\bar{Y} + \bar{X}(\bar{Y} \cdot Y) \qquad \text{(by Theorem 2-13)}$$
$$= 0 \cdot \bar{Y} + \bar{X} \cdot 0 \qquad \text{(by Theorem 2-7)}$$
$$= 0 \qquad \text{(by Theorem 2-3)}$$

---

* This, of course, depends on the uniqueness of the complement, that is, if $X + Z = 1$ and $X \cdot Z = 0$, then $Z = \bar{X}$.

Similarly,

$$
\begin{aligned}
(\bar{X} \cdot \bar{Y}) + (X + Y) &= (\bar{X} \cdot \bar{Y}) + X + Y & \text{(by Theorem 2-12)} \\
&= X + \bar{X} \cdot \bar{Y} + Y & \text{(by Theorem 2-10)} \\
&= X + X\bar{Y} + \bar{X} \cdot \bar{Y} + Y & \text{(by Theorem 2-16)} \\
&= X + (X + \bar{X})\bar{Y} + Y & \text{(by Theorem 2-14)} \\
&= X + 1 \cdot \bar{Y} + Y & \text{(by Theorem 2-6)} \\
&= X + \bar{Y} + Y & \text{(by Theorem 2-4)} \\
&= X + 1 & \text{(by Theorem 2-6)} \\
&= 1 & \text{(by Theorem 2-2)}
\end{aligned}
$$

The hypothesis is proved.

In a similar manner we can show the validity of Theorem 2-21.

We can use the results of the previous example to extend DeMorgan's theorem to more than two variables. Thus,

$$\overline{X + Y + Z} = \bar{X} \cdot (\overline{Y + Z}) = \bar{X} \cdot \bar{Y} \cdot \bar{Z} \tag{2-26}$$

and

$$\overline{XYZ} = \overline{X(YZ)} = \bar{X} + (\overline{YZ}) = \bar{X} + \bar{Y} + \bar{Z} \tag{2-27}$$

An interesting consequence of DeMorgan's theorem is that given a switching function expressed as a combination of a number of switching variables by means of OR, AND, and NOT operations, the complement of the switching function is obtained by replacing in the original expression each OR symbol by AND, each AND symbol by OR, each variable by its *complement*, and each constant by its *complement*. Thus if, for example,

$$A = X \cdot \bar{Y} \cdot (Z + W \cdot \bar{X})$$

then

$$\bar{A} = \bar{X} + Y + \bar{Z} \cdot (\overline{W} + X)$$

Consider next the switching function

$$B = X \cdot \bar{Y} \cdot \overline{(Z + W \cdot \bar{X})}$$

then

$$\bar{B} = \bar{X} + Y + Z + W\bar{X}$$

where $(Z + W\bar{X})$ has been treated as a unit.

An alternate way of proving the theorems is with the aid of Karnaugh maps which provide a graphical representation of switching functions. We discuss the Karnaugh maps in the following section.

2-5　Karnaugh Map

To understand the basic idea behind the Karnaugh map representation of a switching function, consider first an arbitrary switching function $g(X)$ of a single switching variable X. Note that $g(X)$ takes the value $g(1)$ when $X = 1$ and the value $g(0)$ when $X = 0$, where $g(1)$ and $g(0)$ can each have a truth value of 0 or 1. Irrespective of the form of the switching function we can always express $g(X)$ as

$$g(X) = g(1) \cdot X + g(0) \cdot \bar{X} \tag{2-28}$$

The above result can be proved by showing that the left-hand side is equal to the right-hand side for the two values of X. If we set $X = 1$, the right-hand side of Eq. (2-28) becomes

$$g(1) \cdot 1 + g(0) \cdot \bar{1} = g(1) \cdot 1 + g(0) \cdot 0 = g(1)$$

(by Theorems 2-3, 2-4, and 2-1) which is equal to the left-hand side with $X = 1$. If we next set $X = 0$, then the right-hand side of Eq. (2-28) reduces to

$$g(1) \cdot 0 + g(0) \cdot \bar{0} = g(1) \cdot 0 + g(0) \cdot 1 = g(0)$$

(again, by Theorems 2-3, 2-4, and 2-1) and is thus seen to be the same as the left-hand side for $X = 0$.

The above result can be easily generalized to a function of more than one variable. Consider next an arbitrary switching function $g(X, Y)$ of two switching variables X and Y. It can always be expressed as

$$g(X, Y) = g(1, Y) \cdot X + g(0, Y) \cdot \bar{X} \tag{2-29}$$

Equation (2-29) can be proved again in a manner used to prove Eq. (2-28), that is, by showing that the right-hand side is equal to the left-hand side for the two possible values of X. The details of the proof are left as an exercise.

Generalizing further we arrive at the following theorem:

THEOREM 2-22.　A switching function $g(X_1, X_2, \ldots, X_n)$ can always be expressed as

$$g(X_1, X_2, \ldots, X_n) = g(1, X_2, \ldots, X_n) \cdot X_1 + g(0, X_2, \ldots, X_n) \cdot \bar{X}_1 \tag{2-30}$$

Let us apply the above theorem repeatedly on an arbitrary switching function. For simplicity we consider a switching function of two switching variables, $g(X, Y)$. We thus obtain

$$\begin{aligned}
g(X, Y) &= g(X, 1) \cdot Y + g(X, 0) \cdot \bar{Y} \\
&= \{g(1, 1) \cdot X + g(0, 1) \cdot \bar{X}\} \cdot Y + \{g(1, 0) \cdot X + g(0, 0) \cdot \bar{X}\} \cdot \bar{Y} \\
&= g(1, 1) \cdot XY + g(0, 1) \cdot \bar{X}Y + g(1, 0) \cdot X\bar{Y} + g(0, 0) \cdot \bar{X}\bar{Y} \tag{2-31}
\end{aligned}$$

In deriving the last expression we have made use of the distributive theorem 2-14. Equation (2-31) is an interesting result. Since XY, $\bar{X}Y$, $X\bar{Y}$, and $\bar{X}\bar{Y}$ are the four possible minterms of the two switching variables X and Y, Eq. (2-31) shows that any 2-variable switching function can always be expressed as an OR combination of the minterms. Each of the minterms are ANDed with a coefficient which is the truth value of the original switching function evaluated for a particular set of truth values of the switching variables. The particular set of values substituted for the variables are determined by the values for which the minterm takes the value 1. Thus the minterm XY is equal to 1 if and only if $X = 1$ and $Y = 1$. The coefficient in front of this minterm is $g(1, 1)$ which is the truth value of $g(X, Y)$ for $X = 1$ and $Y = 1$. Similarly, $g(0, 1)$ is the truth value of $g(X, Y)$ for $X = 0$ and $Y = 1$, and for these values the minterm $\bar{X}Y$ is 1. If any of these coefficients takes the value 1, then the corresponding minterm is present and if the coefficient is 0, then the corresponding minterm is absent.

Example 2-7. Express $g(X, Y) = X + Y$ as an OR combination of minterms.

From the definition of the OR operation as given in Table 2-1, $g(1, 1) = 1$, $g(0, 1) = 1$, $g(1, 0) = 1$, and $g(0, 0) = 0$. Substituting these values in Eq. (2-31) and using Theorems 2-3 and 2-4 we arrive at

$$g(X, Y) = X + Y = 1 \cdot XY + 1 \cdot \bar{X}Y + 1 \cdot X\bar{Y} + 0 \cdot \bar{X}\bar{Y} \qquad (2\text{-}32)$$
$$= XY + \bar{X}Y + X\bar{Y}$$

The above result can be extended to more than two variables. For example, a 3-variable function $f(X, Y, Z)$ can be expressed as

$$f(X, Y, Z) = f(0, 0, 0) \cdot \bar{X}\bar{Y}\bar{Z} + f(0, 0, 1) \cdot \bar{X}\bar{Y}Z$$
$$+ f(0, 1, 0) \cdot \bar{X}Y\bar{Z} + f(0, 1, 1) \cdot \bar{X}YZ$$
$$+ f(1, 0, 0) \cdot X\bar{Y}\bar{Z} + f(1, 0, 1) \cdot X\bar{Y}Z$$
$$+ f(1, 1, 0) \cdot XY\bar{Z} + f(1, 1, 1) \cdot XYZ \qquad (2\text{-}33)$$

The expansion of a switching function as an OR combination of minterms is known as a *standard sum-of-products* (SSOP) *form, minterm decomposition, sum-of-minterms form,* or a *disjunctive canonical form.*

An alternate canonic expansion of a switching function is as an AND combination of maxterms. For example, a 2-variable switching function $g(X, Y)$ can always be expanded as

$$g(X, Y) = [g(0, 0) + X + Y] \cdot [g(0, 1) + X + \bar{Y}] \cdot [g(1, 0) + \bar{X} + Y]$$
$$\cdot [g(1, 1) + \bar{X} + \bar{Y}] \qquad (2\text{-}34)$$

which is known as a *standard product-of-sums* (SPOS) *form, maxterm decomposition, sum-of-maxterms form,* or a *conjunctive canonical form.* Equation (2-34) can be established with the aid of the following theorem:

THEOREM 2-23.   A switching function $g(X_1, X_2, \ldots, X_n)$ can always be expressed as

$$g(X_1, X_2, \ldots, X_n) = [g(0, X_2, \ldots, X_n) + X_1] \cdot [g(1, X_2, \ldots, X_n) + \bar{X}_1]$$

$$(2\text{-}35)$$

whose proof is left as an exercise. Note that a particular maxterm will be present in the SPOS form of expansion, if the switching function takes the truth value of 0 when the maxterm also has a 0 truth value.

Now we are in a position to understand the Karnaugh map. We shall only discuss here the Karnaugh map representation obtained from the SSOP form. An analogous map based on the SPOS form exists and is not discussed here because the first type of map is more widely used in practice.

Consider first a 2-variable switching function $g(X, Y)$ whose SSOP form is given by Eq. (2-31). A graphical way to represent the SSOP form is as shown in Fig. 2-1, which is the 2-variable Karnaugh map.* Since there are four possible combinations of the truth values of the two switching variables, the 2-variable Karnaugh map has four squares, with each square corresponding to a unique set of values of the switching variables. To obtain the Karnaugh map of a specific switching function, we evaluate the switching function for each combination of the truth values of the switching variables and place this value in its respective box.

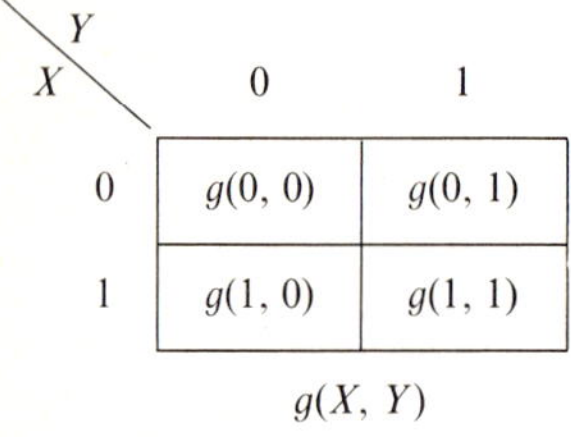

Figure 2-1   A 2-variable Karnaugh map.

Example 2-8.   Obtain the Karnaugh map of $X + Y$.

From Table 2-1 we note that $g(0, 0) = 0$ and hence we place a 0 in the upper left square as shown in Fig. 2-2. In a similar manner, the other three boxes are filled as indicated in Fig. 2-2.

Note that the boxes containing 1's indicate which minterm is present in the SSOP form. Thus, given the Karnaugh map, we can always obtain the SSOP form of representation of a switching function as illustrated in the following example.

---

* The specific ordering scheme used to label each square in a Karnaugh map aids in the simplification of a switching function (see Section 3-10).

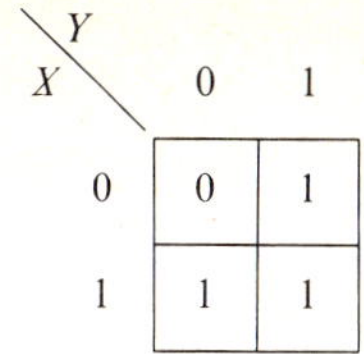

Figure 2-2 Karnaugh map of $X + Y$.

Example 2-9.   Determine the switching function represented by the Karnaugh map of Fig. 2-3.

Note that in this Karnaugh map we have not shown explicitly the 0's, which is the common practice. The upper right box in Fig. 2-3 corresponds to the minterm $\bar{X}Y$ and the lower left box corresponds to the minterm $X\bar{Y}$. Hence the pertinent switching function is given as

$$g(X, Y) = \bar{X}Y + XY$$

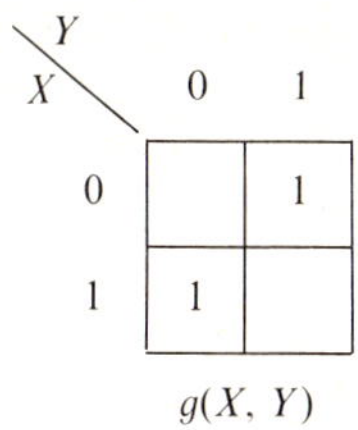

Figure 2-3   A particular 2-variable Karnaugh map.

An alternate way to label the boxes in a Karnaugh map is indicated in Fig. 2-4. Here we also have placed all minterms in their respective boxes to clarify the labeling procedures. If we examine the boxes row-wise, we note that boxes in the bottom row labeled X refer to minterms having the literal X in common. On the other hand, the boxes in the top row (without any label) correspond to minterms with literal $\bar{X}$ in common. Similarly, by examining the boxes column-wise, it is apparent that the boxes in the right column (labeled Y) have the literal Y in their minterms and the boxes in the left column (without the label) have the literal $\bar{Y}$ in their minterms.

	Y	
	$\bar{X}\bar{Y}$	$\bar{X}Y$
X	$X\bar{Y}$	XY

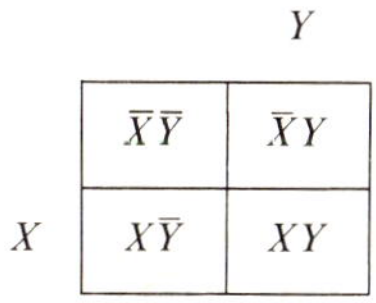

Figure 2-4   An alternate labeling procedure for a 2-variable Karnaugh map.

		Y	Y	
	$\bar{X}\bar{Y}\bar{Z}$	$\bar{X}\bar{Y}Z$	$\bar{X}YZ$	$\bar{X}Y\bar{Z}$
X	$X\bar{Y}\bar{Z}$	$X\bar{Y}Z$	XYZ	$XY\bar{Z}$
		Z	Z	

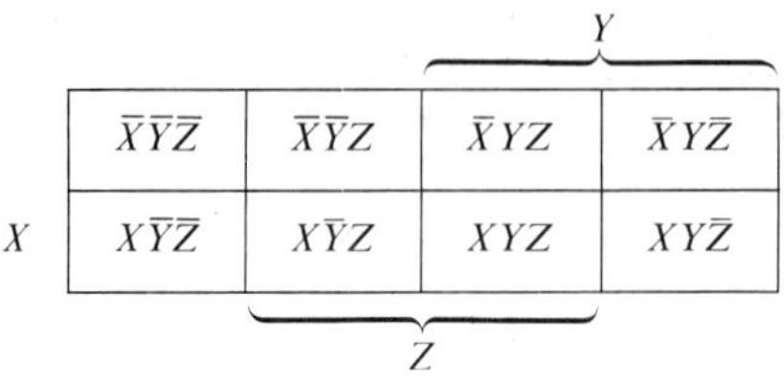

Figure 2-5 A 3-variable Karnaugh map.

The above procedure is also followed in labeling 3-variable and 4-variable Karnaugh maps as shown in Figs. 2-5 and 2-6, respectively. Karnaugh maps for five or more variables can be developed, but they are a little more complicated to use.

We have shown before that in order to obtain the Karnaugh map representation of a switching function we first expand it in a SSOP form which can be obtained by evaluating the truth value of the switching function for all possible truth values of the switching variables. In many cases the SSOP form of expansion can be easily derived with the aid of the switching algebra theorems as illustrated by the next example.

			Y	Y	
	$\bar{W}\bar{X}\bar{Y}\bar{Z}$	$\bar{W}\bar{X}\bar{Y}Z$	$\bar{W}\bar{X}YZ$	$\bar{W}\bar{X}Y\bar{Z}$	
	$\bar{W}X\bar{Y}\bar{Z}$	$\bar{W}X\bar{Y}Z$	$\bar{W}XYZ$	$\bar{W}XY\bar{Z}$	X
W	$WX\bar{Y}\bar{Z}$	$WX\bar{Y}Z$	$WXYZ$	$WXY\bar{Z}$	X
W	$W\bar{X}\bar{Y}\bar{Z}$	$W\bar{X}\bar{Y}Z$	$W\bar{X}YZ$	$W\bar{X}Y\bar{Z}$	
		Z	Z		

Figure 2-6 A 4-variable Karnaugh map.

Example 2-10. Derive the Karnaugh map of $(X + Y)(X + Z)$.

We first express the switching function as a sum-of-products form. Then each product is examined separately. If any of the variables is missing from the product term, the pertinent term is ANDed with an OR combination of the missing variable and its complement. The new expression is expanded as a sum of two modified product terms. This process is continued for each product term or its modified constituents resulting in an expanded sum-of-products form of the switching function where each product contains all variables and is thus a minterm. The resulting expression may contain redundant terms which are removed.

We follow the above procedure for the function of our example and obtain

$$
\begin{aligned}
(X + Y)(X + Z) &= X + XY + XZ + YZ \\
&= X \cdot 1 \cdot 1 + XY \cdot 1 + X \cdot 1 \cdot Z + 1 \cdot YZ \\
&= X(Y + \bar{Y})(Z + \bar{Z}) + XY(Z + \bar{Z}) \\
&\quad + X(Y + \bar{Y})Z + (X + \bar{X})YZ \\
&= XYZ + X\bar{Y}Z + XY\bar{Z} + X\bar{Y}\bar{Z} \\
&\quad + XYZ + XY\bar{Z} + XYZ + X\bar{Y}Z \\
&\quad + XYZ + \bar{X}YZ \\
&= XYZ + X\bar{Y}Z + XY\bar{Z} \\
&\quad + X\bar{Y}\bar{Z} + \bar{X}YZ
\end{aligned}
\tag{2-36}
$$

In deriving the last expression of Eq. (2-36), the following theorems have been used: Theorems 2-14, 2-9, 2-2, 2-6, and 2-8. Equation (2-36) leads to the desired Karnaugh map as sketched in Fig. 2-7.

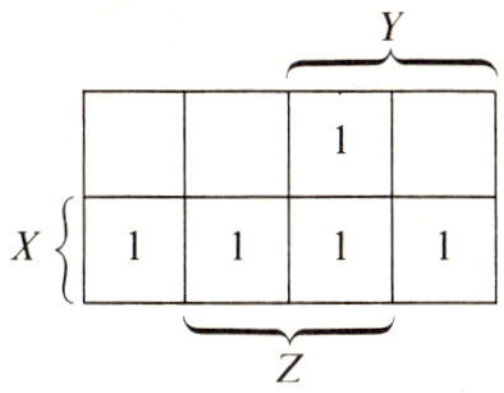

Figure 2-7 The Karnaugh map for $(X + Y)(X + Z)$.

The Karnaugh map can also be used to prove switching algebra theorems.

Example 2-11. Prove Theorem 2-15 using the Karnaugh map.
The SSOP expansion of the left-hand side expression is obtained as

$$
\begin{aligned}
X + YZ &= X \cdot 1 \cdot 1 + 1 \cdot YZ \\
&= X \cdot (Y + \bar{Y}) \cdot (Z + \bar{Z}) + (X + \bar{X}) \cdot YZ \\
&= XYZ + X\bar{Y}Z + XY\bar{Z} + X\bar{Y}\bar{Z} + XYZ + \bar{X}YZ \\
&= XYZ + X\bar{Y}Z + XY\bar{Z} + X\bar{Y}\bar{Z} + \bar{X}YZ
\end{aligned}
\tag{2-37}
$$

which is seen to be the same as that of $(X + Y)(X + Z)$ as given by the right-hand side of Eq. (2-36). Hence the Karnaugh map of $X + YZ$ is also given by Fig. 2-7, proving the theorem.

2-6 Digital Representation of Numbers

The basic purpose of digital electronic systems is to process information. The information, in many applications, is given as numbers, both positive and negative. In the type of digital systems we are considering in this text, a variable can take one of two possible values, 0 and 1. Our interest here, therefore, is to discuss various schemes used in binary systems to represent numbers. We first restrict our attention to the representation of positive numbers. Later, we show how negative numbers can be represented.

Positional Notation of Numbers

The conventional number system with which we are familiar and that we make use of in our daily lives is the decimal number system. Here any number can be represented by making use of a set of ten digits or symbols,

$$0, 1, 2, 3, 4, 5, 6, 7, 8, 9$$

In representing a number greater than 9 or less than 1, we also use the same 10 symbols and in addition use location information. The location of a digit in turn determines the weight or multiplying factor with which the digit is to be multiplied in evaluating the value of the complete number. In the decimal number system, the multiplying factor is a power of 10 with the power depending on the location of the pertinent digit. Thus, for example, a positive decimal number 74.52 is equivalent to

$$(74.52)_{10} = (7 \times 10^1) + (4 \times 10^0) + (5 \times 10^{-1}) + (2 \times 10^{-2})$$
$$= 70.0 + 4.0 + 0.5 + 0.02$$

The digits to the left of the *decimal point* constitute the integer part of the decimal number and are weighted by positive integer powers of 10. Likewise, the digits to the right of the decimal point constitute the fractional part of the number and are weighted by negative integer powers of 10.

If we generalize the above number representation scheme to a system using B symbols, then the representation scheme is said to be in *base* or *radix B*. The dividing part between the integer and fractional parts is then called the *radix point*. The digits to the left of the radix point constituting the integer part are now multiplied by positive integer powers of B, and the digits to the right of the radix point are multiplied by negative integer powers of B. In general, a positive number $\alpha_{m-1}\alpha_{m-2} \cdots \alpha_1\alpha_0 \cdot \alpha_{-1} \cdots \alpha_{-n}$ containing m integer digits and n fractional digits in base or radix B has a decimal value N given by

$$N = \alpha_{m-1}B^{m-1} + \alpha_{m-2}B^{m-2} + \cdots + \alpha_1 B^1 + \alpha_0 B^0 + \alpha_{-1}B^{-1} + \cdots + \alpha_{-n}B^{-n}$$

$$(2\text{-}38)$$

where α_i is a symbol having an integer value between 0 and $B - 1$. α_{m-1} is known as the *most significant digit* and α_{-n} as the *least significant digit*.

Binary Number System

In this system, base $B = 2$ and α_i in Eq. (2-38) can be either 0 or 1. A *binary digit* is often abbreviated as a *bit*. As a result, in Eq. (2-38) with base 2, α_{m-1} is known as the *most significant bit* or simply the *MSB* and α_{-n} is known as the *least significant bit* or simply the *LSB*.

To determine the decimal equivalent of a positive binary number we make use of Eq. (2-38) as explained in the following two examples.

Example 2-12.   Find the decimal equivalent of the binary integer 11001. Using Eq. (2-38) we arrive at

$$11001_2 = (1 \times 2^4) + (1 \times 2^3) + (0 \times 2^2) + (0 \times 2^1) + (1 \times 2^0)$$

$$= (2^4 + 2^3 + 2^0) = 16 + 8 + 1 = 25_{10}$$

Example 2-13.   Find the decimal equivalent of the binary fraction 0.11001. Using Eq. (2-38) we obtain

$$0.11001_2 = (1 \times 2^{-1}) + (1 \times 2^{-2}) + (0 \times 2^{-3}) + (0 \times 2^{-4}) + (1 \times 2^{-5})$$

$$= 2^{-1} + 2^{-2} + 2^{-5} = 0.5 + 0.25 + 0.03125$$

$$= 0.78125_{10}$$

We shall use the subscripts 2 and 10 as illustrated above to indicate whether a number is expressed in the binary system or the decimal system. These subscripts may be dropped if there is no confusion with regard to the system of representation.

The method to convert a positive decimal number to its binary equivalent also follows from Eq. (2-38). We consider the conversion of decimal integers and fractions separately.

To convert a positive decimal integer N greater than 1 to base B, we note from Eq. (2-38) that the sum of all the terms on the right-hand side except α_0 is divisible by B. So if we divide N by B, the remainder would be α_0, the least significant bit. Thus,

$$\frac{N}{B} = r_1 + \frac{\alpha_0}{B}$$

where $r_1 = \alpha_{m-1} B^{m-2} + \alpha_{m-2} B^{m-3} + \cdots + \alpha_1$ is the quotient. Next dividing r_1 by B we get a remainder of α_1,

$$\frac{r_1}{B} = r_2 + \frac{\alpha_1}{B}$$

where $r_2 = \alpha_{m-1} B^{m-3} + \alpha_{m-2} B^{m-4} + \cdots + \alpha_2$ is the quotient. This process is continued until the quotient is less than B.

Example 2-14.   Find the binary equivalent of 107_{10}. We form

$$\frac{107}{2} = 53 + \frac{1}{2} \leftarrow \alpha_0$$

Hence $\alpha_0 = 1$ and $r_1 = 53$. Next

$$\frac{53}{2} = 26 + \frac{1}{2} \leftarrow \alpha_1$$

implying $\alpha_1 = 1$ and $r_2 = 26$. Continuing this process,

$$\frac{26}{2} = 13 + \frac{0}{2} \leftarrow \alpha_2$$

$$\frac{13}{2} = 6 + \frac{1}{2} \leftarrow \alpha_3$$

$$\frac{6}{2} = 3 + \frac{0}{2} \leftarrow \alpha_4$$

$$\frac{3}{2} = 1 + \frac{1}{2} \leftarrow \alpha_5$$

$$r_6$$

Note that the last expression gives $r_6 = 1$ and $\alpha_5 = 1$. Since $r_6 = 1$ it is no longer divisible by 2. Hence $r_6 = \alpha_6 = 1$ is the most significant bit. Thus,

$$107_{10} = 1101011_2$$

To convert a positive decimal number N less than 1 into a number in base B, we also make use of Eq. (2-38). Here now we multiply N by B to yield

$$B \times N = \alpha_{-1} + s_1$$

where s_1 is a decimal number less than 1 and α_{-1} is an integer between 0 and $B - 1$. Note α_{-1} is the most significant digit. Next we multiply s_1 by B obtaining

$$B \times s_1 = \alpha_{-2} + s_2$$

where α_{-2} is the second most significant digit in base B representation and s_2 will be a decimal number less than 1. This process is continued until the desired accuracy is obtained.

Example 2-15. Determine the binary equivalent of 0.3125_{10}.
    First multiply by 2:

$$\alpha_{-1}$$
$$2 \times 0.3125 = 0 + 0.625$$

Hence $\alpha_{-1} = 0$ and $s_1 = 0.625$. Continuing

$$\alpha_{-2}$$
$$2 \times 0.625 = 1 + 0.25$$

indicating $\alpha_{-2} = 1$ and $s_2 = 0.25$. Next

$$2 \times 0.25 = \overset{\alpha_{-3}}{0} + 0.5$$

Thus $\alpha_{-3} = 0$ and $s_3 = 0.5$. Finally,

$$2 \times 0.5 = \overset{\alpha_{-4}}{1} + 0.0$$

implying $\alpha_{-4} = 1$ and $s_4 = 0$. As a result,

$$0.3125_{10} = 0.0101_2$$

Since $s_4 = 0$ in the above example, the process of conversion stops after a finite number of steps. However, for some numbers, the process can go on indefinitely indicating no exact binary equivalent.

Example 2-16.   The conversion of an analog signal voltage into a binary digital form can be obtained by means of an analog-to-digital converter (see Section 8-5). If the output of the converter is limited to 6 bits, determine the digital equivalent of an input analog signal of amplitude 0.3.

Following the procedure outlined above we obtain

$$0.3_{10} \cong 0.010011_2$$

The decimal equivalent of 0.010011_2 is 0.296875_{10}. Thus the error caused by the 6-bit representation is 0.003125. It should be noted that it is not possible to obtain an exact binary equivalent of 0.3_{10}.

Word and Byte

Most practical digital systems, such as the computers, employ a fixed number of bits to represent all data for convenience in storage, transfer, and manipulations. The total number of bits used is usually an integer power of 2, such as 8, 16, 32, and so on. The block of bits representing data is commonly referred to as the *word*. The *word length* or *word size* is given by the number of bits in the word. Often the word size is given in units of 8 bits called the *byte*. Thus a 16-bit word is equivalent to a 2-byte word, a 32-bit word is equivalent to a 4-byte word, and so on.

Octal and Hexadecimal Number Representations

From the hardware implementation point of view, although the binary representation of data is quite convenient, the unusually large number of bits required to represent a moderate-size decimal data is often inconvenient for documentation, display, and other purposes. For example, any decimal positive integer containing 5 digits would require 17 bits for binary representation.

To get around this problem, a number system using a base B which is an integer power of 2 is employed. Two widely used such number systems are described below.

In the *octal* representation of numbers, $B = 2^3 = 8$ and the symbols α_i in Eq. (2-38) can be one of 8 possible decimal digits, 0 through 7. Binary representations of these symbols thus require 3 bits, 000 through 111. As a result conversion of a binary number into an octal representation is achieved quite simply by dividing the integer part and the fractional part of the binary number into groups of 3 bits beginning at the radix point. The decimal equivalent of each binary group represents the corresponding octal digit.

Example 2-17.   Determine the octal representation of 10110001.1111_2. The conversion process is illustrated below.

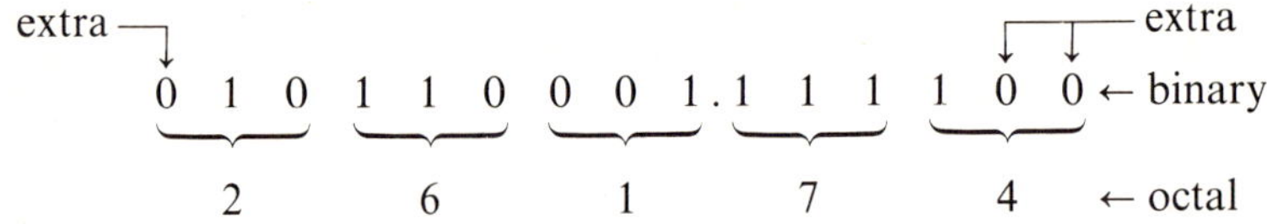

Note that we have added an additional 0 to the left of the MSB of the original binary number and two additional 0's to the right of the LSB to arrive at groupings of 3 bits. From above, the octal representation of the given 12-bit binary number is thus 261.74_8 requiring 5 octal digits.

In the *hexadecimal* representation, $B = 2^4 = 16$. As a result, this system requires 16 different symbols α_i. These symbols are the 10 decimal digits 0 through 9 and six letters of the alphabet A through F where A represents decimal 10, B represents decimal 11, and so on. Conversion of a binary number to a number in hexadecimal form is also very straightforward. Here we divide the binary number into groups of 4 bits beginning at the radix point. The decimal equivalent of each binary group of 4 bits, with the decimal numbers 10 through 15 represented by their alphabet symbols, forms the corresponding hexadecimal digit.

Example 2-18.   Determine the hexadecimal equivalent of 10110001.1111_2. The conversion process is outlined below.

$$1\ \ 0\ \ 1\ \ 1\ \ 0\ \ 0\ \ 0\ \ 1\ .\ 1\ \ 1\ \ 1\ \ 1 \leftarrow \text{binary}$$

$$B \qquad\qquad 1 \qquad\qquad F \qquad \leftarrow \text{hexadecimal}$$

The hexadecimal equivalent is thus $B1.F_{16}$ and requires only 3 digits.

The reverse process of converting an octal or an hexadecimal number to its binary form is extremely simple and should be self-evident. The conversion of an octal or an hexadecimal number into its decimal equivalent can be done by directly evaluating the representation of Eq. (2-38) with appropriate values for B.

Complementary Numbers

In a specific digital system, all binary numbers are in general represented using a fixed number of bits. The total number of bits depends on the size of storage registers being used to store the numbers. Given the size of the register, we can define the *complement* of a binary number. There are two commonly used definitions of complementation.

The *ONEs' complement* of a binary number N is obtained by complementing individual bits of N, that is, by replacing ONE's by ZERO's and vice versa. For example, the ONEs' complement of 110.0101 is 001.1010. If we denote the ONEs' complement of a number N containing m integer bits and n fractional bits as ${}^1N_c$, then mathematically

$$ {}^1N_c = 2^m - N - 2^{-n} \tag{2-39} $$

A more widely used complement number is the *TWO's complement*, which is obtained by forming first the ONEs' complement of the given binary number and then adding a 1 to the LSB. Thus, for example, to determine the TWO's complement of 110.0101, we note that its ONEs' complement is 001.1010; to this we add a 1 to the LSB arriving at the desired result 001.1011. Mathematically, the TWO's complement ${}^2N_c$ of a binary number N containing m integer bits and n fractional bits is given as

$$ \begin{aligned} {}^2N_c &= 2^m - N \\ &= {}^1N_c + 2^{-n} \end{aligned} \tag{2-40} $$

A more convenient way to obtain the TWO's complement is as follows: We examine the binary number to be complemented beginning with its LSB. If it is a 0 we keep it unchanged along with other consecutive 0's to the left. The first 1 encountered while examining from the LSB position is also left unchanged. All other bits are complemented. Consider the binary number 010.0100. Note that the LSB is a 0 followed by a 0 and then a 1 to the left. We keep these bits unchanged. Then we complement the other 4 bits to the left of the first 1. This process leads to the TWO's complement of 010.0100 as 101.1100. This latter algorithm is readily adapted for hardware implementation.

In practice, the ONEs' and TWO's complements are used to represent negative numbers, and are much more suitable for implementing arithmetic operations.

Representation of Negative Numbers

In binary systems, a negative number can be represented three different ways. These are outlined next.

Sign-magnitude Representation.   The simplest way to distinguish between a negative number and a positive number would be to add a sign bit which is placed on the left of the magnitude (essentially the sign bit is now the MSB). According to the standard convention, for a positive number the sign bit is 0 and for a negative number the sign bit is 1. For example, assuming 4 bits are used to represent the magnitude, the number 01.101_2 is equivalent to $+1.625_{10}$ whereas the number 11.101_2 is equivalent to -1.625_{10}.

ONEs' Complement Representation.   An alternate way to obtain a binary representation of a negative decimal number is by forming the ONEs' complement of the positive binary number having the same magnitude and represented in the sign-magnitude form. For example, to obtain the ONEs' complement representation of -1.625_{10} we first observe that the sign-magnitude representation of $+1.625_{10}$ using 4 bits for magnitude is 01.101_2 whose ONEs' complement is 10.010_2. Note that a 1 at the MSB position again indicates that the number is negative.

TWO's Complement Representation.   A more popular binary representation of a negative decimal number is via the TWO's complement of the positive binary number having the same magnitude and represented in sign-magnitude form. Thus the sign-magnitude representation of $+1.625_{10}$ using 4 bits for the magnitude is 01.101_2. Its TWO's complement is 10.011_2 which then represents -1.625_{10}. Here again, a 1 in the MSB position indicates that the number is negative.

To determine the positive number whose TWO's complement is given, we take its TWO's complement. Consider the negative number 11.100_2 in TWO's complement form. We observe that the last 2 bits are 0's which are followed by a 1 on the left. We leave these 3 bits unchanged. Complementing the remaining 2 bits we arrive at 00.100 whose decimal equivalent is $+0.5_{10}$, the desired result.

Note that it is not possible to determine the value of a negative binary number unless the system used to represent the number is known a priori. For example, the number 10.010_2 is equivalent to either -0.25_{10}, -1.625_{10}, or -1.75_{10} depending on whether sign-magnitude, ONEs' complement, or TWO's complement representation is used to represent the negative number.

It should be emphasized here that the three different systems described are used only to represent negative numbers. Positive binary numbers are always represented in sign-magnitude form with a 0 sign bit to indicate the positiveness of the number. In arithmetic operations involving binary numbers, the ONEs' and TWO's complement representation of numbers are usually preferred because the operations can be performed with less hardware.

2-7   Arithmetic Operations

The operations of addition, subtraction, multiplication, and division are basic operations available for arithmetic calculations in almost all digital computers. In addition, some of these operations are often used in special purpose digital systems. We plan now to discuss how these four operations are performed in the binary number system.

Binary Addition and Subtraction

The binary addition procedure is quite similar to the procedure of adding two decimal numbers. Let us illustrate the approach by adding a binary number $A_3 A_2 A_1 A_0$ (addend) to a binary number $B_3 B_2 B_1 B_0$ (augend).* Here A_3 and B_3 are the MSBs and A_0 and B_0 are the LSBs. The sum of these two numbers will in general contain 5 bits which we denote as $S_4 S_3 S_2 S_1 S_0$. In order to form this sum we first add A_0 and B_0 to form the sum S_0 and a carry C_1. Then we add A_1, B_1, and C_1 to form the second digit of sum S_1 and the carry C_2. Next A_2, B_2, and C_2 are added to yield the sum S_2 and the carry C_3. Finally, A_3, B_3, and C_3 are added to yield the sum S_3 and the carry C_4. Since there are no more addend and augend bits left, C_4 is then the most significant bit S_4 in the sum. At any stage in the addition process, 3 binary digits A_i, B_i, and C_i are added to yield the sum S_i and the carry C_{i+1}. The truth table for generating the sum bit S_i and the carry bit C_{i+1} from the 3 binary digits A_i, B_i, and C_i is given in Table 2-12. Observe that when only one of the three variables is 1, their sum S_i is 1 and the carry C_{i+1} is 0. If only two of the three variables are 1, their sum in decimal notation is 2 which, in binary notation is 10, implying that sum S_i is 0 and the carry $C_{i+1} = 1$. Finally, if all three variables are 1, their sum in decimal notation is 3 whose binary equivalent is 11, indicating $S_i = 1$ and $C_{i+1} = 1$. This process of addition is said to occur *serially*.

TABLE 2-12
Truth Table for Full Addition Operation

A_i	B_i	C_i	S_i	C_{i+1}
0	0	0	0	0
0	0	1	1	0
0	1	0	1	0
0	1	1	0	1
1	0	0	1	0
1	0	1	0	1
1	1	0	0	1
1	1	1	1	1

Example 2-19. Determine the sum of $+13_{10}$ and $+7_{10}$ using a 4-bit binary representation of the magnitude.

The binary representation of $+13_{10}$ without the sign bit is 1101_2 and that of $+7_{10}$ is 0111_2. With the aid of the full addition table as given by

---

* Note that $A_3 A_2 A_1 A_0$ and $B_3 B_2 B_1 B_0$ are binary numbers and not switching functions.

Table 2-12, we first add the two LSBs. We then add the next 2 bits of the augend and the addend with the carry generated in the previous step. We continue this process and obtain the final sum in four steps as shown below.

Step 1

```
            1            ← carry
    1   1   0   1        ← +13₁₀
    0   1   1   1        ← +7₁₀
                0        ← partial sum
```

Step 2

```
            1   1        ← carry
    1   1   0   1
    0   1   1   1
            0   0        ← partial sum
```

Step 3

```
        1   1   1        ← carry
    1   1   0   1
    0   1   1   1
        1   0   0        ← partial sum
```

Step 4

```
    1   1   1   1        ← carry
    1   1   0   1
    0   1   1   1
  1 0   1   0   0        ← final sum
```

The bits being added at each step of the addition process are those shown inside the slightly shaded area. Note that a carry bit is generated at each step with the last carry becoming the MSB of the sum. The decimal equivalent of the final sum 10100_2 is $+20_{10}$ as expected.

The above example points out that the addition of two m-bit positive numbers may lead to an $(m + 1)$-bit positive number. In practice, if the sum requires $(m + 1)$ bits with the MSB as 1, we say that an *overflow* has occurred. This generation of an extra bit usually creates an additional problem. In digital system design this problem can be avoided either by appropriate scaling of the augend and addend, or by detecting the overflow and then taking a corrective procedure with the aid of additional logic circuitry.

Binary subtraction can be considered as the addition of a negative number to a positive number. As we shall show now this approach forms the basis of binary subtraction using a TWO's complement arithmetic. As a matter of fact, we can also add two negative numbers rather easily using TWO's complement representation of negative numbers. The following example elaborates this matter.

Example 2-20. Determine the following sums: (a) $8_{10} - 6_{10}$, (b) $6_{10} - 8_{10}$, and (c) $-8_{10} - 6_{10}$. Use a 5 bit representation for both positive and negative numbers with the leftmost bit for the sign.

(a) We observe first that the binary representation of 8_{10} and 6_{10} are 01000_2 and 00110_2, respectively. As a result, in the binary system, the desired sum can be written as

$$S_2 = 01000_2 - 00110_2 \qquad (2\text{-}41)$$

By adding and subtracting the number $2^5{}_{10} = 100000_2$ to the right-hand side of the above expression we arrive at

$$S_2 = 01000_2 - 00110_2 + 100000_2 - 100000_2 \qquad (2\text{-}42)$$

From Eq. (2-42) it can be seen that $(-00110_2 + 100000_2)$ is nothing but the TWO's complement representation of 00110_2 which is 11010_2. Hence, Eq. (2-42) reduces to the form

$$S_2 = 01000_2 + 11010_2 - 100000_2 \qquad (2\text{-}43)$$

The first two 5-bit numbers in the above expression can be added using the binary addition process resulting in a 6-bit number 100010_2 which we substitute in Eq. (2-43) to arrive at

$$S_2 = 100010_2 - 100000_2$$
$$= 00010_2 + 100000_2 - 100000_2 \qquad (2\text{-}44)$$

which is equivalent to 00010_2. Since the sign bit is a 0, the sum is a positive number whose decimal equivalent is $+2_{10}$. Note that in developing the last expression of Eq. (2-44) we have used the fact that $100010_2 = 00010_2 + 100000_2$.

Thus the addition of a negative decimal number to a positive number can be obtained using binary addition technique by adding the binary representation of the numbers with the negative number represented in TWO's complement form, and then dropping the leftmost carry if it is a 1. We summarize the above steps below:

$$
\begin{array}{ccccccl}
 & 0 & 1 & 0 & 0 & 0 & \leftarrow +8_{10} \\
 & 1 & 1 & 0 & 1 & 0 & \leftarrow -6_{10} \\
\hline
1 & 0 & 0 & 0 & 1 & 0 &
\end{array}
$$

drop   sign bit

(b) To form the sum $6_{10} - 8_{10}$ following the above procedure, we convert -8_{10} into its TWO's complement representation. Since $+8_{10}$ is 01000_2, its TWO's complement is 11000_2. Binary representation of $+6_{10}$ is 00110_2. Thus we add 00110_2 and 11000_2 as indicated below:

$$
\begin{array}{ccccc}
0 & 0 & 1 & 1 & 0 \quad \leftarrow +6_{10} \\
1 & 1 & 0 & 0 & 0 \quad \leftarrow -8_{10} \\
\hline
1 & 1 & 1 & 1 & 0 \\
\end{array}
$$

$\uparrow$ sign bit

Since the sign bit is a 1, the sum represents a negative number in TWO's complement form. Its decimal equivalent can be shown to be equal to -2_{10}.

(c) TWO's complement respresentations of -8_{10} and -6_{10} are 11000_2 and 11010_2, respectively. We form their sum next:

$$
\begin{array}{cccccc}
& 1 & 1 & 0 & 0 & 0 \quad \leftarrow -8_{10} \\
& 1 & 1 & 0 & 1 & 0 \quad \leftarrow -6_{10} \\
\hline
1 & 1 & 0 & 0 & 1 & 0 \\
\end{array}
$$

drop $\quad$ sign bit

We drop the leftmost 1. The remainder is 10010 which has a 1 at the MSB position indicating the sum to be a negative number. Its TWO's complement is 01110 and, as a result, $-8_{10} - 6_{10} = -14_{10}$.

It should be pointed out here that the sum of two negative numbers represented in TWO's complement form will not yield a correct result if an overflow occurs, as the following example illustrates.

Example 2-21. Determine the sum $-9_{10} - 10_{10}$ using a 4-bit representation of the magnitude.

The TWO's complement representation of -9_{10} is 10111_2 and that of -10_{10} is 10110_2. Their sum using Table 2-12 is as follows:

$$
\begin{array}{cccccc}
& 1 & 0 & 1 & 1 & 1 \\
& 1 & 0 & 1 & 1 & 0 \\
\hline
1 & 0 & 1 & 1 & 0 & 1 \\
\end{array}
$$

If we drop the 1 on the left, then the remainder is identified as a positive number whose decimal equivalent is $+13_{10}$. Thus the result of the addition process is not -19_{10}.

The reason behind this discrepancy is that to correctly represent the magnitude of the sum we need 5 bits. Hence to obtain the correct result we should represent both augend and the addend using a 5-bit representation of the magnitude. The revised calculations are outlined below:

$$
\begin{array}{rccccccc}
-9_{10} & & 1 & 1 & 0 & 1 & 1 & 1 \\
-10_{10} & & 1 & 1 & 0 & 1 & 1 & 0 \\
\hline
-19_{10} & 1 & 1 & 0 & 1 & 1 & 0 & 1 \\
\end{array}
$$

drop   sign bit

We now drop the leftmost 1. The sign bit has a 1 indicating the sum is a negative number in TWO's complement form. Its decimal equivalent is -19_{10}, as expected.

Binary Multiplication and Division

The multiplication of two binary numbers follows rules similar to the decimal system. If we represent by P_i the product of two binary digits A_i and B_i, then it is seen that $P_i = 1$ if and only if both A_i and B_i are 1; otherwise $P_i = 0$. Consequently, the switching function P_i can be expressed as $A_i \cdot B_i$. The following example illustrates the binary multiplication procedure.

Example 2-22. Form the product of $+11_{10}$ and $+5_{10}$ using the binary multiplication method.

The binary representations of $+11_{10}$ and $+5_{10}$ without the sign bits are 1011_2 and 101_2, respectively. Let us consider 1011_2 as the multiplicand and 101_2 as the multiplier. Since the multiplier consists of 3 bits, the multiplication process essentially consists of three steps. In the first step of the multiplication process we multiply the multiplicand by the LSB of the multiplier to form the partial product. In the next step, the multiplicand is multiplied by the bit to the left of LSB. The result is shifted to the left by 1 bit position and added to the previous partial product to form the accumulated partial product. In the last step, the multiplicand is multiplied by the MSB of the multiplier. The result is shifted to the left by 2 bits position and added to the accumulated partial product obtained at the end of second step giving the final product. The various steps are outlined in detail below.

Step 1

$$
\begin{array}{ccccl}
 & 1 & 0 & 1 & 1 & \leftarrow \text{multiplicand } +11_{10} \\
\times & & 1 & 0 & 1 & \leftarrow \text{multiplier } +5_{10} \\
\hline
 & 1 & 0 & 1 & 1 & \leftarrow \text{first partial product}
\end{array}
$$

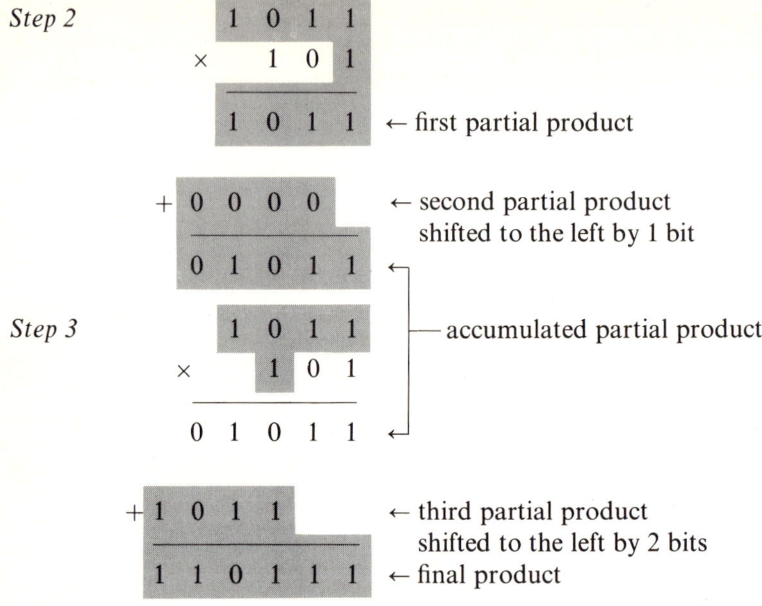

The bits being multiplied at each step of the multiplication process, their product, and the partial products are shown inside the lightly shaded areas. The decimal equivalent of the final product 110111_2 is $+55_{10}$, as expected.

If we denote symbolically the multiplicand and the multiplier by $A_3 A_2 A_1 A_0$ and $B_2 B_1 B_0$, then the first partial product obtained at the end of Step 1 is given by $(A_3 \times B_0)$, $(A_2 \times B_0)$, $(A_1 \times B_0)$, $(A_0 \times B_0)$. Since $B_0 = 1$, the first partial product is identical to $A_3 A_2 A_1 A_0$. In the next step, $A_3 A_2 A_1 A_0$ is multiplied by $(B_1 \times 2)$ which is same as multiplying $A_3 A_2 A_1 A_0$ by B_1 and shifting the result to the left by 1 bit position. B_1 being a 0, the second partial product has all 0's. The shifted second partial product is then added to the first partial product to generate the accumulated partial product given symbolically by $[(A_3 \times B_1), (A_2 \times B_1), (A_1 \times B_1), (A_0 \times B_1), 0] + [(A_3 \times B_0), (A_2 \times B_0), (A_1 \times B_0), (A_0 \times B_0)]$. In our example, the accumulated partial product is 01011_2, obtained by adding $00000_2 + 1011_2$. Finally, in the third step, $A_3 A_2 A_1 A_0$ is multiplied by $(B_2 \times 2^2)$ or equivalently, $A_3 A_2 A_1 A_0$ is multiplied by B_2 and shifted to the left by 2 bits. B_2 being a 1, the third partial product is identical to $A_3 A_2 A_1 A_0$. The shifted third partial product is added to the accumulated partial product of the previous step to generate the final product, given symbolically by $[(A_3 \times B_2), (A_2 \times B_2), (A_1 \times B_2), (A_0 \times B_2), 0, 0] + [\{(A_3 \times B_1), (A_2 \times B_1), (A_1 \times B_1), (A_0 \times B_1), 0\} + \{(A_3 \times B_0), (A_2 \times B_0), (A_1 \times B_0), (A_0 \times B_0)\}]$.

We also observe that the final answer contains 6 bits. In general, a product of a p-bit number with a q-bit number will have $(p + q)$ bits.

The method followed to divide a binary number (dividend) by another binary number (divisor) is essentially the same as that followed in the decimal number system. The method is illustrated next with the aid of an example for a positive integer-valued dividend and divisor with the latter being smaller in magnitude than the former.

Example 2-23.   Divide $+38_{10}$ by $+7_{10}$ using the binary division method.

The binary representations of the dividend 38_{10} and the divisor 7_{10} without the sign bits are 100110_2 and 111_2, respectively. Binary division is performed by repeated subtraction of the divisor from the dividend. In the first step, we compare the magnitude of the divisor with that of the most significant 3 bits of the dividend. Since the divisor is greater we cannot subtract it from the most significant 3 bits of the dividend. Next we subtract the divisor from the most significant 4 bits of the divisor and place a 1 in the MSB position of the quotient. The remainder after first subtraction is 10_2. We add an LSB to the remainder which is the fifth bit from the left of the dividend to give us a new remainder of 101_2. In the third step, we compare the magnitude of this remainder with that of the divisor. Since the divisor is greater we cannot subtract it from the remainder of the previous step. Thus we place a 0 in the quotient and bring in the sixth bit from the left of the dividend (which happens to be its LSB) and obtain the modified remainder of 1010_2. In the last step, the dividend is subtracted from the modified 4-bit remainder of the previous step and a 1 is placed to the right of 0 in the quotient. The result of the subtraction is 11_2 which is the final remainder. Since all of the bits of the dividend have been compared, the result of the division process yields a quotient of 101_2 and a remainder of 11_2. All the steps involved in the division process are described below.

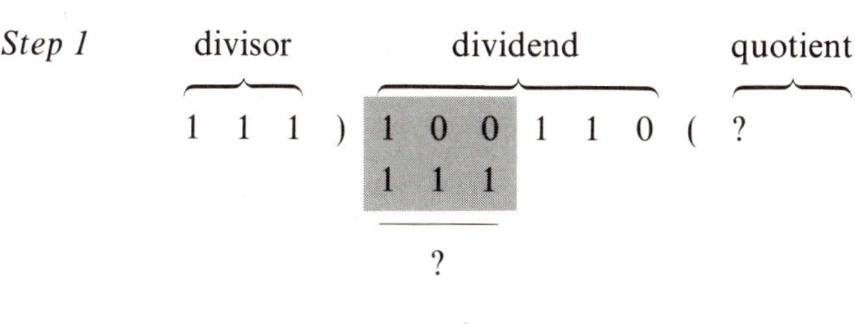

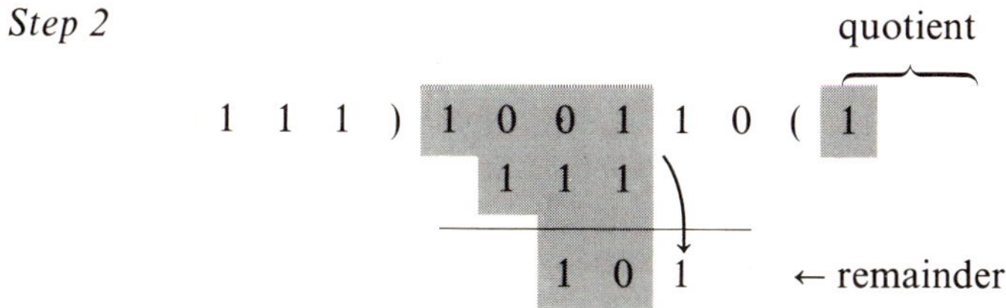

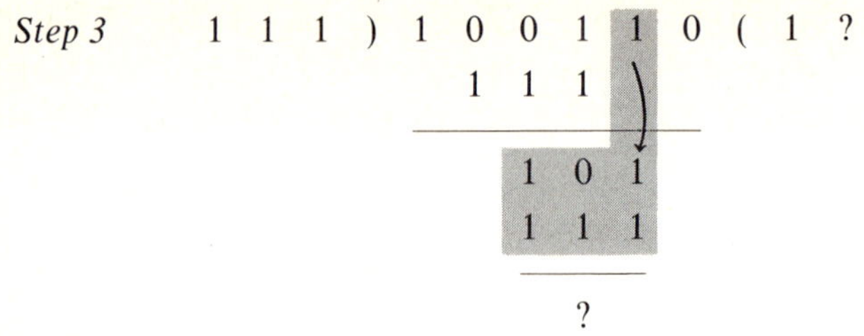

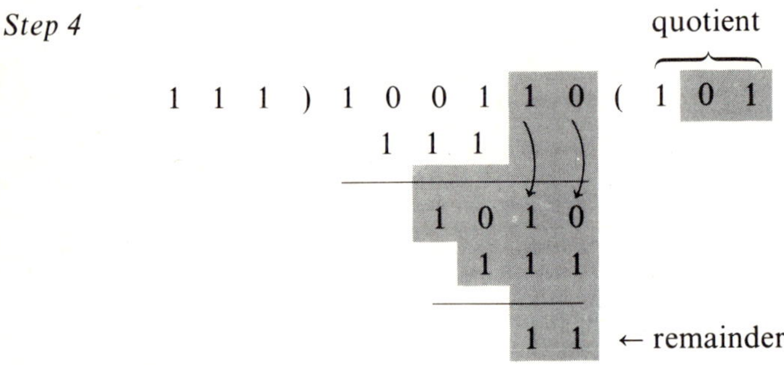

The numbers and bits involved at each step of the division process are shown in the lightly shaded area.

2-8   Coding in Digital Systems

Digital systems usually use the binary number representation and arithmetic because of the compatibility of the binary representation with the conventional digital circuitry. In a number of applications, however, it is often convenient for the user to make use of the decimal representation and arithmetic. Since commonly used digital circuitry are binary in nature, it is necessary to code decimal numbers using binary digits and then design the proper circuits for manipulating these coded decimal numbers. Such coded representations are used in displaying numbers as in digital voltmeters and digital frequency meters. Binary coded decimal numbers are also used in a number of digital computers specifically designed for business applications such as payroll, accounting, and tax calculations. In these machines, the arithmetic operations are also performed in the decimal system. In this section we briefly review a number of such codes.

Natural Binary Coding

Positive and negative decimal numbers can be represented using the conventional binary coding discussed earlier in Section 2-6. This type of coding is commonly used in scientific computers and most other digital systems. Table 2-13 illustrates such coding schemes with 4 bits. Using 4 bits we can represent 16 decimal numbers.

In *offset binary*, the normal binary representation of positive decimal numbers is essentially used; however, about half of 2^m numbers generated by m bits are used to represent negative numbers and the remainder represents positive numbers, as illustrated in Table 2-13 for $m = 4$. Here all ZERO's represent the most negative number and all ONE's represent the most positive number. Note that a negative number is identified by a 0 in the MSB position and a positive number by a 1 in the MSB position. Also, there is no representation for the decimal number $+8$. The offset binary coding is not suitable for arithmetic computations; it is primarily used in bipolar digital-to-analog conversion.

TABLE 2-13

Representation of Positive and Negative Decimal Numbers Using Straight Binary Coding

Decimal Number	Sign-Magnitude	Offset Binary	ONEs' Complement	TWO's Complement
-8		0000		1000
-7	1111	0001	1000	1001
-6	1110	0010	1001	1010
-5	1101	0011	1010	1011
-4	1100	0100	1011	1100
-3	1011	0101	1100	1101
-2	1010	0110	1101	1110
-1	1001	0111	1110	1111
-0	1000	1000	1111	0000
$+0$	0000		0000	
$+1$	0001	1001	0001	0001
$+2$	0010	1010	0010	0010
$+3$	0011	1011	0011	0011
$+4$	0100	1100	0100	0100
$+5$	0101	1101	0101	0101
$+6$	0110	1110	0110	0110
$+7$	0111	1111	0111	0111
$+8$				

In sign-magnitude, ONEs' complement, and TWO's complement coding, the leftmost bit is the sign bit—a 0 for a positive number and a 1 for a negative number. The remaining 3 bits have been used to code the magnitude of the decimal number. Both in sign-magnitude and in ONEs' complement coding, the decimal number 0 does not have a unique representation. This aspect may lead to problems in some digital hardware. On the other hand, in TWO's complement coding and offset binary coding, the decimal 0 has a unique representation.*

---

* By complementing the leftmost bit, a number in offset binary code can be converted to its representation in TWO's complement form and vice versa.

Decimal numbers with more than one digit can be coded using additional bits by simple extension of that shown in Table 2-13. Thus, for example, using 5 bits, we can code decimal numbers in the range -31 to $+31$. For example, the decimal number -25_{10} is represented as 100111_2 in TWO's complement coding. In some applications, it is preferable to code each decimal digit in a decimal number separately using a group of bits. This is discussed next.

Binary-Coded-Decimal (BCD) Representation

To represent the ten decimal digits 0 through 9, we need at least 4 bits. However, we can have $2^4 = 16$ different combinations of these 4 bits, of which ten combinations can be used to code the 10 decimal digits and the remaining 6 remain superfluous. The number of ten combinations out of 16 possible representations is $16!/6! \cong 2.8 \times 10^{10}$, which is roughly the number of possible BCD codes. Of these possible codes there are about 7.6×10 unique codes.[1] We discuss here only a few of the commonly used BCD codes.

If we denote the 4 bits as $A_3 A_2 A_1 A_0$, then by assigning different weights w_i to each bit A_i, a number of codes can be generated. The decimal equivalent of the coded digit $A_3 A_2 A_1 A_0$ is thus given by

$$n = w_3 A_3 + w_2 A_2 + w_1 A_1 + w_0 A_0 \qquad (2\text{-}45)$$

For example, if we choose $w_3 = 8$, $w_2 = 4$, $w_1 = 2$, and $w_0 = 1$, we arrive at

$$n = 8A_3 + 4A_2 + 2A_1 + A_0 \qquad (2\text{-}46)$$

This particular BCD code is more commonly known as the *8-4-2-1 code*. There are 16 other such weighted codes with positive weights, that is, $w_i > 0$. Table 2-14 shows two such weighted codes along with the 8-4-2-1 code.

Often, unweighted codes are preferred for coding decimal numbers. One such code is the *excess-3 code* shown in the last column of Table 2-14. This code is obtained from the 8-4-2-1 code by adding the binary equivalent of 3 to each coded representation. Thus the excess-3 representation of 8_{10} is $1000_2 +$

TABLE 2-14
Several BCD Codes

Decimal Number	8-4-2-1 Code	5-2-1-1 Code	3-3-2-1 Code	Excess-3 Code
0	0000	0000	0000	0011
1	0001	0001	0001	0100
2	0010	0011	0010	0101
3	0011	0110	0011	0110
4	0100	0111	0101	0111
5	0101	1000	1010	1000
6	0110	1001	1100	1001
7	0111	1100	1101	1010
8	1000	1110	1110	1011
9	1001	1111	1111	1100

$0011_2 = 1011$. An interesting feature of this code is that the NINE's complement of the decimal digit* is obtained by complementing each bit. For example, the NINE's complement of 8_{10} is 1_{10}, whose excess-3 representation is obtained by complementing the representation of 8_{10}. As a result design of digital circuits for performing decimal arithmetic using this code is much simpler than with the weighted codes.

Unit-Distance Codes

All of the codes discussed so far have one major drawback which arises due to the change of more than one bit in going from the coded representation of certain numbers to that of their consecutive numbers. For example, in the BCD code, as we change from the decimal number 7 to the decimal number 8, all four bits change states. A similar situation occurs in *5-2-1-1 code* in going from decimal 4 to decimal 5, and so on, for the other codes. In using codes to read digitally the angle of a shaft or in similar applications, multiple changes may lead to errors in reading the information. This problem can be avoided by using the *Gray code* shown in Table 2-15. Here only one bit changes from a number to its adjacent number and such a code is usually called a unit-distance code which is no longer a weighted code. Many other unit-distance codes can be generated simply by following adjacent squares in a Karnaugh map so that all squares are traversed in the process.

An interesting feature of the Gray code can be observed if we examine carefully the entries of Table 2-15. We first observe that the top half of all the

TABLE 2-15
A Unit-Distance Code

Decimal Number	Gray Code
0	0000
1	0001
2	0011
3	0010
4	0110
5	0111
6	0101
7	0100
8	1100
9	1101
10	1111
11	1110
12	1010
13	1011
14	1001
15	1000

---

* NINE's complement of x_{10} is defined as $(9 - x)_{10}$.

numbers (above the dotted line) have a 0 in the MSB position whereas the remaining numbers have a 1 at the same position. Next, if we ignore the MSB, then the entries in the bottom half are sort of "mirror images" of the entries in the top half. A similar feature is seen to hold for the top half (and also the bottom half) numbers if we divide them in the middle. Finally, if we group the numbers in two, similar property holds for each adjacent pair of numbers. Because of this property, the Gray code is also known as the *reflected binary code.*

Example 2-24. Represent the decimal number 629 using the various codes discussed above.

Consider first the 8-4-2-1 code. From Table 2-13, the coded form of the decimal number 6 is 0110. Similarly the coded forms of the other two decimal numbers 2 and 9 are respectively 0010 and 1001. Hence the BCD representation of 629 is

$$0110 \quad 0010 \quad 1001 \quad \text{(BCD code)}$$

where the first 4 bits represent the most significant digit 6, and so on.

In a similar manner, the representations of 629 in other codes can be found and are given below:

$$
\begin{array}{llll}
1001 & 0011 & 1111 & \text{(5-2-1-1 code)} \\
1100 & 0010 & 1111 & \text{(3-3-2-1 code)} \\
1001 & 0101 & 1100 & \text{(excess-3 code)}
\end{array}
$$

2-9 Summary

The fundamental concepts and the mathematical foundation of digital systems are discussed in this chapter. Digital systems process information which is presented as discrete variables commonly known as switching variables. Thus the input, output, and internal variables in digital systems are switching variables which, unlike variables in analog circuits and systems, can assume only one of two possible values (Section 2-1). The two possible values are usually designated as 0 (logical ZERO) and 1 (logical ONE). Since the designation is arbitrary, there are two possible assignment schemes known as positive logic and negative logic assignments (Section 2-2). The output switching variables of a logic circuit are then functions of the input switching variables and possible internal switching variables. A function of switching variables, usually known as a switching function, can be represented by making use of the basic operations such as AND, OR, and NEGATION as discussed in Section 2-3. The theory, operation, and design of logic circuits hinge upon the switching algebra theorems, reviewed in Section 2-4. No attempt has been made to present the switching algebra in a formal rigorous manner. A convenient graphical representation of the switching function is with the aid of a Karnaugh map described in Section 2-5. Such maps are widely used in simplifying the expression of a switching function as will be outlined in the next chapter. Section 2-6 discusses the representation of numbers in digital systems; in particular, the various methods of representing negative numbers. Section 2-7 is concerned with the basic approaches followed in performing arithmetic operations. Finally, Section 2-8 covers the coding of decimal numbers.

References

1. A. Barna and D. I. Porat, *Integrated Circuits in Digital Electronics*, Wiley-Interscience, New York, 1973.
2. L. W. Bell, *Digital Concepts*, Publication No. 062-1030-00, Tektronix, Inc., Beaverton, Oregon, June 1969.
3. H. J. Beuscher, A. H. Budlong, M. B. Haverty, and G. Waldbaum, *Electronic Switching Theory and Circuits*, Van Nostrand Reinhold, New York, 1971.
4. T. L. Booth, *Digital Networks and Computer Systems*, John Wiley & Sons, New York, 1971.
5. D. L. Dietmeyer, *Logic Design of Digital Systems*, Allyn and Bacon, Boston, Mass., 1971.
6. F. J. Hill and G. R. Peterson, *Introduction to Switching Theory and Logical Design*, John Wiley & Sons, New York, 1968.
7. V. T. Rhyne, *Fundamentals of Digital System Design*, Prentice-Hall, Englewood Cliffs, N.J., 1973.
8. H. Torng, *Introduction to the Logical Design of Switching Circuits*, Addison-Wesley, Reading, Mass., 1964.
9. W. E. Wickes, *Logic Design with Integrated Circuits*, John Wiley & Sons, New York, 1968.
10. The Application Engineering Department Staff, *Digital-to-Analog Converter Handbook*, Hybrid Systems Corporation, Burlington, Mass., 1970.

Problems

2-1 The RTL logic circuits have the following allowable ranges:

$$0.9 \text{ V} < \text{HIGH} \leq 3.6 \text{ V}$$

$$0 \text{ V} \leq \text{LOW} < 0.3 \text{ V}$$

Using the positive logic assignment and 0, 1 labeling, determine the truth values of an output signal X whose measured values are:

(a) 1.8 V   **(b)** 3.5 V   **(c)** 0.15 V
(d) 0.299 V   **(e)** 0.31 V   **(f)** 0.8 V

2-2 Repeat Problem 2-1 with negative logic assignment.

2-3 Determine the truth values of the following:
(a) $1 + 0 + (1 \cdot 0)$ **(b)** $1 \cdot 0 \cdot (1 + 0)$
(c) $\bar{1} + 0 + (1 \cdot \bar{0})$ **(d)** $1 \cdot (\overline{\bar{0} + 1 \cdot \bar{1}}) \cdot \bar{0}$

2-4 Prove the following theorems using truth tables:
(a) Theorems 2-1, 2-2, 2-3, and 2-4
(b) Theorems 2-6 and 2-7
(c) Theorems 2-5 and 2-9
(d) Theorems 2-12 and 2-13

2-5 Develop the truth table for the following switching functions:
(a) $XY\bar{Z}W + XY(\bar{Z} + X\bar{Y})$ (b) $\overline{XY\bar{Z}W + XY(\bar{Z} + X\bar{Y})}$
(c) $XY\bar{Z}\overline{W} + XY(\bar{Z} + X\bar{Y})$

2-6 Using truth tables show that
(a) $X + YZ \neq (X + Y) \cdot Z$ (b) $\overline{X + Y} \neq \bar{X} + \bar{Y}$ (c) $\overline{XY} \neq \bar{X} \cdot \bar{Y}$

2-7 Identify the minterms and maxterms for a switching function of the four switching variables, W, X, Y, and Z in the following list of expressions:
(a) $XY\bar{Z}W$ (b) $XW\bar{Z}\overline{W}$ (c) $X + YZ\overline{W}$
(d) $XZXW\bar{Y}$ (e) $X + Y + Z + W$ (f) $XY\overline{W}$
(g) $\bar{X} + Y + \bar{Z} + W$

2-8 For each of the minterms identified in Problem 2-7, determine the truth values of the switching variables for which the minterm takes the truth value of 1.

2-9 Prove the following theorems using previous switching algebra theorems:
(a) Theorem 2-16, (b) Theorem 2-17, (c) Theorem 2-18, (d) Theorem 2-21.
Indicate the theorems used in the proofs.

2-10 Prove Theorem 2-23.

2-11 In addition to the OR, AND operations discussed in Section 2-3, 14 other operations can be defined to combine two switching variables. Using a truth table define these new operations.

2-12 Prove the following theorems using switching algebra theorems:
(a) *Theorem 2-24.* $XY + YZ + \bar{X}Z = XY + \bar{X}Z$
(b) *Theorem 2-25.* $(X + Y)(\bar{X} + Z) = XZ + \bar{X}Y$
(c) *Theorem 2-26.* $XY + \bar{X}Y = Y$
(d) *Theorem 2-27.* $(X + Y)(\bar{X} + Y) = Y$
Indicate the theorems used in the proofs.

2-13 Prove the following identities:
(a) $X \cdot g(X, Y) = X \cdot g(1, Y)$ (b) $\bar{X} \cdot g(X, Y) = \bar{X} \cdot g(0, Y)$
(c) $X + g(X, Y) = X + g(0, Y)$ (d) $\bar{X} + g(X, Y) = \bar{X} + g(1, Y)$
where $g(X, Y)$ is a switching function.

2-14 Determine the Karnaugh map representation of the switching functions of Problem 2-5.

2-15 Prove Theorems 2-24, 2-25, 2-26, and 2-27 using Karnaugh maps.

2-16 Repeat Problem 2-6 using the Karnaugh map.

2-17 Determine the Karnaugh map representations of the following 3-variable switching functions:
(a) $\bar{X} + \bar{Y} + Z$ (b) $XY + YZ + \bar{X}\bar{Z}$
(c) $\bar{X}Y + Y\bar{Z} + X\bar{Z}$ (d) $\bar{X}(X + Y) + Z$

   (e) $(X + Y + Z)(X + \bar{Y} + Z)(\bar{X} + Y + Z)$ (f) $X + Y\bar{Z}$
   (g) $XY + \bar{Y}Z$

2-18 Prove the following identities using the Karnaugh maps:
   (a) $X(\bar{X} + Y) = XY$ (b) $\overline{X\bar{Y}}(X + Y) = X\bar{Y} + \bar{X}Y$
   (c) $X(X + Y)Y = XY$ (d) $\overline{X + Y} + XY = (X + \bar{Y})(\bar{X} + Y)$
   (e) $(X + Y)(Y + Z)(Z + X) = XY + YZ + ZX$
   (f) $(X + Y)(Y + Z)(Z + \bar{X}) = (X + Y)(Z + \bar{X})$
   (g) $XY + \bar{X}(Z + \bar{Y}) = (X + Z)Y + \bar{X}\bar{Y}$
   (h) $(X + Y + Z + W)(X + Y + Z + \bar{W})(\bar{X} + Y + Z + W) = Y + Z + XW$
   (i) $\bar{X}\bar{Y} + X\bar{Z} + Y\bar{W} + \bar{Z}\bar{W} = \bar{X}\bar{Y} + X\bar{Z} + Y\bar{W}$

2-19 Prove the above identities using switching algebra theorems.

2-20 Using DeMorgan's theorem and Theorems 2-24 and 2-26, prove the following:

$$(\bar{X} + \bar{Y})(\bar{Y} + \bar{Z})(X + \bar{Z}) = (\bar{X} + \bar{Y})(X + \bar{Z})$$

$$(\bar{X} + \bar{Y})(X + \bar{Y}) = \bar{Y}$$

2-21 Determine the binary equivalents of the following positive decimal numbers (limit fractional part to 6 bits)
   (a) 101   (b) 0.33   (c) 75   (d) 33.33
   (e) $\sqrt{2}$ (f) 129.15   (g) 0.625   (h) 0.75

2-22 Determine the decimal equivalent of the following positive binary numbers (leftmost bit is the sign bit):
   (a) 00.111   (b) 011.111   (c) 010101
   (d) 0111111   (e) 0110110   (f) 0101.101

2-23 Determine the ONEs' complement of the binary equivalent of the positive decimal numbers listed in Problem 2-21.

2-24 Determine the ONEs' complement of the binary numbers of Problem 2-22.

2-25 Determine the TWO's complement of the binary equivalent of the positive decimal numbers of Problem 2-21.

2-26 Determine the TWO's complement of the binary numbers of Problem 2-22.

2-27 Perform the following arithmetic operations using TWO's complement representation of negative numbers and binary addition:
   (a) $17_{10} + 12_{10}$ (b) $17_{10} - 12_{10}$
   (c) $12_{10} - 17_{10}$ (d) $-12_{10} - 17_{10}$
   (e) $23_{10} + 7_{10}$ (f) $23_{10} - 7_{10}$
   (g) $7_{10} - 23_{10}$ (h) $-23_{10} - 7_{10}$

2-28 Develop the rules for forming the difference of two binary numbers using the ONEs' complement of the negative number.

2-29 Determine the decimal equivalents of the following octal and hexadecimal numbers (indicated by the subscripts):
   (a) 637.25_8 (b) 637.25_{16} (c) $2B9E.7C3_{16}$ (d) 54.2137_8

2-30 Convert the following positive numbers into octal numbers and hexadecimal numbers:
   (a) 637.25_{10} (b) 1011.0101_2 (c) 89.372_{10} (d) 110.11011_2

2-31 Develop the following weighted BCD codes:
   (a) 6-3-1-1   **(b)** 3-3-2-1   **(c)** 2-4-2-1

2-32 Develop the Gray code for representing decimal numbers 0 through 7 using 3 bits.

2-33 Develop an alternative code to represent decimal numbers 0 through 15 using 4 bits such that only one bit changes in the coded representations of two consecutive numbers. (Consider 0 and 15 to be consecutive numbers in developing the code.)

CHAPTER 3

Combinational Digital Circuits

The input, output, and internal signals of a digital circuit are switching variables. Each variable at any instant can have one of two possible values—logical ZERO and logical ONE. The purpose of a digital circuit is to perform some arithmetic or logical operations on the input signal(s) and develop the appropriate output signal(s). The input and output signals are often called *data*. There are basically two modes of data transmission. In the serial mode, each data is sent 1 bit at a time over a single wire, whereas, in the parallel mode, all the bits of a single data are sent at the same time using multiple wires. As a result, any digital operation can be performed serially if the data involved are available in a serial mode or can be performed, instead, in a parallel fashion if the data are available in a parallel mode. In serial operation the wires and logic circuits can be time-shared which, in turn, minimizes the cost of data transmission and processing. On the other hand, in parallel operation, data transmission and processing can be performed at the highest possible rate but at a higher cost as a result of a larger number of wires and more complex logic circuits.

There are two types of digital circuits. In the first type, known as *combinational circuits*, the truth value of an output variable at a certain instant of time is determined completely by the truth values of the input switching variables at the same instant. In the second type of digital circuits, the output depends on both the present and the past values of inputs and some additional internal switching variables. Such circuits are more commonly known as *sequential circuits*. In both types of digital circuits, the output signals can thus be expressed as switching functions of input, and possibly some internal switching variables.

In this chapter, we discuss primarily the analysis and design of combinational circuits. The sequential circuits are treated in the following chapter.

3-1   A Simple Design Example

Before we begin a formal study of combinational digital circuits it is instructive to consider a simple but typical design problem. The problem is to design a circuit to count the votes cast by four shareholders of a small company. The total number of outstanding shares of the company is 100, and the owner of one share is entitled to one vote at the stockholders' meeting. The four shareholders of the company and the number of shares they own are as follows: A—40 shares, B—30 shares, C—20 shares, and D—10 shares. At the stockholders' meeting, it is necessary to have 60 percent of the vote cast in favor of a measure to have it passed. It can be seen that a measure passes if either A *and* B (independent of the votes of C and D) or A *and* C (independent of the votes of B and D) or B *and* C *and* D (independent of the action of A) vote in favor of the measure.

We design the voting machine using switches, a battery, and a bulb. Shareholders indicate their actions by positioning a set of switches—the switches are moved to the ON position for voting " yes," and they are moved to the OFF position for voting " no." After all shareholders have moved their respective switches, the state of the bulb indicates the outcome of the voting. If 60 percent of the votes cast are in favor of the motion, then the bulb is lit, otherwise it remains in the OFF state. For convenience, we denote the switching variable characterizing the state of the switches under the control of the shareholder A by A, and so on, for the others. Likewise we denote the switching variable characterizing the state of the bulb by L.

Before we develop the realization of the voting machine, let us first examine the implementations of the AND and OR operations by switches, as shown in

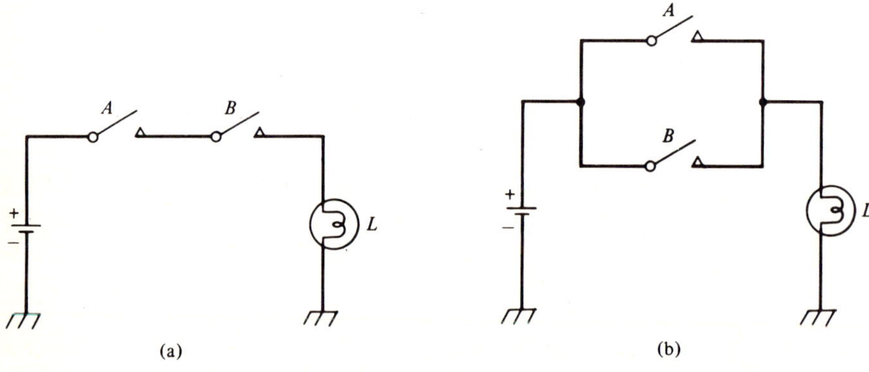

(a)   (b)

Figure 3-1   (a) Implementation of AND operation by switches, (b) implementation of OR operation by switches.

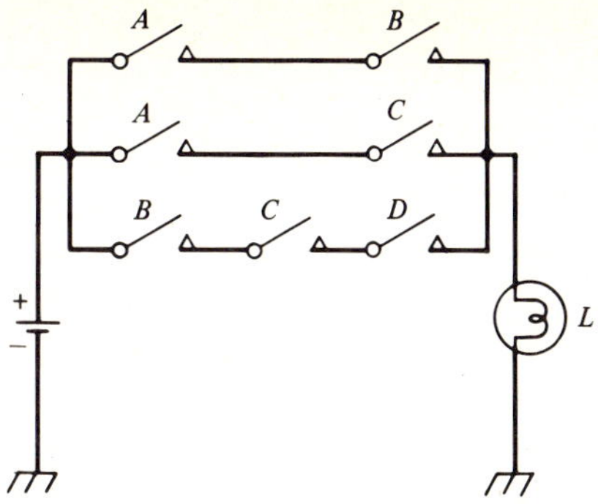

Figure 3-2 Realization of the voting machine.

Fig. 3-1. Note that in Fig. 3-1(a), if both A and B are ON, then L is ON; otherwise L is OFF. Hence this circuit implements $L = AB$. Similarly, in Fig. 3-1(b), if either A or B or both are "ON," then L is ON, otherwise L is in the OFF state. As a result this circuit implements $L = A + B$.

Now going back to the operation of the voting machine it is evident that L is ON if either "A and B are ON" or "A and C are ON" or "B and C and D are ON." Hence the expression for the switching function L is

$$L = AB + AC + BCD \qquad (3\text{-}1)$$

and the corresponding circuit realization is as shown in Fig. 3-2.

It is apparent from the above discussion that the first step in the design of a combinational circuit is to develop the expression for the pertinent switching function(s) from the problem statement. The switching function is then implemented in hardware form in the second step. For our design example of this section, the expression for the switching function was developed more or less in a heuristic fashion and no attempt was made to obtain an expression that would require fewer components for implementation. Later in this chapter we outline a systematic approach to derive the switching function and also outline a technique to simplify a switching function.

3-2 Logic Gates

As indicated earlier, the implementation of a combinational circuit requires the realization of the switching functions describing the operation of the circuit. To this end we need circuits that can implement the logical operations needed to combine the input switching variables. Three basic switching operations were described earlier in Section 2-3. Several other types of switching operations will be described later in this section. In Fig. 3-1 we illustrated the implementation of AND and OR operations using switches. Electronic circuits implementing these logical operations are more commonly known as *logic gates* and various types of IC gates are widely available as off-the-shelf items.

Basic Gates

The digital circuit implementing the OR operation is called an *OR gate*; its symbolic representation is shown in Fig. 3-3(a). The implementation of the AND operation is achieved by the *AND gate*. Its symbolic representation is sketched in Fig. 3-3(b). Finally, the NEGATION operation is performed by an *INVERTER gate*, shown symbolically in Fig. 3-3(c). This latter gate is also called a *NOT gate*. The OR and the AND gate are available with two or more input terminals.

As an illustration of the implementation of a switching function using the above basic gates, consider the following example.

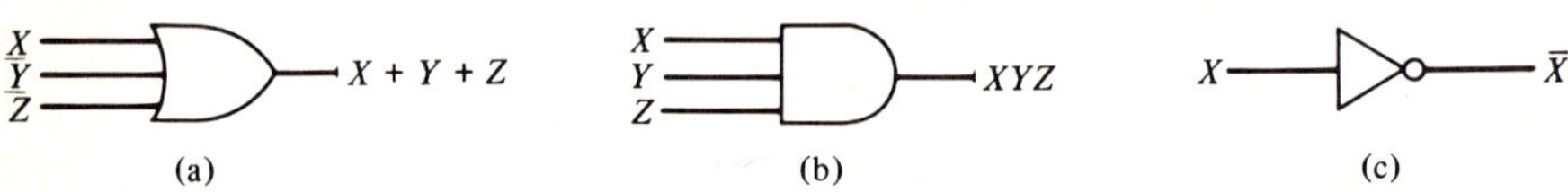

(a)   (b)   (c)

Figure 3-3 Symbolic representations of the three basic gates: (a) OR gate, (b) AND gate, (c) NOT gate.

Example 3-1. Realize the switching function

$$C = \overline{X + Y\overline{Z}} + Z \tag{3-2}$$

We first implement the product $Y\overline{Z}$ using a 2-input AND gate. The output of this AND gate and the input variable X are fed into a 2-input OR gate which is followed by a NOT gate, generating $\overline{X + Y\overline{Z}}$. The output of the NOT gate and the input variable Z are combined by an OR gate to implement the switching function C. The complete realization is shown in Fig. 3-4(a). Note that if the input variable Z is not available in its complemented form, an additional NOT gate will be needed to produce $\overline{Z}$.

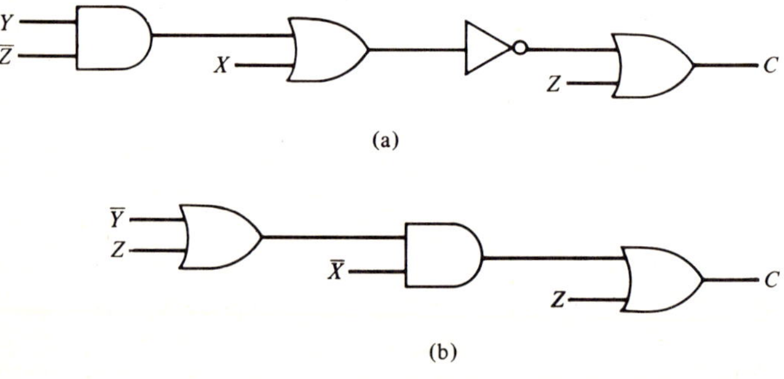

Figure 3-4 Circuit realization of a typical switching function (a) for Eq. (3-2), (b) for Eq. (3-3).

An alternate realization can be obtained by rewriting Eq. (3-2) with the aid of DeMorgan's theorem:

$$C = \bar{X} \cdot (\bar{Y} + Z) + Z \tag{3-3}$$

Implementation of Eq. (3-3) is sketched in Fig. 3-4(b).

The schematic diagram showing the interconnection of the gates implementing a switching function, as shown in Fig. 3-4, is known as a *logic diagram.*

In Section 3-7 we outline a systematic method to develop an expression for a switching function in a form which lends itself easily to implementation using the three basic gates. However, the three basic gates do not form a complete set, whereas the NOT gate with either the AND or the OR gate forms a *complete set.* This can be proven by showing that the AND gate can be implemented using an OR gate and three NOT gates. Similarly an OR gate can be realized using an AND gate and three NOT gates (Problem 3-1).

Gate Packages

In practice, a number of gates are fabricated on a single chip and then mounted in a package. The total number of gates assembled in a package depends on the number of pin connections available for input and output terminals, which in turn depends on the type of package being used. For example, in a 14-pin DIP, one pin is used for power supply connection, and another pin is used for ground connection. Thus the number of pins available for input and output terminal connections is 12. If two gates with single output each are assembled in the package, the number of pins available for input connections is 10, which can be divided between the two gates in various combinations. A commonly used combination is 4 inputs for each gate (two pins are then left unconnected). Another combination is three inputs for one gate and two inputs for the other gate.

The number of gates in a single package and the number of input terminals of each gate are used to designate the total package. For example, a *dual 4-input* gate package contains two gates with 4 inputs each. A *dual 3-2-input* gate package contains two gates, one gate with 3 inputs and the other gate with 2 inputs. Figure 3-5 shows some typical gate packages: a *quad 2-input* OR gate package, a *triple 3-input* AND gate package, and a *hex* INVERTER package.

Other Types of Gates

IC logic gates implementing operations other than the three basic operations discussed earlier are also available. We describe next several such gates.

NOR Gate.   The NOR operation is defined as

$$A = \overline{X + Y} \tag{3-4}$$

which is the complement of the OR operation. Thus the truth value of $\overline{X + Y}$ is 1 if and only if both X and Y have truth values of 0; otherwise $\overline{X + Y}$ is 0.

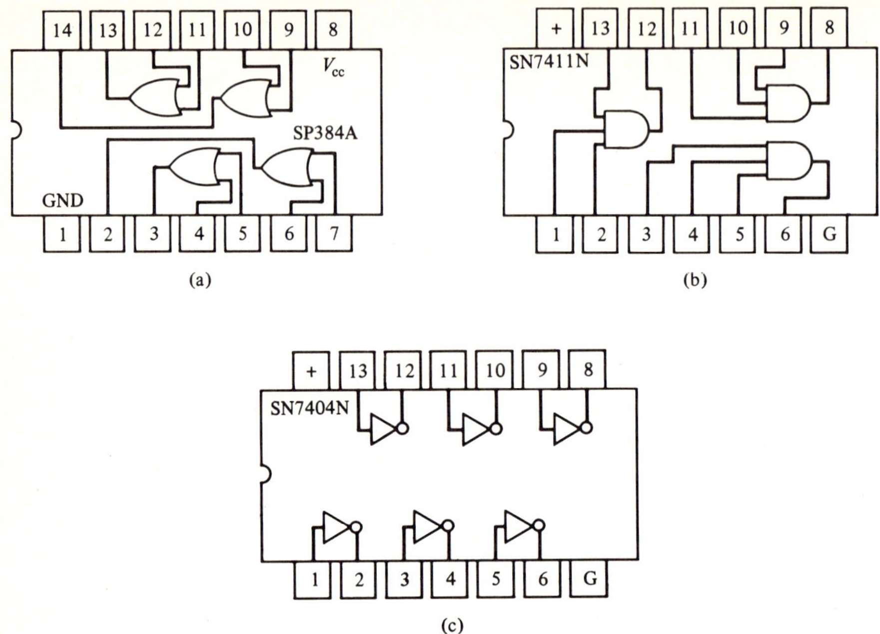

Figure 3-5 Top views of typical IC logic-gate packages: (a) quad 2-input OR gate ICP—SP384A, (b) triple 3-input AND gate ICP—SN7411N, (c) hex INVERTER ICP—SN7404N.

The truth table for the NOR operation is given in Table 3-1. The symbolic representation of the NOR gate is shown in Fig. 3-6(a). It follows from the definition that the NOR gate can be considered as an OR gate followed by an INVERTER, as shown in Fig. 3-6(b). An alternate realization of the NOR gate shown in Fig. 3-6(c) is arrived at by noting that $\overline{X + Y} = \overline{X} \cdot \overline{Y}$ by De-Morgan's theorem. This equivalence leads to the second symbolic representation sketched in Fig. 3-6(d).

It should be noted that the NOR gate by itself forms a complete set and, as a consequence, any switching function can always be realized using only NOR gates. This fact is proved in Section 3-8.

TABLE 3-1
The Truth Table for
the NOR Operation

X	Y	$\overline{X + Y}$
0	0	1
0	1	0
1	0	0
1	1	0

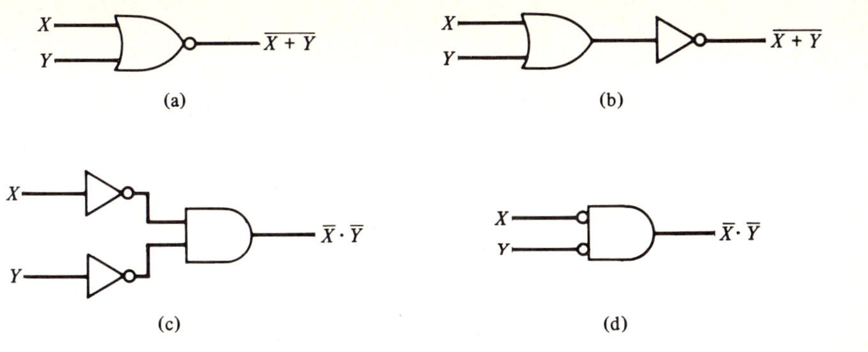

Figure 3-6   The NOR gate and its equivalent representations.

Some IC realizations provide simultaneously both the OR and NOR outputs. Such gates, often providing multiple outputs, are usually called OR–NOR gates. A typical OR–NOR gate is shown in Fig. 3-7.

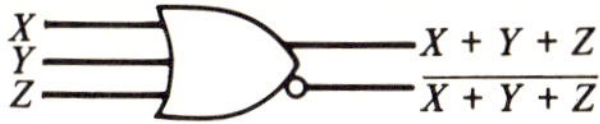

Figure 3-7   An OR–NOR gate.

NAND Gate.   Another useful operation is the NAND operation, defined as

$$A = \overline{X \cdot Y} \tag{3-5}$$

which takes the truth value of 0 if and only if both X and Y are 1; otherwise it is 1. Its corresponding truth table is shown in Table 3-2. The symbolic representation of the NAND gate is shown in Fig. 3-8(a). The two circuit realizations of the NAND gate indicated in Figs. 3-8(b) and 3-8(c) using the basic gates should be self-evident. Figure 3-8(c) leads to the alternate symbolic representation sketched in Fig. 3-8(d). Multi-output AND–NAND gates are also available. As shown later in Section 3-8, the NAND gate is also a complete set.

TABLE 3-2
The Truth Table for the
NAND Operation

X	Y	$\overline{XY}$
0	0	1
0	1	1
1	0	1
1	1	0

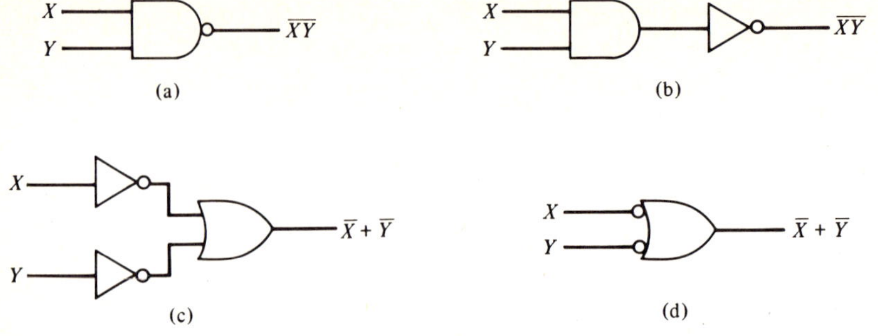

Figure 3-8 The NAND gate and its equivalent representations.

Exclusive-OR Gate. This gate implements the exclusive-OR operation de-noted symbolically as

$$A = X \oplus Y \tag{3-6}$$

Its operation is defined as follows: "$X \oplus Y$ is 1 if either X or Y is 1 (but not both); otherwise it is 0." The truth table for this operation is given in Table 3-3.

It can be easily shown that we can express

$$X \oplus Y = X\bar{Y} + \bar{X}Y = (\bar{X} + \bar{Y})(X + Y) \tag{3-7}$$

TABLE 3-3
The Truth Table for the
Exclusive-OR Operation

X	Y	$X \oplus Y$
0	0	0
0	1	1
1	0	1
1	1	0

The logic symbol of the exclusive-OR gate is sketched in Fig. 3-9. The exclusive-OR operation is commutative and associative, that is,

$$X \oplus Y = Y \oplus X \tag{3-8}$$

$$(X \oplus Y) \oplus Z = X \oplus (Y \oplus Z) = X \oplus Y \oplus Z \tag{3-9}$$

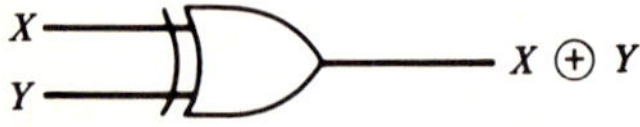

Figure 3-9 The exclusive–OR gate symbol.

Some useful properties and identities using this operation are summarized in Table 3-4, the proofs of which are left as an exercise (Problems 3-8). Additional identities involving this operation are to be found in Problems 3-9, 3-10, 3-11, 3-14, and 3-15.

TABLE 3-4
Useful Identities Using the Exclusive-OR Operation

$X \oplus 0 = X$	$X \oplus XY = X\bar{Y}$
$X \oplus 1 = \bar{X}$	$X \oplus \bar{X}Y = X + Y$
$X \oplus \bar{X} = 1$	$X \oplus (X + Y) = \bar{X}Y$
$X \oplus X = 0$	$X \oplus (\bar{X} + Y) = \bar{X} + \bar{Y}$

Coincidence Gate.   The coincidence operation, which is symbolically denoted as

$$A = X \odot Y \tag{3-10}$$

is defined as follows: "$X \odot Y$ is 1 if X and Y have the same truth values, otherwise it is 0." The corresponding truth table is given in Table 3-5. Comparison of this table with Table 3-3 shows that

$$X \odot Y = \overline{X \oplus Y} \tag{3-11}$$

TABLE 3-5
The Truth Table for the Coincidence Operation

X	Y	$X \odot Y$
0	0	1
0	1	0
1	0	0
1	1	1

which leads to the alternate name *exclusive-NOR* given to this operation. Two commonly used logic symbols for the gate implementing this operation are sketched in Fig. 3-10. The coincidence gate is also known as the *equality* and *comparator* gate. IC gates providing both exclusive-OR and exclusive-NOR outputs are also available.

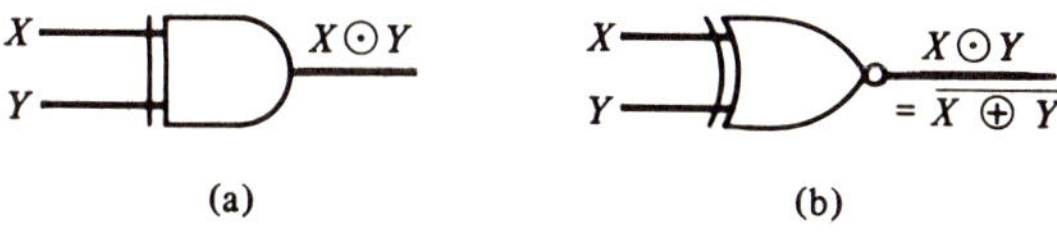

Figure 3-10   The coincidence gate symbols.

The coincidence operation is again commutative and associative. It follows from Eqs. (3-11) and (3-7) that

$$X \odot Y = XY + \bar{X}\bar{Y} = (\bar{X} + Y)(X + \bar{Y}) \tag{3-12}$$

Table 3-6 lists some interesting properties and identities using the coincidence operation. Some additional identities are given in Problems 3-14 through 3-17.

TABLE 3-6
Useful Identities Using the Coincidence Operation

$X \odot 0 = \bar{X}$	$X \odot XY = \bar{X} + Y$
$X \odot 1 = X$	$X \odot \bar{X}Y = \bar{X}\bar{Y}$
$X \odot \bar{X} = 0$	$X \odot (X + Y) = X + \bar{Y}$
$X \odot X = 1$	$X \odot (\bar{X} + Y) = XY$

AND–OR–INVERT (A-O-I) and AND–OR (A-O) Gates. In contrast to the previously discussed gates, these recently introduced types of gates do not implement a single binary operation but, instead, realize switching functions. The A-O-I gate implements a switching function expressed in a complemented sum-of-products form. Essentially, it consists of several multi-input AND gates whose outputs are combined by a NOR gate. An A-O-I gate is usually specified in terms of the total number of AND gates in the package and the number of input terminals of each AND gate. For example, Fig. 3-11 shows the schematic of a dual 2-wide 2-input A-O-I gate, with each A-O-I gate implementing a switching function of the form

$$\overline{A_1 B_1 + A_2 B_2}$$

where A_1, A_2, B_1, and B_2 are the input variables. Similarly, a 4-wide 4-2-3-2-input A-O-I gate package contains 4 AND gates with 4, 2, 3, and 2 inputs, respectively, implementing a switching function of the form

$$\overline{A_1 B_1 C_1 D_1 + A_2 B_2 + A_3 B_3 C_3 + A_4 B_4}$$

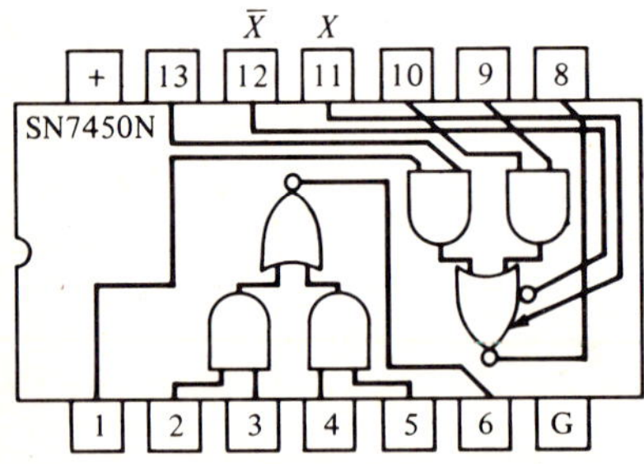

Figure 3-11   Top view of SN7450N—dual 2-wide two-input exp. A-O-I gate ICP.

Even though an A-O-I gate appears like a connection of several gates, its properties are similar to that of a single gate.

The A-O gate implements a switching function expressed in a sum-of-products form and has properties similar to the A-O-I gate.

Expander. Often an A-O-I gate may be provided with extra terminals to connect additional logic circuits called *expanders* to increase the complexity of the A-O-I operation. These expansion input terminals (labeled X and $\bar{X}$ in the upper A-O-I gate of Fig. 3-11) are essentially at the input of the NOR gate, allowing the connection of more AND gates (available in the expander package) to the NOR gate. Some A-O-I gates have two expansion input terminals, one for the input in normal form and the other for the input in its complemented form. The corresponding expander logic circuits must thus be able to provide outputs in both forms. A typical connection of this type of A-O-I gate with an expander is shown in Fig. 3-12. Other types of A-O-I gates provide only one expansion terminal and their corresponding expanders do not need to provide outputs in both forms. In general, the expansion input terminals, if not used, should be left open. Expander units are available in packages in various forms, such as dual 4-input AND gates, triple 3-input, quad 3-2-2-3 inputs, and so on.

Buffer. In practice, an IC gate, due to its output current limitations, is unable to drive a large number of other gates connected to its output. This problem is usually circumvented by connecting a *buffer* between the driving gate and the driven gates. The buffer is designed specifically to deliver large output current. There are basically two types of buffers. In one type, the logic level at the output of the buffer is the same as that at the input, and consequently its introduction into the system does not affect the overall logic implementation. The other type is a NOT gate. If the latter type is used, then either two NOT

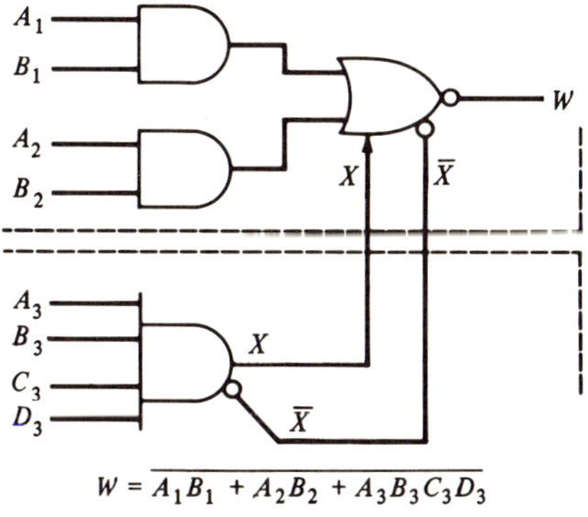

$$W = \overline{A_1 B_1 + A_2 B_2 + A_3 B_3 C_3 D_3}$$

Figure 3-12 An A-O-I gate with an expander element.

gates are connected in cascade to insure no change in logic implementation, or appropriate changes in logic implementation are made so that a single NOT gate can be used as a buffer.

3-3 Analysis of Combinational Circuits

A combinational circuit implemented using any one or more of the gates described in the previous section can be analyzed very easily to determine the expression for the output switching function. The following two examples illustrate the analysis procedure.

> ***Example 3-2.*** Analyze the logic diagram of Fig. 3-13 and determine the expression for the switching function A.
>
> For convenience in analysis, we denote the output of each gate by a letter symbol and write down the expressions for the output of each gate:

$$B = X + \bar{Y}$$
$$C = BZ$$
$$D = C\bar{X}$$
$$A = Y + D$$

Combining the above equations and eliminating the internal variables, we obtain

$$A = Y + C\bar{X} = Y + (BZ)\bar{X} = Y + [(X + \bar{Y})Z]\bar{X} \qquad (3\text{-}13)$$

The expression on the right-hand side of Eq. (3-13) is the desired answer.

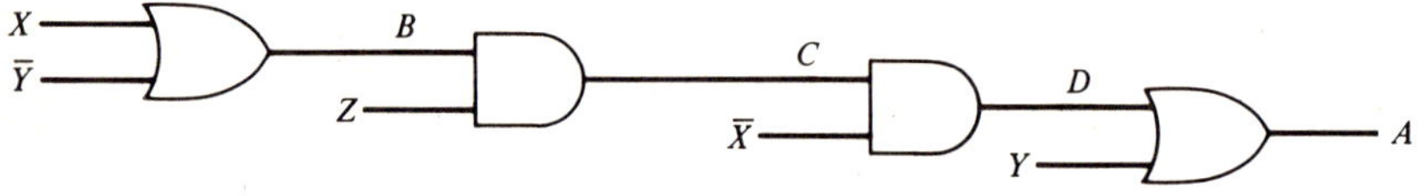

Figure 3-13 A combinational circuit.

> ***Example 3-3.*** Analyze the combinational circuit of Fig. 3-14.
>
> We first label the outputs of the three internal gates as X, Y, and Z, as shown in the figure. These output variables are then expressed as functions of their input variables:

$$X = \overline{AB\bar{C}}$$
$$Y = \bar{A} + B$$
$$Z = X \oplus Y$$
$$F = Z \odot Y$$

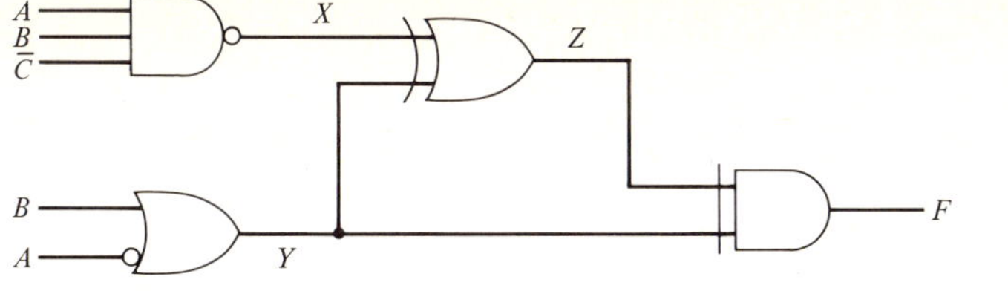

Figure 3-14 Another example of a combinational circuit.

Substituting the first three equations in the last one, we arrive at

$$F = (X \oplus Y) \odot (\bar{A} + B)$$
$$= [(\overline{AB\bar{C}}) \oplus (\bar{A} + B)] \odot (\bar{A} + B) \qquad (3\text{-}14)$$

Figure 3-14 illustrates another commonly used feature in drawing logic diagrams whereby circles at the inputs are used to denote NEGATIONs. Note that the OR gate on the lower left-hand side inverts A and combines it with B by an OR operation, that is, output of this gate is $\bar{A} + B$.

3-4 Characteristics of Practical Logic Circuits

IC logic circuits are in general nonideal elements. A thorough understanding of the practical limitations of these circuits is necessary to ensure that a given design of a digital system built using these elements will function properly. We outline next some of the important characteristics of practical logic circuits.

Input and Output Limitations

The number of unique inputs available to a specific logic gate is known as its *fan-in* limit. In the case of IC gates, this limitation is a direct result of the packaging considerations. In some cases, expanders can be connected to circumvent the fan-in limitation of a given gate package.

Often in a digital system, the output of a logic circuit is fed into several similar logic elements. The maximum number of other logic units (load) a specified logic circuit can drive without deterioration of its performance is, known as its *fan-out* limit. Consider, for example, the combinational circuit of Fig. 3-15. Here the AND gate labeled a must be able to drive the two OR gates labeled c and d. Similarly the AND gate labeled b must be able to drive the three OR gates labeled c, d, and e and the INVERTER f.

The output of a gate that is at logical-0 level draws current from the loads; it supplies current to the loads when it is at logical-1 level. The fan-out of a gate is a relative measure of its output current with respect to the load current requirements and is essentially given by the number of loads that can be connected at the output. Since the number of allowable loads in the output at logical-0 level is in general different from the number of allowable loads

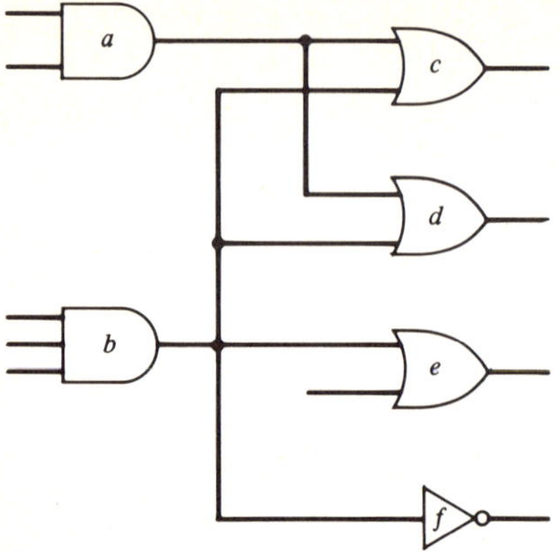

Figure 3-15   A typical combinational circuit connection.

when the output is at logical-1 level, the fan-out is usually specified under worst-case conditions.

Example 3-4.   A gate in the TTL 74LS family driving gates in the TTL 74 family can accept a load of 10 gates at logical-1 level and a load of 5 gates at logical-0 level. Thus the (worst-case) fan-out of the driving gate in this case is 5.

Manufacturers often assign numbers to the input and output terminals of a gate, which can be used to determine its fan-out. These numbers, called *load factors*, are determined from actual input and output current requirements of the driving and the driven gates, and are normalized with respect to the output currents of the driving gate. The load factors are usually indicated by numbers inside parentheses next to the terminals, as shown in Fig. 3-16.

Thus the NAND gate of Fig. 3-16(a) or the exclusive-OR gate of Fig. 3-16(b) has an output loading factor of 5 and, as a result, can drive 5 other gates with unity loading factors. However, either of these gates can be fanned only to two exclusive-OR gates (or buffers) having input loading factors of 2, and at most, one other gate with a unity input loading factor. On the other hand, the buffer of Fig. 3-16(c) can drive 25 gates with unity loading factors. The number of logic circuits connected directly to the output of the buffer will be less than 25 if these logic circuits have input loading factors greater than 1. The output

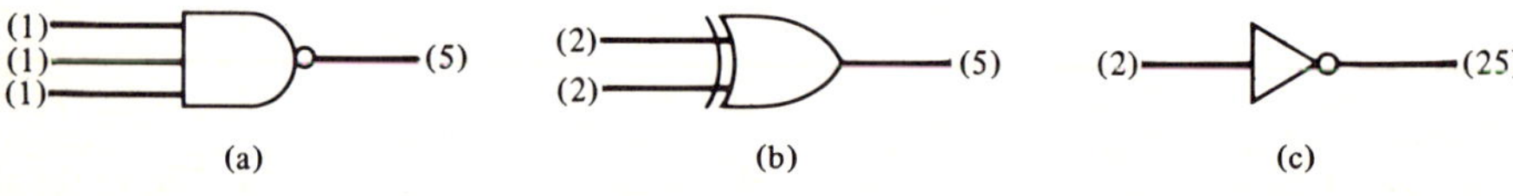

Figure 3-16   Illustration of loading factors.

loading factor is used to indicate the fan-out under the assumption that all of the driven logic circuits have unity loading factors. It should be noted that the above comments apply only for interconnection of gates and buffers of the same type or family, which is usually the practice. When different types of gates are interconnected, the actual values of the input and output currents should be used to compute the fan-out.

Logic Levels and Noise Margin

As indicated above, a digital system is designed by interconnecting a number of logic circuits. For the interconnected system to work properly, it is necessary that the output logic levels of the driving circuit be recognized correctly by the input of the driven circuits or, in other words, they be compatible with the input logic levels. The actual logic levels are functions of circuit component values, temperature, supply voltage, and fan-out. The magnitude of the difference between the two output logic levels is known as the *logic swing*. The worst-case values of these logic levels are usually specified by the manufacturer to take into account the unavoidable manufacturing tolerances of component values, supply voltage variations, and maximum fan-out. The pertinent logic levels are defined as follows:

V_{IL},　　the maximum input voltage level recognized as a logical 0
V_{IH},　　the minimum input voltage level recognized as a logical 1
V_{OL},　　the maximum output voltage level with the gate at logical-0 state
V_{OH},　　the minimum output voltage level with the gate at logical-1 state

Thus, for example, for a NOT gate, if V_I and V_O denote the actual input and output voltage levels under any normal operating condition, then if $V_I \leq V_{IL}$, it is guaranteed that $V_O \geq V_{OH}$, and if $V_I \geq V_{IH}$, then it is guaranteed that $V_O \leq V_{OL}$. Note that any input voltage in the range $V_{IL} < V_I < V_{IH}$ (the uncertainty region) cannot be properly recognized by the gate.

To understand the problem of compatibility, consider the case of Fig. 3-17, in which the NOT gate a is driving the NOT gate b. If gate a is at logical-0 state, then its output voltage $V_a \leq V_{OL}$. But for V_a to be recognized as a logical 0 by gate b, we must ensure that $V_a \leq V_{IL}$, and as a consequence we must have

$$V_{OL} < V_{IL}$$

If, on the other hand, gate a is at logical-1 state, then $V_a \geq V_{OH}$. Gate b will recognize this level as logical 1 if $V_a \geq V_{IH}$, which in turn implies that we must have

$$V_{OH} > V_{IH}$$

The difference between V_{IL} and V_{OL} and the difference between V_{OH} and V_{IH} are known as *noise margins*. More specifically, we define the two noise margins as

$$\Delta 0 = V_{IL} - V_{OL}$$
$$\Delta 1 = V_{OH} - V_{IH}$$

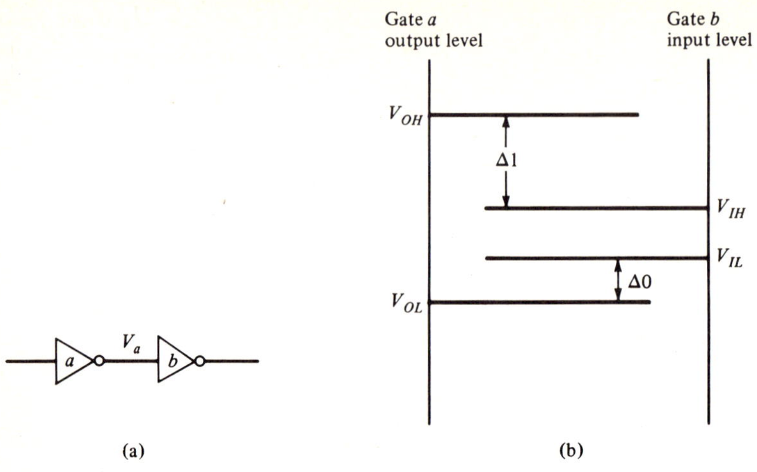

Figure 3-17   Illustration of concept of noise margins.

The noise margin is alternately called the *noise immunity* and denoted by *NI*, with the subscript *L* for low level and the subscript *H* for high level, to designate specifically the two margins.

Example 3-5.   A typical IC gate has the following voltage level specifications at 75°C:

$$V_{IL} = 0.95 \text{ V}, \qquad V_{IH} = 1.80 \text{ V}$$
$$V_{OL} = 0.50 \text{ V}, \qquad V_{OH} = 2.50 \text{ V}$$

Then its low-level and high-level noise immunities are computed as follows:

$$\Delta 0 = 0.95 - 0.50 = 0.45 \text{ V}$$
$$\Delta 1 = 2.50 - 1.80 = 0.70 \text{ V}$$

In practice, noise voltages caused by various sources, such as switching and powerline transients, coupling between signal leads, and so on, may be added to the signals at various points in the system. To prevent the occurrence of false logic signals, it is preferable to use logic circuits with high noise margins.

Dynamic Response Characteristics

A logic circuit does not respond immediately to a change of status of the input signals as a result of the finite switching speed of the transistors and the charging and discharging of circuit capacitances. Consequently the dynamic response of a logic circuit is of considerable importance to the system designer. Such a response is characterized normally by three parameters—rise time, fall time, and delay. Since the voltage waveforms in a digital circuit do not change values instantaneously, a typical input waveform will be as shown at the top of Fig. 3-18. The output waveform of an INVERTER to this input is as shown at the

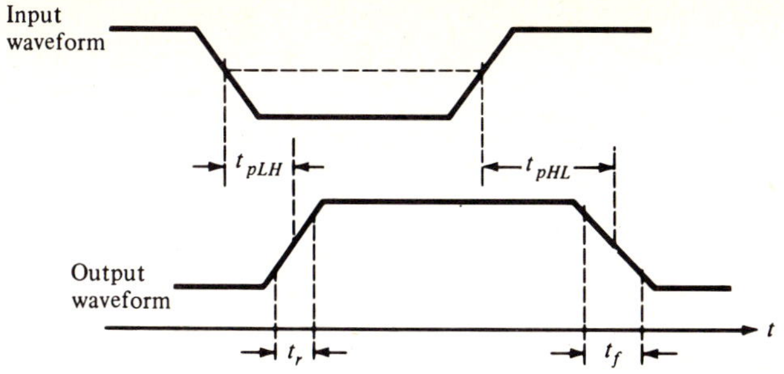

Figure 3-18 Typical input and output waveform of a NOT gate.

bottom of Fig. 3-18. The delay time between the switching threshold level of the activating negative-going edge of the input waveform and the switching threshold level of the corresponding edge of the output waveform is denoted as t_{pLH}. Likewise the delay between the switching threshold level of the activating positive-going edge of the input and the switching threshold level of the corresponding edge of the output waveform is denoted as t_{pHL}. In general the two delay times are not equal, and an average of the two,

$$t_{pd} = (t_{pLH} + t_{pHL})/2$$

is used to define the *propagation delay time* of the logic circuit. The total capacitance encountered by the logic circuit is dependent upon the number of inputs and the number of other circuits connected to the output, that is, upon the fan-in and fan-out. Thus the use of a large number of expanders to increase the fan-in increases the input capacitance of the gate, leading to an increase in the propagation delay. In addition to its dependence on fan-in and fan-out, the propagation delay is also a function of temperature and supply voltage. It should be noted that the switching threshold levels of the input and the output waveforms are usually taken to be of the same value. However, the exact value used to measure the propagation delay for a given type of IC logic varies from manufacturer to manufacturer.

The other two parameters of interest in the response characteristics of a gate are the rise time and the fall time. The time taken by the output voltage waveform to rise from a **LOW** level to a **HIGH** level is its *rise time* t_r. It is measured by the time taken to traverse from 10 to 90 percent of the logic swing as depicted in Fig. 3-18. Similarly, the time taken to traverse from 10 to 90 percent of the voltage swing in the falling edge is the *fall time* t_f (see Fig. 3-18). The rise time, fall time, and the delay are usually specified in nanoseconds.

Power Dissipation

In practice, a digital system contains a large number of gates whose power is usually supplied by a single power supply. Moreover a single package may contain from several gate circuits to several hundreds of gate circuits fabricated relatively close to each other. If the power dissipations of individual gate packages are high, the total power dissipation of the system may become too high and may require external cooling facilities to maintain the gate temperature within the allowable operating range. Moreover power supply and distribution costs also become higher. The *average power dissipation* pd_{av} of the package is an important design factor and is defined as the average of the power dissipation with all gate inputs at logical-1 level and that with all gate inputs at logical-0 level.

By adjusting the operating points of the transistors in the circuit, the power dissipated in a gate can be decreased. However, a decrease in the dissipated power is associated with an increase in the propagation delay of the gate. Hence, a more appropriate performance measure of a gate is its *speed-power product* in joules, which is given by the product of the propagation delay and the power dissipation. A gate with a low speed-power product is then highly desirable.

3-5 Logic Families

Several different types of IC logic circuits have been introduced since their first entry in 1958. Their developments have been dictated by the needs of the logic designer and depend primarily on the applications involved. For example, digital systems used in industrial environments with high noise levels need logic gates with very high noise margins. Similarly, logic circuits used in applications with stringent space requirements such as in satellites, calculators, and watches should preferably be low-power gates. Speed is a concern in many fast computers, and here, logic gates with very low propagation delays are desirable.

At present, the available IC logic circuits can be classified into seven basic groups or families: resistor-transistor logic (RTL), diode-transistor logic (DTL), high-threshold logic (HTL), transistor-transistor logic (TTL), emitter-coupled logic (ECL), metal-oxide semiconductor (MOS) logic, and complementary metal-oxide semiconductor (CMOS) logic circuits.

Some of these groups are further divided into subgroups by trading off their speed and power requirements. Different manufacturers producing IC logic packages belonging to one of these groups usually follow the same functional approach. Consequently, there are usually some similarities in the actual performance of identical logic packages produced by different manufacturers, and in some cases, they are compatible in the sense that they can be interchanged in a given design without any appreciable change in the operation of the whole digital system. However, it should be noted that each manufacturer in general uses different internal circuits and fabrication techniques to produce his IC packages. As a result the characteristics and features of IC logic circuits of each manufacturer in most cases are distinctly different. Detailed description of the logic circuits of each manufacturer is beyond the scope of this book;

however, in this section, we provide an outline of the gross features of each of the seven basic groups.

A Comparison[19–21]

It is difficult to compare in depth the seven logic families mentioned above without getting involved with the development of their circuit configurations and their analysis, and the associated fabrication procedures. Some general comments can, however, be made regarding their important characteristics, such as fan-out, output impedance, noise immunity, supply voltage requirements, power dissipation, propagation delay, specified temperature range, and cost. These have been summarized in Table 3-7. Another parameter included in this table is the clock rate for flip-flops. This latter item is of concern in the design of sequential circuits, the subject of Chapter 4.

The RTL circuit was one of the earliest types of IC logic introduced on the market. Because of its fairly low noise margin and low fan-out, it is rarely used in new systems. Some of its attractive features are very low cost and ease of interface with circuits built with discrete components. A low-power version of the RTL circuits with very good speed-power product is available.

The DTL circuit has a somewhat higher noise margin than the RTL circuits, but it is still lower than most other logic families. It also has a fairly high propagation delay. However, it has a good fan-out, relatively low power dissipation, ease of interface with discrete logic, and is compatible with TTL circuits.

The HTL circuit has been specifically designed with rather high noise immunities and high threshold. It requires higher supply voltages. The logic swing is typically 14 V. It is the slowest of all logic families and, as a result, is insensitive to short noise pulses. It can interface readily with discrete circuits, linear circuits, and electromechanical components. It is somewhat more costly than most others and has a high power dissipation.

The most widely used IC logic is the TTL circuit. It is characterized with a fairly low propagation delay along with a moderate power dissipation and, as a result, exhibits an attractive speed-power product. Its noise immunity is very good compared to all other logic types except the HTL. In addition, it has a high fan-out. A drawback of the TTL circuit is the unequal power supply current drains at the two logic levels which results in "spikes" in the power supply current during gate switchings. These spikes may appear as noise spikes in the output voltage waveform, causing false read-out if the power supplies are not properly bypassed by capacitors connected to ground. The TTL is available in basically three different forms: medium-speed TTL, high-speed TTL, and Schottky TTL. The characteristics of the first two versions are provided in Table 3-7. The characteristics of Schottky TTL are essentially similar to that of the high-speed TTL except the former has a very low propagation delay, of the order of 2 nsec. Its rise and fall times are also quite low, in the range of 2–3 nsec.

The emitter-coupled logic (ECL), also called current-mode logic (CML), is the highest speed logic available in the market today. Some of its attractive

TABLE 3-7
Comparison Chart of the Major IC Digital Logic Families[a]

Parameters	RTL	Low-Power RTL	DTL	HTL	12-nsec TTL	5-nsec TTL	4-nsec ECL	2-nsec ECL	1-nsec ECL	CMOS
Typical high-level Z_o, ohms	640	3.6 kΩ	6 or 2 kΩ	15 or 1.5 kΩ	70	10	15	6	6	1.5 kΩ
Fan-out	5	4	8	10	10	10	25	25 inputs or 50 Ω	10 low-Z inputs or 50 Ω	50 or higher
Specified temperature range, °C	−55 to 125 0 to 75 15 to 55	−55 to 125 0 to 75 15 to 55	−55 to 125 0 to 75	−30 to 75	−55 to 125 0 to 70	−55 to 125 0 to 75	−55 to 125 0 to 75	−55 to 125 0 to 75	0 to 75	−55 to 125
Supply voltage	3.0 V ± 10% 3.6 V ± 10%	3.0 V ± 10% 3.6 V ± 10%	5.0 V ± 10%	15 ± 1 V	5.0 V ± 10% 5.0 V ± 5%	5.0 V ± 10% 5.0 V ± 5%	−5.2 V +20% −10%	−5.2 V +20% −10%	−5.2 V ± 10%	4.5 to 16 V
Typical power dissipation per gate	12 mW	2.5 mW	8 or 12 mW	55 mW	12 mW	22 mW	40 mW	55 mW plus load	55 mW plus load	0.01 mW static ≈1 mW at 1 MHz
Immunity to external noise	Nominal	Fair	Good	Excellent	Very good	Very good	Good	Good	Good	Very good
Noise generation	Medium	Low–medium	Medium	Medium	Medium–high	High	Low	Low–medium	Medium	Low–medium
Propagation delay per gate, nsec	12	27	30	90	12	6	4	2	1	70
Typical clock rate for flip-flops, MHz	8	2.5	12 to 30	4	15 to 30	30 to 60	60 to 120	200	400	5
Cost per function	Low	Low	Low	Medium	Low	Medium	Low	Medium	High	Medium to high

[a] Copyright © 1970 by The Institute of Electrical and Electronics Engineers, Inc. Reprinted, by permission, from *IEEE Spectrum*, Vol. 7, no. 12, December 1970, pp. 30–42.

features are high fan-out, low output impedance, single-supply operation, temperature-independent noise immunity. Some of its drawbacks are lower noise immunity than TTL and HTL circuits, higher power dissipation, and the necessity of using logic level *translators* for interfacing with the previously discussed logic circuits. Some other attractive features of ECL are pointed out later in this section.

There are two types of MOS logic circuits—P-channel metal-oxide semiconductor (PMOS) circuits and N-channel metal-oxide semiconductor (NMOS) circuits. The CMOS logic circuit is usually preferred over the MOS circuit, as the former exhibits considerably lower power dissipation, shorter propagation delay, and also shorter rise and fall times. MOS circuits have high input and output impedances and can have high fan-out at the sacrifice of speed. MOS circuits are also characterized by very high propagation delays. Both MOS and CMOS circuits can be fabricated with a very high packing density, which makes them attractive for large-scale integration of complex logic circuits. These circuits also exhibit very large logic swings and reasonably large noise margins.

Logic Flexibility

An additional feature often considered by logic designers in choosing a suitable logic family is its versatility in meeting the system requirements. Some of the important factors used in determining the versatility are described next.

One such factor is the *wired-logic* capability. This allows the implementation of additional logic without extra gates by simply joining together the outputs of certain logic circuits. In the case of some types of gates, this results in the realization of an OR function, leading to the name " implied-OR " wired logic. In the case of some other types of gates, tying together results in the realization of an " implied-AND " wired logic (Fig. 3-19). The implied-AND connection

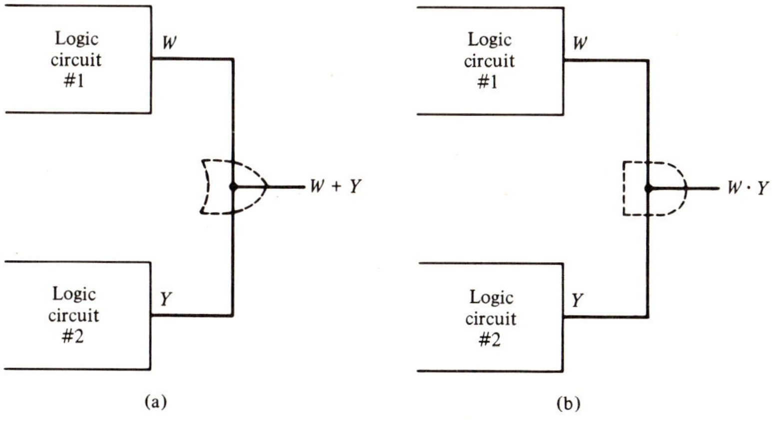

Figure 3-19 Wired logic implementation scheme.

is possible with some gates in the RTL family and most gates in the DTL and the HTL families. On the other hand, all gates in the ECL family permit implied-OR connection. In general, the wired-logic capability is not available with the standard line of TTL circuits. However, a modified TTL family capable of implied-AND connection has been made available recently.

Another factor of interest to the designer is the variety of logic functions of a given family available in a single package. The number of packages needed to implement a digital system will be fewer if greater varieties are available as off-the-shelf items. For example, availability of all types of gates (AND, OR, NAND, NOR, NOT, exclusive-OR, exclusive-NOR, and A-O-I) makes the logic implementation process much simpler. In this respect, the TTL circuits are most attractive. In addition to these gates, a very large variety of complex logic circuits are available in the TTL family, making it the most versatile logic family. Complex functions are also available in ECL form.

Many designs require a logic variable in both direct and complemented forms. In these designs, gates with outputs in both forms will be attractive because they eliminate the need for additional NOT gates. ECL gates provide this feature which, along with the implied-OR wired-logic capability, makes these gates quite versatile. An example of their versatility is illustrated in Fig. 3-20. An implementation of the three output switching functions of this figure using 3-input OR gates and NOT gates would have required a total of eight gates (Problem 3-24).

Integrated-Injection Logic

The integrated-injection logic (I^2L) is a recently introduced logic family. It is most attractive for integrating complex logic circuits in MSI and LSI forms. Only a handful of I^2L circuit packages are currently available on the market. One of its most attractive features is its speed-power product, which can go as low as 10^{-13} J. I^2L circuits can operate faster than N-channel MOS logic

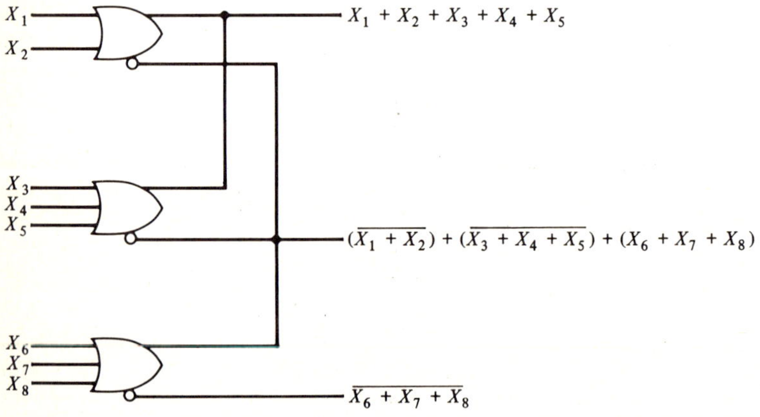

Figure 3-20   Illustration of ECL versatility.

circuits and have a power dissipation less than CMOS circuits. A packing density of about 250 gates/mm^2 has been achieved in this type of logic. The rather high packing density is achieved by letting one semiconductor region of the chip function as part of a number of circuit components. This sharing feature is usually called merging; hence the name merged-transistor logic (MTL) is also given to I^{2}L circuits. The merging feature significantly reduces the number of interconnections between components on the chip, which in turn extends the reliability of the IC.

Tri-State Logic

An interesting modification of the TTL logic gates is the recently introduced tri-state logic (TSL) circuits. As the name implies, the output of such a gate has a third possible stable "state" in addition to the conventional logical-1 and logical-0 states. In this third state, the gate essentially is in a logically "inert" state with its output exhibiting a very high impedance which makes the gate appear physically disconnected from the circuit. The transfer of the gate to the high-impedance state is achieved by a special control input line called "disable."

A symbolic representation of a TSL INVERTER is shown in Fig. 3-21(a); an equivalent representation of which is as shown in Fig. 3-21(b). Here, if the DISABLE input is LOW, the "equivalent switch" at the output remains closed, and the gate operates as a normal NOT gate with output Y equal to $\bar{X}$. However, if the DISABLE input is set to HIGH, the "equivalent switch" opens, "disconnecting" the output from other circuits connected to it. Note that in some TSL gates the control input line is normally at HIGH level and they are "disabled" when the control input goes to LOW level.

TSL gates are available in multigate packages and are used primarily in *bussing*. The disable feature allows the connection of a number of logic circuits to a common bus via TSL gates which can be controlled to ensure that only a single logic circuit is used for data transfer at one time. An example of such a connection is shown in Fig. 3-22. Note that the TSL buffers shown here are activated with a LOW "disable" input. When the control signal is LOW, the lower TSL gate is disabled, and then Logic Circuit #1 can transmit data via

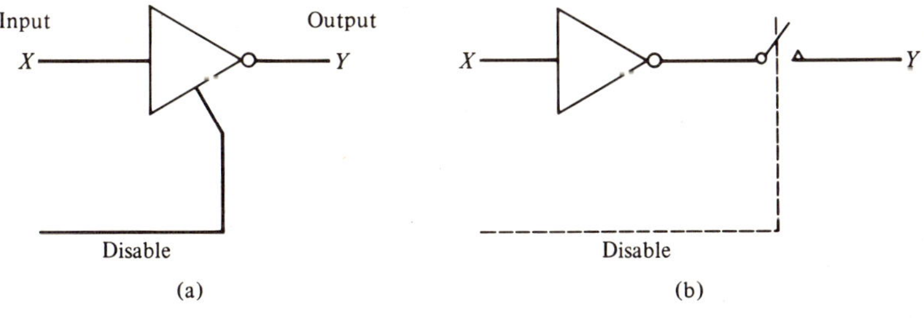

Figure 3-21   TSL inverter: (a) logic diagram, and (b) equivalent circuit.

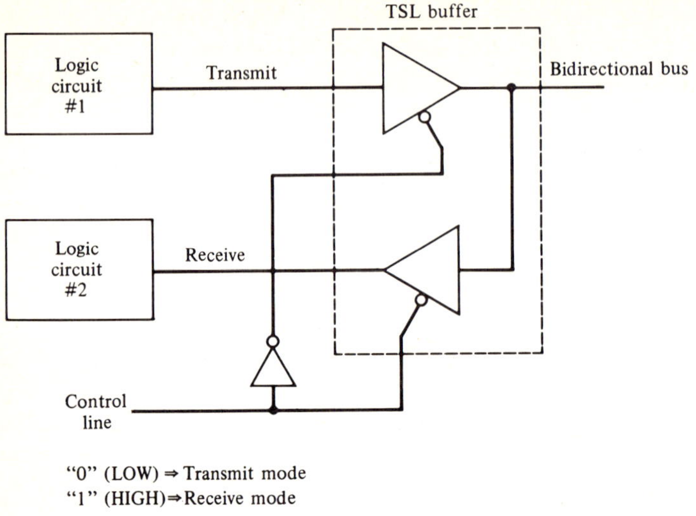

Figure 3-22   Bussing with TSL gates.

the bidirectional bus. On the other hand, when the control input is HIGH, the upper TSL buffer is disabled and then the Logic Circuit #2 is ready to receive data from the bidirectional bus.

Typically as many as 128 TSL gates may be connected to a single bus and still be able to drive 3 TTL loads. The TSL gate is designed to switch to the high-impedance state faster than the transition times between the HIGH and LOW states. This feature prevents the transmit and receive TSL gates being physically connected to the bus simultaneously.

3-6   Design Considerations

In the implementation of a combinational circuit, it is necessary to take into account the nonideal characteristics of the gates used. There are several additional factors that influence the proper operation of the circuit. The question of whether the design is the " best " design is much more difficult to answer. The difficulty in answering this question is due to the fact that different criteria lead to different designs, and it is not always possible to meet all desired criteria. However, an understanding of the practical constraints will provide some guidelines in choosing an " optimum " design that may be satisfactory for the application concerned.

Nonuniqueness of Realization
In general, the realization of a specified switching function is not unique. Thus, often there is more than one realization adding flexibility to the design procedure. A comparison of each realization can then lead to the best circuit.

There are many ways to obtain alternate equivalent realizations. One simple approach is to rewrite the expression for the switching function by making use of the switching algebra theorems of Tables 2-7 and 2-8. This approach was illustrated earlier in Example 3-1. Another example is given below.

Example 3-6.   Let us obtain an alternate realization of the circuit of Fig. 3-13. The pertinent switching function is given in Eq. (3-13), which we modify with the aid of Theorems 2-14, 2-13, and 2-7 as

$$A = Y + X\bar{X}Z + \bar{X}\bar{Y}Z = Y + \bar{X}\bar{Y}Z \tag{3-15}$$

Finally, by applying Theorem 2-18, we simplify it further to

$$A = Y + \bar{X}Z \tag{3-16}$$

whose implementation is indicated in Fig. 3-23.

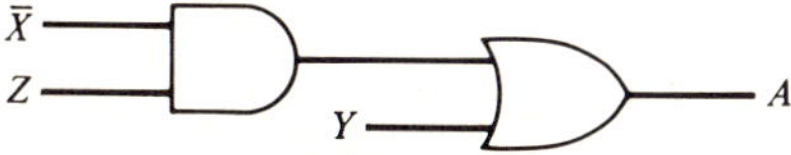

Figure 3-23   Simplified equivalent realization of the circuit of Fig. 3-13.

Levels of Realization

If all the paths from the output of the combinational circuit to all the input terminals of the circuit are traced, the path encountering the maximum number of gates determines the number of *levels* in the realization. For example, the circuit of Fig. 3-13 is a four-level realization, the circuit of Fig. 3-4(b) is a three-level realization, and the circuit of Fig. 3-23 is a two-level realization. The first gate encountered in tracing the paths from the output is said to be at the first logic level, the second gate in the path is at the second level, and so on. In general, a gate at one level in one path will be at the same level with respect to other paths. However, in some cases, this may not be true, as illustrated by the logic diagram of Fig. 3-24. Here the AND gate appears at the second level with respect to one path, but is at the third level with respect to the path going through the NOR gate. Hence, this circuit should be considered as a three-level realization.

The type of logic connection depicted in Fig. 3-24 may cause the occurrence of a false read-out due to the finite propagation delay in gates.[22] To illustrate

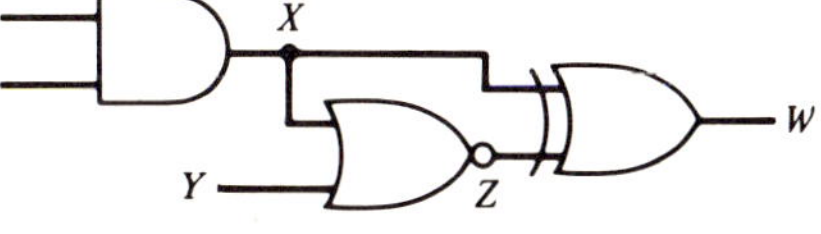

Figure 3-24   A combinational circuit with a gate belonging to two different levels.

this problem, let us denote the propagation delay of the NOR gate by τ_2 and that of the exclusive-OR gate by τ_3, with τ_3 less than τ_2. Let us assume that the output X of the AND gate is at logical-1 level and the input Y of the NOR gate is at logical-0 level. The output Z of the NOR gate is then 0, resulting in the output W of the exclusive-OR gate being at logical-1 level. Now, if the gates are all ideal with no propagation delays, then changing X to 0 will set Z to 1 and, as a result, W will be maintained at logical-1 level. However, because of the gate delays, the output of the combinational circuit will exhibit a negative-going pulse, as shown in Fig. 3-25. Note that the transition of X to 0 occurring at time t_a causes a transition in the output Z of the NOR gate, after a delay of τ_2 sec at time t_c. Since the transition of X to 0 occurs before the change of Z to 1, both X and Z being at 0 levels cause a change in the output W of the exclusive-OR gate after only a delay of τ_3 sec at time t_b. Finally, the transition of Z at t_c causes a second change of voltage level of W at t_d, after another delay of τ_3 sec. The overall result is the occurrence of a negative going pulse

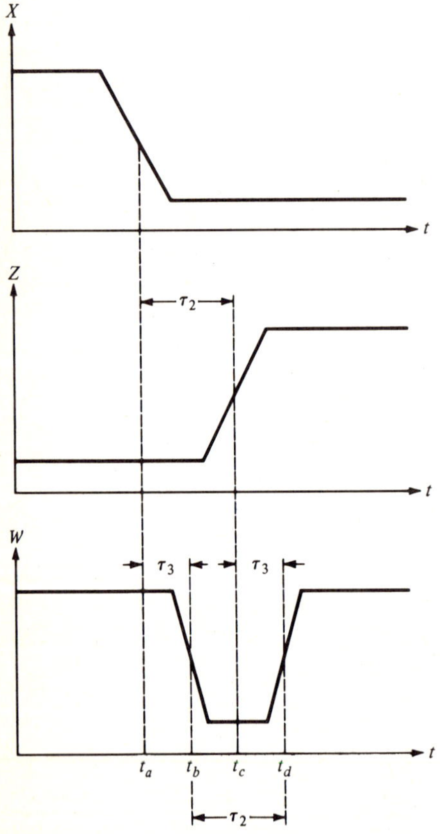

Figure 3-25　Illustration of propagation delay hazards.

of width τ_2 sec at the output of the exclusive-OR gate, instead of W maintaining its state at logical-1 level. This type of unintentional change in voltage levels is called a *delay hazard*.

Design Guidelines

In implementing a design, all of the above factors and the constraints imposed by the gates must be taken into account. In addition, the costs of implementation is an additional factor to be considered by the designer. One way to estimate the cost is by counting the number of gates in the logic diagram. For example, the design of Fig. 3-23 has the least number of gates in comparison to the original design of Fig. 3-13. However, a digital IC logic package usually contains more than one gate, and the cost of a digital IC logic package is roughly the same whether it contains one gate or more than one. Consequently, in estimating the cost of implementation of a circuit, it is more relevant to estimate the number of packages needed rather than the number of individual gates. Since the number of leads in a package is fixed, a rough estimate of the cost of implementation can be obtained by counting the number of leads of all gates.

If switching speed is of interest, then it can be maximized by decreasing the number of levels in the realization. Note now that the two-level realization of Fig. 3-23 is attractive from this viewpoint in comparison to the original four-level design given in Fig. 3-13.

In some cases, direct implementation of a logic diagram may not be possible because of the fan-in and fan-out limitations of the available gates. The fan-in limitations may be eliminated in some designs with the aid of expanders. In other cases, additional gates may be used to solve this problem. Consider, for example, the implementation of a design requiring a 3-input AND gate, whereas the available AND gates have only 2 inputs. The 3-input AND gate can be replaced by 2-input AND gates as outlined in Fig. 3-26(a). This type of fan-in extension increases the number of levels in the realization. Problems 3-2 and 3-3 consider other approaches to fan-in extension of gates.

The fan-out limitation can be circumvented in a number of ways. One approach is to connect a buffer with higher fan-out at the output of the gate in question. Another approach is to connect a number of identical gates in parallel, as shown in Fig. 3-26(b). This latter approach increases the loading (i.e., the output current requirements) of the driving gate.

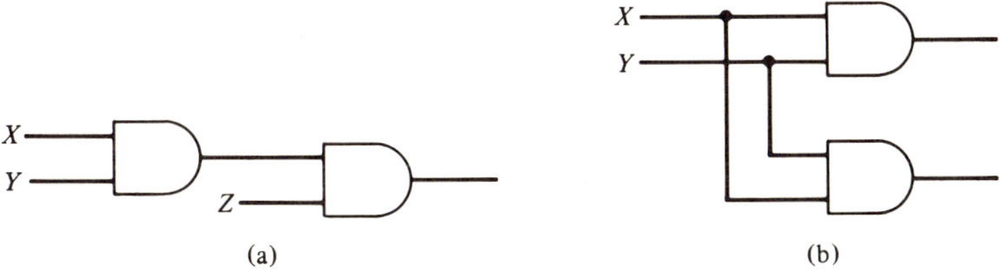

Figure 3-26 Fan-in and fan-out extensions.

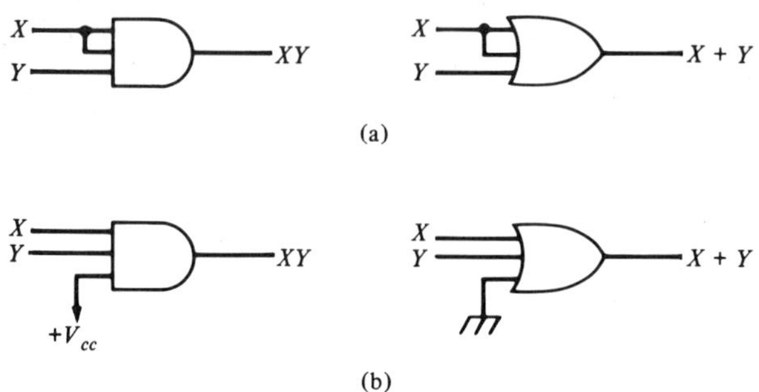

Figure 3-27   Techniques for converting a 3-input gate to a 2-input gate.

The reverse problem of converting a gate with many inputs to a gate with fewer inputs is a simpler problem. Some IC gates are designed to work with unused input terminals left open. However, inputs left open lower the noise immunity of the gate through extraneous signal pickup and also reduce operating speeds caused by an increase in input capacitances. The recommended approach in most cases is to tie together the unused input terminals with one of the active input terminals, as illustrated in Fig. 3-27(a). The tying of two or more inputs again increases the loading of the driving gate. An alternate approach, shown in Fig. 3-27(b), is to connect the unused input terminals to HIGH or LOW dc voltage levels, depending on the type of gate. These two logic levels are easily obtained from the dc supply voltages and/or ground, and the appropriate connection depends on the logic family to which the gate belongs. For example, in the case of TTL gates, the HIGH level is the 5-V supply and the low level is the system ground for positive logic. Since the 5-V supply may exhibit transients, it is preferable to connect the unused input terminal either to the supply through a 1-kΩ resistor or to the HIGH level output of an unused gate.

3-7   Synthesis of Switching Functions

Before a digital combinational circuit can be implemented, we must have the mathematical expression for the switching function combining the switching variables. In this section, we present a systematic approach to develop the switching function from a given truth table. Using this method, any switching function can be synthesized—the final expression using only OR, AND, and NEGATION operations. We illustrate the procedure with the aid of several examples.

Example 3-7.   Design a single-output, 3-input combinational circuit whose output is 1 if the total number of 1's appearing at the input is two; otherwise it is 0.

Let us designate the output variable as A and the 3 input variables as X, Y, and Z. The first step in the synthesis procedure is to develop the truth table. In the second step, the corresponding Karnaugh map is derived from which the expression for the output switching function is obtained by inspection. In many cases, it may be possible to directly arrive at the Karnaugh map without recourse to a truth table.

TABLE 3-8
The Truth Table for the Combinational Circuit of Example 3–7

X	Y	Z	A
0	0	0	0
0	0	1	0
0	1	0	0
0	1	1	1
1	0	0	0
1	0	1	1
1	1	0	1
1	1	1	0

The truth table for the problem stated in this example is shown in Table 3-8. The corresponding Karnaugh map is as indicated in Fig. 3-28. The expression for the switching function in SSOP form is obtained by inspection of the Karnaugh map as

$$A = X\bar{Y}Z + XY\bar{Z} + \bar{X}YZ \tag{3-17}$$

An implementation of the above is shown in Fig. 3-29, where we have assumed that the input variables are also available in their complemented forms.

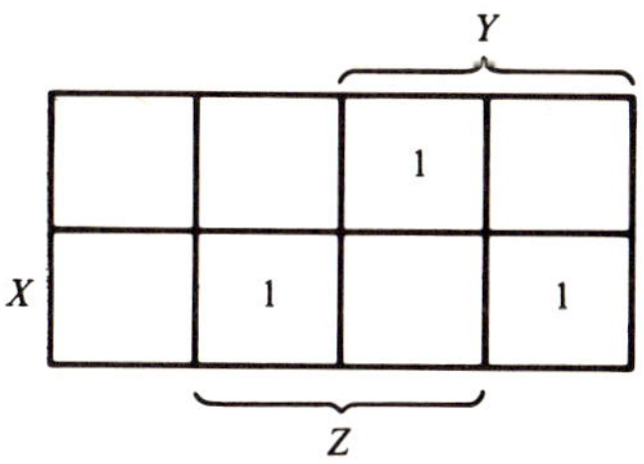

Figure 3-28   The Karnaugh map for Example 3-7.

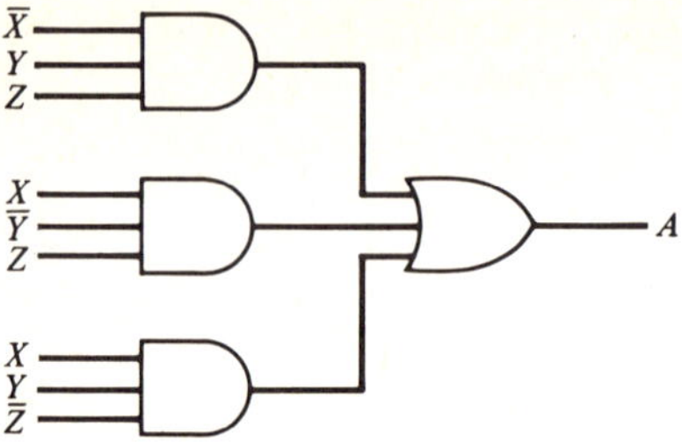

Figure 3-29   Realization of the circuit of Example 3-7.

The circuit based on SSOP-type expansion can be readily implemented by an A-O gate package, possibly with additional expanders.

Two alternate realizations of the above circuit are developed in the following two examples.

Example 3-8.   Obtain a realization of the circuit of Example 3-7 using an A-O-I gate.

We first observe from the Karnaugh map of Fig. 3-28 that the complement of the pertinent switching function can be expressed in the SSOP form as

$$\bar{A} = \bar{X}\bar{Y}\bar{Z} + \bar{X}\bar{Y}Z + \bar{X}YZ + X\bar{Y}\bar{Z} + XYZ \qquad (3\text{-}18)$$

This implies that we can write

$$A = \overline{(\bar{X}\bar{Y}\bar{Z} + \bar{X}\bar{Y}Z + \bar{X}Y\bar{Z} + X\bar{Y}\bar{Z} + XYZ)} \qquad (3\text{-}19)$$

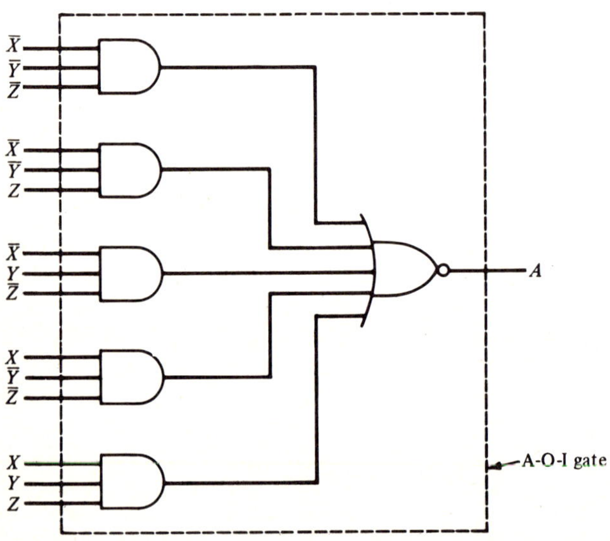

Figure 3-30   A-O-I gate realization of the circuit of Example 3-7.

The expression for A as given above is in appropriate form for implementation using an A-O-I gate (possibly with expanders) as sketched in Fig. 3-30.

Example 3-9. Obtain a realization of the circuit of Example 3-7 using a SPOS-type expression for the output.

The Karnaugh map of Fig. 3-28 can also be used to obtain an expression for the output switching function A as a product of maxterms. The basis for this is the expression for $\bar{A}$ as a sum of minterms, as given in Eq. (3-18) and its subsequent complemented form in Eq. (3-19). If we now apply DeMorgan's theorem to the right-hand side of Eq. (3-19) we arrive at

$$A = \overline{(\bar{X}\bar{Y}\bar{Z})} \cdot \overline{(\bar{X}\bar{Y}Z)} \cdot \overline{(\bar{X}Y\bar{Z})} \cdot \overline{(X\bar{Y}\bar{Z})} \cdot \overline{(XYZ)}$$

which further reduces to

$$A = (X + Y + Z)(X + Y + \bar{Z})(X + \bar{Y} + Z)(\bar{X} + Y + Z)(\bar{X} + \bar{Y} + \bar{Z})$$

$$(3\text{-}20)$$

by a second application of DeMorgan's theorem. The expression on the right-hand side of Eq. (3-20) is the desired result, and its implementation is shown in Fig. 3-31.

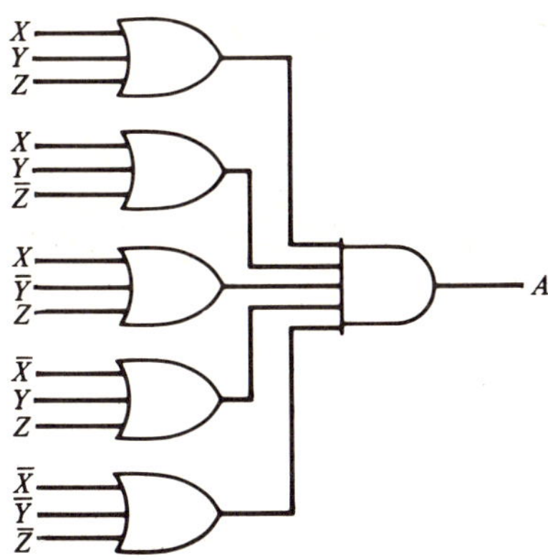

Figure 3-31   A third realization of circuit of Example 3-7.

For convenience, the minterms and the maxterms are usually designated symbolically as m_i and M_i, respectively, as shown in Table 3-9 for the 3-variable case. The subscript i refers to the decimal equivalent of the number formed by the truth values of the input variables for which the minterm takes the value 1 and the maxterm takes the value 0.

TABLE 3-9
Symbolic Notations for Minterms and Maxterms

Row Number	$X\ Y\ Z$	Minterm		Maxterm	
		Expression	Symbol	Expression	Symbol
0	0 0 0	$\bar{X}\bar{Y}\bar{Z}$	m_0	$X + Y + Z$	M_0
1	0 0 1	$\bar{X}\bar{Y}Z$	m_1	$X + Y + \bar{Z}$	M_1
2	0 1 0	$\bar{X}Y\bar{Z}$	m_2	$X + \bar{Y} + Z$	M_2
3	0 1 1	$\bar{X}YZ$	m_3	$X + \bar{Y} + \bar{Z}$	M_3
4	1 0 0	$X\bar{Y}\bar{Z}$	m_4	$\bar{X} + Y + Z$	M_4
5	1 0 1	$X\bar{Y}Z$	m_5	$\bar{X} + Y + \bar{Z}$	M_5
6	1 1 0	$XY\bar{Z}$	m_6	$\bar{X} + \bar{Y} + Z$	M_6
7	1 1 1	XYZ	m_7	$\bar{X} + \bar{Y} + \bar{Z}$	M_7

Using this short-hand notation we can write the switching function of Eq. (3-17) in SSOP form as

$$A = m_5 + m_6 + m_3 \tag{3-21}$$

or simply as

$$A = \Sigma m(5, 6, 3) \tag{3-22}$$

Similarly, the SPOS expression of Eq. (3-20) can be rewritten as

$$A = M_0 M_1 M_2 M_4 M_7 \tag{3-23}$$

or simply as

$$A = \Pi M(0, 1, 2, 4, 7) \tag{3-24}$$

It should be noted that whether we express the switching function as a sum-of-products form or product-of-sums form, the implementation is always a two-level realization as long as the variables are available in both direct and complemented form.

An interesting relation between the maxterms and minterms via their symbolic notations is obtained from an examination of Table 3-9. Consider any minterm in the table, say m_3. Then, if the input switching variables are X, Y, and Z, we observe from Table 3-9 that

$$m_3 = \bar{X}YZ$$

Taking the complement of the above we arrive at

$$\bar{m}_3 = \overline{\bar{X}YZ} = X + \bar{Y} + \bar{Z} = M_3$$

It can be easily verified that

$$\bar{m}_i = M_i$$

for $i = 0, 1, \ldots, 7$. In fact, the above result is true in general and holds for the n-variable case.

3-8   Implementation Using NAND, NOR Gates

In the previous section we outlined two approaches to synthesize a switching function which made use of the three basic operations. The function synthesized in this manner was shown to lead to a two-level realization using only AND and OR gates, provided the switching variables are available in both direct and complemented forms. If the variables are not available in their complemented forms, then they, of course, can be generated using INVERTERs. As a result, the three basic gates are sufficient to generate any switching functions. Of these three types of gates, the NOT gate with either the AND or the OR gate are all that are necessary to implement a switching function (see Problem 3-1).

We shall now show that using either NAND gates or NOR gates exclusively, we can implement any switching function. This can be proved by constructing AND, OR, and NOT gates using the NAND (NOR) gates only. Figure 3-32 shows the implementation of the three basic gates using NAND gates. Likewise, Fig. 3-33 shows how the three basic gates can be implemented using the NOR gates only. In each of these figures, we have used the commonly used notation of representing a NOT gate by either a single-input NAND gate or a single-input NOR gate.

Since any switching function can always be realized using the three basic gates, implementation of the basic gates proves the completeness of the NAND and NOR operations. That is, any switching function can be implemented

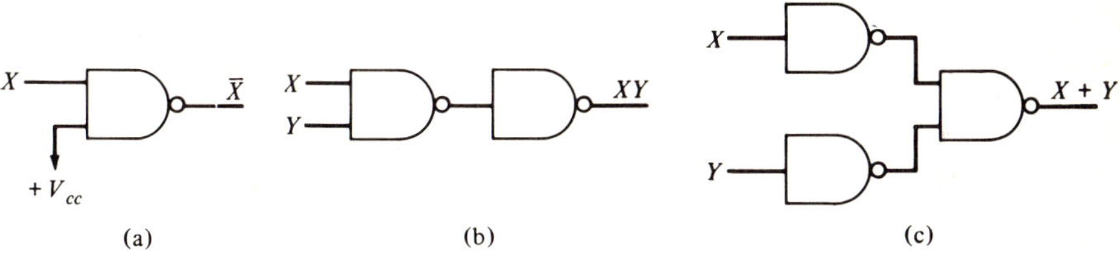

      (a)   (b)   (c)

Figure 3-32   Implementation of (a) INVERTER, (b) AND, and (c) OR gates using only NAND gates.

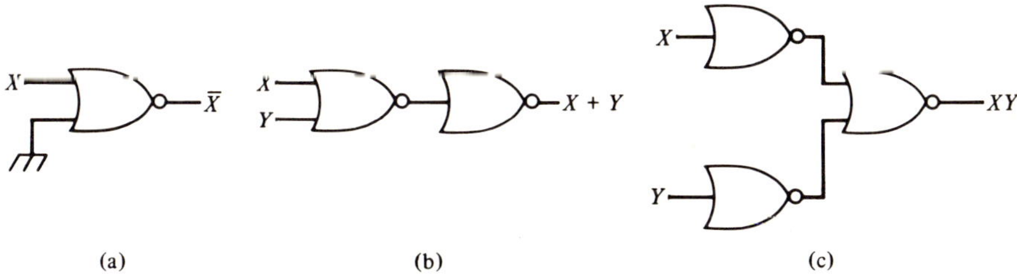

      (a)   (b)   (c)

Figure 3-33   Implementation of (a) INVERTER, (b) OR, and (c) AND gates using only NOR gates.

using either NAND or NOR gates. However, from a practical point of view, it is usually not economical to convert a realization containing the basic gates to an all-NAND or all-NOR realization by replacing each basic gate by its equivalent shown in Fig. 3-32 or in Fig. 3-33. In some cases, it is possible to obtain an economical realization; the procedure is outlined in the next section.

3-9   All-NAND (NOR) Logic Circuits

In the previous section we showed that any switching function can be implemented using either only NOR gates or only NAND gates. We now consider the analysis of combinational circuits implemented using either NAND or NOR gates only. We show that the switching functions of this type of circuit can also be written down by inspection, as can be done with the AND–OR-type circuits.

From Fig. 3-6(b) we see that the NOR gate is equivalent to an OR gate with an INVERTER at its output. Alternately, from Fig. 3-6(c), the NOR gate can also be considered as an AND gate with inverted inputs. Consider now the two-level NOR logic circuit of Fig. 3-34(a). If we replace the NOR gate in the second level by its equivalent representation of Fig. 3-6(b) and the NOR gate of the first level by its equivalent representation of Fig. 3-6(c), then the circuit of Fig. 3-34(a) leads to an equivalent circuit shown in Fig. 3-34(b), which in turn reduces to that of Fig. 3-34(c). From Fig. 3-34(c) we can write down by inspection the expression for the output switching function A as

$$A = (X + Y)\overline{W} \qquad (3\text{-}25)$$

If we now compare Fig. 3-34(c) with Fig. 3-34(a), we note that the first (odd) level NOR gate behaves like an AND gate with inputs inverted and the second (even) level NOR gate behaves like an OR gate with output inverted. Since the two NOR gates are in series, the INVERTER at the output of the OR gate at the second (even) level cancels the INVERTER at the top input of the AND gate at the first (odd) level. However, the variable W goes into the NOR gate

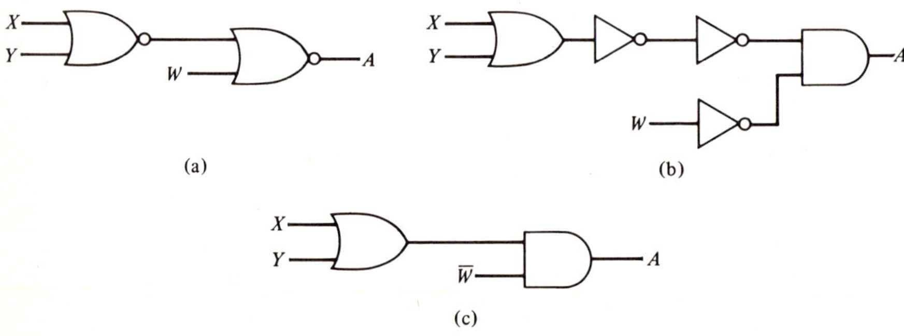

Figure 3-34   Conversion of an all-NOR realization to a realization containing OR and AND gates.

at an odd level, which means W must be inverted before being fed to the AND gate at the odd level.

We can generalize the previous results and state that to convert a NOR logic into an AND–OR-type circuit, (a) *replace each NOR gate at even levels by an OR gate*; (b) *replace each NOR gate at odd levels by an AND gate*; and (c) *invert all switching variables entering the combinational circuit at odd levels.*

Example 3-10. Analyze the circuit of Fig. 3-35(a) and determine the expression for the output.

Note that gate number 2 acts as an even-level gate when its output is considered as input to gate number 4 and acts as an odd-level gate when its output is considered as input to gate number 3. Treating its dual role separately, we obtain the equivalent NOR realization shown in Fig. 3-35(b). Conversion of this equivalent realization is then carried out following the rules outlined above, which leads to the equivalent AND–OR realization sketched in Fig. 3-35(c). The expression for the output switching function A can now be written down by inspection as

$$A = (\overline{X}\,\overline{Y} + \overline{Z})(Z(X + Y) + \overline{X}) \tag{3-26}$$

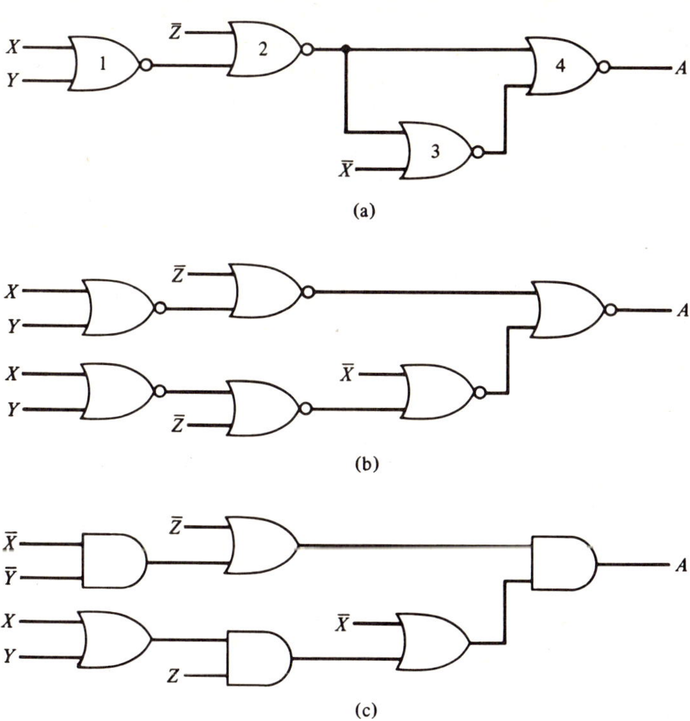

(a)

(b)

(c)

Figure 3-35 A second example showing the steps in the conversion of an all-NOR realization to an OR–AND realization.

For analyzing NAND logic, we can formulate similar rules. The rules are: (a) *Replace each NAND gate at even levels by an AND gate*; (b) *replace each NAND gate at odd levels by an OR gate*; and (c) *invert all switching variables entering the combinational circuit at odd levels.*

We can use the idea conveyed in analysis to synthesize a switching function using NAND or NOR gates. The following example illustrates the procedure.

Example 3-11.   Obtain an all-NAND and an all-NOR realization of the combinational circuit of Example 3-7.

The AND–OR logic implementation of the pertinent switching function is given in Fig. 3-29. Since the OR gate is at the odd level and the AND gates are at an even level, the circuit can be readily converted to a NAND logic as shown in Fig. 3-36(a). The OR–AND implementation of the same switching function appears in Fig. 3-31. Here the AND gate is at an odd level and the OR gates are at an even level. Thus this circuit can be converted readily into a circuit employing NOR gates as shown in Fig. 3-36(b). In both of these realizations, the input switching variables do not appear directly at the input terminals of the gate at the odd level. Hence no inversion of inputs is necessary.

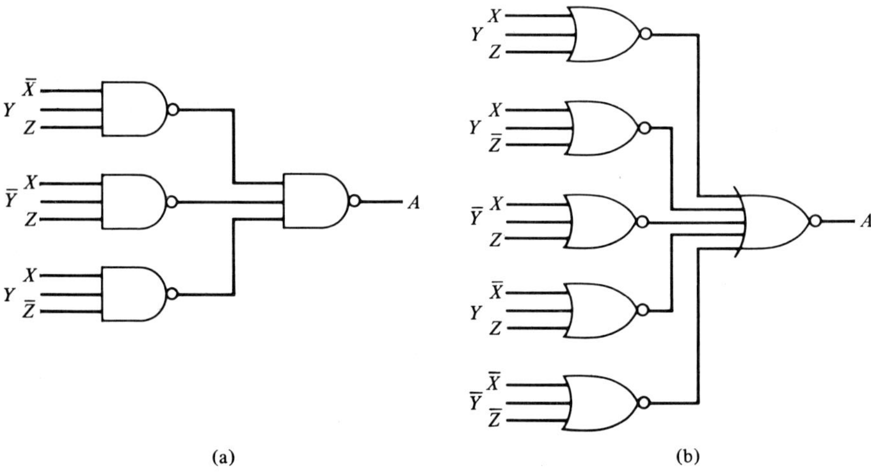

(a)   (b)

Figure 3-36   (a) An all-NAND realization and (b) an all-NOR realization of the circuit of Example 3-7.

3-10   Simplification of Switching Functions

One aim of the circuit designer is to minimize the cost of implementing a given switching function. In most cases, this can be achieved by minimizing the number of gates used in the realization. However, as pointed out earlier, if IC gate packages are used, minimization of individual gates in the realization may not always mean a minimization of IC packages. In spite of this, it is still instructive to attempt a simpler realization with fewer gates.

To illustrate the idea behind simplification consider, for example, the implementation of the following switching function:

$$A = XZ + X\bar{Z} + Y\bar{X} + Y(\overline{\bar{X} + Y}) \tag{3-27}$$

A realization of the above assuming that the variables are not available in their complement form is indicated in Fig. 3-37(a), requiring nine gates. A simpler realization is obtained if we simplify the expression for the switching function, making use of the switching algebra theorems listed in Tables 2-7 and 2-8. The steps in the simplification process are outlined below:

$$
\begin{aligned}
A &= XZ + X\bar{Z} + Y\bar{X} + Y(\overline{\bar{X} + Y}) \\
  &= X(Z + \bar{Z}) + Y\bar{X} + Y(\overline{\bar{X} + Y}) && \text{(by Theorem 2-14)} \\
  &= X + Y\bar{X} + Y(\overline{\bar{X} + Y}) && \text{(by Theorems 2-6 and 2-4)} \\
  &= X + Y\bar{X} + Y(X\bar{Y}) && \text{(by Theorem 2-20)} \\
  &= X + Y\bar{X} && \text{(by Theorems 2-13 and 2-7)} \\
  &= X + Y && \text{(by Theorem 2-18)} \tag{3-28}
\end{aligned}
$$

Implementation of the last expression in Eq. (3-28) requires one OR gate only, as shown in Fig. 3-37(b).

In general, the algebraic approach outlined above is not used in simplifying a switching function because one may not be sure whether the simplest form in expressing the switching function has been obtained by this approach. A

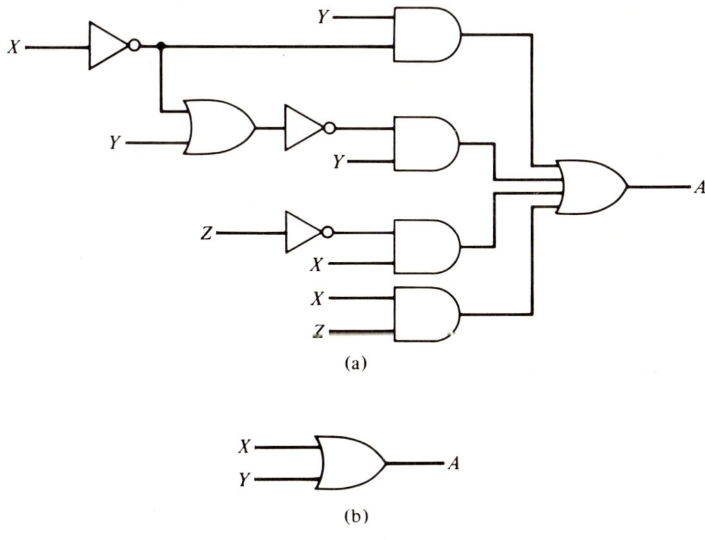

(a)

(b)

Figure 3-37 Two equivalent realizations of the switching function of Eq. (3-27).

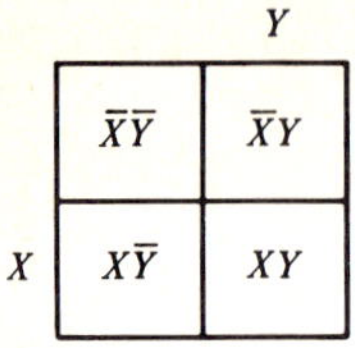

$$Y$$

$\overline{X}\overline{Y}$	$\overline{X}Y$
$X\overline{Y}$	XY

X

Figure 3-38   A 2-variable Karnaugh map.

more reliable and convenient approach to the simplification of a switching function of up to 6 variables is with the aid of a *Karnaugh map*, which is discussed next.

Consider, first, the simplification of a 2-variable function. Figure 3-38 shows a 2-variable Karnaugh map showing all the minterms in their respective boxes. If we examine any two adjacent boxes in this map, we will find that their corresponding minterms differ in only one literal, with the remaining literal being identical in both. Thus, in the map of a particular function, if a 1 is present in two adjacent boxes, then their corresponding minterms can be combined to eliminate the literal which differs in form. To illustrate this idea, consider the Karnaugh map of a switching function A_1, as shown in Fig. 3-39(a). Note that there are 1's in two adjacent boxes. As can be seen from the map in this case, the literal Y differs in the two corresponding minterms. Hence it can be eliminated by combining the two minterms, leading to the literal X. We can also show this algebraically:

$$A_1 = X\overline{Y} + XY = X(Y + \overline{Y}) = X \cdot 1 = X \qquad (3\text{-}29)$$

Example 3-12.   Simplify the 2-variable switching function A_2 represented by the Karnaugh map of Fig. 3-39(b).

Note here that the lower-right box containing a 1 is adjacent to two boxes with 1's in them. Its corresponding minterm can thus be combined with the minterms of both adjacent boxes. It is convenient to show this combination by placing a loop around the 1's in the adjacent boxes, as shown in the figure. Loop *a* covers the two lower boxes whose minterms have literal X in common, and as a result this combination yields X. Loop *b* covers the

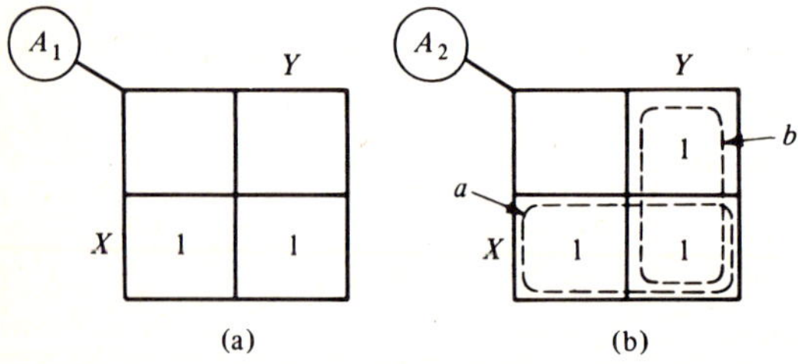

Figure 3-39   Karnaugh map representations of two 2-variable switching functions.

two boxes on the right whose minterms have the literal Y in common, and the resultant combination leads to Y. Thus by inspection

$$A_2 = X + Y$$

Note that the minterm XY has been used twice to aid in the simplification. This follows from Theorem 2-8 by which we can write $XY = XY + XY$.

To verify above result algebraically, we observe

$$A_2 = X\bar{Y} + XY + \bar{X}Y$$
$$= X\bar{Y} + XY + XY + \bar{X}Y$$
$$= X(\bar{Y} + Y) + (X + \bar{X})Y$$
$$= X \cdot 1 + 1 \cdot Y$$
$$= X + Y$$

as expected.

We extend the above idea to the simplification of a 3-variable function. Here, as before, the minterms corresponding to adjacent boxes contain one literal that changes from normal to complement form, and in combination this literal disappears, leaving a 2-variable product. The way the map has been developed, the boxes in the upper-right and upper-left corners are "adjacent." Similarly, the boxes in the lower-right and lower-left corners are also "adjacent." In addition to the elimination of one literal by combining the minterms of two adjacent boxes, we can also eliminate two literals by combining the minterms of four adjacent boxes. This is illustrated in the following example.

Example 3-13.   Simplify the 3-variable function B mapped in Fig. 3-40.

The four minterms inside loop a have the literal $\bar{Z}$ appearing unchanged. The other two literals change forms. Hence combination of these four minterms lead to $\bar{Z}$. We can also show this algebraically. We have

$$\bar{X}\bar{Y}\bar{Z} + X\bar{Y}\bar{Z} + \bar{X}Y\bar{Z} + XY\bar{Z} = (\bar{X} + X)\bar{Y}\bar{Z} + (\bar{X} + X)Y\bar{Z}$$
$$= \bar{Y}\bar{Z} + Y\bar{Z} = (\bar{Y} + Y)\bar{Z} = \bar{Z}$$

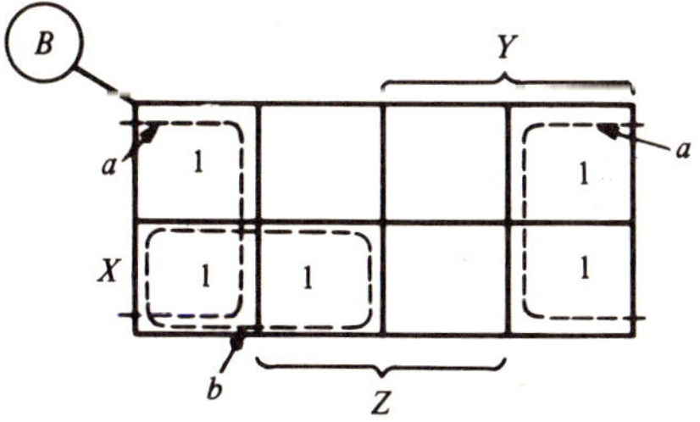

Figure 3-40   The Karnaugh map representation of a 3-variable function.

The minterms inside loop b yield $X\bar{Y}$ after combination. Hence the simplified form of B is given by

$$B = \bar{Z} + X\bar{Y}$$

Example 3-14. Simplify the following 3-variable switching function:

$$W = \bar{A}\bar{B}\bar{C} + \bar{A}\bar{B}C + \bar{A}BC + A\bar{B}\bar{C} + A\bar{B}C + ABC \qquad (3\text{-}30)$$

The map of W is first constructed as shown in Fig. 3-41. We then combine the four minterms inside loop a which yields $\bar{B}$. Similarly the minterms inside loop b yield C after simplification. As a result W simplifies to

$$W = \bar{B} + C$$

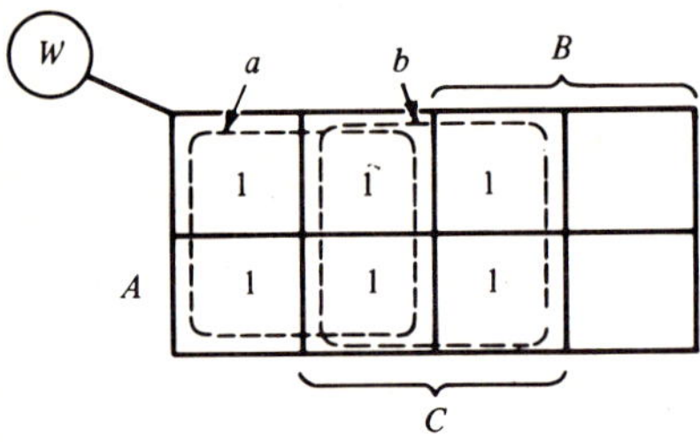

Figure 3.41   Another 3-variable function.

The above procedure is further extended to the four-variable case. Here, in addition to two adjacent boxes and four adjacent boxes, we can also have eight adjacent boxes. We consider an example to illustrate the procedure.

Example 3-15. Simplify the following 4-variable switching function:

$$D(W, X, Y, Z) = \Sigma m(0, 2, 3, 5, 6, 7, 8, 10, 11, 13, 14, 15) \qquad (3\text{-}31)$$

The map of D is as indicated in Fig. 3-42. The eight minterms inside loop a combine to yield Y.

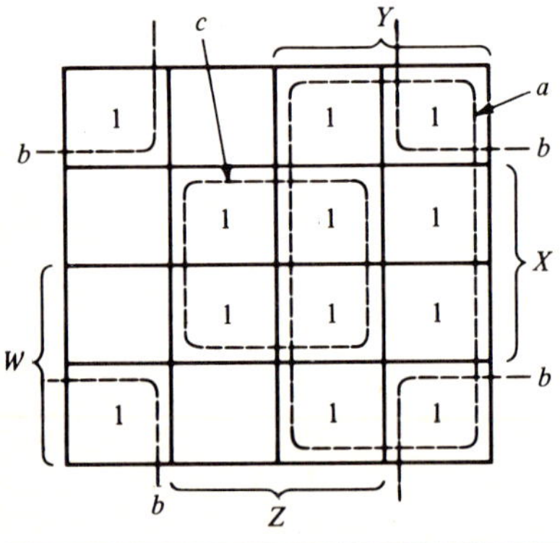

Figure 3-42   A 4-variable function.

The four minterms at the corners inside loop b are combined to form $\overline{X}\overline{Z}$. Finally, the four minterms inside loop c at the center lead to the product XZ. Thus the simplified SOP form of D is given as

$$D = Y + \overline{X}\overline{Z} + XZ \tag{3-32}$$

The final aim of the simplification process is to arrive at an equivalent expression in SOP form with fewest literals. This is achieved by combining the 1's in the map in as few loops as possible, with each loop containing as many adjacent boxes as possible. However, the numbers of adjacent boxes (with 1's) inside a loop must be an integral power of 2.

The map method of simplification can be applied to switching functions of up to 6 variables without too much difficulty; however, in the cases of 5-variable and 6-variable Karnaugh maps, "adjacency" of the boxes are not that straightforward to locate. For switching functions of more than 6 variables, the Karnaugh maps become too unwieldy, and as a result, the map method is rarely used in these cases. An alternate tabular method of simplification due to Quine and McCluskey[1, 5, 6] is not restricted to the number of variables and is preferable there. This method can be easily programmed on a digital computer.

As indicated earlier, the simplification of a switching function using any one of the approaches indicated above leads to an expression of the switching function in SOP form with the fewest literals. As a result, the corresponding realization using AND and OR gates (assuming all variables are available in direct and complement forms) requires the fewest gates. Since any switching function can be implemented using either the NAND or the NOR gate only, it is of interest to obtain a realization using the fewest of these gates. However, to this end, no systematic method has yet been developed. Catalogs of minimum NAND (NOR) gate realizations have been developed by exhaustive analysis procedures and can be consulted if necessary. These tables are usually restricted to switching functions of few variables. One such table is that due to Hellerman[17] which lists minimum gate realizations of switching functions of up to 3 variables using either NAND or NOR gates for cases in which the variables are assumed not to be available in complement form.

3-11   NAND and NOR Trees[18]

In Section 3-7 we outlined a method to express any switching function as a "sum" of minterms or as a "product" of maxterms. Combinational circuits generating all minterms or all maxterms can thus be employed for direct implementations of arbitrary switching functions. For n-input variables, all minterms can be generated using 2^n n-input AND gates and n NOT gates. Likewise all maxterms can be generated using 2^n n-input OR gates and n NOT gates. Alternate approaches to generation of all minterms and maxterms are via the NOR and NAND tree, respectively.

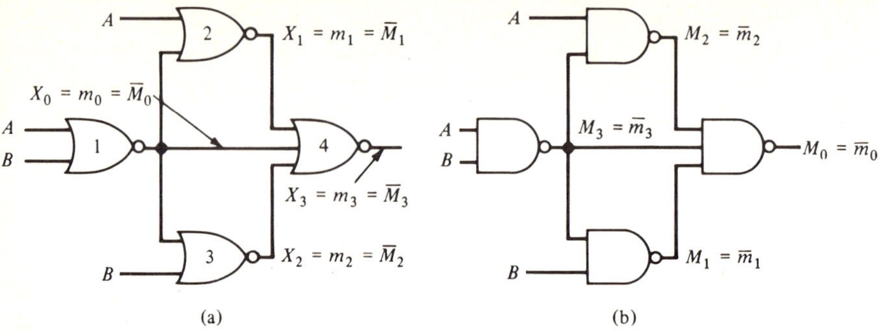

Figure 3-43   (a) 2-variable NOR-tree and (b) 2-variable NAND-tree.

Figures 3-43(a) and 3-43(b) show, respectively, a 2-variable NOR tree and a 2-variable NAND tree. Analysis of the circuit of Fig. 3-43(a) yields the expressions for the outputs of all NOR gates as

$$X_0 = \overline{A + B} = \overline{A}\overline{B} = m_0 = \overline{M}_0$$
$$X_1 = \overline{\overline{A}\overline{B} + A} = (A + B)\overline{A} = \overline{A}B = m_1 = \overline{M}_1$$
$$X_2 = \overline{\overline{A}\overline{B} + B} = (A + B)\overline{B} = A\overline{B} = m_2 = \overline{M}_2 \tag{3-33}$$
$$X_3 = \overline{\overline{A}\overline{B} + \overline{A}B + A\overline{B}} = (A + B)(A + \overline{B})(\overline{A} + B) = AB = m_3 = \overline{M}_3$$

In a similar manner, it can be verified by direct analysis that the outputs of the NAND gates in the NAND tree of Fig. 3-43(b) are indeed the maxterms as labeled.

We illustrate next the use of the above trees in the synthesis of switching functions.

Example 3-16.   Realize the coincidence operation using a NOR tree.

From Eq. (3-12) we observe that the SPOS-type expression for the coincidence operation is given as

$$A \odot B = (\overline{A} + B)(A + \overline{B}) = M_1 M_2$$

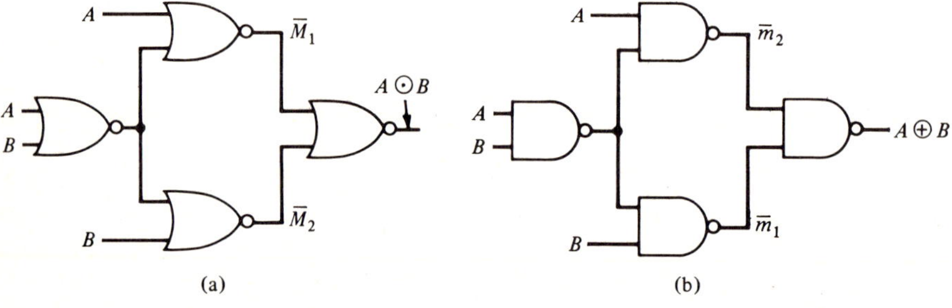

Figure 3-44   (a) All-NOR realization of the coincidence operation, (b) all-NAND realization of the exclusive-OR operation.

We rewrite the above using Theorems 2-5 and 2-21 as

$$A \odot B = \overline{M_1 M_2} = \overline{M_1} + \overline{M_2} \tag{3-34}$$

An implementation of the above is readily obtained by NORing the outputs of NOR gates 2 and 3 of Fig. 3-43(a). Since generation of m_3 is not needed, NOR gate 4 can be a 2-input gate. This leads to the final realization sketched in Fig. 3-44(a) which can be implemented using a quad 2-input NOR-gate package.

Example 3-17.   Obtain an all-NAND implementation of the exclusive-OR operation.

   We note from Eq. (3-7) that the exclusive-OR operation can be expressed in SSOP form as

$$A \oplus B = \bar{A}B + A\bar{B} = m_1 + m_2$$

The above form can be rewritten as

$$A \oplus B = \overline{\overline{m_1 + m_2}} = \overline{\overline{m_1} \cdot \overline{m_2}} \tag{3-35}$$

with the aid of Theorems 2-5 and 2-20. A realization of the above using the NAND tree of Fig. 3-43(b) is shown in Fig. 3-44(b). An implementation of this exclusive-OR gate requires a quad 2-input NAND-gate package.

The above ideas can be extended to the realization of switching functions of more than two variables. A 3-variable NAND tree is shown in Fig. 3-45. Analysis of this maxterm generator is left as an exercise (Problem 3-42). The corresponding 3-variable NOR-tree minterm generator is quite similar and its development is also left as an exercise (Problem 3-43). The following example illustrates the application of the three-variable maxterm generator of Fig. 3-45.

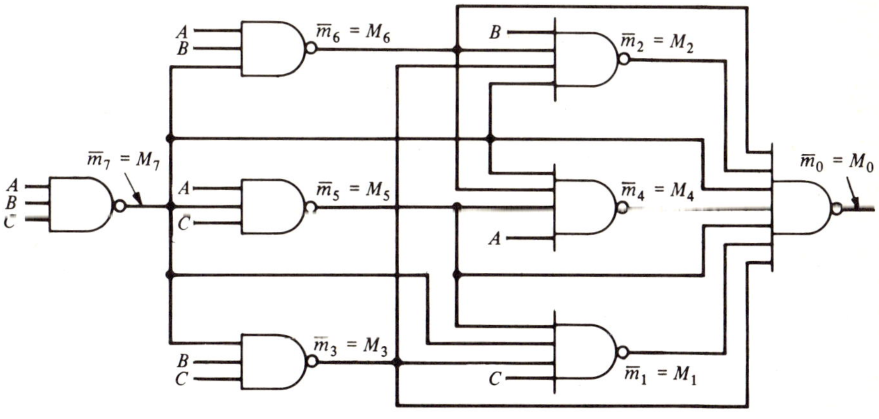

Figure 3-45   A 3-variable NAND tree.

Example 3-18.   Implement the combinational circuit of Example 3-7 using a 3-variable NAND tree.

The expression for the switching function in SSOP form is given in Eq. (3-21), which we rewrite as

$$A = \overline{m_5 + m_6 + m_3} = \overline{m}_5 \cdot \overline{m}_6 \cdot \overline{m}_3 \tag{3-36}$$

The corresponding NAND-tree-based realization is shown in Fig. 3-46. Note that this circuit requires two less NAND gates than the one depicted in Fig. 3-36(a) if variables are not available in complement form.

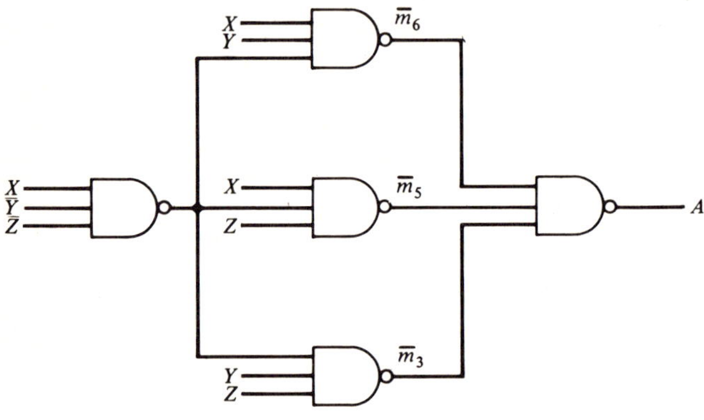

Figure 3-46   A NAND-tree based realization of the circuit of Example 3-7.

Another application of the NAND and NOR trees is in the implementation of multi-output combinational circuits. This is illustrated later in Section 3-13.

3-12   Arithmetic Circuits

We now consider the design of several useful combinational circuits. Some of these circuits will be designed directly, without recourse to the truth tables or Karnaugh maps to illustrate alternate approaches to combinational circuit design. In this section we will restrict our attention to arithmetic circuits. Code conversion circuits are considered in the following section.

Digital circuits implementing binary arithmetic operations are used as basic units in the design of digital computers and many other digital systems. Some of the commonly used arithmetic units are comparators, adders, subtractors, multipliers, and dividers. We discuss here the design of some of these units. For a more complete and detailed discussion the reader is referred to the texts referred to at the end of this chapter.

Comparators

Comparators are used to compare the magnitudes of two binary numbers. We illustrate the development of these circuits when both numbers are positive and in natural binary code. Consider first the design of circuits for comparing two 1-bit numbers A_0 and B_0. The comparator, in its general form, has three outputs which we denote by $f_{(A_0 > B_0)}$, $f_{(A_0 = B_0)}$, and $f_{(A_0 < B_0)}$, with only one of the outputs being at logical-1 level at any time, in accordance with the following convention:

(i) $f_{(A_0 > B_0)} = 1$ if and only if $A_0 > B_0$,

(ii) $f_{(A_0 = B_0)} = 1$ if and only if $A_0 = B_0$,

(iii) $f_{(A_0 < B_0)} = 1$ if and only if $A_0 < B_0$.

The truth table describing the 1-bit comparator is as shown in Table 3-10. The expressions for the three output switching functions are readily derived from the truth table as

$$f_{(A_0 > B_0)} = A_0 \overline{B}_0 \tag{3-37}$$

$$f_{(A_0 = B_0)} = \overline{A}_0 \overline{B}_0 + A_0 B_0 = A_0 \odot B_0 \tag{3-38}$$

$$f_{(A_0 < B_0)} = \overline{A}_0 B_0 \tag{3-39}$$

It can be seen from the above that we can rewrite $f_{(A_0 = B_0)}$ as

$$f_{(A_0 = B_0)} = \overline{A_0 \overline{B}_0 + \overline{A}_0 B_0} = \overline{f_{(A_0 > B_0)} + f_{(A_0 < B_0)}} \tag{3-40}$$

An implementation of Eqs. (3-37), (3-39), and (3-40) yields the logic diagram for the 1-bit comparator as sketched in Fig. 3-47.

Next we consider the design of a 2-bit comparator to compare two 2-bit numbers, $A_1 A_0$ and $B_1 B_0$, with A_1 and B_1 as the MSBs. Using the previous convention we label the three outputs as $f_{(A_1 A_0 > B_1 B_0)}$, $f_{(A_1 A_0 = B_1 B_0)}$, and $f_{(A_1 A_0 < B_1 B_0)}$ with the condition that only one of the outputs can be 1 at any time.

Now $A_1 A_0$ will be greater than $B_1 B_0$ if $A_1 > B_1$. If $A_1 = B_1$, then $A_1 A_0$ will be greater than $B_1 B_0$ if $A_0 > B_0$. If these conditions are not met, then $A_1 A_0$ is not greater than $B_1 B_0$. As a result the switching function $f_{(A_1 A_0 > B_1 B_0)}$ is given as

$$f_{(A_1 A_0 > B_1 B_0)} = f_{(A_1 > B_1)} + \left[f_{(A_1 = B_1)} \cdot f_{(A_0 > B_0)} \right] \tag{3-41}$$

Using the results of Eq. (3-37) in the above, we arrive at

$$f_{(A_1 A_0 > B_1 B_0)} = f_{(A_1 > B_1)} + \left[f_{(A_1 = B_1)} \cdot (A_0 \overline{B}_0) \right] \tag{3-42}$$

TABLE 3-10
Truth Table for 1-Bit Comparator

Inputs		Outputs		
A_0	B_0	$f_{(A_0 > B_0)}$	$f_{(A_0 = B_0)}$	$f_{(A_0 < B_0)}$
0	0	0	1	0
0	1	0	0	1
1	0	1	0	0
1	1	0	1	0

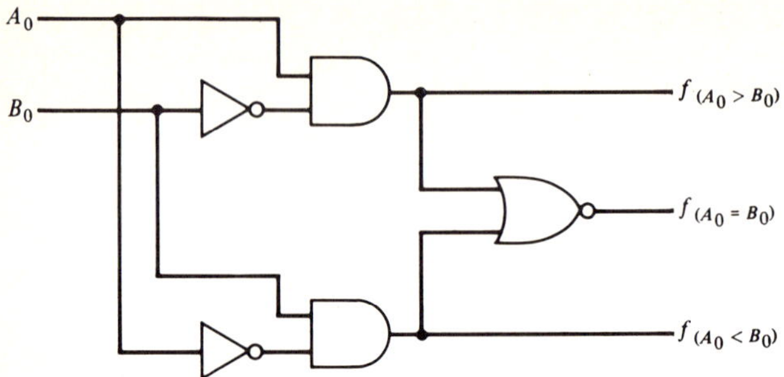

Figure 3-47   A 1-bit comparator.

In a similar manner we can show that

$$f_{(A_1A_0 < B_1B_0)} = f_{(A_1 < B_1)} + [f_{(A_1 = B_1)} \cdot (\bar{A}_0 B_0)] \tag{3-43}$$

It also follows from our discussion of the 1-bit comparator design that

$$f_{(A_1 A_0 = B_1 B_0)} = \overline{f_{(A_1A_0 > B_1B_0)} + f_{(A_1A_0 < B_1B_0)}} \tag{3-44}$$

An implementation of Eqs. (3-42), (3-43), and (3-44) shown in Fig. 3-48 is the desired realization of the 2-bit comparator. Note that this is a 6-level realization and the results of MSB and LSB comparisons have to "ripple" through a number of other gates before correct logic levels appear at the outputs.

To design a *simultaneous-carry comparator* one needs to express Eqs. (3-41)–(3-44) in either SOP or POS forms and implement the resultant expressions using at most a 3-level realization.

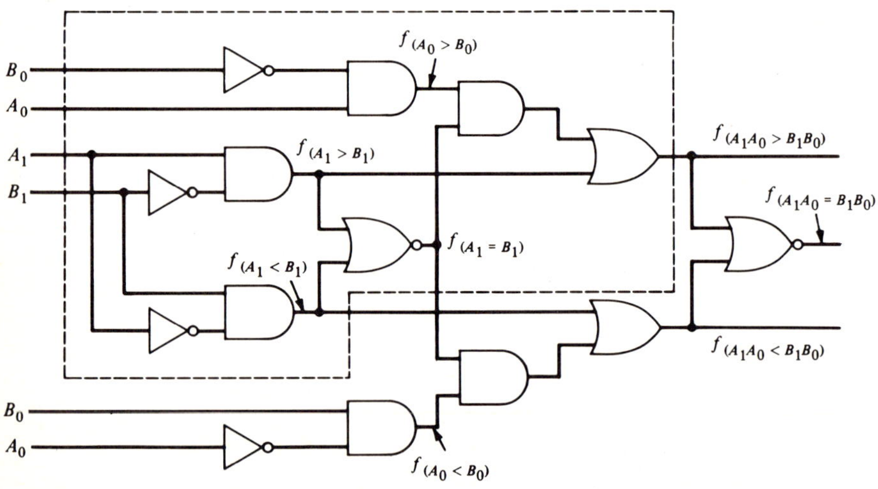

Figure 3-48   A 2-bit ripple comparator.

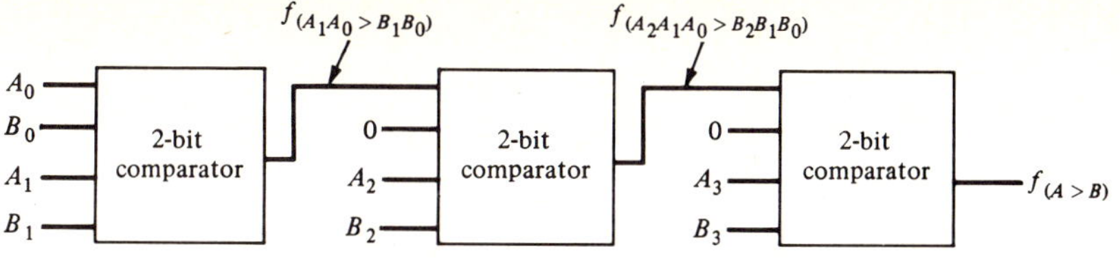

Figure 3-49   Implementation of a 4-bit comparator using three 2-bit comparators.

It is possible to extend the design principles indicated above to design an n-bit comparator where $n > 2$. However, since 2- to 4-bit comparators are available in IC form, it is simpler to design higher-order comparators using these lower-order comparators. For example, Fig. 3-49 outlines the implementation of a 4-bit comparator with only " greater than " output using three 2-bit comparators. A possible implementation of the 2-bit " greater than " comparator is that shown inside the dotted box in Fig. 3-48.

Adder

The method of adding two binary numbers was explained in Section 2-7 where we observed that if the binary numbers are added in a serial fashion, at any stage of the addition process, only three bits, A_i, B_i, and C_i need to be added to generate the sum bit S_i and the carry bit C_{i+1}. A digital circuit implementing this operation is called a *full adder* and is described by the truth table of Table 2-12. The full-addition process can be broken down into two half-addition processes, as illustrated next.

A *half-adder* circuit, shown in block-diagram form in Fig. 3-50, adds two bits X and Y to generate a sum bit S and a carry bit C. The corresponding truth table is as shown in Table 3-11, from which we observe that

$$S = X\bar{Y} + \bar{X}Y = X \oplus Y \tag{3-45}$$

$$C = XY \tag{3-46}$$

The above expressions can be implemented in a number of forms. A novel implementation of the half adder is obtained if we examine the NOR realization of the exclusive-OR gate shown in Fig. 3-51. (The development of this circuit is left as an exercise.) We note that the intermediate variable W shown in the diagram is precisely XY, or C, while the final output is S. Hence, this circuit also

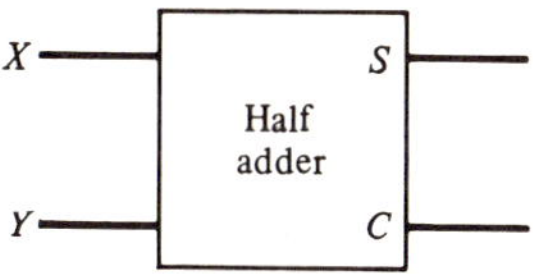

Figure 3-50   Symbolic representation of a half-adder.

TABLE 3-11
The Truth Table for a
Half Adder

X	Y	S	C
0	0	0	0
0	1	1	0
1	0	1	0
1	1	0	1

can be used as a half adder. Alternate half-adder realizations is found in Problems 3-51 and 3-52. The half adder, also known as a *modulo-2 adder*, is available in IC form.

Going back to the full-addition process, we observe from Table 2-12 that we can write

$$S_i = \bar{A}_i \bar{B}_i C_i + \bar{A}_i B_i \bar{C}_i + A_i \bar{B}_i \bar{C}_i + A_i B_i C_i \tag{3-47a}$$

$$C_{i+1} = \bar{A}_i B_i C_i + A_i \bar{B}_i C_i + A_i B_i \bar{C}_i + A_i B_i C_i \tag{3-47b}$$

which can be rewritten as

$$S_i = A_i \oplus B_i \oplus C_i \tag{3-48a}$$

$$C_{i+1} = A_i B_i + C_i(A_i \oplus B_i) \tag{3-48b}$$

Implementation of the above with two half adders is shown in Fig. 3-52, which also shows two additional output terminals whose purpose will be explained later. A symbolic representation of the full adder is as indicated in Fig. 3-53. The full adder is also available in IC form.

A simple way to realize an *n*-bit adder is to connect *n* full adders in cascade in a ripple-carry adder configuration. Since there is no carry bit for the addition of the LSBs, we can use a half-adder for the generation of the LSB of the sum. The complete realization of a 4-bit ripple-carry adder is shown in Fig. 3-54. Note that in such an adder circuit, the correct sum is available only after the carry signals have *rippled* through all higher-order full adders. This leads to a very

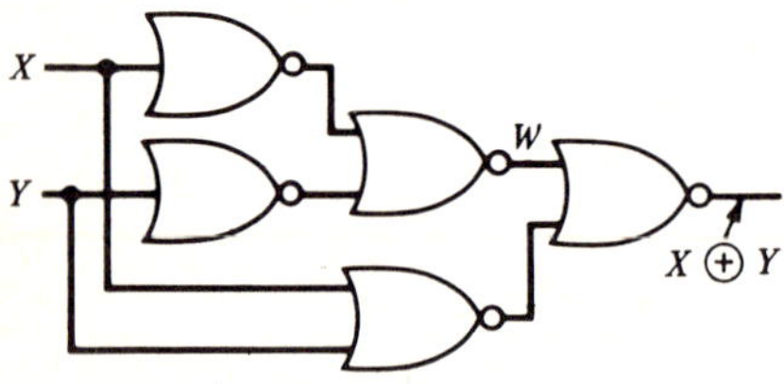

Figure 3-51   All NOR realization of the exclusive-OR operation.

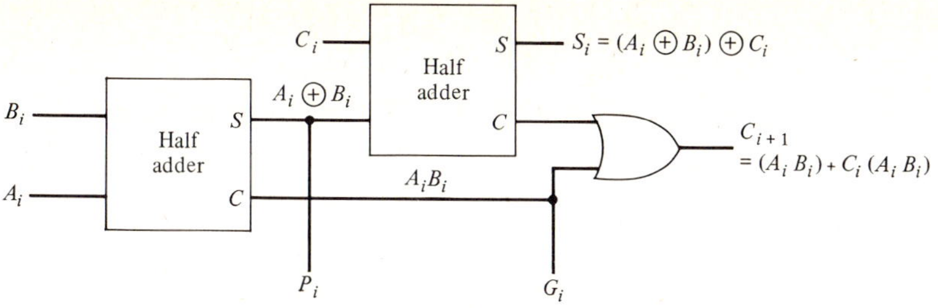

Figure 3-52   Implementation of a full adder with two half-adders.

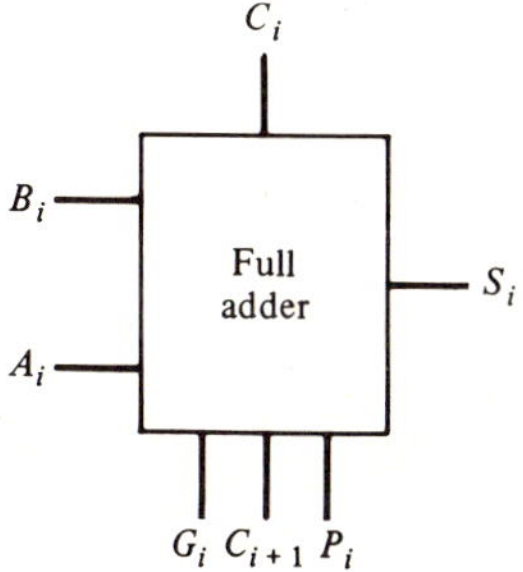

Figure 3-53   Symbolic representation of a full adder.

long propagation delay for large values of n. The propagation delay can be minimized by generating the individual carry bits simultaneously with the aid of a *carry-look-ahead* circuit.

To understand the basic principle behind the carry-look-ahead circuit, let us rewrite the expression of Eq. (3-48b) for carry C_{i+1} as

$$C_{i+1} = G_i + C_i P_i \qquad (3\text{-}49)$$

where $G_i = A_i B_i$ denotes the *carry-generate* variable and $P_i = A_i \oplus B_i$ denotes the *carry-propagate* variable, since the latter " propagates " the carry bit C_i.

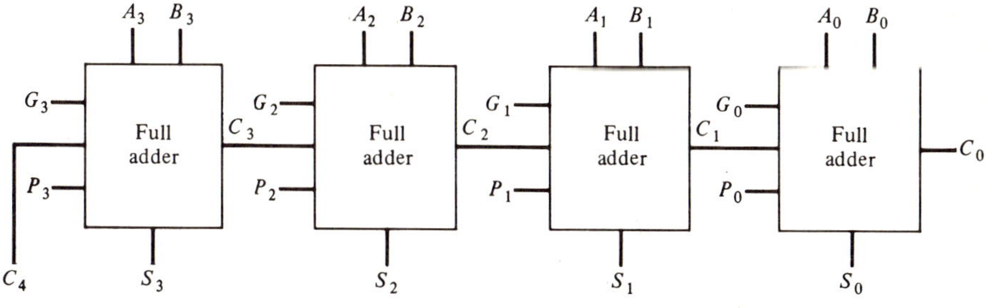

Figure 3-54   A 4-bit ripple-carry adder.

These variables are the extra switching variables shown in the full-adder circuit of Fig. 3-52. For $i = 1, 2$, and 3, we arrive at

$$
\begin{aligned}
C_1 &= G_0 + C_0 P_0 = G_0 \\
C_2 &= G_1 + C_1 P_1 = G_1 + G_0 P_1 \\
C_3 &= G_2 + C_2 P_2 = G_2 + (G_1 + G_0 P_1)P_2 \\
    &= G_2 + G_1 P_2 + G_0 P_1 P_2
\end{aligned}
\tag{3-50}
$$

where we have made use of $C_0 = 0$. Two-level realization of the carry bits using the carry-generate and carry-propagate variables is thus straightforward. A block-diagram representation of a 4-bit adder using a carry-look-ahead circuit is thus as shown in Fig. 3-55; a symbolic representation of a 4-bit adder is shown in Fig. 3-56.

It was illustrated earlier in Example 2-18 that binary subtraction can be readily achieved using an adder with the negative numbers represented in 2's-complement form. A simple algorithm for obtaining the 2's complement of a binary number was also advanced in Section 2-6. An implementation of this algorithm is shown in Fig. 3-57 for a 5-bit number. The input is $A_4 A_3 A_2 A_1 A_0$ and the complemented output is $B_4 B_3 B_2 B_1 B_0$, with A_0 and B_0 being the LSBs. In this circuit, the LSB A_0 has to propagate through a number of gates before the MSB of the complemented output can be generated. A 2-level implementa-

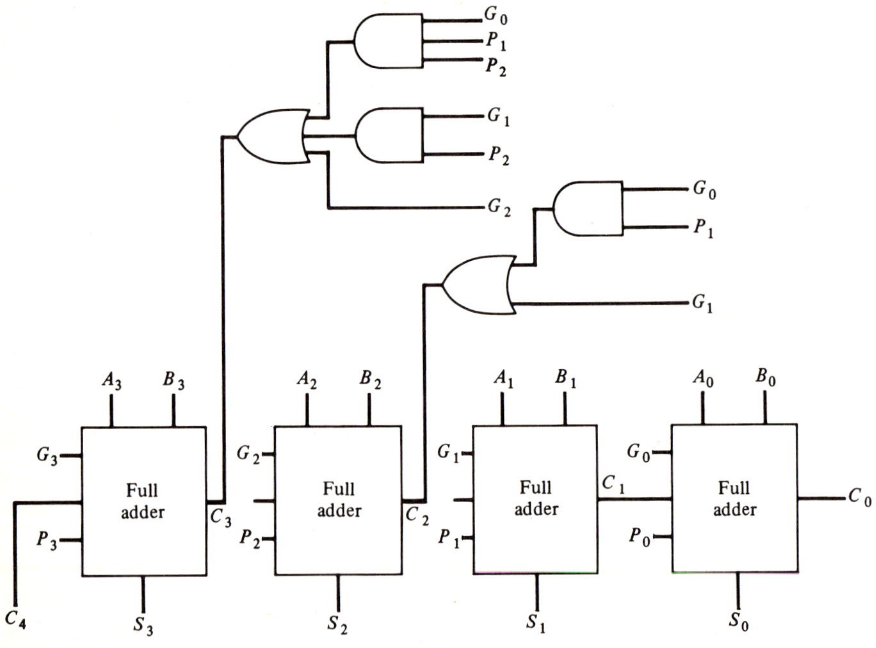

Figure 3-55   A 4-bit adder with carry-look-ahead.

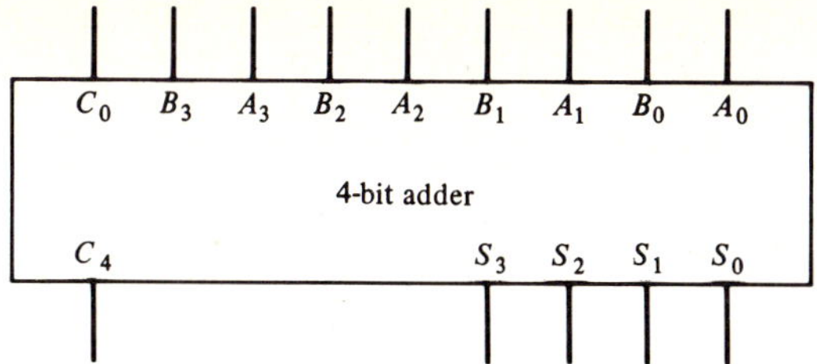

Figure 3-56 Symbolic representation of a 4-bit adder.

tion of the same circuit requiring OR gates with larger fan-in requirements can easily be obtained (Problem 3-54).

The design of a full adder/subtractor unit is considered in Problem 3-55.

Multiplier

The third type of arithmetic circuit we plan to discuss is the binary multiplier. Its realization is based on the multiplication process explained in Section 2-7. Let us develop the realization of a 4-bit-by-4-bit multiplier. Let the multiplicand and the multiplier be denoted as $X_3 X_2 X_1 X_0$ and $Y_3 Y_2 Y_1 Y_0$, respectively. Their product P will be an 8-bit number $P_7 P_6 P_5 P_4 P_3 P_2 P_1 P_0$. We can express the product as

$$P = (X_3 X_2 X_1 X_0)Y_0 + (X_3 X_2 X_1 X_0)Y_1 \cdot 2 + (X_3 X_2 X_1 X_0)Y_2 \cdot 2^2$$
$$+ (X_3 X_2 X_1 X_0)Y_3 \cdot 2^3 \tag{3-51}$$

Implementation of Eq. (3-51) is achieved as follows. In the first step, the multiplicand is multiplied by Y_0. Next the multiplicand is multiplied by Y_1, shifted to the left by one bit, and added to the previous product. In the third step, the multiplicand is multiplied by Y_2, shifted to the left by 2 bits, and added to the result of the second step. Finally the product of the multiplicand and Y_3 is formed,

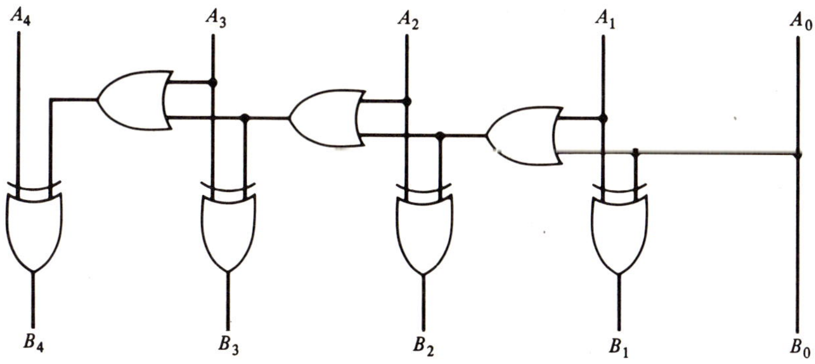

Figure 3-57 Generation of TWO's complement.

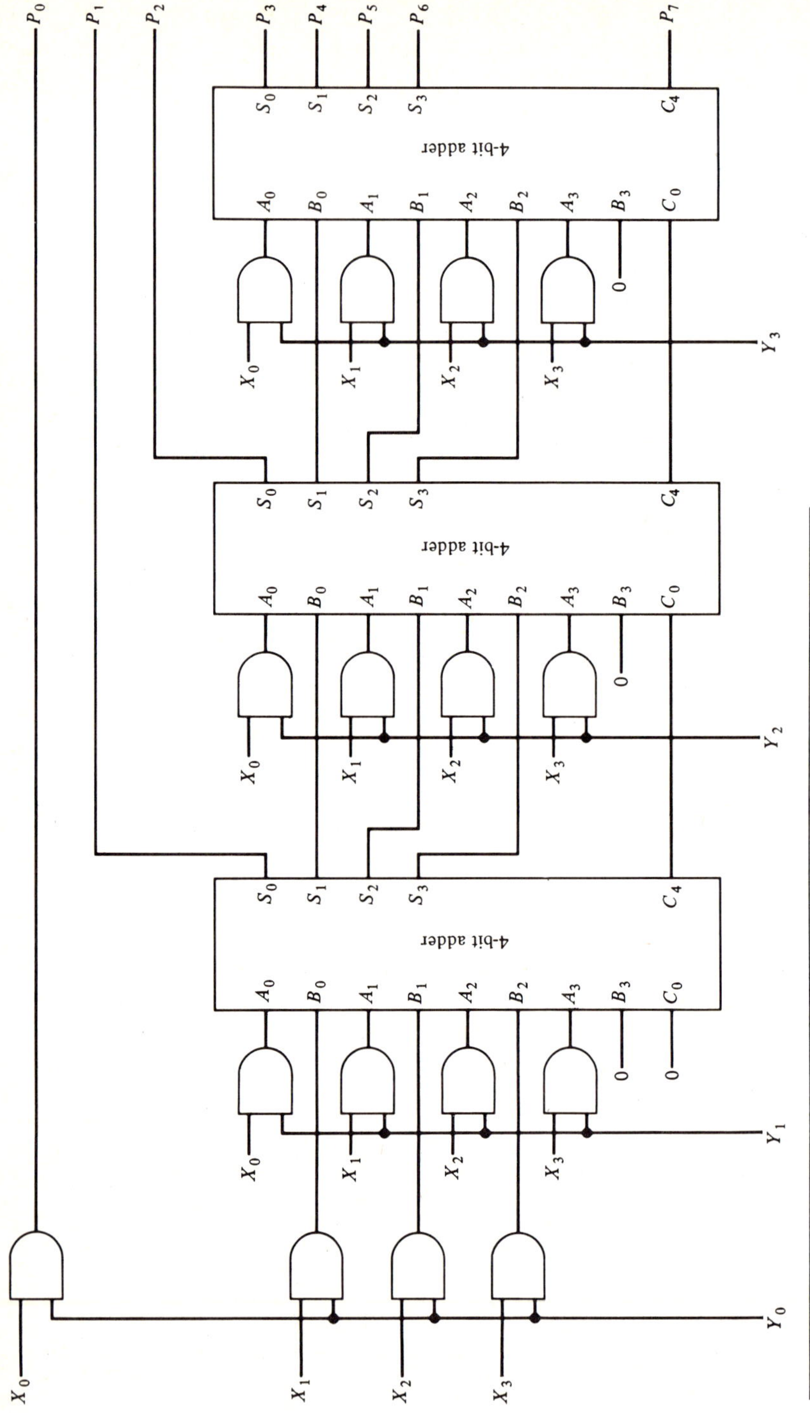

Figure 3-58 A 4-bit × 4-bit multiplier.

shifted to the left by 3 bits, and added to the result of the previous step. The complete realization is illustrated in Fig. 3-58.

An alternate realization of a multiplier to be discussed later is with the aid of read-only memories.

3-13 Code Conversion and Error Detection Circuits

A variety of codes for coding binary numbers were discussed in Section 2-8. In the design of digital systems, it is often necessary to convert a number represented in one code into a representation in another code. In this section we outline the design of several code-conversion circuits. In addition, we outline a simple coding scheme along with its combinational circuit implementation for the detection of errors in binary data.

Binary-to-Gray Code and Gray-to-Binary Code Converters

For simplicity, we consider the conversion of 4-bit numbers. We denote the number coded in natural binary code as $B_3 B_2 B_1 B_0$ with B_3 as the MSB, and the number coded in Gray code as $G_3 G_2 G_1 G_0$ with G_3 as the MSB. The relation between the binary and Gray codes is given by the truth table of Table 3-12. If we are developing a binary-to-Gray code converter, then the input variables of the converter are B_3, B_2, B_1, and B_0 and the output variables are G_3, G_2, G_1, and G_0. In the case of a Gray-to-binary code converter, their roles are reversed.

TABLE 3-12
Truth Table Relating Binary and Gray Codes

Binary Code				Gray Code			
B_3	B_2	B_1	B_0	G_3	G_2	G_1	G_0
0	0	0	0	0	0	0	0
0	0	0	1	0	0	0	1
0	0	1	0	0	0	1	1
0	0	1	1	0	0	1	0
0	1	0	0	0	1	1	0
0	1	0	1	0	1	1	1
0	1	1	0	0	1	0	1
0	1	1	1	0	1	0	0
1	0	0	0	1	1	0	0
1	0	0	1	1	1	0	1
1	0	1	0	1	1	1	1
1	0	1	1	1	1	1	0
1	1	0	0	1	0	1	0
1	1	0	1	1	0	1	1
1	1	1	0	1	0	0	1
1	1	1	1	1	0	0	0

In order to develop the design equation for realizing a binary-to-Gray code converter, we shall have to develop the expressions for the switching functions G_i in terms of the input variables B_i. To this end we can use the method outlined in Section 3-7. However, a simpler realization can be obtained as a result of some special relations between the two codes.

We observe from Table 3-12 that the relations between B_i and G_i are simply given as

$$
\begin{aligned}
G_3 &= B_3 \\
G_2 &= B_3 \oplus B_2 \\
G_1 &= B_2 \oplus B_1 \\
G_0 &= B_1 \oplus B_0
\end{aligned}
\tag{3-52}
$$

An implementation of the 4-bit binary-to-Gray code converter based on Eq. (3-52) is shown in Fig. 3-59.

For the reverse process, the following relations hold:

$$
\begin{aligned}
B_3 &= G_3 \\
B_2 &= G_3 \oplus G_2 = B_3 \oplus G_2 \\
B_1 &= G_3 \oplus G_2 \oplus G_1 = B_2 \oplus G_1 \\
B_0 &= G_3 \oplus G_2 \oplus G_1 \oplus G_0 = B_1 \oplus G_0
\end{aligned}
\tag{3-53}
$$

A realization of a Gray-to-binary code converter based on Eq. (3-53) is sketched in Fig. 3-60.

An extension of the above method to implement an n-bit converter should be evident.

BCD Converters

Next we consider the design of converters where either the input or the output or both are in some type of BCD code. The design considered makes use of the synthesis and the simplification methods of Sections 3-7 and 3-8, respectively.

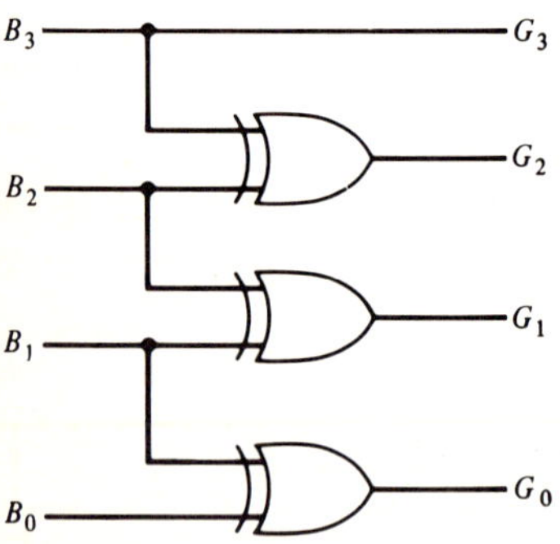

Figure 3-59 A binary-to-Gray-code converter.

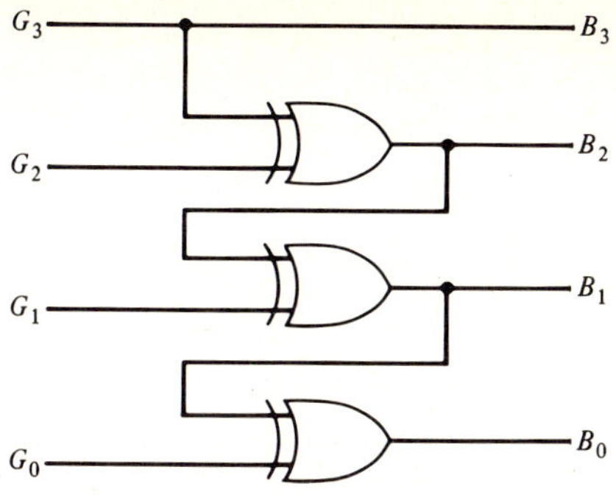

Figure 3-60   A Gray-code-to-binary converter.

Example 3-19.   Design a code converter to translate a 4-bit number in 5-2-1-1 code to a number in 8-4-2-1 code.

Let us denote the number in 5-2-1-1 code as $A_3 A_2 A_1 A_0$ with A_3 as the MSB. Likewise, let us denote the number in 8-4-2-1 code as $B_3 B_2 B_1 B_0$ with B_3 as the MSB. From Table 2-14 we obtain first the truth table of Table 3-13 relating the output variables B_3, B_2, B_1, and B_0 to the input variables A_3, A_2, A_1, and A_0.

From Table 3-13 we arrive at the expressions for the four output variables as a sum of minterms as given below:

$$B_3 = m_{14} + m_{15}$$
$$B_2 = m_7 + m_8 + m_9 + m_{12}$$
$$B_1 = m_3 + m_6 + m_9 + m_{12}$$
$$B_0 = m_1 + m_6 + m_8 + m_{12} + m_{15}$$

(3-54)

TABLE 3-13
Truth Table Relating 5-2-1-1 and 8-4-2-1 Codes

5-2-1-1 Code				8-4-2-1 Code			
A_3	A_2	A_1	A_0	B_3	B_2	B_1	B_0
0	0	0	0	0	0	0	0
0	0	0	1	0	0	0	1
0	0	1	1	0	0	1	0
0	1	1	0	0	0	1	1
0	1	1	1	0	1	0	0
1	0	0	0	0	1	0	1
1	0	0	1	0	1	1	0
1	1	0	0	0	1	1	1
1	1	1	0	1	0	0	0
1	1	1	1	1	0	0	1

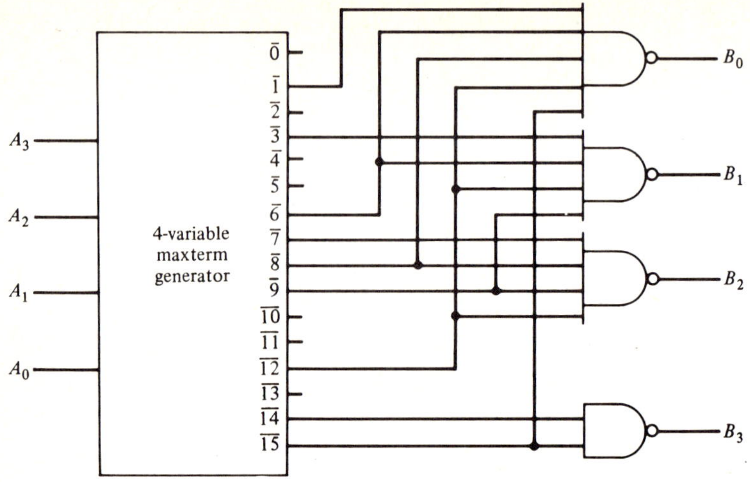

Figure 3-61   Realization of the code converter of Example 3-19 using a 4-variable NAND tree.

An implementation of the above using a 4-variable maxterm generator is sketched in Fig. 3-61. The maxterm generator can be designed as a 4-variable NAND tree.

An alternate realization of the above code converter which makes use of the simplification scheme based on Karnaugh maps is illustrated next. As seen from the truth table of Table 3-13, there are six possible input logic level combinations that are guaranteed not to appear at the converter input. These combinations are given as follows:

$$
\begin{array}{llll}
A_3 = 0, & A_2 = 0, & A_1 = 1, & A_0 = 0 \\
A_3 = 0, & A_2 = 1, & A_1 = 0, & A_0 = 0 \\
A_3 = 0, & A_2 = 1, & A_1 = 0, & A_0 = 1 \\
A_3 = 1, & A_2 = 0, & A_1 = 1, & A_0 = 0 \\
A_3 = 1, & A_2 = 0, & A_1 = 1, & A_0 = 1 \\
A_3 = 1, & A_2 = 1, & A_1 = 0, & A_0 = 1
\end{array}
\tag{3-55}
$$

and are known as *don't care* conditions. Such don't care conditions often can be taken advantage of in simplifying a switching function expression.

The Karnaugh map representations of the four switching functions are next developed as shown in Fig. 3-62. In these maps we show the don't care input conditions as *d*. It does not matter whether we include or do not include their corresponding minterms in the expressions for the switching functions. If a don't care minterm is included in the switching function expression, its truth value is always 0, as the input switching variables will never assume the truth values to make the minterm take the value of 1.

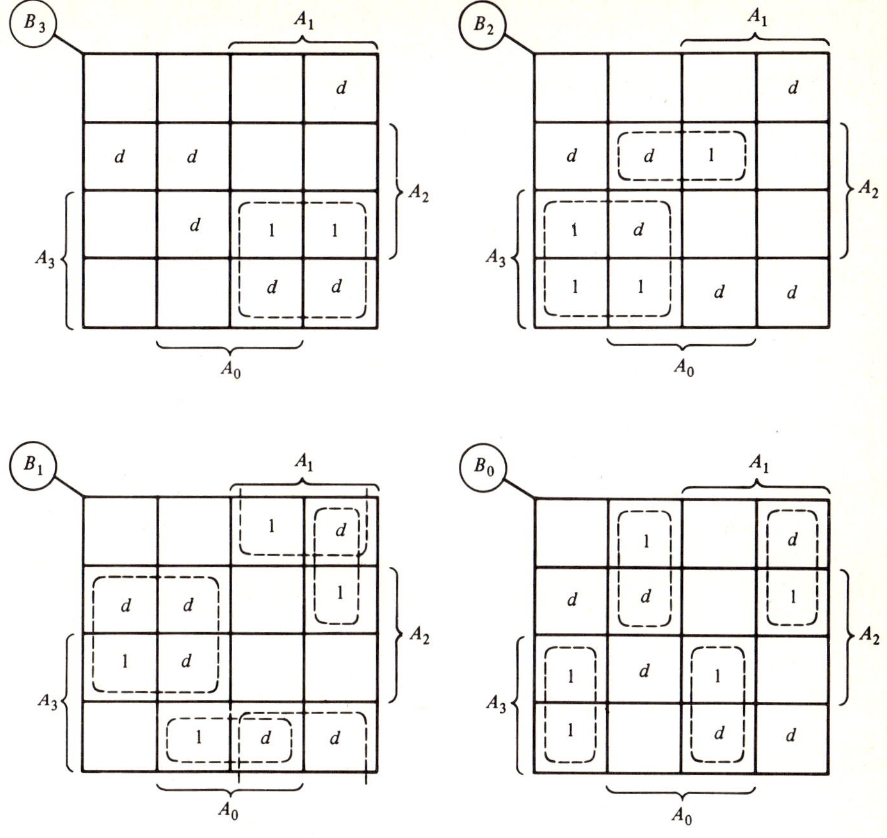

Figure 3-62   Karnaugh map representation of output variables.

By combining the minterms inside the loops shown, we arrive at

$$
\begin{aligned}
B_3 &= A_3 A_1 \\
B_2 &= A_3 \overline{A}_1 + \overline{A}_3 A_2 A_0 \\
B_1 &= A_2 \overline{A}_1 + \overline{A}_2 A_1 + A_3 \overline{A}_2 A_0 + \overline{A}_3 A_1 \overline{A}_0 \\
B_0 &= A_3 \overline{A}_1 \overline{A}_0 + A_3 \overline{A}_1 A_0 + \overline{A}_3 A_1 \overline{A}_0 + A_3 A_1 A_0
\end{aligned}
\tag{3-56}
$$

Circuit implementation of the above is left as an exercise. It should be noted that in the implementation of this converter, the total number of gates can further be minimized by sharing certain gates between several switching functions.

An alternate approach to code-converter realization is with the aid of read-only memories. We discuss this approach later in Section 3-15.

Error Detection Circuits

Often due to component failure or noise pickup, digital data received for some particular operation may not be the same as the data originally transmitted.

For example, a binary data 1101 when received may be read as 0101 if the most significant bit changed from 1 to 0. This type of single-bit error can be detected by adding an additional *parity* bit to the number. The truth value of the parity bit is determined by the condition that the total number of 1's in the data, including the parity bit, be even. This type of coding is called *even-parity* coding. Thus the binary number 1101 when coded with even parity would be 11101, where the leftmost bit is now the parity bit having a truth value of 1 to make the total number of 1's in the coded number an even number. Suppose now this coded number is received as 10101. We check the total number of 1's in the received number and find it to be an odd number, concluding that the data received are not correct. Alternatively, *odd-parity* coding may be used to make the total number of 1's an odd number.

If $X_{n-1}X_{n-2} \cdots X_1 X_0$ represents the n-bit number and P is the parity bit, then it can be easily shown that the expression $X_{n-1} \oplus X_{n-2} \oplus \cdots \oplus X_1 \oplus X_0 \oplus P$ must be ZERO for even-parity check and ONE for odd-parity check. A simple implementation of parity-check circuits is using exclusive-OR gates.

Generation of the parity bit is also quite straightforward. For even-parity check, the parity bit is given as

$$P = X_{n-1} \oplus X_{n-2} \oplus \cdots \oplus X_1 \oplus X_0$$

and for odd-parity check, it is

$$P = 1 \oplus X_{n-1} \oplus \cdots \oplus X_1 \oplus X_0$$

The parity bit can be added to any binary data including BCD data. For example, Table 3-14 shows a 7-4-2-1 BCD code to which a parity bit has been added for even parity. Note that this particular BCD code uses 5 bits (including the parity bit) to represent a decimal digit with each representation containing exactly two 1's and three 0's. (The representation of decimal zero is not in usual 7-4-2-1 code.) The resultant code is known as *2-out-of-5 code* and can detect all single-bit errors and up to 40 percent of double-bit errors. (Why?)

TABLE 3-14
Two-Out-of-Five BCD Code

Decimal Equivalent	Coded Representation				
	7	4	2	1	Parity Bit
0	1	1	0	0	0
1	0	0	0	1	1
2	0	0	1	0	1
3	0	0	1	1	0
4	0	1	0	0	1
5	0	1	0	1	0
6	0	1	1	0	0
7	1	0	0	0	1
8	1	0	0	1	0
9	1	0	1	0	0

3-14   Multiplexers and Decoders

In addition to the arithmetic circuits, code converters, and error-detection circuits, there are several other types of special combinational circuits that find applications in digital system designs. Two such circuits are the multiplexer and the decoder. Both of them are available as MSI circuit packages in various forms.

Digital Multiplexer

The multiplexer (usually abbreviated as MUX) is a multi-input, single-output combinational circuit with additional control terminals. Usually the input lines are called the *data* lines and the control lines are called the *data select* or *address* lines. The status of the data select lines determine which one of the input lines is "connected" to the output. By connection we mean that the truth value of the output is the same as the truth value of the selected input data line. If the number of data select lines is k, then the number of data input lines is 2^k, and the

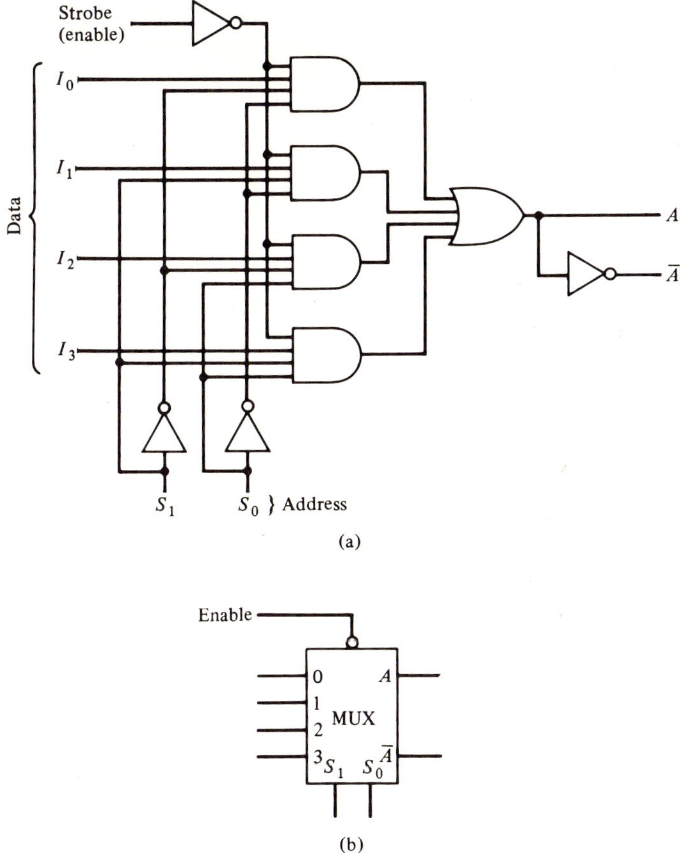

Figure 3-63   A 4-input multiplexer: (a) logic diagram, (b) schematic.

device is called a 2^k-input multiplexer. Often a multiplexer is provided with an additional control terminal called *strobe* or *enable* line. The multiplexer is enabled or activated if the strobe input assumes a particular logic level (HIGH or LOW, depending on the design of the circuit). The strobe input can be used to multiplex input lines at desired time intervals.

Figure 3-63 shows the logic diagram and logic symbol of a 4-input multiplexer with an additional strobe or enable input. Its corresponding truth table is indicated in Table 3-15. Note that the multiplexer is enabled only if the strobe input is LOW (logical 0). If the strobe input is HIGH, then output A of the multiplexer is set LOW, independent of the status of the data input lines. We can write the expression for the output switching function from Table 3-15 (assuming strobe to be LOW) as

$$A = I_0 \bar{S}_0 \bar{S}_1 + I_1 \bar{S}_0 S_1 + I_2 S_0 \bar{S}_1 + I_3 S_0 S_1 \qquad (3\text{-}57)$$

IC multiplexers are available as quad 2-input, dual 4-input, 8-input, and 16-input multiplexer packages. However, multiplexers with any other desired number of inputs can be designed by connecting multiplexers with fewer inputs in a tree-like structure. Figure 3-64 shows the logic diagram of a 4-input multiplexer designed using three 2-input multiplexers. The operation of this circuit can be verified by constructing the truth table as shown in Table 3-16. Note that the selected subscripting is such that when the control input (with S_0 as the MSB and S_1 as the LSB, as suggested by their positions in Table 3-16) is binary code $N_1 N_2$, the multiplexer causes the selection of input I_n where n is decimal equivalent of $N_1 N_2$. Thus, if $S_0 = 1$ and $S_1 = 0, n = 2_{10}$, and input I_2 is selected. Following a similar procedure, a 16-input multiplexer unit can be designed using five 4-input multipliers as illustrated in Fig. 3-65. Verification of this circuit is left as an exercise.

One application of the multiplexer is in sending many signals over a single wire by time multiplexing. We postpone the discussion on multiplexing until

TABLE 3-15
Truth Table for a 4-Input Multiplexer with Strobe Input

			Inputs				Output
$S_0{}^a$	S_1	Strobe	I_0	I_1	I_2	I_3	A
d	d	1	d	d	d	d	0
0	0	0	0	d	d	d	0
0	0	0	1	d	d	d	1
0	1	0	d	0	d	d	0
0	1	0	d	1	d	d	1
1	0	0	d	d	0	d	0
1	0	0	d	d	1	d	1
1	1	0	d	d	d	0	0
1	1	0	d	d	d	1	1

$^a d$ denotes don't care condition.

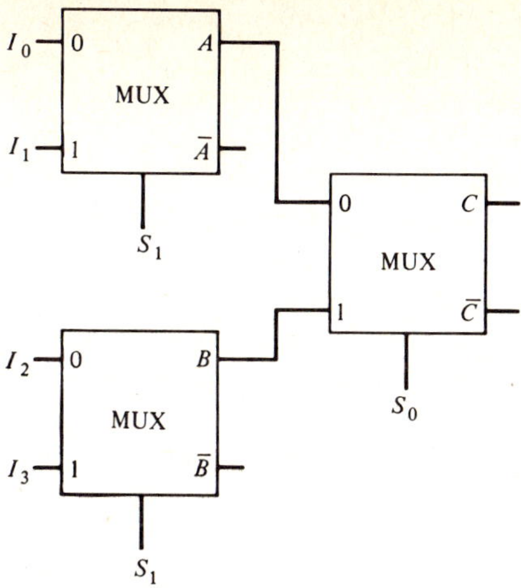

Figure 3-64 A 4-input multiplexer tree.

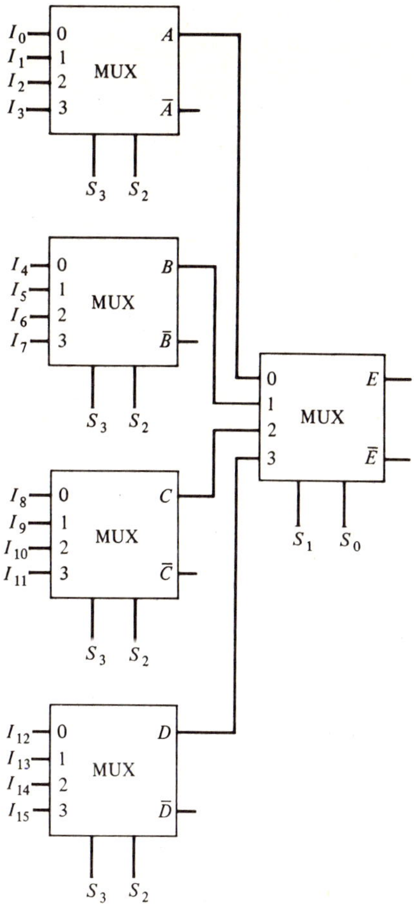

Figure 3-65 A 16-input multiplexer tree.

the end of this section. Another elegant application of multiplexers is in the implementation of an arbitrary switching function as illustrated in the next two examples. Since multiplexers are available as single MSI circuit packages, in many cases, this type of combinational circuit design may be quite economical.

TABLE 3-16
The Truth Table of the 4-Input Multiplexer Tree of Fig. 3-64

S_0	S_1	A	B	C
0	0	I_0	I_2	$A = I_0$
0	1	I_1	I_3	$A = I_1$
1	0	I_0	I_2	$B = I_2$
1	1	I_1	I_3	$B = I_3$

Example 3-20. Let us implement the combinational circuit of Example 3-7 using a single 4-input multiplexer.

Let the input variables X and Y be used as data select line variables S_0 and S_1, respectively. If we substitute this identification in Eq. (3-17) we arrive at

$$A = S_0 \bar{S}_1 Z + S_0 S_1 \bar{Z} + \bar{S}_0 S_1 Z \tag{3-58}$$

Now comparing Eq. (3-58) with Eq. (3-57) we note that they will be identical if we set

$$I_0 = 0, \qquad I_1 = Z, \qquad I_2 = Z, \qquad I_3 = \bar{Z} \tag{3-59}$$

The corresponding realization is then as shown in Fig. 3-66. Other single multiplexer realizations can be obtained by choosing either X and Z or Y and Z as data select line variables.

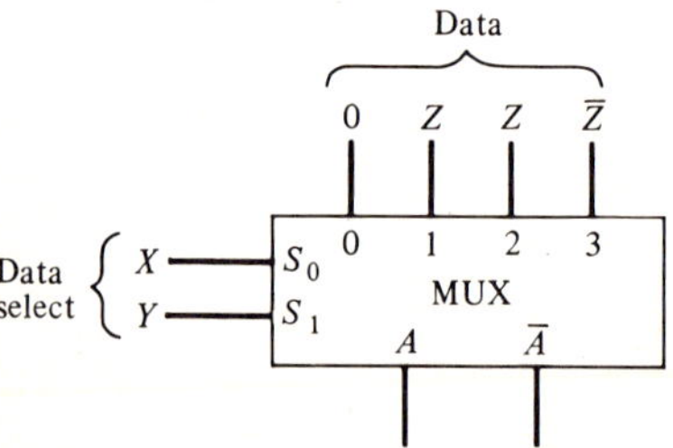

Figure 3-66 Realization of the circuit of Example 3-7 using a single 4-input multiplexer.

In general, a 3-variable switching function can be implemented using a 4-input multiplexer by applying two variables to the data select lines and by applying the third variable or binary constants 0 and 1 to the data input lines in an appropriate manner. In a similar manner, a 4-variable switching function can be implemented using an 8-input multiplexer. Here the three variables are connected to the three data select lines. The fourth variable and binary constants are applied to the data input lines in an appropriate fashion. Switching functions of five or more variables can be realized with higher-order multiplexers. We illustrate in the next example the realization of a 4-variable switching function employing an 8-input multiplexer.

Example 3-21.   Realize the voting machine of Fig. 3-2 using a single multiplexer.

The output of an 8-input multiplexer with control inputs S_0, S_1, and S_2 is given as

$$L = \bar{S}_0 \bar{S}_1 \bar{S}_2 I_0 + \bar{S}_0 \bar{S}_1 S_2 I_1 + \bar{S}_0 S_1 \bar{S}_2 I_2 + \bar{S}_0 S_1 S_2 I_3$$
$$+ S_0 \bar{S}_1 \bar{S}_2 I_4 + S_0 \bar{S}_1 S_2 I_5 + S_0 S_1 \bar{S}_2 I_6 + S_0 S_1 S_2 I_7 \qquad (3\text{-}60)$$

where $\{I_k\}$ denote input data variables. Let us use the variables A, B, and C of the voting machine as the address inputs S_0, S_1, and S_2, respectively. Using this identification we rewrite Eq. (3-1) as

$$L = S_0 S_1 + S_0 S_2 + S_1 S_2 D \qquad (3\text{-}61)$$

Before we can compare Eq. (3-61) with Eq. (3-60) to identify the data input variables $\{I_k\}$, we have to expand the right-hand side of Eq. (3-61) so that each product term contains each one of the address input variables or their complements. Using Theorems 2-4, 2-6, 2-14, and 2-8 we rewrite Eq. (3-61) as

$$L = S_0 S_1 (S_2 + \bar{S}_2) + S_0 (S_1 + \bar{S}_1) S_2 + (S_0 + \bar{S}_0) S_1 S_2 D$$
$$= S_0 S_1 S_2 + S_0 S_1 \bar{S}_2 + S_0 S_1 S_2 + S_0 \bar{S}_1 S_2 + S_0 S_1 S_2 D + \bar{S}_0 S_1 S_2 D$$
$$= S_0 S_1 S_2 + S_0 S_1 \bar{S}_2 + S_0 \bar{S}_1 S_2 + \bar{S}_0 S_1 S_2 D \qquad (3\text{-}62)$$

We now compare Eqs. (3-62) and (3-60) and identify the address input variables:

$$I_0 = 0, \quad I_1 = 0, \quad I_2 = 0, \quad I_3 = D, \quad I_4 = 0, \quad I_5 = 1, \quad I_6 = 1, \quad I_7 = 1.$$
$$(3\text{-}63)$$

The corresponding realization is sketched in Fig. 3-67.

Decoders

A *decoder* is a multi-input, multi-output combinational MSI circuit, one or more outputs of which are activated (usually to logical 0 with the remaining outputs at opposite state) for a particular combination of input truth values. In the

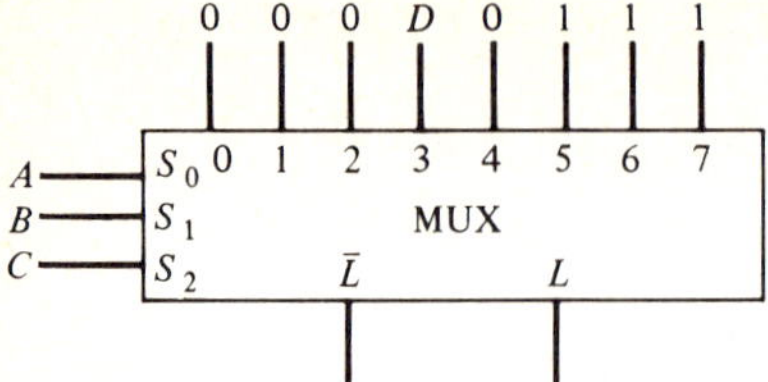

Figure 3-67   Realization of the voting machine of Fig. 3-2 using a single multiplexer.

general case, a decoder has 2^n output lines if n is the number of input lines. Some decoder circuits are designed to drive displays, lamps, relays, and other special devices in addition to the usual logic gate inputs, and are more commonly known as *decoder/drivers*.

Figure 3-68 shows a 3-to-8-line decoder where each decoded output assumes a logical-0 value only for a unique combination of input values, in accordance with the truth table of Table 3-17. As can be seen from the truth table, this decoder is essentially a maxterm generator with the output labeled $\bar{k}$, realizing

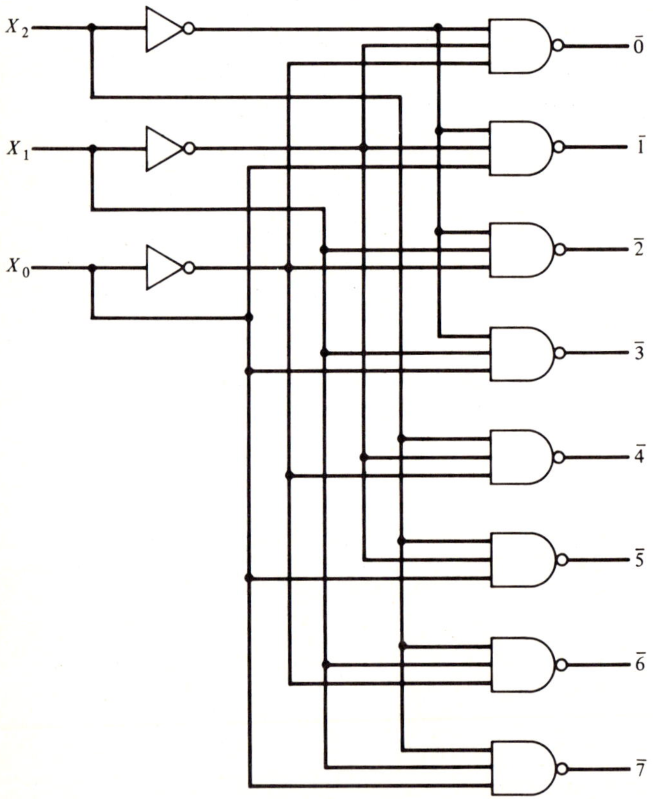

Figure 3-68   A 3-to-8-line decoder.

TABLE 3-17
Truth Table of the 3-to-8-Line Decoder

Inputs			Outputs							
X_2	X_1	X_0	$\bar{0}$	$\bar{1}$	$\bar{2}$	$\bar{3}$	$\bar{4}$	$\bar{5}$	$\bar{6}$	$\bar{7}$
0	0	0	0	1	1	1	1	1	1	1
0	0	1	1	0	1	1	1	1	1	1
0	1	0	1	1	0	1	1	1	1	1
0	1	1	1	1	1	0	1	1	1	1
1	0	0	1	1	1	1	0	1	1	1
1	0	1	1	1	1	1	1	0	1	1
1	1	0	1	1	1	1	1	1	0	1
1	1	1	1	1	1	1	1	1	1	0

the maxterm $M_k = \overline{m}_k$. As a result, such a decoder can also be implemented with the aid of 3-variable NAND tree. Commonly available MSI packages are dual 4-output, single 8-output, and single 16-output decoder circuits.

Like the multiplexer, some decoders have an additional control (enable) line for strobing which permits decoding if and only if the strobe input is at a particular logic level (HIGH or LOW, depending on the design used). Strobe input may be used to enable decoding at desired time intervals. A simple way to strobe a decoder is by connecting a separate strobe circuit, such as that shown in Fig. 3-69 for a 3-input decoder, at the input. Alternatively, the strobe may be connected directly to an additional input of each of the output gates of the decoder.

A simple application of an n-line-to-2^n-line decoder is in the implementation of any n-variable switching function by combining the appropriate outputs of the decoder. The combination is performed with the aid of a NAND or an AND gate if the decoder is of maxterm generator type. An OR gate or a NOR gate is used to combine the outputs of a minterm generator type decoder. This approach to switching function implementation is illustrated in Fig. 3-70 which shows the realization of the switching function of Eq. (3-17) of Example 3-7.

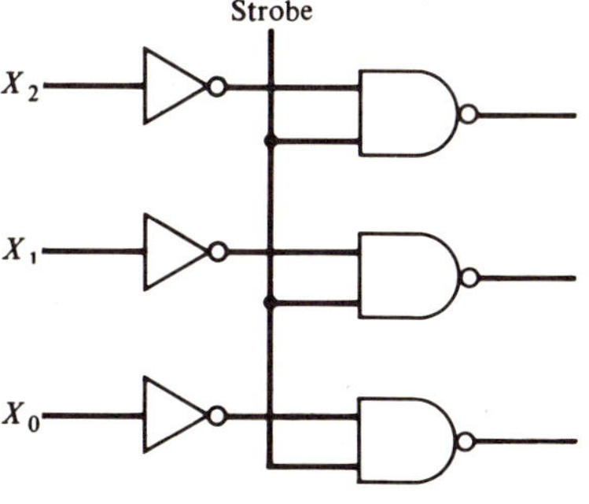

Figure 3-69   A strobe circuit for a 3-input decoder.

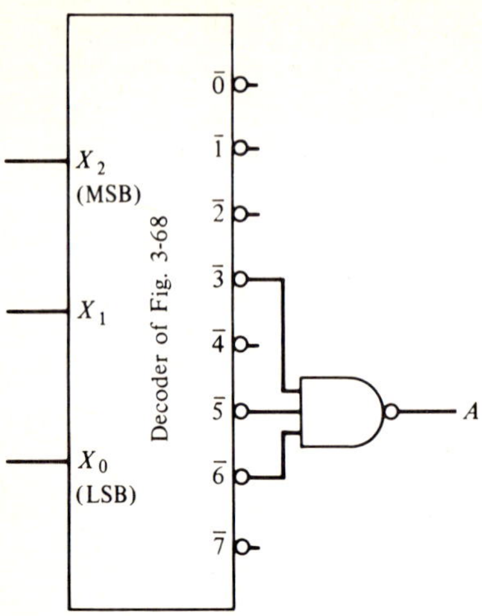

Figure 3-70   Implementation of a 3-variable switching function.

The above types of decoders can also be used in implementing multi-output combinational circuits. Another application of a decoder is as a demultiplexer to be discussed later in this section.

Decoding of BCD data is performed with the aid of a BCD-to-decimal decoder. There are two types of such decoders. If the decoder is designed without considering the six don't care input conditions, then it can be used to reject false input conditions. On the other hand, if the don't care input conditions are used to simplify the expressions for the output switching functions, the resultant decoder does not have the false data rejection feature. It should be noted that a 4-to-16-line decoder can be used also as a BCD-to-decimal decoder for any type of BCD codes by suitably selecting 10 out of the 16 outputs of the decoder (Problem 3-67).

A commonly used decoder is the BCD-to-7-segment decoder/driver circuit[12] to drive 7-segment LED or similar display devices. As the name implies, such a

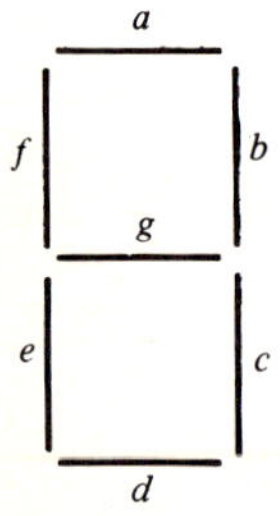

Input state	0000	0001	0010	0011	0100	0101	0110	0111	1000	1001
Lighted segments										
Segments turned off	g	a, d e, f, g	c, f	e, f	a, d e	b, e	a, b	d, e f, g		d, e

(a)   (b)

Figure 3-71   (a) A 7-segment display device, (b) display arrangements for 8-4-2-1 code.

display device consists of seven illuminated segments as shown in Fig. 3-71(a). In most cases, all segments are initially in the illuminated condition. By turning off appropriate segments, specific decimal digits are displayed [Fig. 3-71(b)]. For example, the number 4 is displayed by turning off segments a, d, and e. The pertinent truth table describing the operation of the decoder is shown in Table 3-18. As can be seen from Table 3-18, there are six don't care input conditions:

$$1010, \quad 1011, \quad 1100, \quad 1101, \quad 1110, \quad 1111$$

These don't care conditions can be used to simplify the expression for the output switching functions in complemented form. The final results are

$$\bar{a} = \bar{A}C + A\bar{B}\bar{C}\bar{D}$$
$$\bar{b} = A\bar{B}C + \bar{A}BC$$
$$\bar{c} = \bar{A}B\bar{C}$$
$$\bar{d} = ABC + A\bar{B}\bar{C} + \bar{A}BC \qquad\qquad (3\text{-}64)$$
$$\bar{e} = A + \bar{B}C$$
$$\bar{f} = AB + B\bar{C} + A\bar{D}\bar{C}$$
$$\bar{g} = ABC + \bar{B}\bar{C}\bar{D}$$

The derivation of Eq. (3-64) is left as an exercise. A 2-level all-NAND realization of Eq. (3-64) is straightforward.

An alternate realization of a BCD-to-7-segment decoder can be obtained by combining the outputs of a BCD-to-decimal decoder with false data rejection as illustrated in Fig. 3-72.

TABLE 3-18
BCD-to-7-Segment Decoder Truth Table

Inputs				Decimal Equivalent	Outputs						
D	C	B	A		a	b	c	d	e	f	g
0	0	0	0	0	1	1	1	1	1	1	0
0	0	0	1	1	0	1	1	0	0	0	0
0	0	1	0	2	1	1	0	1	1	0	1
0	0	1	1	3	1	1	1	1	0	0	1
0	1	0	0	4	0	1	1	0	0	1	1
0	1	0	1	5	1	0	1	1	0	1	1
0	1	1	0	6	0	0	1	1	1	1	1
0	1	1	1	7	1	1	1	0	0	0	0
1	0	0	0	8	1	1	1	1	1	1	1
1	0	0	1	9	1	1	1	0	0	1	1

   ↑   ↑
MSB   LSB

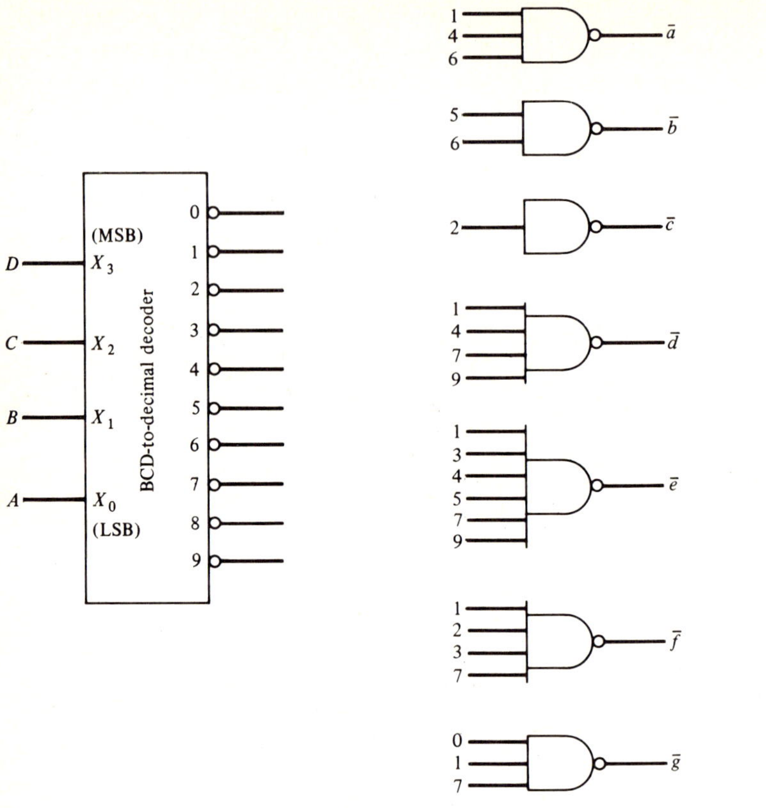

Figure 3-72   Implementation of a BCD-to-7-segment decoder.

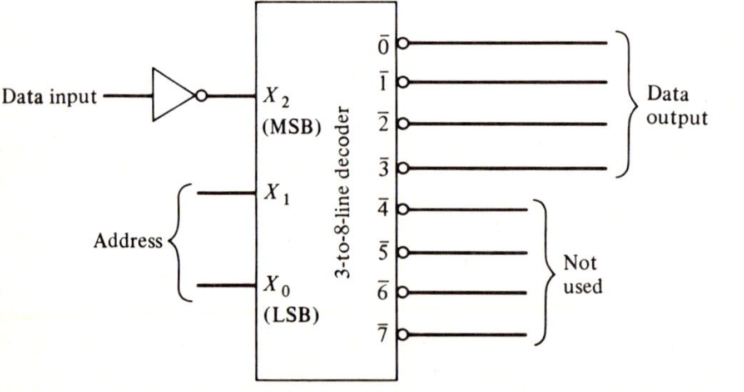

Figure 3-73   A 3-to-8-line decoder used as a 4-output demultiplexer.

Demultiplexer

The function of a demultiplexer is exactly opposite to that of a multiplexer. The demultiplexer has one data input line, k address lines, and 2^k output lines. The status of the address lines determines which one of the output lines will be at the same logic level as that of the data input line.

A decoder can be directly used as a demultiplexer by using one of its input lines as the data input line and the remaining input lines as address inputs. Furthermore only half of the output lines of the decoder are used as demultiplexer outputs. For example, if we examine the truth table of Table 3-17 and

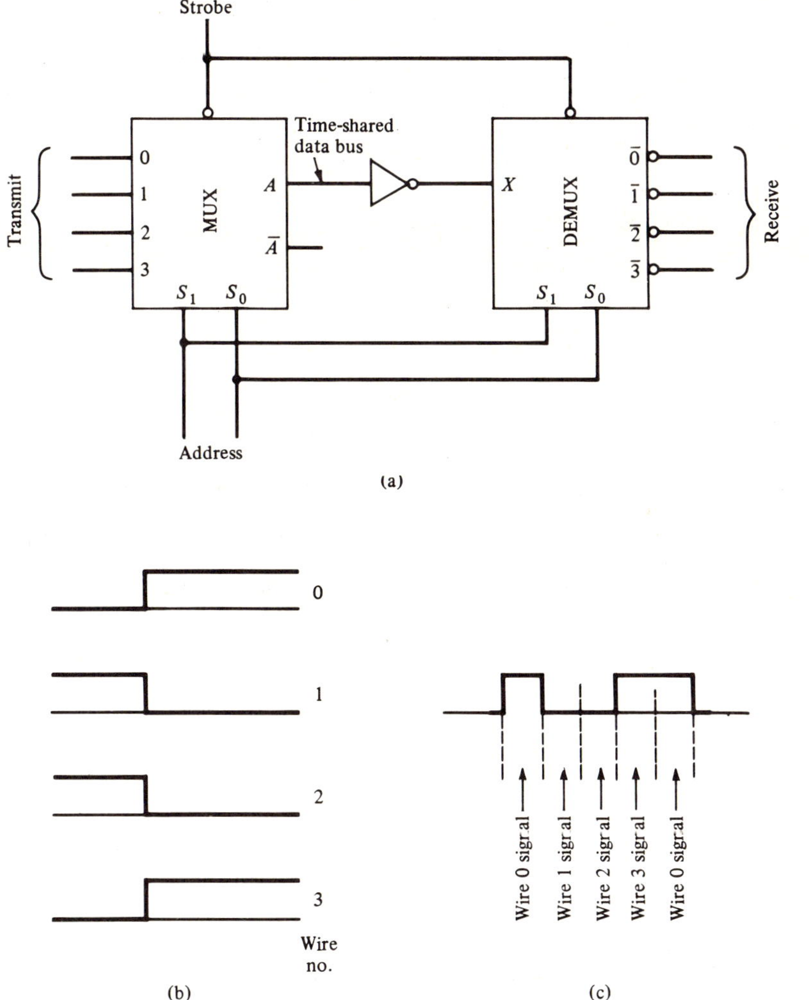

Figure 3-74 Time-multiplexing with a multiplexer and a demultiplexer: (a) schematic, (b) transmit waveform, and (c) output of the multiplexer for transmit waveform in (b).

restrict our attention to the first four output columns, we observe that output $\bar{0}$ assumes the same truth value as that of input X_2 when $X_1X_0 = 00$, with the remaining three outputs remaining always at a logical 1. Likewise, output $\bar{1}$ assumes the truth value of X_2 when $X_1X_0 = 01$ with outputs $\bar{0}$, $\bar{2}$, and $\bar{3}$ remaining at logical 1. Similarly, with $X_1X_0 = 10$ and $X_1X_0 = 11$, outputs $\bar{2}$ and $\bar{3}$, respectively, assume the same truth values as those of input X_2. The decoder of Fig. 3-68 can thus be used as a 4-output demultiplexer as shown in Fig. 3-73. Note that in this application, the last four output terminals of the decoder are not used.

As indicated earlier, the multiplexer and the demultiplexer are used in single-wire transmission of many signals by time multiplexing. Figure 3-74 illustrates the time sharing of a single wire among four input signals. There are various applications of the time-sharing approach to system design. One application was the aircraft music distribution system described in Section 1-4. Another application would be the time sharing of a large computer among a number of users connected through teletypes or other peripheral input/output devices.

3-15 Read-Only Memories and Programmable-Logic Arrays

A number of LSI digital circuit packages have been introduced in recent years. Of these, two very commonly used circuits are the *read-only memory* (abbreviated as ROM) and the *programmable logic array* (abbreviated as PLA). These two circuits are essentially multi-input, multi-output combinational circuits where each output variable is a function of a number of input variables. Moreover, each of these are designed to implement a large variety of multi-input, multi-output combinational circuits. Often, a single ROM or a PLA package can replace complex circuits realized with a large number of SSI digital circuits.

Read-Only Memories

The ROM consists of an address decoder and an array of semiconductor devices (memory array) that are suitably connected at the last fabrication step to realize a prescribed set of switching functions (Fig. 3-75). The process of making the desired connection is known as *programming*.

The operation of a ROM is determined by the number of " words " and the number of " bits " in each word. Each word is selected with the aid of the decoder by a unique combination of input bits, more commonly known as address bits. If the number of input lines is n, then the total number of words in the ROM is 2^n. If b is the number of bits in each word, then the total number of bits in the ROM, $2^n \times b$, is often used to indicate the size of the ROM. It should be noted that ROMs with 16,384 bits are not uncommon.

To understand the operation of a ROM consider a 4-word × 3-bit ROM shown symbolically in Fig. 3-76. The four words of the ROM are uniquely selected by a 2-bit address through a decoder. In this figure we have assumed that the decoder selects a *word line* with an active HIGH (1) on that line and

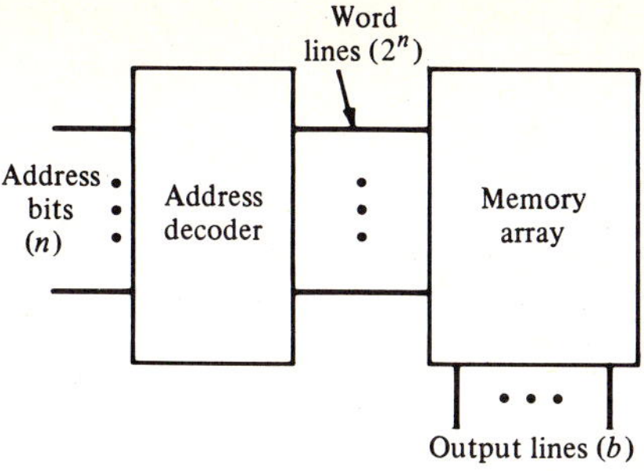

Figure 3-75 A general block diagram of a ROM.

with the other three word lines remaining LOW (0). As shown in the figure, the output lines of the ROM intersect the word lines at right angles. An intersection shown symbolically with a heavy dot indicates a " connection " of the pertinent word line with the output line through a semiconductor device. This connection is such that if the respective word line is HIGH, it raises the potential of the corresponding output line to HIGH. Or, in other words, the switching variable designating this output line is assigned a truth value of 1. If, on the other hand, none of the word lines connected to this output line is HIGH, the potential of the output line remains LOW indicating a 0 truth value for corresponding switching variable. For example, in the ROM shown in Fig. 3-76, $B = 1$ if $X_1 = 0$ and $X_0 = 1$. B is also 1 for $X_1 = 1$ and $X_0 = 0$. B is 0 for the other two input combinations. Thus it is seen that the output line B has implemented the switching function $\bar{X}_1 X_0 + X_1 \bar{X}_0 = X_1 \oplus X_0$. Likewise it can be seen that the output line A is implementing the switching function $\bar{X}_1 X_0 + X_1 \bar{X}_0 + X_1 X_0 = X_1 + X_0$ and the output line C is implementing the function $\bar{X}_1 \bar{X}_0 + X_1 X_0 = X_1 \odot X_0$. Note that here each word line implements a minterm.

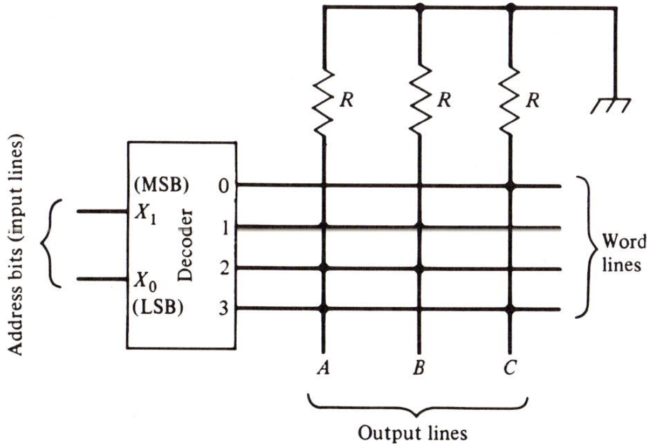

Figure 3-76 Symbolic representation of a 4×3 ROM.

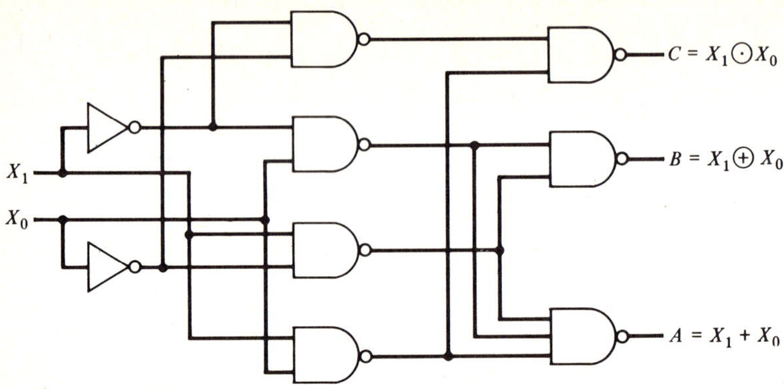

Figure 3-77   Equivalent circuit of the ROM of Fig. 3-76.

An equivalent representation of the ROM of Fig. 3-76 is thus as shown in Fig. 3-77. If we implement the combinational circuit of Fig. 3-77 using IC gates, it requires three IC packages, whereas the ROM realization requires one package.

An alternate interpretation of the ROM operation is by considering the output as a 3-bit word. Thus, if the input bits are $X_1 = 0$ and $X_0 = 1$, the output bit pattern is $A = 1$, $B = 1$, and $C = 0$; similarly for the other three input combinations. The truth table of Table 3-19 describes the four different bit patterns that can be selected by addressing the ROM appropriately.

The ROM is thus ideally suited for use as a multi-input, multi-output combinational circuit such as code converters or as a look-up table such as generating trigonometric functions. Other applications of ROMs include process control, character generation, and in microprogramming. We illustrate two simple applications next.

ROM Applications

Many arithmetic operations can be conveniently implemented by precalculating the results of the operations for all possible values of the operand, storing them in some type of memory, and fetching one particular answer for a prescribed

TABLE 3-19
Truth Table for the ROM Circuit
of Fig. 3-76

X_1	X_0	A	B	C
0	0	0	0	1
0	1	1	1	0
1	0	1	1	0
1	1	1	0	1

operand by addressing the memory locations appropriately. To illustrate such an application, we consider in the next example the design of a 2-bit × 2-bit multiplier.

Example 3-22.   Let the two numbers be denoted as $A_1 A_0$ and $B_1 B_0$ which we assume for simplicity to be positive integers. Their product will be a 4-bit number which we denote as $P_3 P_2 P_1 P_0$. The truth table describing the multiplication operation is given in Table 3-20. This multiplier can be implemented using a 16-word × 4-bit ROM as shown in Fig. 3-78.

TABLE 3-20
Truth Table for a 2-Bit Multiplier

A_1	A_0	B_1	B_0	P_3	P_2	P_1	P_0
0	0	0	0	0	0	0	0
0	0	0	1	0	0	0	0
0	0	1	0	0	0	0	0
0	0	1	1	0	0	0	0
0	1	0	0	0	0	0	0
0	1	0	1	0	0	0	1
0	1	1	0	0	0	1	0
0	1	1	1	0	0	1	1
1	0	0	0	0	0	0	0
1	0	0	1	0	0	1	0
1	0	1	0	0	1	0	0
1	0	1	1	0	1	1	0
1	1	0	0	0	0	0	0
1	1	0	1	0	0	1	1
1	1	1	0	0	1	1	0
1	1	1	1	1	0	0	1

The size of the ROM required for generating the product can be determined readily. If the number of multiplier bits is N_1 and the number of multiplicand bits is N_2, then it can be shown that the ROM size is given by $2^{(N_1 + N_2)} \times (N_1 + N_2)$ bits. For $N_1 = N_2 = 8$, the number of ROM bits is 1,048,576 which is considerably larger than that currently available. This problem can be avoided by breaking up the product of two numbers into a sum of four products involving fewer bits. For example, the product of two 8-bit numbers can be obtained using four 256 × 8 bit ROM and six 4-bit adders.

Other popular look-up table applications of ROMs include generating trigonometric functions, exponentials, logarithm and antilogarithms, integer and fractional powers, and so on.

ROMs are also ideally suited for implementing code conversion algorithms. Several off-the-shelf ROM based single chip code converters are readily available. One such converter is the Motorola MC4001 package[16] shown in Fig. 3-79.

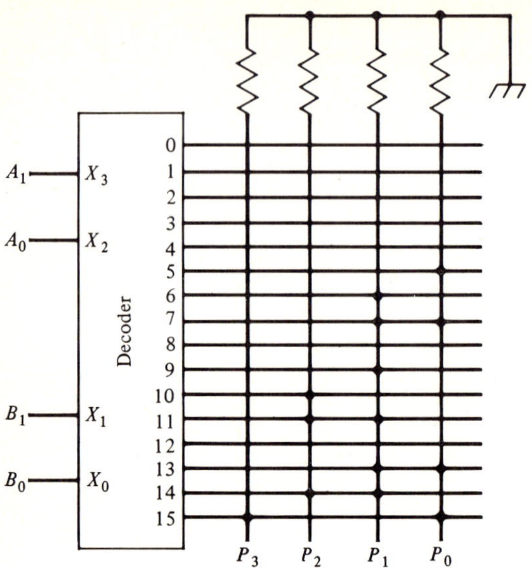

Figure 3-78   ROM implementation of a 2-bit multiplier.

It can be used for either binary-to-BCD (8-4-2-1) or BCD (8-4-2-1)-to-binary conversions. It has four input lines A_0, A_1, A_2, A_3, and eight output lines B_0, $B_1, \ldots, B_7$, and is designed from the 128-bit (16-word × 8-bit) ROM package MCM4000. The truth table describing the MC4001 package is shown in Table 3-21.

The MC4001 can be directly used in converting a 5-bit binary number (less than decimal 20) to BCD form which is often needed in the design of desk calculators and similar applications. The 5-bit binary number is obtained from

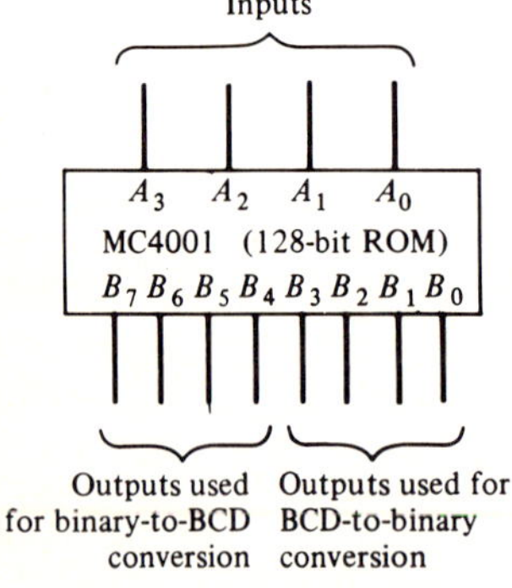

Figure 3-79   Schematic representation of MC4001 code converter. (Courtesy Motorola Semiconductor Products Inc.)

TABLE 3-21
MC4001 Truth Table[a]

Decimal Equivalent	Inputs				Outputs Binary ⟶ BCD				Outputs BCD ⟶ Binary			
	A_3	A_2	A_1	A_0	B_7	B_6	B_5	B_4	B_3	B_2	B_1	B_0
0	0	0	0	0	0	0	0	0	0	0	0	0
1	0	0	0	1	0	0	0	1	0	0	0	1
2	0	0	1	0	0	0	1	0	0	0	1	0
3	0	0	1	1	0	0	1	1	0	0	1	1
4	0	1	0	0	0	1	0	0	0	1	0	0
5	0	1	0	1	1	0	0	0	0	1	0	1
6	0	1	1	0	1	0	0	1	0	1	1	0
7	0	1	1	1	1	0	1	0	0	1	1	1
8	1	0	0	0	1	0	1	1	0	1	0	1
9	1	0	0	1	1	1	0	0	0	1	1	0
10	1	0	1	0	1	0	0	0	0	1	1	1
11	1	0	1	1	1	0	0	1	1	0	0	0
12	1	1	0	0	1	1	1	0	1	0	0	1
13	1	1	0	1	1	1	1	1	0	0	1	0
14	1	1	1	0	1	0	1	1	0	0	1	0
15	1	1	1	1	0	1	0	0	0	0	0	0

[a] Courtesy Motorola Semiconductor Products Inc.

adding two BCD numbers using four binary full adders. The appropriate connections for this application is shown in Fig. 3-80 whose validity can be checked by forming the corresponding truth table (Problem 3-74). Note that the LSBs of the binary number and its BCD equivalent are the same.

By connecting a number of MC4001 packages, a code converter to convert binary or BCD number of lengths greater than 4 bits can be constructed. The

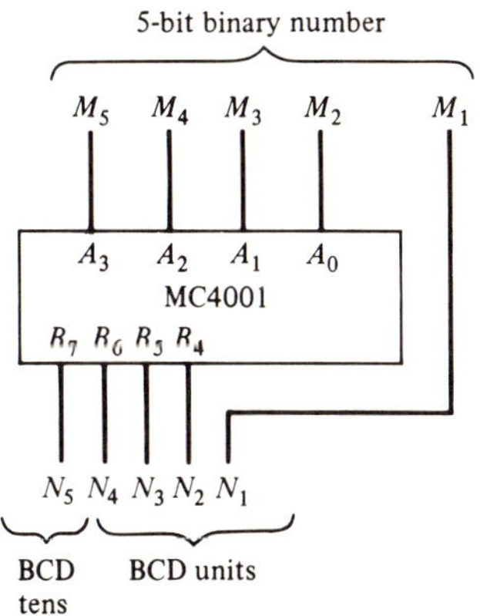

Figure 3-80   Conversion of a 5-bit binary number (less then decimal 20) to its BCD equivalent. (Courtesy Motorola Semiconductor Products Inc.)

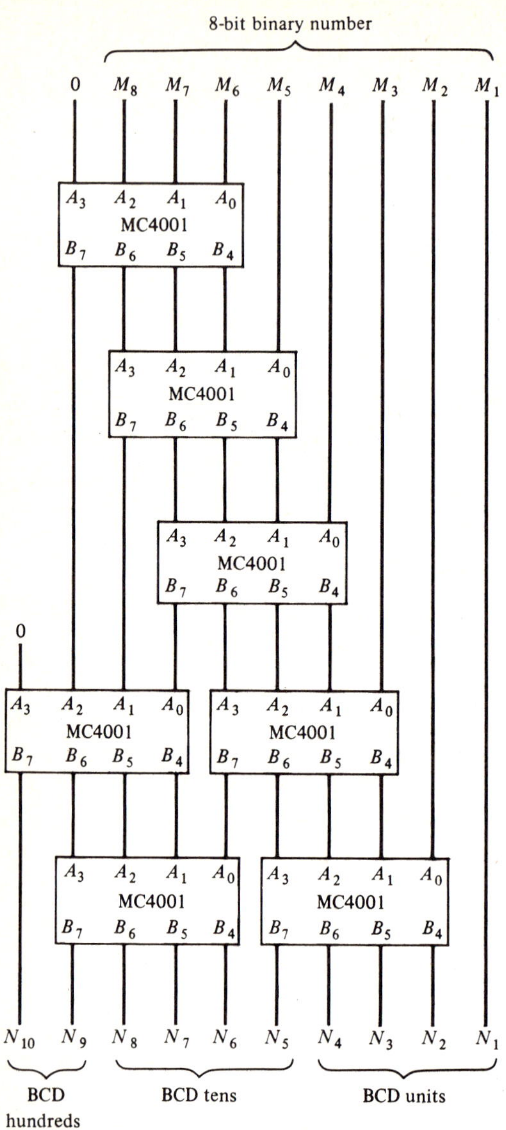

Figure 3-81   An 8-bit binary-to-BCD converter. (Courtesy Motorola Semiconductor Products Inc.)

design of an 8-bit binary-to-BCD converter using seven MC4001s is shown in Fig. 3-81. The associated algorithm used for the design and its development are explained in details in Reference 16, where also the design of converters to convert a BCD number of length greater than 4 to an equivalent binary number is considered.

Another novel application of the MC4001 is as a full adder and a full subtractor. The connection for full-adder operation is sketched in Fig. 3-82. From Table 3-21 we obtain

$$B_6 = \bar{A}_3 A_2 \bar{A}_1 \bar{A}_0 + A_3 \bar{A}_2 \bar{A}_1 A_0 + A_3 A_2 \bar{A}_1 \bar{A}_0 + A_3 A_2 \bar{A}_1 A_0 + A_3 A_2 A_1 A_0$$

$$B_4 = \bar{A}_3 \bar{A}_2 \bar{A}_1 A_0 + \bar{A}_3 \bar{A}_2 A_1 A_0 + \bar{A}_3 A_2 A_1 \bar{A}_0 + A_3 \bar{A}_2 \bar{A}_1 \bar{A}_0 + A_3 \bar{A}_2 A_1 A_0$$
$$ + A_3 A_2 \bar{A}_1 A_0 + A_3 A_2 A_1 \bar{A}_0$$

But as shown in Fig. 3-82, $A_3 = 1$, $A_2 = \bar{A}_i$, $A_1 = B_i$, and $A_0 = \bar{C}_i$. Hence the above two expressions simplify to

$$B_6 = A_i \bar{B}_i \bar{C}_i + \bar{A}_i \bar{B}_i C_i + \bar{A}_i B_i \bar{C}_i + \bar{A}_i B_i \bar{C}_i \qquad (3\text{-}65)$$

$$B_4 = A_i \bar{B}_i C_i + A_i B_i \bar{C}_i + \bar{A}_i \bar{B}_i \bar{C}_i + \bar{A}_i B_i C_i \qquad (3\text{-}66)$$

Comparing Eqs. (3-65) and (3-66) with Eqs. (3-47) and (3-48) it can be seen that $B_6 = \overline{C_{i+1}}$ and $B_4 = \bar{S}_i$. The full-subtractor connection is described in Problem 3-75.

ROM Programming

In practice, a semiconductor device is present at each intersection of the word lines and the output lines. At a particular intersection, if the device is " *connected* " between the respective word line and the output line (indicated by a heavy dot), a logical 1 is said to be *stored* in that *cell*. On the other hand, if the device is not connected (indicated by the absence of a dot), a logical 0 is said to be stored in the cell. The storing of the desired bit pattern in the memory array is thus accomplished by making the corresponding connections or breaking the connecting paths, as the case may be, and is known as ROM programming.

If a ROM with a particular bit pattern is desired, the necessary programming is usually performed by the manufacturer on a custom order basis. The process involves the development of a metallization mask containing the bit pattern from the prescribed ROM truth table which is then used to provide the necessary connection paths in the last step of the ROM fabrication. This type of custom ROM manufacture is quite economical if a large number of identically programmed ROMs are required.

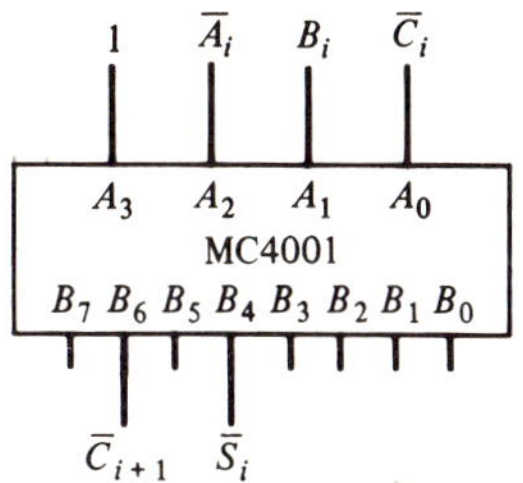

Figure 3-82 Implementation of a full adder using one MC4001 code converter. (Courtesy Motorola Semiconductor Products Inc.)

Certain types of ROMs, known as *programmable* ROMs (or PROMs), are field programmable allowing the user to write the desired bit pattern at his own location. In one class of PROMs, a metallic link is used as a connecting path in each cell. Any one of these links can be fused (i.e., broken) by suitably applying a very high current pulse. In another class of PROMs, a conductive path in a cell is developed through an *avalanche-induced migration process* by applying suitably a current through that particular cell.

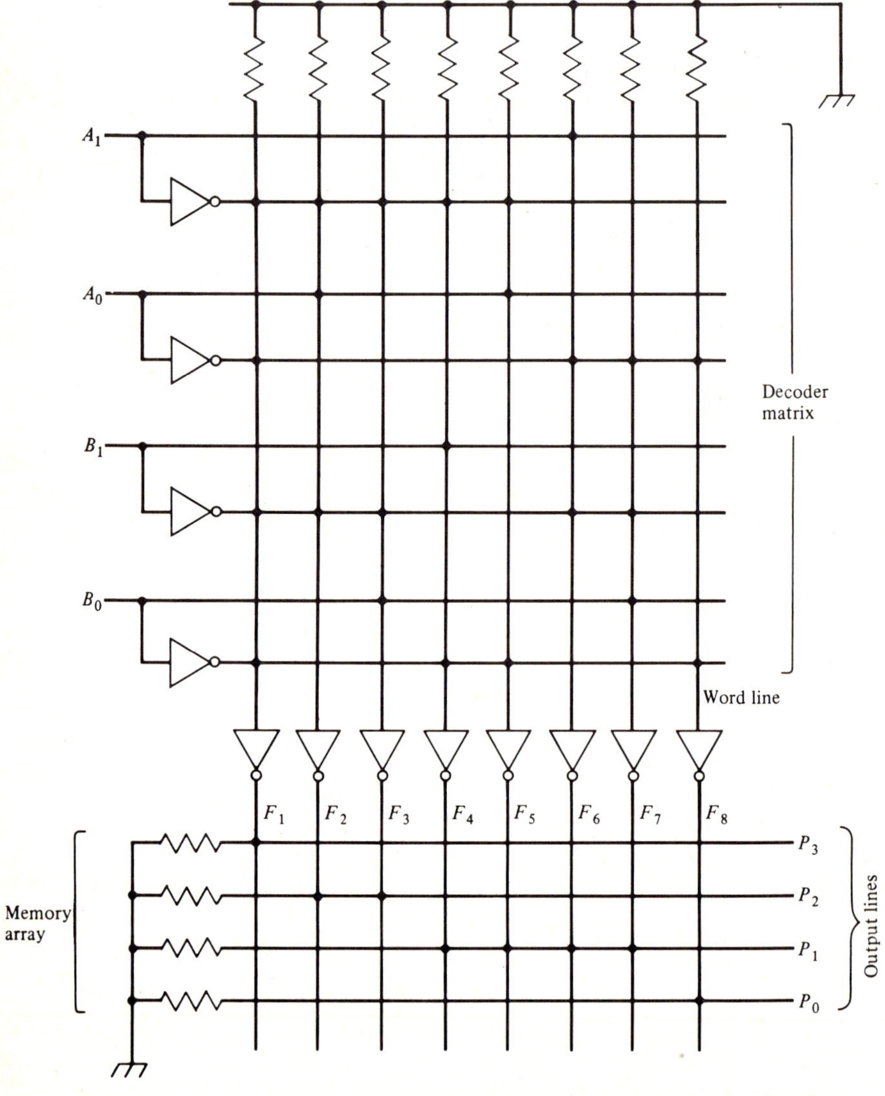

Figure 3-83   PLA implementation of a 2-bit multiplier.

The bit pattern written in a ROM by the IC manufacturer or in a PROM by the user is permanent. If it is desired to change the bit pattern, the old ROM has to be discarded and an entirely new ROM or a PROM has to be programmed. In many situations, when a digital system is under a development stage, it may be necessary to experiment with ROMs with different bit patterns. For such applications, reprogrammable ROMs have been introduced. In these ROMs the bit pattern is erased by illuminating the ROM with an ultraviolet light which returns all cells to the logical-0 state. Next the ROM is programmed by applying a large voltage to appropriate output lines while the pertinent words are selected one at a time.

Programmable-Logic Arrays

A basic difficulty associated with the ROM is that its memory array size increases very rapidly with an increase in the input lines, as the number of inputs determine uniquely the number of word lines with each word line corresponding to a unique minterm. However, in many applications not all minterms are used in generating the output functions. For example, in the 2-bit multiplier of Fig. 3-78, only nine word lines out of a total of 16 are being used. This fact can be used to advantage to reduce the memory array size and is accomplished by making the decoder matrix also programmable which is the basis of the programmable logic array (PLA). The PLA is thus a modified version of a ROM with a programmed decoder matrix. Here the number of word lines is usually considerably less than 2^n where n is the number of inputs.

A schematic diagram of a possible PLA implementation of the 2-bit multiplier is shown in Fig. 3-83. As in the case of the ROM, a heavy dot at an intersection of two lines indicates a connection between these lines through a semiconductor device. The vertical lines are called the word lines, the top eight horizontal lines are the input lines, and the bottom four horizontal lines are the output lines. The word lines are connected to ground (logical-0 level) through resistors. If any one or more of the input lines " connected " to a word line (shown by heavy dots) are at logical-1 levels, then the potential at the input of the inverter connected to the word line is forced to be at logical-1 level and, as a result, the output of the inverter assumes a logical-0 level. Otherwise, the inverter output remains at a logical-1 level. For example, the word line $\overline{F}_1$ is connected to the input lines $\overline{A}_1, \overline{A}_0, \overline{B}_1,$ and $\overline{B}_0$. Hence $\overline{F}_1$ is 1 if and only if $\overline{A}_1 = 1$ or $\overline{A}_0 = 1$ or $\overline{B}_1 = 1$ or $\overline{B}_0 = 1$. Or, in other words, $F_1 = \overline{\overline{A}_1 + \overline{A}_0 + \overline{B}_1 + \overline{B}_0} = A_1 A_0 B_1 B_0$. Similarly we obtain for the other word lines,

$$F_2 = A_1 \overline{A}_0 B_1, \qquad F_3 = A_1 B_1 \overline{B}_0, \qquad F_4 = A_1 \overline{B}_1 B_0$$

$$F_5 = A_1 \overline{A}_0 B_0, \qquad F_6 = \overline{A}_1 A_0 B_1, \qquad F_7 = A_0 B_1 \overline{B}_0$$

$$F_8 = A_0 B_0$$

Note that each word line here implements a product term which may or may not be a minterm.

Each output line is connected to ground through a resistor and is normally at logical-0 level, unless one or more word lines connected to it (shown by a heavy dot) are at logical-1 level. Thus it follows that for the PLA circuit of Fig. 3-83 we have

$$P_3 = F_1 = A_1 A_0 B_1 B_0$$
$$P_2 = F_2 + F_3 = A_1 \bar{A}_0 B_1 + A_1 B_1 \bar{B}_0 \tag{3-67}$$
$$P_1 = F_4 + F_5 + F_6 + F_7 = A_1 \bar{B}_1 B_0 + A_1 \bar{A}_0 B_0 + \bar{A}_1 A_0 B_1 + A_0 B_1 \bar{B}_0$$
$$P_0 = F_8 = A_0 B_0$$

It can be easily shown that the switching function expressions given above are those obtained from Table 3-20 after simplification.

It should be noted that the size of the memory array in the PLA implementation is now 32 bits which is half of that in the ROM implementation. A PLA is programmed in a manner similar to that used for a ROM.

3-16 Summary

The primary purpose of this chapter is to outline the techniques for analyzing and synthesizing combinational circuits. The components of a combinational circuit are logic gates. These gates provide the operations needed to combine the switching variables to form the switching functions characterizing the operation of the combinational circuits. The basic gates (AND, OR, and INVERTER gates) are described in Section 3-2. Several additional logic gates (NOR, NAND, exclusive-OR, coincidence, AND–OR–INVERT, and AND–OR gates) are also introduced in this section. The logic circuit analysis method is outlined in Section 3-3. Practical IC gates are nonideal devices and characteristics of such gates are discussed in Section 3-4. A review of the classification of IC logic circuits is provided in Section 3-5. Section 3-6 considers the effects of the nonideal characteristics of gates on the complete design of a combinational circuit. Section 3-7 outlines methods of developing a switching function from the truth table or a Karnaugh map description. The synthesis methods are illustrated with the aid of several examples in this section and in later sections. The realization of a switching function using the basic gates is illustrated in Sections 3-2 and 3-7. It is shown in Section 3-8 that any switching function can be implemented using exclusively either NOR gates or NAND gates. Section 3-9 considers the analysis of all-NAND and all-NOR logic circuits and conversion of logic circuits containing basic gates into all-NAND (NOR) circuits. A Karnaugh-map-based approach is described in Section 3-10 to obtain a simplified expression for a switching function. Simplification in general leads to a realization with fewer gates. Realization of switching function via the NAND- and NOR-tree minterm and maxterm generators is illustrated in Section 3-11. The following three sections describe the design of a number of useful combinational circuits. Some of these designs have been developed using the synthesis method of Section 3-7. Alternate approaches have been followed to develop the remaining circuits. Several arithmetic circuits, such as comparators, adders, and multipliers, are considered in Section 3-12. Code converters and simple error detection circuits are described in Section 3-13. Section 3-14 outlines the development of multiplexers, decoders, and demultiplexers. Often these devices can be used to implement complex switching functions rather easily. Finally Section 3-15 describes the read-only memories and the programmable-logic arrays, and outlines several applications of these LSI circuits.

References

1. D. L. Dietmeyer, *Logic Design of Digital Systems*, Allyn and Bacon, Boston, Mass., 1971.
2. W. E. Wickes, *Logic Design with Integrated Circuits*, John Wiley & Sons, New York, 1968.
3. "Application Memos," Signetics Corporation, Sunnyvale, Calif., September 1969.
4. L. W. Bell, "Digital Concepts," Publication No. 062-1030-00, Tektronix, Inc., Oregon, June 1969.
5. F. J. Hill and G. R. Peterson, *Introduction to Switching Theory and Logical Design*, John Wiley & Sons, New York, 1968.
6. H. Torng, *Introduction to the Logical Design of Switching Circuits*, Addison-Wesley, Reading, Mass., 1964.
7. "Technical Data," Motorola MRTL Digital Integrated Circuits, Motorola Semiconductor Products, Inc., Phoenix, Ariz.
8. A. Barna and D. I. Porat, *Integrated Circuits in Digital Electronics*, Wiley-Interscience, New York, 1973.
9. V. T. Rhyne, *Fundamentals of Digital System Design*, Prentice-Hall, Englewood Cliffs, N.J., 1973.
10. G. K. Kostopoulos, *Digital Engineering*, Wiley-Interscience, New York, 1975.
11. T. R. Blakeslee, *Digital Design with Standard MSI and LSI*, Wiley-Interscience, New York, 1975.
12. R. L. Morris and J. R. Miller, Eds., *Designing with TTL Integrated Circuits*, McGraw-Hill Book Co., New York, 1971.
13. J. D. Lenk, *Handbook of Logic Circuits*, Reston Publishing Co., Reston, Va., 1972.
14. H. V. Malmstadt and C. G. Enke, *Digital Electronics for Scientists*, W. A. Benjamin, New York, 1969.
15. J. L. Anderson, "Multiplexers Double as Logic Circuits," *Electronics*, October 27, 1969, pp. 100–105.
16. J. Linford, "Code Conversion with Semiconductor Read Only Memories," Application Note AN-506, Motorola Semiconductor Products, Inc., Phoenix, Ariz.
17. L. Hellerman, "A Catalog of Three-variable OR-invert and AND-invert Logical Circuits," *IEEE Trans. Electronic Computers*, Vol. EC-12, June 1963, pp. 198–223.
18. G. A. Maley and J. Earle, *The Logic Design of Transistor Digital Computers*, Prentice-Hall, Englewood Cliffs, N.J., 1963.
19. L. S. Garrett, "Integrated-circuit Digital Logic Families—Part I," *IEEE Spectrum*, Vol. 7, October 1970, pp. 46–58.
20. L. S. Garrett, "Integrated-circuit Digital Logic Families—Part II," *IEEE Spectrum*, Vol. 7, November 1970, pp. 63–72.
21. L. S. Garrett, "Integrated-circuit Digital Logic Families—Part III," *IEEE Spectrum*, Vol. 7, December 1970, pp. 30–42.

22. H. Taub and D. L. Schilling, *Digital Integrated Electronics*, McGraw-Hill Book Co., New York, 1976.

23. T. L. Booth, *Digital Networks and Computer Systems*, John Wiley & Sons, New York, 1971.

Problems

3-1 Realize an AND gate using an OR gate and NOT gates. Realize an OR gate using an AND gate and NOT gates.

3-2 Implement a 32-input NAND circuit using a single 8-input NAND gate and four dual 4-input AND gate packages.

3-3 Implement a 32-input OR circuit using a single 8-input NAND gate and four quad 2-input NOR gate packages.

3-4 Realize a 5-input NOR circuit using a quad 2-input NOR gate package and a hex INVERTER package.

3-5 Realize $F = XYZ + X\bar{Y}Z + \bar{X}YZ + XY\bar{Z}$ using 2-input gates only.

3-6 Realize $A = X \oplus Y \oplus Z \oplus W \oplus V$ in three different ways using only a quad exclusive-OR gate package.

3-7 Verify the two different expressions for the exclusive-OR operation given in Eq. (3-7).

3-8 Verify the exclusive-OR identities of Table 3-4.

3-9 Show that if $X \oplus Y = 1$, then $X \oplus 1 = Y$ and $Y \oplus 1 = X$.

3-10 Prove the following identities: **(a)** $\bar{X} \oplus \bar{Y} = X \oplus Y$, **(b)** $X \oplus Y = \overline{\bar{X} \oplus Y}$, **(c)** $X \oplus (X \oplus Y) = Y$, and **(d)** $(X \oplus Y) \oplus (\bar{X} \oplus \bar{Y}) = 0$.

3-11 Show that if $X \oplus Y \oplus Z \oplus W = 0$, then **(a)** $X \oplus Z \oplus W \oplus Y = 0$, and **(b)** $0 \oplus Y \oplus X \oplus W = Z$.

3-12 Verify the two different expressions for the coincidence operation given in Eq. (3-12).

3-13 Verify the exclusive-NOR identities of Table 3-6.

3-14 Show that $X \oplus Y \oplus XY = X + Y = X \odot Y \odot XY$.

3-15 Prove the identities: $X \odot Y = \bar{X} \oplus Y = X \oplus \bar{Y}$.

3-16 Show that if $X \odot Y \odot Z = 1$, then **(a)** $Y \odot 1 \odot Z = X$ and **(b)** $Y \odot Z = X \odot 1$.

3-17 Show that $X \odot Y \odot Z = X \oplus \bar{Y} \odot Z = X \oplus \bar{Y} \oplus \bar{Z}$.

3-18 Using a 2-wide, 2-input A-O-I gate implement the following gates: **(a)** NOR gate, **(b)** NAND gate, and **(c)** NOT gate.

3-19 Implement a 3-input OR gate using a 3-wide 2-input A-O gate.

3-20 Realize an exclusive-OR gate using three 2-input gates.

3-21 Realize a exclusive-NOR (coincidence) gate using only a quad NOR gate package.

3.22 Realize an exclusive-OR circuit and a exclusive-NOR circuit using only a 2-wide, 2-input A-O-I gate. Assume that inputs are available in *double rail* form, that is, in both normal and complemented forms.

3-23 Analyze the combinational circuits of Fig. 3-84 and determine the expressions for the output variables (without simplification) and the corresponding truth tables.

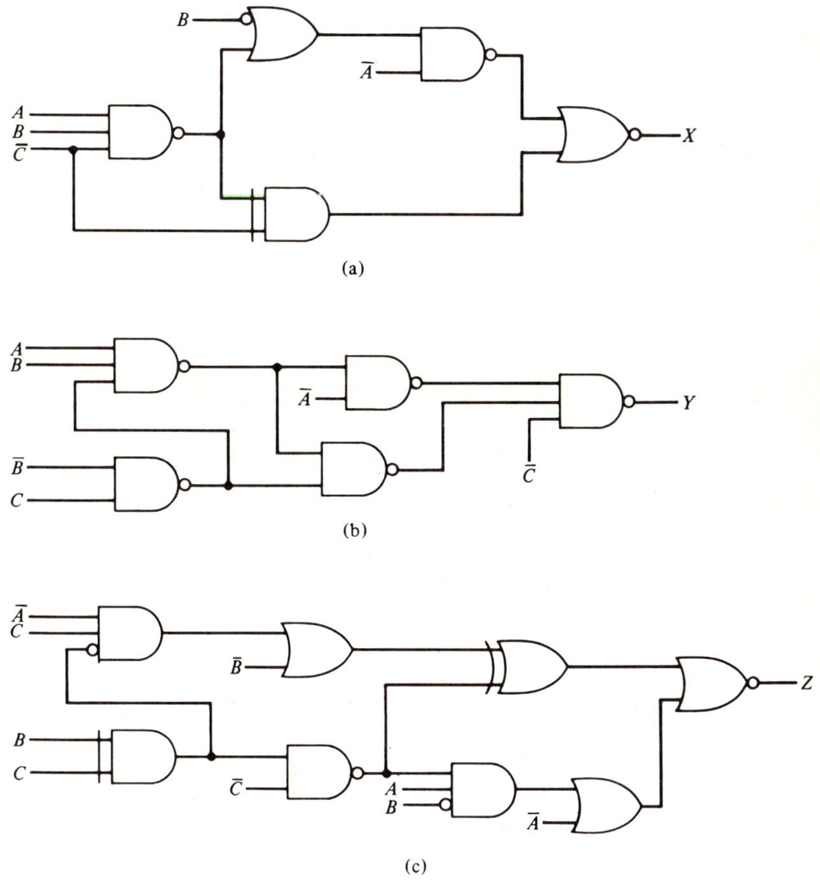

Figure 3-84

3-24 Obtain an equivalent realization of the combinational circuit of Fig. 3-20 using 3-input OR gates and NOT gates.

3-25 Determine the levels of realization of each of the circuits of Fig. 3-84.

3-26 A combinational circuit is to be designed which examines a 4-bit number in natural binary code and produces an output of 1 if and only if the decimal equivalent of the number is less than 13 but greater than 5. Develop the pertinent truth table describing the circuit and obtain a logic circuit implementation in SSOP form.

3-27 Develop an SSOP-type realization of a combinational circuit which examines a 4-bit number to determine if it is in 3-3-2-1 code. The output of the circuit is a 1 if and only if the input is an invalid combination not recognizable by this code.

3-28 Develop alternative realizations of the circuits of Problems 3-26 and 3-27 based on the SPOS-type expansions of the output variables.

3-29 Implement the circuits of Problems 3-26 and 3-27 using A-O-I gates and expanders.

3-30[23] A candy vending machine accepts only quarters, dimes, and nickels. The cost of any one of the candies is 15 cents. The machine dispenses a candy if either three nickels, or a dime and a nickel, or two dimes, or a quarter are inserted. A logic circuit is to be designed to activate a relay which dispenses a candy if and only if correct combination of coins have been used. Let $N_1 N_0$ denote the number of nickels, $D_1 D_0$ denote the number of dimes, and Q denote the number of quarters. Develop the appropriate truth table and a logic circuit implementation in SSOP form. For simplicity, assume the machine does not have facilities to return change.

3-31 Develop a 4-variable "dissent circuit" for which the output is a 1 if and only if the input variables are not all identical. Use only basic gates in the implementation.

3-32 Design a 5-variable "majority voter circuit" to indicate whether a majority of its inputs are 1's.

3-33 A 2-function logic unit shown symbolically in Fig. 3-85 implements one of two logic operations depending on the status of a function selector input terminal S. Develop the realizations for the following 2-function logic units:
(a) $A = X + Y$ if $S = 0$, $A = XY$ if $S = 1$,
(b) $A = X \oplus Y$ if $S = 0$, $A = X + Y$ if $S = 1$,
(c) $A = X \odot Y$ if $S = 0$, $A = XY$ if $S = 1$.

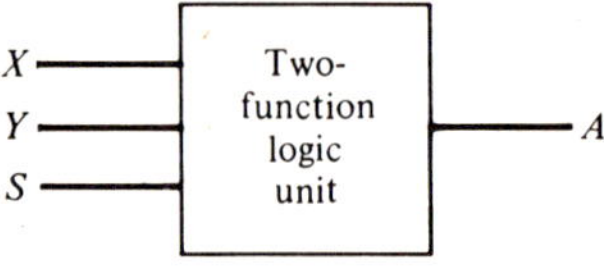

Figure 3-85.

3-34 Develop a realization of a 4-function logic unit characterized by Table 3-22 where S_1 and S_0 are the function selector inputs.

TABLE 3-22

S_1	S_0	A
0	0	$X + Y$
0	1	XY
1	0	$X + \bar{Y}$
1	1	$\bar{X}\bar{Y}$

3-35 Design a combinational circuit with fewest gates which receives a 4-bit binary number in natural binary code and produces an output of 1 if and only if the decimal equivalent of the binary number is nonzero and a multiple of three. The inputs are in normal form.

3-36 A 4-bit BCD number in 8-4-2-1 code is to be checked to determine whether the 4 bits represent a valid BCD number. Develop a minimum-gate-count combinational checking circuit whose output will be a 1 if the input is not a valid BCD number; otherwise it will be a 0.

3-37 Develop all-NAND realizations of the circuits of Problem 3-26, 3-27, and 3-30–3-36.

3-38 Develop all-NOR realizations of the circuits of Problems 3-26, 3-27, and 3-30–3-36.

3-39 Realize the circuits of Problems 3-26, 3-27, and 3-30–3-32, using a minimum number of logic gates.

3-40 Obtain a simplified expression in SOP form with fewest literals for each of the following three-variable switching functions:

(a) $f_a = \bar{X}(\bar{Y} + Z) + X\bar{Y}$
(b) $f_b = (X \odot Y)\bar{Z} + Z(Y + X\bar{Y})$
(c) $f_c = B(A \oplus C) + A\bar{C}$
(d) $f_d = A(B \oplus C) + A(C + B) + A\bar{B}\bar{C} + \bar{A}\bar{B}C$

3-41 Simplify the following 4-variable switching functions and obtain an SOP-type expression with fewest literals in each case:

(a) $g_a(X, Y, Z, W) = m_2 + m_3 + m_5 + m_7 + m_{11} + m_{13} + m_{14} + m_{15}$
(b) $g_b(A, B, C, D) = \Sigma m(0, 1, 5, 7, 8, 10, 14, 15)$
(c) $g_c(A, B, C, D) = \bar{B}\bar{C}D + CD + BC\bar{D}$
(d) $g_d(A, B, C, D) = \Sigma m(1, 5, 7, 6, 12, 13, 11, 15)$
(e) $g_e(W, X, Y, Z) = m_2 + m_7 + m_8 + m_9 + m_{10} + m_{11} + m_{13} + m_{15}$
(f) $g_f(A, B, C, D) = \bar{A}\bar{B}\bar{C} + \bar{A}\bar{C}D + \bar{A}BD + A\bar{B}\bar{C} + \bar{B}C\bar{D} + \bar{B}CD$

3-42 Analyze the 3-variable NAND tree of Fig. 3-45.

3-43 Develop the logic diagram of the 3-variable NOR tree minterm generator.

3-44 Develop all-NAND realizations of the circuits of Problems 3-26, 3-27, and 3-30–3-36 using a NAND tree maxterm generator.

3-45 Develop all-NOR realizations of the circuits of Problems 3-26, 3-27, and 3-30–3-36 using a NOR tree minterm generator.

3-46 Design a 1-bit comparator with minimum-gate-count to compare two input bits A and B. The comparator has three outputs W_1, W_2, and W_3 which are normally at logical-0 level except for the following input conditions:

$$W_1 = 1 \qquad \text{for} \qquad A > B$$
$$W_2 = 1 \qquad \text{for} \qquad A = B$$
$$W_3 = 1 \qquad \text{for} \qquad A < B$$

3-47 Realize a 2-bit equality detector circuit using the SSOP expansion of Eq. (3-44). Simplify the realization.

3-48 Realize a 2-bit simultaneous carry comparator.

3-49 Verify the 4-bit comparator realization of Fig. 3-49.

3-50 Design a 7-bit comparator using two 4-bit comparators. Modify the circuit using an additional 4-bit comparator to realize a 10-bit comparator.

3-51 Realize a half adder by implementing Eqs. (3-45) and (3.46). The inputs are available in normal forms only.

3-52 Obtain a simplified realization of a half adder by expressing the sum output S of Eq. (3-45) in a SPOS form.

3-53 Show that the all-NAND circuit of Fig. 3-86 is a full adder.

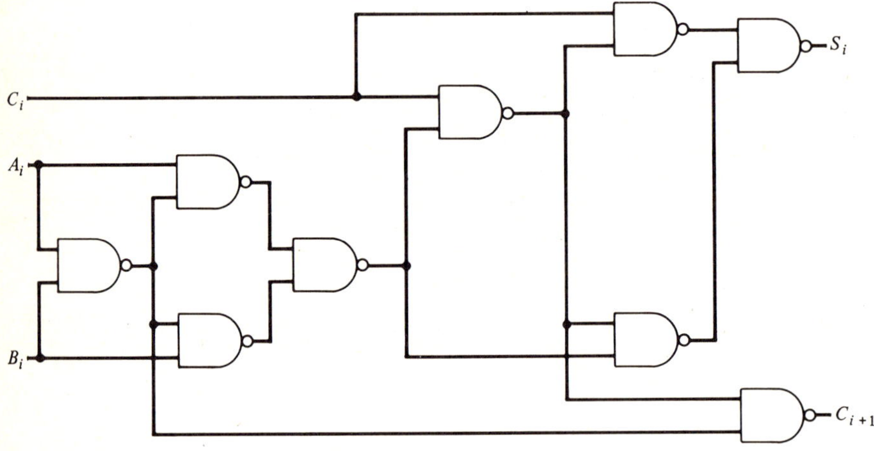

Figure 3-86   Full-adder implementation using only NAND gates.

3-54 Develop a two-level realization of a 5-bit TWO's complement logic circuit.

3-55* A combinational full-adder/subtractor unit shown symbolically in Fig. 3-87 is to be designed. When the control input E is 0, the circuit acts as a full adder by adding the input bits X, Y, and carry-in bit W and produces the sum bit S/D and carry-out C/B. On the other hand, with E set to a 1, the circuit behaves as a full subtractor, and subtracts X from Y with borrow-in bit W producing the difference bit S/D and borrow-out bit C/B. The truth table of the adder/subtractor unit is given in Table

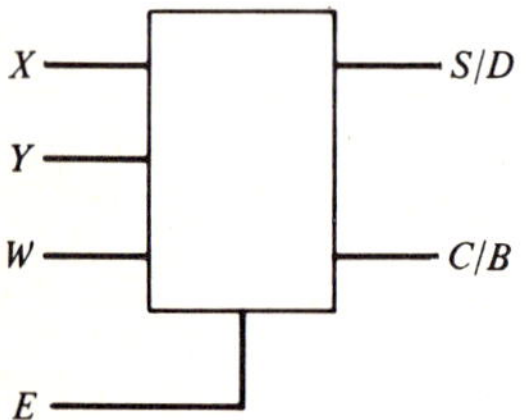

Figure 3-87

---

* Problems 3-55 and 3-56 are based on Robert A. Gabel, "A Logic Design Example: The Full Adder/Subtractor," *IEEE Trans. on Education*, Vol. 18, August 1975, pp. 171–173.

3-23. Develop a realization using only two input gates and having the shortest propagation delay (assume all gates have identical delays).

TABLE 3-23

E	X	Y	W	S/D	C/B
d^a	0	0	0	0	0
0	0	0	1	1	0
d	0	1	0	1	0
0	0	1	1	0	1
0	1	0	0	1	0
d	1	0	1	0	1
0	1	1	0	0	1
d	1	1	1	1	1
1	0	0	1	1	1
1	0	1	1	0	0
1	1	0	0	1	1
1	1	1	0	0	0

$^a d =$ don't care.

3-56* The circuit of Problem 3-55 is to be designed so that the cost of production of a single unit is a minimum. Available gates in the stock room are given below with their per package cost shown inside parentheses:

   quad 2-input OR (60¢), quad 2-input AND (50¢)
   quad 2-input NOR (60¢), quad 2-input NAND (60¢)
   dual 4-input AND (50¢), dual 4-input NAND (60¢)
   hex INVERTER (75¢)

3-57 Design a dual-purpose 4-bit code converter unit which operates as a binary-to-Gray code converter when the control bit S is a ZERO and as a Gray-to-binary code converter when the control bit S is a ONE.

3-58 Verify the relations between the natural binary- and Gray-coded numbers as given by Eq. (3-52) and (3-53).

3-59 Implement the 4-bit code converter to convert from 5-2-1-1 code to 8-4-2-1 code using Eq. (3-56).

3-60 Design a 4-bit converter to convert 8-4-2-1 code to 5-2-1-1 code. Utilize the don't care input conditions to simplify the design.

3-61 Design a 4-bit BCD (8-4-2-1) to excess-3 code converter. Make use of the don't care input conditions to simplify the design.

3-62 Develop a 9-bit parity detector/generator circuit using full adders only.

3-63 Explain why the 2-out-of-5 code of Table 3-14 can detect up to 40 percent of double-bit errors.

3-64 Realize the three 2-function logic units of Problem 3-33 using 4-input multiplexer packages.

---

* See footnote p. 150.

3-65 Realize the 3-variable switching functions of Problem 3-40 using 4-input multiplexer packages.

3-66 Realize the circuits of Problems 3-26, 3-27, 3-31, 3-34, 3-35, and 3-36 using 8-input multiplexer packages.

3-67 Convert a 4-to-16-line decoder to a BCD (5-2-1-1)-to-decimal decoder.

3-68 Design a BCD (8-4-2-1)-to-decimal decoder without false data rejection feature.

3-69 Realize the 3-variable switching functions of Problem 3-40 using the 3-to-8-line decoder of Fig. 3-68 and a NAND gate.

3-70 Verify the expressions for the outputs of the BCD-to-7-segment decoder as given by Eq. (3-64) and obtain a 2-level realization of the decoder.

3-71 Show in block-diagram form the design of an 8-bit $\times$ 8-bit multiplier unit which uses four 256-word $\times$ 8-bit ROMS, and five 4-bit adders.

3-72* Determine the size of the ROM to implement a sin X look-up table. Consider X limited to first quadrant values from 0 to 90°. X is available with a resolution of 0.01° and an accuracy of 16 bits is desired in the function evaluation. How would you implement the sin X look-up table with four ROMs of smaller size?

3-73† Show the schematic of a ROM implementation of a 4-input, 4-output combinational circuit described by:

$$W = B\bar{C}\bar{D} + A\bar{C}\bar{D} + AB\bar{C}D + A\bar{B}CD$$

$$X = \bar{A}BCD + \bar{A}C\bar{D} + ABC\bar{D} + A\bar{C}D$$

$$Y = AB + AC + BC$$

$$Z = \bar{A}B\bar{C}D + \bar{A}BC\bar{D} + \bar{A}C\bar{D} + \bar{B}C$$

3-74 Verify the 5-bit binary-to-BCD conversion scheme using MC4001 shown in Fig. 3-80.

3-75 Show that the circuit of Fig. 3-88 is a full subtractor, where B_i is the borrow bit, D_i is the difference bit, Y_i is the minuend bit, and X_i is the subtrahend bit.

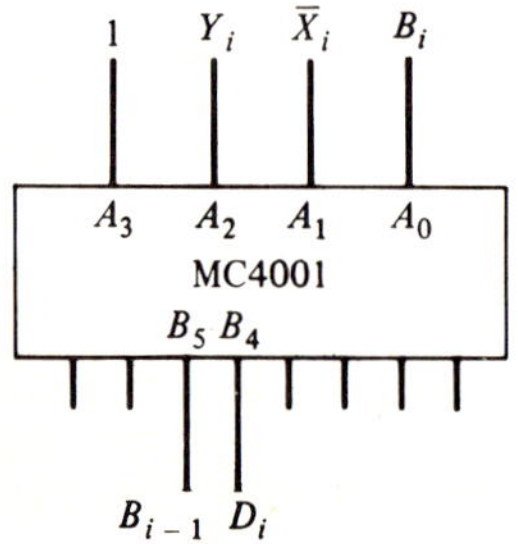

Figure 3-88 The MC4001 as a full subtractor. (Courtesy Motorola Semiconductor Products Inc.)

---

* A. Hemmel, "Making Small ROMs Do Math Quickly, Cheaply, and Easily," *Electronics*, Vol. 43, no. 10, May 11, 1970, pp. 104–111.

† W. I. Fletcher and A. M. Despain, "Simplify Combination Logic Circuits," *Electronic Design*, Vol. 13, June 24, 1971, pp. 72–73.

CHAPTER 4

Sequential Digital Circuits

The logic circuits discussed in the previous chapter have the common property that the output logic levels at any instant of time depend only on the input logic levels at the same instant (assuming no propagation delay from the input to the output) as the output switching variables are given by combination of the input variables using various switching operations. These circuits, known as *combinational circuits*, constitute a significant part of most digital systems. However, in many digital system designs, it is necessary to store past input and output values, as the outputs depend on these, along with the present values of the inputs. This second type of logic circuit is known as the *sequential circuit* and is the subject of discussion of this chapter.

To illustrate the need for a sequential circuit, consider the problem of estimating periodically as accurately as possible the position and velocity of an airplane.[13] These parameters are in turn determined from a sequence of range and angle measurements obtained from a periodic scanning of a radar antenna on the airplane. However, due to a number of factors, such as wind twisting the antenna, the measurements include disturbance effects which need to be removed before the position and velocity are computed. Thus, for example, the observed sequence of angle measurements x_n is in fact given by

$$x_n = a_n + d_n \qquad n = 1, 2, 3, \ldots \qquad (4\text{-}1)$$

where a_n is the signal representing the true angle during the nth scan and d_n is the noise caused by external factors. One way of reducing the effect of d_n to obtain a better estimate y_n of the angle is to pass the angle data sequence through a moving-average filter characterized by the equation

$$y_n = \sum_{i=0}^{M} h_i x_{n-i} \tag{4-2}$$

where the filter coefficients h_i are suitably chosen a priori. To implement the moving-average filter of Eq. (4-2) one needs to store M past values of the measured angle sequence $x_{n-1}, x_{n-2}, \ldots, x_{n-M}$. As a result, a digital implementation would require the design of a sequential circuit. Note that in addition to the storage registers storing past input values, the circuit requires multiplication and addition facilities. A simple example of a moving-average filter is

$$y_n = (x_n + x_{n-1})/2 \tag{4-3}$$

Problem 4-1 illustrates the dramatic reduction of the effect of disturbance using a filter described by Eq. (4-3) when the signal representing the actual angle does not change rapidly as a function of n, whereas the noise does.

A typical block diagram of a 3-input, 4-output sequential circuit is shown in Fig. 4-1. Here the output variables Y_1, Y_2, Y_3, and Y_4 are functions of the present values of the input X_1, X_2, and X_3, and past values of Y_3 and Y_4, and intermediate variables S_1 and S_2.

In general there are two classes of sequential circuits: synchronous and asynchronous. In a *synchronous sequential circuit*, a series of equally spaced pulses generated by a timing circuit known as *clock* is fed into the sequential circuit at appropriate places to insure that all pertinent operations are initiated only by the synchronizing clock pulse. The clock pulses also eliminate the effect of unequal propagation delays of the components of a sequential circuit and connecting lines. The synchronizing process guarantees the operation of the sequential circuit in an orderly and systematic fashion. On the other hand, an *asynchronous sequential circuit* is designed to operate without the synchronizing clock pulses. Here the signal indicating the completion of one operation is used to initiate the next operation in sequence and, as a result, the time taken for each

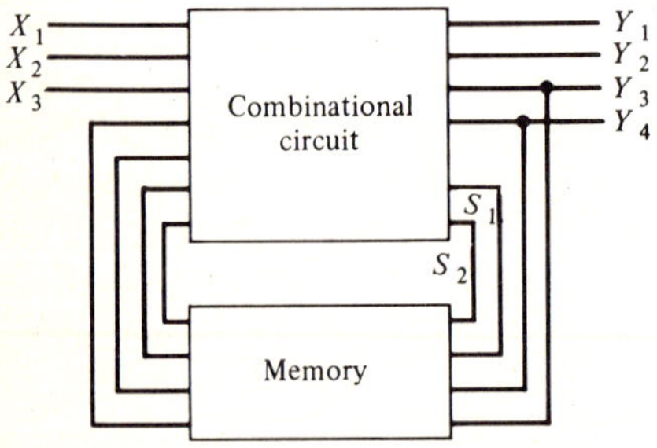

Figure 4-1   Block diagram of a typical sequential circuit.

operation is usually different from others. The majority of sequential circuits in use are the synchronous type as it is much easier to analyze and design such circuits. We shall restrict our attention in this chapter primarily to this type.

4-1   Flip-Flops

The two basic components in a sequential digital circuit or system are the logic gate and the binary *storage* or *memory* element. The various types of logic gates have been described in Section 3-2. In this section we plan to discuss the flip-flop, the most widely used binary memory device, which is readily available in IC forms. Later in this chapter, we outline systematic methods for the analysis and synthesis of sequential circuits containing gates and flip-flops. It should be noted that there are also other types of storage devices, such as magnetic cores, disks, tapes, and so on, discussion of which are beyond the scope of this book.

In general, a flip-flop has two output terminals and two or more input terminals, as shown symbolically in Fig. 4-2. As indicated in the figure, the logic level of one of the output terminals is the complement of that of the other output terminal. The *state* of the flip-flop is determined by the logic levels of the output terminals. The flip-flop is said to be in logical-1 state if the Q output is HIGH (i.e., at logical-1 level), and it is at logical-0 state if the Q output is LOW (i.e., at logical-0 level). Both the states of the flip-flop are stable. This means that the flip-flop when in one state remains in that state if there is no change in the input logic levels. A transition to the other state may occur only when there is a prescribed change in the logic levels at one or more of the input terminals.

The two inputs labeled A and B are synchronous inputs, and they work in conjunction with the clock input labeled CP. A transition at the data inputs has no effect on the state of the flip-flop unless the clock input is at appropriate logic level. The clock input in the flip-flop shown in Fig. 4-2 has to go to a high level (i.e., logical 1) to enable flip-flop state transitions. In some flip-flops, state transitions take place with the clock input going low (i.e., logical-0 level) which is

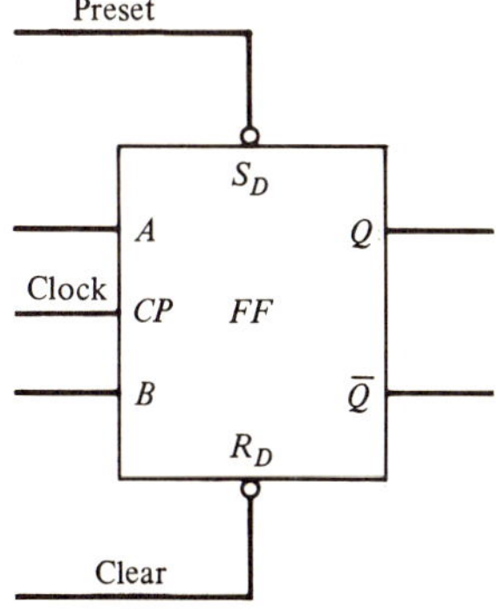

Figure 4-2   Symbolic notation of a typical flip-flop.

indicated by a circle at the clock terminal in the symbolic notations of the flip-flops. The synchronous inputs of a flip-flop are often called the *data* inputs.

The *preset* and *clear* inputs of a flip-flop are asynchronous and they operate independent of other inputs. In the flip-flop shown in Fig. 4-2, these inputs are normally at high or logical-1 levels. A low input at the clear terminal *clears* or *resets* the flip-flop to a logical-0 state, and a low input at the preset terminal *sets* the flip-flop to a logical-1 state. A common use of these inputs is to initialize the flip-flop to a known state, independent of the clock, before it accepts the data inputs. The preset and clear inputs are often called the *direct set* and *direct reset* inputs, respectively. In the symbolic notation of the flip-flops, these terminals are labeled S_D and R_D, as shown in Fig. 4-2.

The R–S Latch and Flip-Flop

The R–S latch is a 2-input, 2-output asynchronous binary storage component shown symbolically in Fig. 4-3. The input variables are labeled R and S, and the state of the flip-flop is denoted by Q.

The operation of the R–S latch is best explained by a truth table showing the transition of states as a function of the R–S inputs. Such a truth table, usually known as a *transition table*, is shown in Table 4-1. In Table 4-1 the input variables and the state at some arbitrary time $t = t_n$ are specifically shown by the use of subscripts n. Q_{n+1} indicates the next state at a later time $t = t_{n+1}$.

The normal or "resting" condition of the latch is $R = 0, S = 0$, shown in the top two lines of the table. Note that the output Q does not change with the input and may be equal to either 1 or 0. That is, it can be in either of its two stable states. A 1-pulse applied to the S-input terminal "sets" the Q output to assume the value of 1 a short time later, independent of its original value. This is shown in the third and fourth lines of the truth table. On the other hand, a 1-pulse applied to the R terminal "resets" the latch to the ZERO state (i.e., Q goes to 0) independent of its original value, as shown in the fifth and sixth lines.

Application of a 1-pulse to both the S and R terminals generally causes undesirable effects and hence is not allowed. The difficulty arises in trying to get the condition in which $R = 1, S = 1$ to the condition in which $R = 0, S = 0$. Even if the external signals change at the same time, unequal circuit delays will cause either the condition $R = 0, S = 1$ or the condition $R = 1, S = 0$ to occur for a brief period. Hence, when the resting condition is reached, the output Q may be a 1 (if $R = 0, S = 1$ occurred) or may be a 0 (if the $R = 1, S = 0$ occurred). This uncertainty in the final state of the latch is eliminated by requiring that the

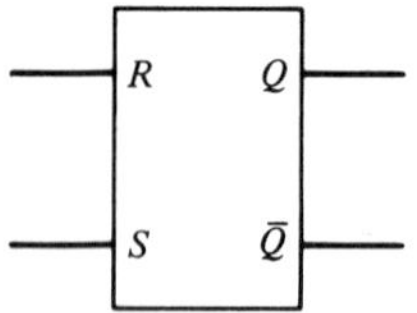

Figure 4-3 Symbolic representation of an R–S latch.

TABLE 4-1
**Truth Table for an R-S Latch
and Flip-Flop**

R_n	S_n	Q_n	Q_{n+1}
0	0	0	0
0	0	1	1
0	1	0	1
0	1	1	1
1	0	0	0
1	0	1	0
1	1	0	?
1	1	1	?

input combination $R = S = 1$ be prohibited. Mathematically this means that
the relation

$$R_n S_n = 0 \tag{4-4}$$

has to be maintained at all times.

Making use of Eq. (4-4) we can write the expression for Q_{n+1} in terms of R_n,
S_n, and Q_n as

$$Q_{n+1} = S_n + \bar{R}_n Q_n \tag{4-5}$$

Two different implementations of the R–S latch are shown in Fig. 4-4. The
operations of both of these circuits hinge upon the finite propagation delay inherent in a practical gate. Let us analyze the implementation of Fig. 4-4(a).
Analysis of the second realization follows similar lines.

A convenient way to analyze sequential circuits is to assume a consistent set
of input and output signal levels and then to observe the change in the signal
levels at pertinent terminals with changes in the input signal levels. As an aid to
the analysis, plots of various signals as a function of time, known as *timing
diagrams*, are often made. We follow this procedure for the circuit of Fig. 4-4(a),
assuming positive logic assignment scheme.

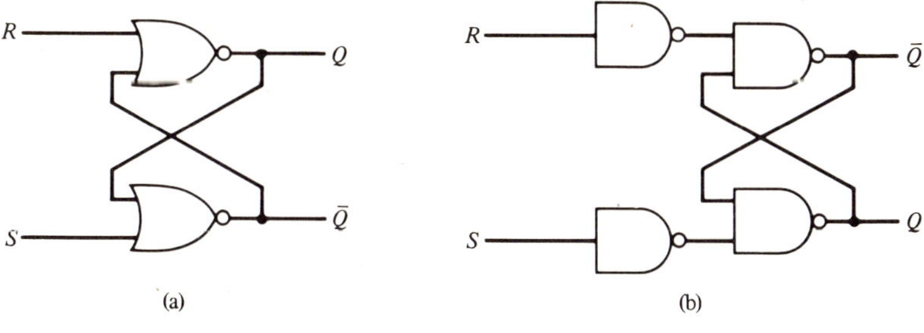

Figure 4-4   Implementations of an R–S latch using (a) NOR gates and (b) NAND gates.

Let the propagation delay of each NOR gate be τ sec. Assume next that at time $t = t_1$, $R = 0$, $S = 0$, $Q = 0$, and $\bar{Q} = 1$. Since S and Q are both 0's, the output $\bar{Q}$ of the bottom NOR gate is 1, as assumed. Likewise, $R = 0$ and $\bar{Q} = 1$ ensure that the output Q of the top NOR gate remains at 0, consistent with our earlier assumption. At time $t = t_2$ (with $t_2 > t_1$), let S go to 1. Since Q is still 0, $\bar{Q}$ goes to 0 after a delay of τ sec. Now the inputs to the top NOR gate are $R = 0$ and $\bar{Q} = 0$. Hence, after an additional delay of τ sec, Q goes to 1. At this time both the inputs to the bottom NOR gate are 1, implying that $\bar{Q} = 0$ is maintained. If at some later time $t = t_3$, S goes to 0, inputs to the bottom NOR gate would be $Q = 1$ and $S = 0$, indicating that $\bar{Q} = 0$ will be maintained. If, at $t = t_4$, R goes to 1, after a delay of τ sec, Q will go to 0. Since $S = 0$ now, $\bar{Q}$ will become 1 after a delay of τ sec. $R = 1$ and $\bar{Q} = 1$ will maintain $Q = 0$. Likewise, $Q = 0$ and $S = 0$ will maintain $\bar{Q} = 1$. R going to 0 will not change the status of Q and $\bar{Q}$. The corresponding timing diagram is indicated in Fig. 4-5. It should be noted that $R = 1$ and $S = 1$ will force both the output terminals to go to 0. If both of the inputs are made 0, status of the output terminals will not be known, as it depends on which of the input terminals become 0 first. It is evident that the operation of this circuit is described by the truth table of Table 4-1; that is, the circuit behaves as an R–S latch.

The R–S latch can be converted to a clocked R–S flip-flop by ANDing the R and S inputs with clock pulses. Figure 4-6 shows the R–S flip-flop derived from the circuit of Fig. 4-4(a), along with its symbolic notation. The modification of the R–S latch of Fig. 4-4(b) to a clocked R–S flip-flop is left as an exercise (Problem 4-4).

Other Types of Flip-Flops

There are three other types of basic flip-flops which we describe next. In the *J–K flip-flop* both the input variables can take the value of 1 at the same time, in contrast to the R–S flip-flop. The J–K flip-flop is symbolically represented as sketched in Fig. 4-7, where for simplicity the asynchronous clear and preset inputs have not been shown. The input data variables are denoted by J and K. Representing the state of the flip-flop by Q, the operation of the flip-flop can be

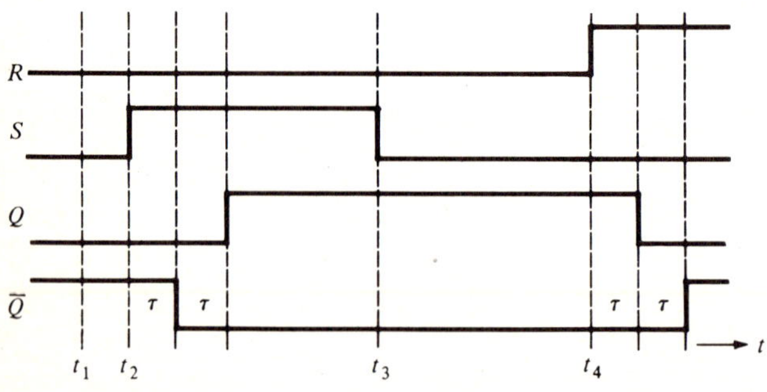

Figure 4-5   The R–S latch timing diagram.

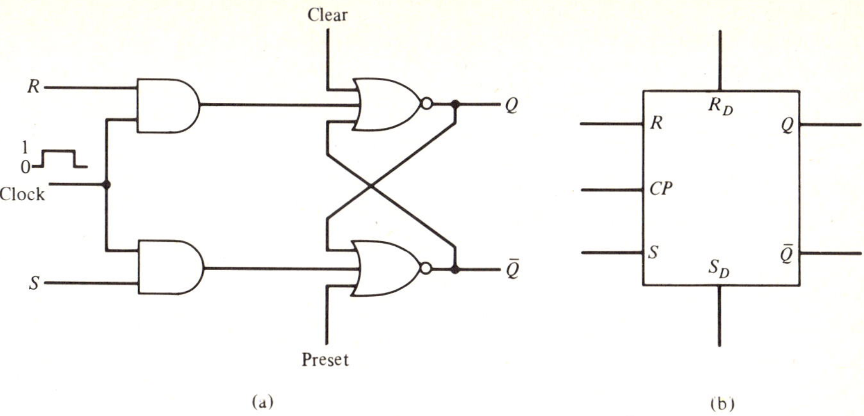

Figure 4-6 Clocked R–S flip-flop using NOR gates.

explained by its transition table given in Table 4-2. Note that the state transition in a clocked flip-flop occurs only when the clock pulse allows it. For the J–K flip-flop of Fig. 4-7, this transition would be when the clock pulse becomes positive. From this table we obtain readily the expression for the next state Q_{n+1} in terms of the state Q_n and the inputs J_n and K_n at time $t = t_n$:

$$Q_{n+1} = J_n \bar{Q}_n + \bar{K}_n Q_n \tag{4-6}$$

Note that the J–K flip-flop behaves like an R–S flip-flop (with J corresponding to S and K corresponding to R) for all input combinations except for the case $J_n = 1$, $K_n = 1$ in which situation the flip-flop changes state (changes Q) from its preceding state.

The two other basic flip-flops are the T flip-flop and the D flip-flop, with each having a single clocked data input terminal. The T or *toggle flip-flop* changes states for each pulsing of the data input variable which happens during a clock pulse. The symbolic representation of the T flip-flop is indicated in Fig. 4-8 where again for simplicity the asynchronous clear and preset inputs have not been shown. The transition table of Table 4-3 describes the operation of this flip-flop. Mathematically we can describe the operation of this device as

$$Q_{n+1} = \bar{T}_n Q_n + T_n \bar{Q}_n = T_n \oplus Q_n \tag{4-7}$$

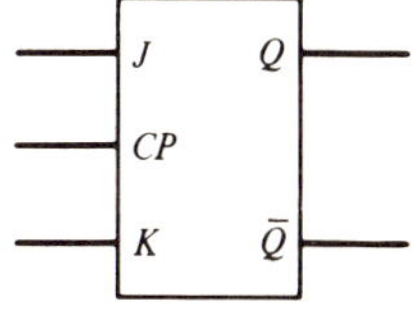

Figure 4-7 Symbolic representation of a J–K flip-flop.

TABLE 4-2
**Transition Table of a *J-K*
Flip-Flop**[a]

J_n	K_n	Q_n	Q_{n+1}
0	0	0	0
0	0	1	1
0	1	0	0
0	1	1	0
1	0	0	1
1	0	1	1
1	1	0	1
1	1	1	0

[a] Note transition occurs only when $CP = 1$. If $CP = 0$, there is no transition.

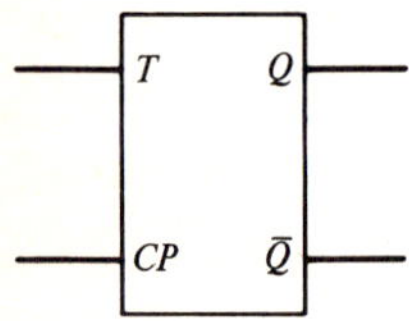

Figure 4-8   Symbolic representation of a *T* flip-flop.

In a *D or delay flip-flop*, irrespective of the present state, the next state is always identical to the present logic level of the data input if fed in conjunction with a clock pulse. The symbolic representation of the *D* flip-flop without the asynchronous preset and clear inputs is sketched in Fig. 4-9. Its operation is described by the transition table of Table 4-4. Essentially this flip-flop is a delaying element which delays the input signal by the amount of the clock period.

Any one of the above four flip-flops can be modified with possibly additional logic circuitry to work as one of the remaining three flip-flops. This is illustrated later in this section.

TABLE 4-3
**Transition Table
for a *T* Flip-Flop**[a]

T_n	Q_n	Q_{n+1}
0	0	0
0	1	1
1	0	1
1	1	0

[a] Note transition occurs only when $CP = 1$.

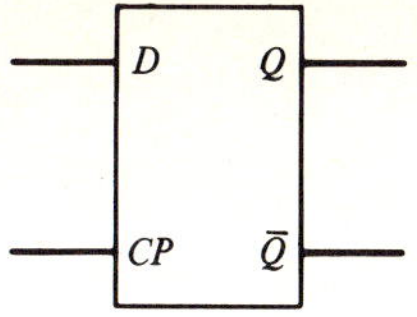

Figure 4-9   Symbolic representation of a D flip-flop.

TABLE 4-4
Transition Table for
a D Flip-Flop[a]

D_n	Q_n	Q_{n+1}
0	0	0
0	1	0
1	0	1
1	1	1

[a] Note that transition occurs only when $CP = 1$.

Flip-Flop Clocking Techniques

In general, the flip-flops are used as memories in the design of sequential circuits (Fig. 4-1). As a result, in many cases, the outputs of the flip-flops are fed back to the inputs. This type of feedback may lead to undesirable responses, as illustrated by the simple sequential circuit of Fig. 4-10. Assume that in the

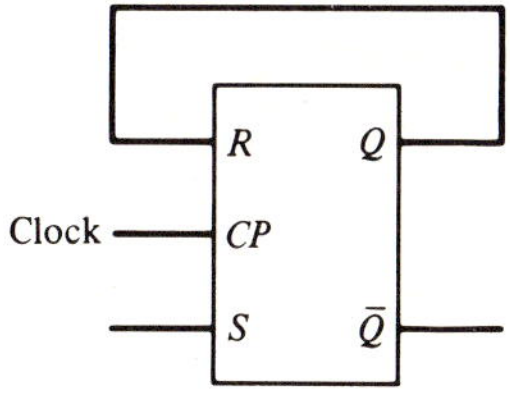

Figure 4-10   Illustration of the racing problem.

absence of the clock pulse, the SET terminal is 1 and the state of the flip-flop is ZERO, that is, $S = 1$ and $Q = 0$. Since $Q = 0$, R is also 0. Thus when the clock pulse is applied, the next state is 1, making R also 1. Now both S and R are 1, indicating that Q will go to 0.* This makes R equal to 0. But S being 1, Q will be *set* to 1, that is, the flip-flop state will start oscillating between ONE and ZERO. Or, in other words, the circuit of Fig. 4-10 will become a free-running

* We are assuming here that the R–S flip-flop is designed using cross-coupled NOR gates.

multivibrator (see Section 4-11 for the definition). This oscillation occurs only if the clock is set at 1 for too long. Thus, for this type of circuit to operate properly, it is necessary that the clock pulse width be so short that by the time the flip-flop has changed states and has made corresponding changes in the status at its inputs, the clock is no longer present. On the other hand, the clock pulse width must be sufficiently long for the flip-flop to be activated by narrow input pulses.

One way to avoid this *racing* problem is to use two flip-flops in cascade in what is known as the *master-slave* (MS) configuration.[1] To explain the operation of the MS flip-flop, consider first the circuit of Fig. 4-11(a), where two R–S flip-flops have been connected in cascade. The flip-flop on the left is enabled by the clock pulses C_A, and the flip-flop on the right is enabled by the clock pulses C_B. The waveforms of the two clock pulses are shown in Fig. 4-11(b). When $C_A = 1$ at $t = t_1$, the leftmost flip-flop responds to the truth values of the set and reset inputs, S_A and R_A, respectively. Since at that time $C_B = 0$, the outputs of this flip-flop do not affect the state of the right flip-flop. Next, at $t = t_2$, $C_B = 1$, and $C_A = 0$, implying the flip-flop at the left remains at the state determined previously by the truth values of R_A and S_A at $t = t_1$ and will not change even though R_A and S_A status may have changed. However, C_B being 1 now, the state of the flip-flop on the right assumes the state of the flip-flop on the left. The flip-flop on the left is known as the *master* flip-flop and the one on the right is called the *slave* flip-flop. Thus when $C_A = 1$ and $C_B = 0$, the slave is disconnected from the master and the master responds to the input SET and RESET signals. Then when $C_A = 0$ and $C_B = 1$, the external inputs are disconnected from the master and the state of the master is transferred to the slave.

In practice the two-phase clock signals are combined into one waveform of the form shown in Fig. 4-12 at the bottom where the gates are enabled by the

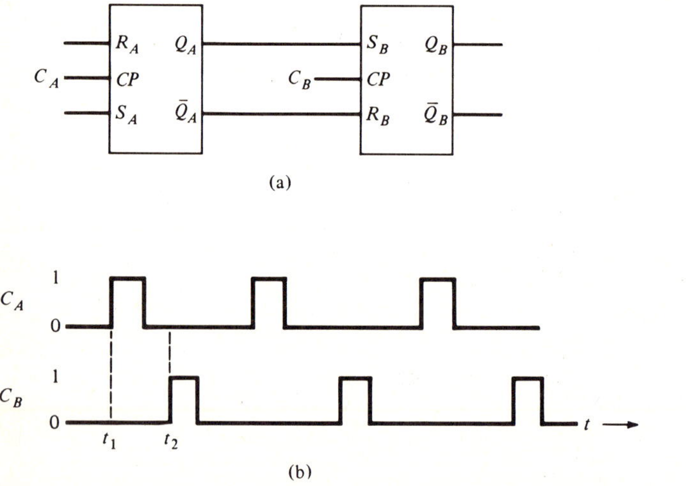

Figure 4-11 Illustration of master-slave flip-flop concept.

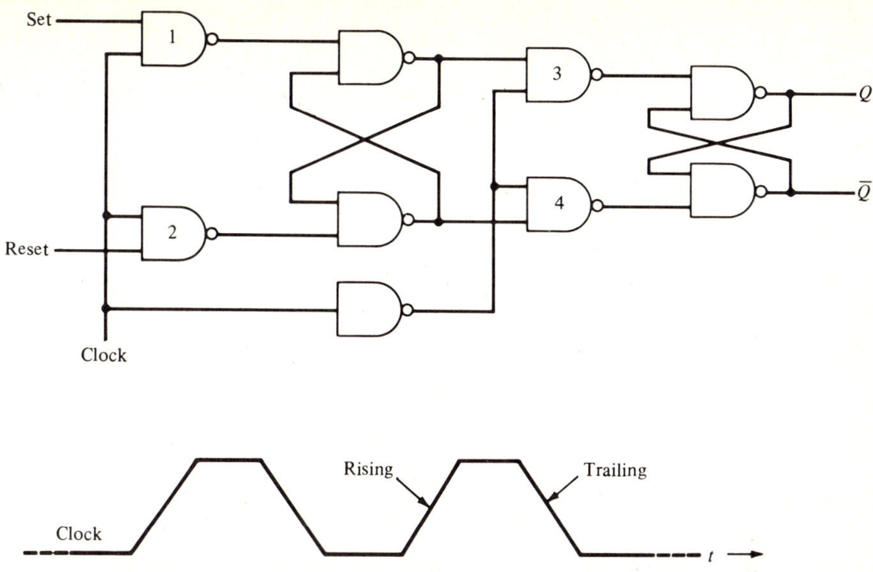

Figure 4-12 Realization of the master-slave R–S flip-flop.

rising and the trailing edges of the clock pulse. An implementation of the MS flip-flop using NAND gates is also shown in this figure. Note that if all the NAND gates are enabled only by the rising edge of the signals at their input, then the leading edge of the clock pulse actuates the gates 1 and 2, and the trailing edge actuates the gates 3 and 4.

An alternate way to avoid the racing problem is by a d-c or *edge-triggered* clock. Here the latch is activated at a particular voltage level of an internally generated narrow pulse obtained from the main clock input. The voltage level recognized is either the positive-going (rising) edge or the negative-going (trailing) edge, depending on the actual design of the flip-flop.

Note that in a MS flip-flop, both transitions of the clock are used to operate the device. On the other hand, in an edge-triggered flip-flop, either the positive-going or the negative-going transition is utilized. As a result, the latter type of flip-flop can be operated at a speed higher than the former type.

Relation Between the Flip-Flops

As indicated earlier, any one of the flip-flops discussed earlier can be converted into another type of flip-flop with additional circuitry. We describe some of these modifications here.

The J–K flip-flop can be used as a T flip-flop by connecting the two inputs together as shown in Fig. 4-13(a). This imposes the constraint $K_n = J_n = T_n$ which, when substituted in Eq. (4-6), leads to Eq. (4-7). Similarly either a J–K or an R–S flip-flop can be converted to a D flip-flop using an additional NOT gate, as shown in Fig. 4-13(b).

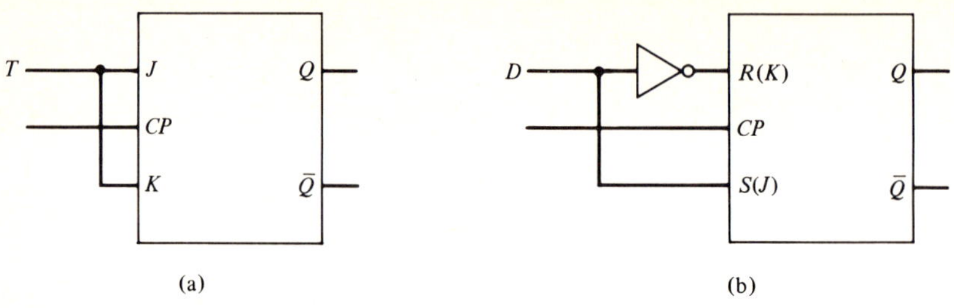

Figure 4-13 (a) Realization of a T flip-flop from a J–K flip-flop, (b) realization of a D flip-flop from an R–S or a J–K flip-flop.

An R–S master-slave or edge-triggered type flip-flop can be converted to a J–K flip-flop with the aid of two additional AND gates as shown in Fig. 4-14. It is seen from the circuit diagram that when the clock is 1 we have

$$R_n = K_n Q_n$$
$$S_n = J_n \bar{Q}_n \tag{4-8}$$

Substituting above in Eq. (4-5) we arrive at the expression for the next state as

$$Q_{n+1} = S_n + \bar{R}_n Q_n = J_n \bar{Q}_n + (\overline{K_n Q_n})Q_n$$
$$= J_n \bar{Q}_n + \bar{K}_n Q_n \tag{4-9}$$

which is the same as Eq. (4-6).

Figure 4-15 illustrates two possible schemes for converting a D flip-flop to a T flip-flop. Likewise, two different circuit arrangements to convert an R–S flip-flop to a T flip-flop are sketched in Fig. 4-16. Note that in the implementation schemes of Figs. 4-15(a) and 4-16(a), the clock pulse is used to toggle the flip-flops. Verification of the operation of these four circuits is left as an exercise.

IC Flip-Flop Packages

The flip-flops available as IC packages are usually the R–S, J–K, and D types, with one to six flip-flops per package. Most of the clocked IC flip-flops are

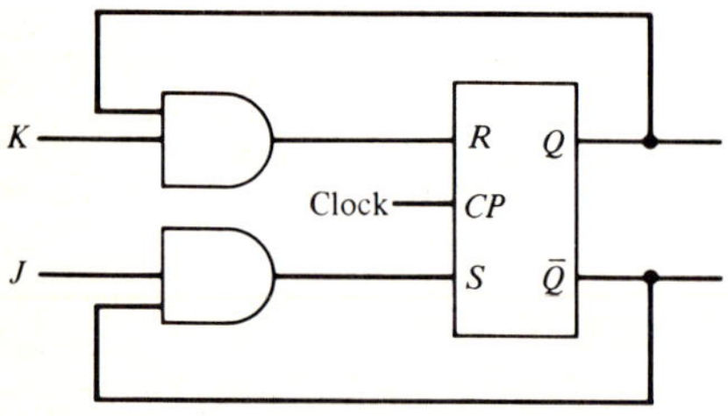

Figure 4-14 Implementation of a J–K flip-flop using an R–S flip-flop.

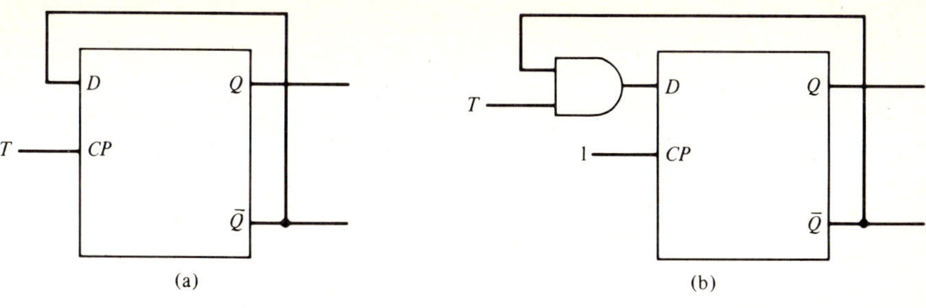

Figure 4-15 Implementation of a T flip-flop using a D flip-flop.

equipped with two sets of input terminals. One set of these consists of synchronous input terminals that are enabled with the clock pulses. The other set contains a preset and a clear, asynchronous input terminals which are not gated by the clock, and can thus be used to establish any desired state in the flip-flop independent of the clock if necessary. Clocked R–S and J–K flip-flops are available usually in both forms, edge-triggered and master-slave, whereas the D flip-flop is available with edge-triggered clocking. In addition to the flip-flops, multiple R–S NOR and NAND latches are also marketed.

Several typical IC packages are shown in Fig. 4-17. In these figures, the terminals labeled "R_D" are the clear input terminals and the terminals labeled "S_D" are the preset input terminals. As can be seen, these terminals are activated with a ZERO. Some flip-flops have logic gates at the input terminals of the flip-flops to allow the flip-flop's input signal to be generated by an AND combination of several switching variables. For example, the flip-flop package of Fig. 4-17(b) has three J and three K inputs. One each of the J and K inputs is inverted before ANDing. Some flip-flop packages contain A–O gates connected to the flip-flop inputs to provide additional flexibility to the system designer.

Characteristics of Practical Flip-Flops[9]

An IC flip-flop belonging to a particular logic family has characteristics almost identical to those of the IC gates in the same family. An overview of these

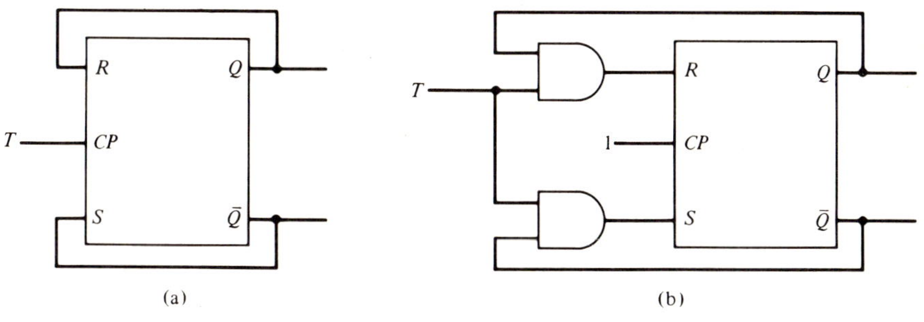

Figure 4-16 Conversion of an R–S flip-flop to a T flip-flop.

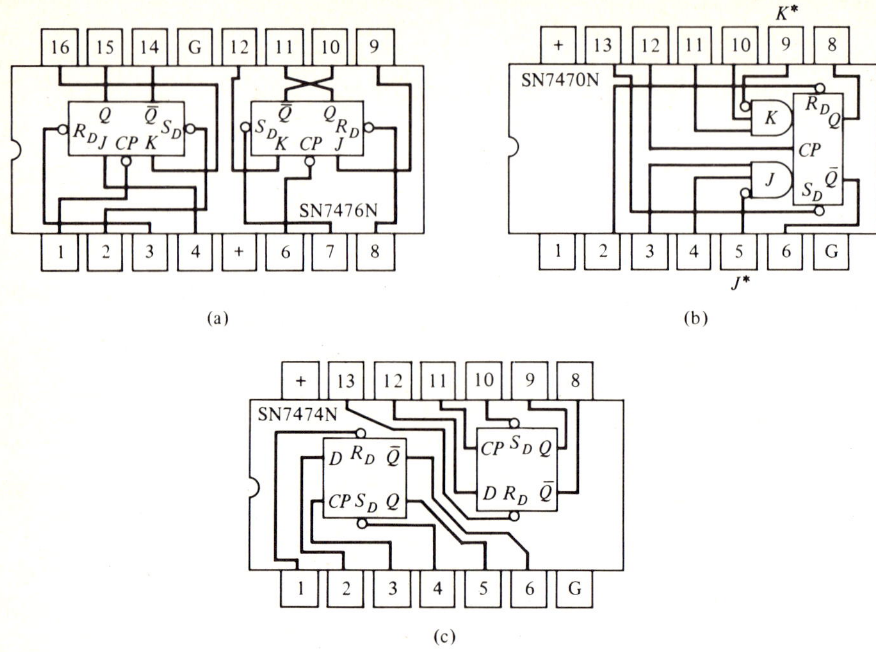

(a)

(b)

(c)

Figure 4-17   Top views of (a) SN7476N—Dual J–K master-slave flip-flop ICP with preset and clear, (b) SN7470N—dc clocked J–K flip-flop ICP, and (c) SN7474N—Dual D edge triggered flip-flop ICP.

characteristics has been provided in Section 3-5. However, there are some additional parameters characterizing an IC flip-flop which are normally found on the manufacturer's specification sheet. These parameters are of interest in selecting an appropriate flip-flop for a given application. We discuss these parameters next.

In order to operate reliably, every flip-flop requires the widths of the clock pulse, preset pulse, and the clear pulse to be greater than some minimum values. The minimum values are denoted as $t_{p\text{(clock)}}$, $t_{p\text{(preset)}}$, and $t_{p\text{(clear)}}$, and are given in nanoseconds. The width is measured as the time between the threshold point (usually 50 percent) of the leading and the trailing edges of the pertinent pulse.

Two important parameters characterizing the data input pulse are its set-up time and hold time. The *set-up time* is defined as the time interval for which the data input must remain stable *before* the flip-flop is activated by the clock pulse. As shown in Fig. 4-18, the set-up time (denoted by t_s) is measured as the time between the threshold points (usually 50 percent) of the transition edge of the data pulse and the activating edge of the clock pulse. For TTL master-slave J–K flip-flops, the set-up should be equal to or greater than $t_{p\text{(clock)}}$, the minimum allowable width of the clock pulse. Or, in other words, the J and K data inputs must be held stable until the clock falls. The *hold time* is defined as the time interval for which the data input must be held stable *after* the flip-flop has

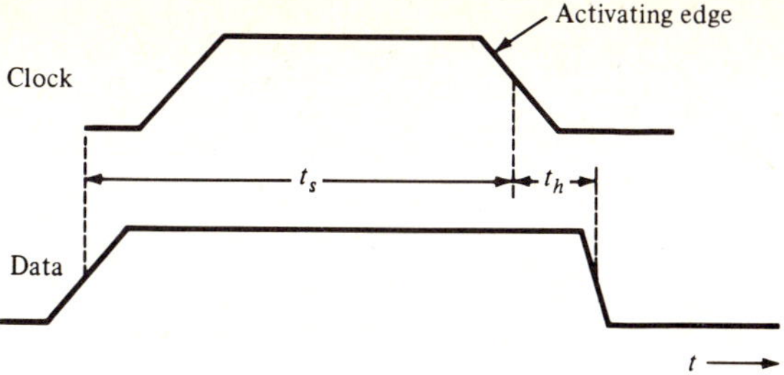

Figure 4-18 Illustration of set-up time and hold time measurements.

been activated by a clock pulse. As indicated in Fig. 4-18, the hold time (denoted by t_h) is measured as the time between the threshold points (usually 50 percent) of the activating edge of the clock and the data pulse edge returning to other state. In most TTL flip-flops, the hold time is zero. Both hold time and set-up time are measured in nanoseconds.

Another parameter of concern is the propagation delay, which is the time required for the flip-flop to switch from one state to another. Two different types of propagation delay times are defined for a flip-flop. The time required for the Q or $\bar{Q}$ output to switch from a logical-1 to a logical-0 state is denoted by t_{pHL} and that for Q or $\bar{Q}$ output to switch from a logical-0 state to a logical-1 state is denoted by t_{pLH}. Usually, the designer is interested in the average worst case or average maximum propagation delay given by

$$t_{pd(avg)} = (t_{pLH(max)} + t_{pHL(max)})/2 \tag{4-10}$$

where the maximum values of t_{pLH} and t_{pHL} are given by the manufacturer on the specification sheet. For a clocked flip-flop with asynchronous preset and clear inputs, the propagation delay time is defined with respect to both the activating clock input and the asynchronous inputs. As shown in Fig. 4-19, the propagation delay time with respect to the clock is the time interval between the threshold levels (usually 50 percent) of the activating edge of the clock pulse and

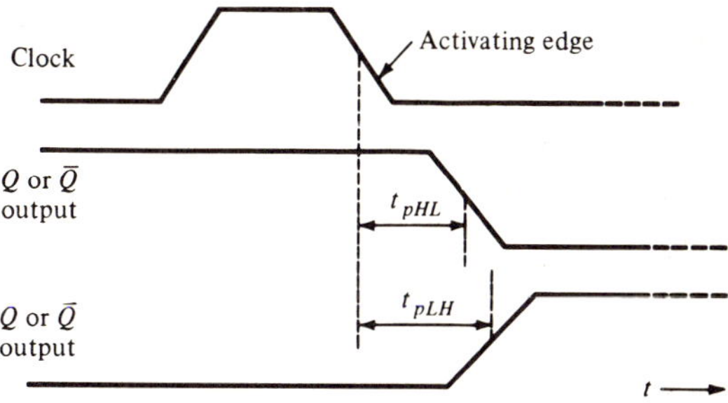

Figure 4-19 Illustration of flip-flop propagation delay measurements.

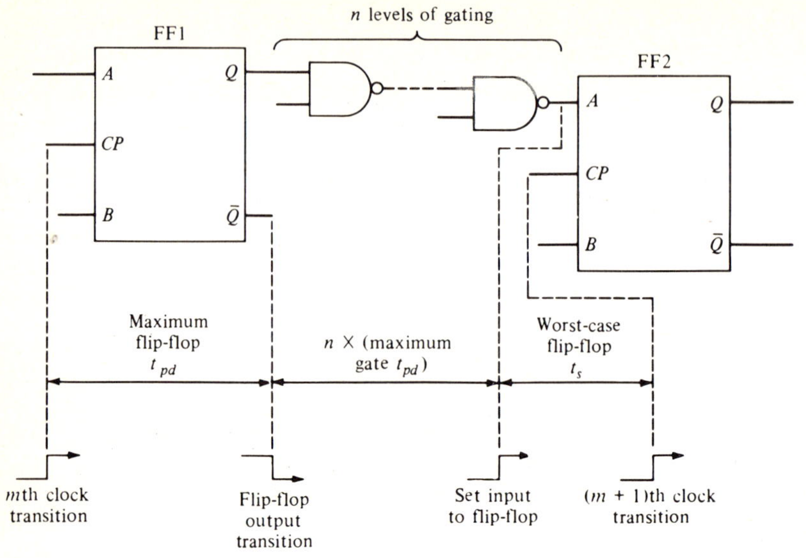

Figure 4-20   Determination of maximum clock rate.

the transition edge of the output voltage levels. A similar scheme is used to measure the propagation delay with respect to the asynchronous inputs.

In the design of a synchronous sequential circuit, another parameter of considerable interest is the *maximum clock rate* at which the circuit can be operated reliably. This clock rate is a function of the propagation delays of the flip-flop and logic gates and the set-up time of the flip-flop. It is determined as follows. First, the signal path containing highest levels of gating (i.e., with longest delay) between the output of one flip-flop to the input of another flip-flop is identified. Let the longest delay signal path be modeled by the logic diagram of Fig. 4-20, showing two flip-flops with n levels of gating between them. If the flip-flop 1 changes state in response to the mth clock transition, this change has to propagate through the n-level combinational circuit and then permit flip-flop 2 to establish its new data input. As a result, the minimum clock period T_{min} (the time interval between the threshold levels of two consecutive activating edges) is determined by the following relation:

$$T_{min} = (\text{maximum flip-flop } t_{pd}) + n(\text{maximum gate } t_{pd})$$
$$+ (\text{worst-case flip-flop } t_s) \qquad (4\text{-}11)$$

The T_{min} is expressed in nanoseconds. Its reciprocal is the maximum clock frequency f_{max}, usually expressed in megahertz (MHz).

Example 4-1. Determine the maximum clock rate for a sequential circuit designed using SN54/7476 dual J–K flip-flops and SN54/7400 quad 2-input NAND gates. The path with longest delay has five levels of gating.

The pertinent timing parameters of the logic elements from the manufacturer's specification sheet are

Flip-flop:
$$t_{pHL(max)} = 40 \text{ nsec}$$
$$t_{pLH(max)} = 25 \text{ nsec}$$
$$t_{s(worst\text{-}case)} = 20 \text{ nsec}$$

NAND gate:
$$t_{pHL(max)} = 15 \text{ nsec}$$
$$t_{pLH(max)} = 22 \text{ nsec}$$

Using the maximum values of the propagation delays, we determine T_{min} as

$$T_{min} = 40 + (5 \times 22) + 20 = 170 \text{ nsec}$$

Taking its reciprocal, we obtain the maximum clock rate as 5.88 MHz.

In many applications, it is more realistic to use the average propagation delay to determine the maximum clock rate since, on the average, there will be equal numbers of 1-to-0 and 0-to-1 transitions. The average propagation delays of the flip-flop and the NAND gate are 32.5 and 18.5 nsec, respectively. Using these figures, the value of T_{min} becomes 145 nsec which implies a maximum clock rate of 6.89 MHz.

It should be noted that in Table 3-7 comparing the various logic families, the figures given for maximum clock rate are typical values. In most applications, the actual computed value of the clock rate will be about 30–40 percent of these values.

Electrical characteristics of the IC flip-flops are described in manufacturer's specification sheets, which should be consulted to select an appropriate flip-flop package for a given application. As an example of manufacturer's specifications, consider the Texas Instruments SN54/7470 positive edge-triggered TTL dual J–K flip-flop package. The temperature range of the 54 series TTL ICs is -55–$125°C$, and that of the 74 series is 0–70°C. The recommended supply voltage (V_{cc}) range for the 5470 is 4.5–5.5 V, and that for the 7470 is 4.75–5.25 V. The nominal value of V_{cc} for both is 5 V. Other pertinent parameters are

Input set-up time, $t_s = 20$ nsec
Input hold time, $t_h = 5$ nsec
Minimum pulse width of clock at high level = 20 nsec
Minimum pulse width of clock at low level = 30 nsec
Minimum pulse widths of preset or clear pulse at low level = 25 nsec
Maximum clock frequency, $f_{max} = 20$ MHz
Maximum propagation delay t_{pLH} (or t_{pHL}) from preset (or clear) to output = 50 nsec
Typical propagation delay from clock to output, $t_{pLH} = 27$ nsec and $t_{pHL} = 18$ nsec
Minimum high-level input voltage, $V_{IH} = 2$ V
Maximum low-level input voltage $V_{IL} = 0.8$ V
Minimum high-level output voltage, $V_{OH} = 2.4$ V
Maximum low-level output voltage, $V_{OL} = 0.4$ V

4-2   Analysis of Sequential Circuits

The type of sequential circuits we are considering in this book are formed by interconnecting combinational logic circuits and flip-flops. Let the complete circuit contain n flip-flops. Since each of the flip-flops can be in one of two possible states, the total number of states of the sequential circuit is 2^n. Application of the input signals with the clock present moves the flip-flop from one state to another, producing an output signal. A convenient way to describe the operation of a sequential circuit is with the aid of a *state table*—a truth table showing the next state and the output as a function of the present state and the input. A pictorial representation of the state table is the *state diagram* where the states are represented by circles with labels identifying the states and the transition of states are shown by directed paths. The paths are labeled in the format $XY----/AB----$, where $XY----$ represents the current values of the input variables, and $AB----$ are the corresponding output variable values.

The systematic process of deriving the state table and the state diagram of a sequential circuit is the object of analysis. We outline next the analysis procedure with the aid of an example.

Example 4-2. Determine the state table and the state diagram of the sequential circuit of Fig. 4-21, the input of which is X and the output is W.

The circuit contains two flip-flops and, as a result, the total number of stable states of the sequential circuit is at most four. If we use subscript n to denote the various switching variables and the states of the flip-flops at time

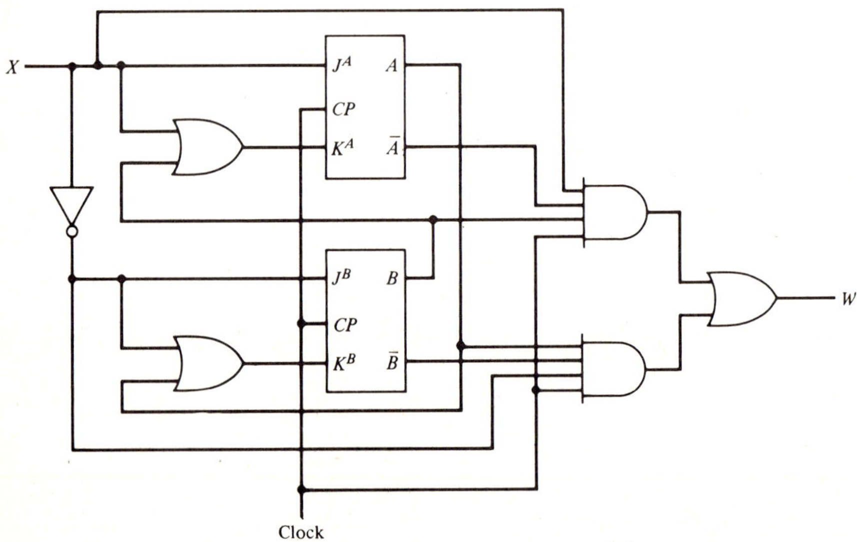

Figure 4-21   A 4-state sequential circuit.

$t = t_n$, then the possible four states of the sequential circuit at $t = t_n$ are described by

$$A_n = 0, B_n = 0; \quad A_n = 0, B_n = 1; \quad A_n = 1, B_n = 0; \quad A_n = 1, B_n = 1 \quad (4\text{-}12)$$

It follows from the circuit diagram that

$$J_n{}^A = X_n; \quad K_n{}^A = X_n + B_n$$
$$J_n{}^B = \bar{X}_n; \quad K_n{}^B = \bar{X}_n + A_n \qquad (4\text{-}13)$$

Let us determine the state table of the circuit. Assume that initially the sequential circuit is in the state $A_n = 0$, $B_n = 0$, as shown by the first two entries in the first row of Table 4-5.

TABLE 4-5
State Table for the Sequential Circuit of Figure 4-21

Present State		Next State				Output	
		$X_n = 0$		$X_n = 1$		$X_n = 0$	$X_n = 1$
A_n	B_n	A_{n+1}	B_{n+1}	A_{n+1}	B_{n+1}	W_n	W_n
0	0	0	1	1	0	0	0
0	1	0	0	1	1	0	1
1	0	1	1	0	0	1	0
1	1	0	0	0	0	0	0

If the input $X_n = 0$, then using Eq. (4-13) we obtain $J_n{}^A = 0$, $K_n{}^A = 0$ and $J_n{}^B = 1$ and $K_n{}^B = 1$. Now using this information and the transition table of the J–K flip-flop as given by Table 4-2, we observe that A_{n+1} will be 0 and B_{n+1} will be 1, as recorded in the third and fourth entries in the first row of Table 4-5. On the other hand, if $X_n = 1$, then Eq. (4-13) implies $J_n{}^A = 1$, $K_n{}^A = 1$ and $J_n{}^B = 0$, $K_n{}^B = 0$. With the aid of Table 4-2 we then deduce $A_{n+1} = 1$ and $B_{n+1} = 0$, as recorded in Table 4-5. This procedure is repeated by assuming that the state of the sequential circuit to be one of the remaining states described by Eq. (4-12) and determining the next states for the case of $X_n = 0$ and $X_n = 1$ separately. The complete state transition is as indicated by the first six columns of Table 4-5 and can be readily verified (Problem 4-12).

To determine the output value we observe from Fig. 4-21 that

$$W_n = \bar{A}_n B_n X_n + A_n \bar{B}_n \bar{X}_n$$

For $A_n = 0$, $B_n = 0$, and $X_n = 0$, W_n is evaluated from the above equation and is found to be equal to 0. Similarly, the output is computed for other combinations of these variables as indicated by Table 4-5 and are recorded in the last two columns of this table.

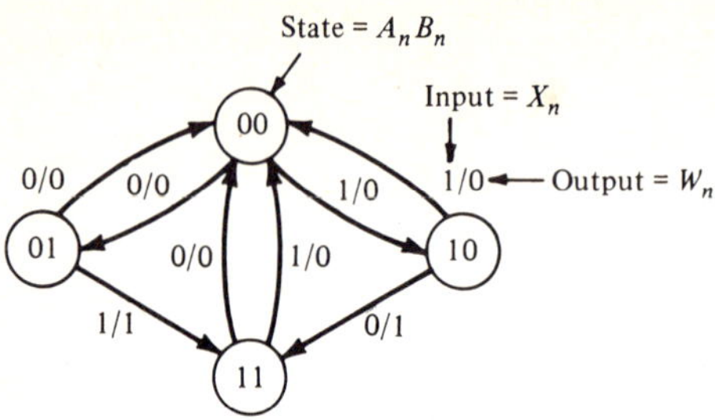

Figure 4-22 State diagram describing the operation of the sequential circuit of Fig. 4-21.

This completes the determination of the state table of the circuit of Fig. 4-21.

The state diagram for this sequential circuit can be constructed from the information given by the state table. The pertinent state diagram is sketched in Fig. 4-22. The four states are represented by four circles. Each state is designated by a 2-bit binary number, with the leftmost bit representing the state of the flip-flop A and the rightmost bit representing the state of the flip-flop B. Thus the state "01" implies $A_n = 0$ and $B_n = 1$, and so on. From Table 4-5 we observe that $X = 0$ moves the state "00" to the state "01" and produces an output $W = 0$. This is indicated by an arrow going from the circle marked "00" to the circle marked "01" with the notation "0/0" next to the path. In a like manner, all the other connections are obtained.

4-3 State Table Formulation and Simplification

The objective of the analysis procedure is to arrive at a state table or a state diagram description of a given sequential circuit. The reverse process to obtain a realization to perform a set of specific operations in sequence is known as synthesis. A systematic procedure to synthesize a sequential circuit is outlined in the next section. To follow this procedure, it is necessary to have a state table description of the circuit. The purpose of this section is to describe the process of developing a state table from the problem statement. Since the number of flip-flops needed to implement the circuit is directly related to the total number of states in the state table, it is often prudent to attempt to simplify the state table description with the aim of obtaining an equivalent state table description with fewer states. This, in turn, would lead to a realization with fewer flip-flops.

Both the state table formulation and the simplification procedures are illustrated with the aid of examples.

Example 4-3. Develop the state diagram and the state table characterizing a single-input, single-output sequential circuit which examines an input binary sequence serially and generates an output sequence in accordance with the following rule: The output bit is 1 if and only if the present input bit is 1 and the previous input bit is 0; otherwise the output is 0.

Time t_n	•	•	•	t_0	t_1	t_2	t_3	t_4	t_5	t_6	t_7	t_8	t_9	•	•	•	•
Input X_n	•	•	•	0	0	1	1	0	0	0	0	1	1	•	•	•	•
Output W_n	•	•	•	0	0	1	0	0	0	0	0	1	0	•	•	•	•

Figure 4-23   A typical input–output sequence for the circuit of Example 4-3.

Let $\{X_n\}$ and $\{W_n\}$ denote the input and output sequences, respectively. A typical input and output bit pattern for the desired sequential circuit is then as shown in Fig. 4-23.

In order to determine an appropriate state diagram, let us assume that the circuit has already examined two consecutive ZEROs in the input sequence and is in state a. If the next input is a 0, the circuit returns to state a and produces an output of 0. If, instead, the next input is a 1, it goes to state b producing an output of 1. Pictorially, this is represented as indicated in Fig. 4-24(a). From state b, the circuit goes to another state d producing an output of 0 if the following input is a 0. Otherwise it moves to a fourth state c for an input of 1 and produces an output of 0. This is shown in Fig. 4-24(b). We follow this procedure and obtain the state transitions from states c and d and arrive at the complete state diagram as sketched in Fig. 4-24(c).

The state diagram of Fig. 4-24(c) leads to the equivalent state-table description given by Table 4-6. Note that the state-table representation is slightly different from that used in Example 4-1. Both representations are used in practice.

Example 4-4.   Develop the state diagram and the state table for a sequential circuit which examines a 3-bit sequence $X_{n-2} X_{n-1} X_n$ sequentially and produces an output of 1 if and only if the input sequence is either 011 or 111. Otherwise the output is 0. An output of 0 is produced after the first and second

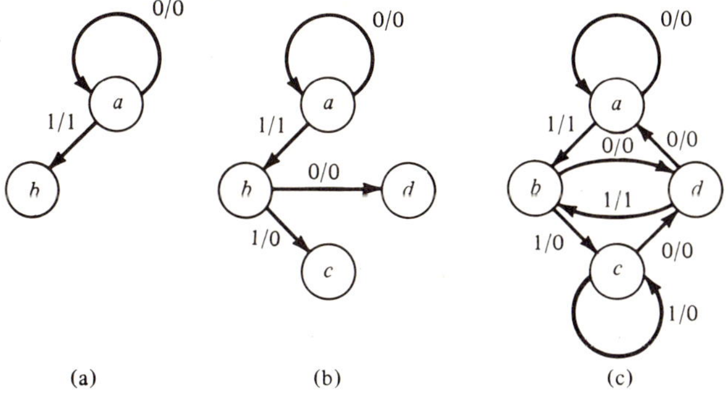

(a)   (b)   (c)

Figure 4-24   Development of the state diagram of the sequential circuit of Example 4-3.

TABLE 4-6
**State-Table Description of the Sequential
Circuit of Example 4-3**

Present State	$X_n = 0$		$X_n = 1$	
	Next State	W_n	Next State	W_n
a	a	0	b	1
b	d	0	c	0
c	d	0	c	0
d	a	0	b	1

bits have been examined. The circuit goes back to the initial state after 3 bits
have been examined so that it will be ready to examine another 3-bit sequence.

Assume that the circuit is in initial state a. After examining the first input
bit, the circuit goes to state b if the input bit is a 0; otherwise it goes to state
c if the input bit is a 1. In either case an output of 0 results. From state b the
next state is d if the input bit is a 0 or is e if the input bit is a 1. Similarly, from
state c the next state would be either f or g depending on whether the second
bit is a 0 or a 1. After the second input bit has been received, the output is
still a 0. The partial state diagram developed so far is as shown in Fig. 4-25(a).
When the third input bit is received, the circuit is in either state d or e or f
or g. After the third bit has been examined the circuit must reset itself in
order to be ready to examine another set of 3 bits. Hence, from either one of
the last four states, the next state must be a, independent of the third input
bit. Now if the machine is in state e and the third bit received is a 1, then the
sequence examined has been 011, implying that the output must be a 1. Likewise, if the circuit is in state g and the input bit is a 1, the output must be a 1
as the input sequence examined in this case has been 111. For all other

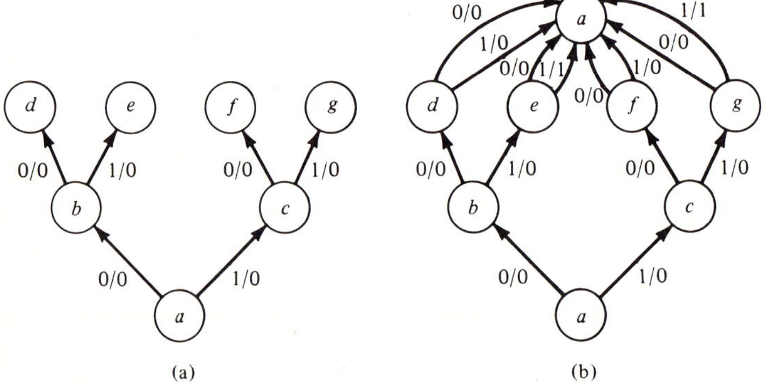

Figure 4-25   Development of a state diagram of the sequential circuit of Example 4-4.

TABLE 4-7
State Table Description of the Sequential Circuit of Example 4-4

Present State	$X_n = 0$		$X_n = 1$	
	Next State	Y_n	Next State	Y_n
a	b	0	c	0
b	d	0	e	0
c	f	0	g	0
d	a	0	a	0
e	a	0	a	1
f	a	0	a	0
g	a	0	a	1

situations, the output is a 0. The complete state diagram is sketched in Fig. 4-25(b). The corresponding tabular representation of the state transitions is given by the state table of Table 4-7, where we indicate the output as Y_n.

Once the state table has been derived, the next step in most design problems is to simplify the state table. Recall that a sequential circuit containing n flip-flops has 2^n states. Alternately, if the total number of states needed is N, then the minimum number of flip-flops needed is

$$n = \log_{10} N / \log_{10} 2 = 3.32(\log_{10} N) \qquad (4\text{-}14)$$

If the number computed above is not an integer, the next largest integer is chosen as n. By simplification, we mean the elimination of redundant states and development of a state table description of the sequential circuit with fewer states. Since the minimum number of flip-flops needed to implement a circuit is directly related to the total number of states, in accordance with Eq. (4-14), minimization of states would thus lead to a realization which is less complex, and hence cheaper.

Example 4-5. Determine the minimum number of flip-flops needed to construct the sequential circuit of Example 4-4.

From the state diagram of Fig. 4-25(b), we note that the total number of states in the diagram is 7. Therefore,

$$n = 3.32 (\log_{10} 7) = 2.8057 \qquad (4\text{-}15)$$

which is a fraction. The next largest integer is 3. Thus three flip-flops will be needed to design this sequential circuit.

There are several techniques that can be applied to simplify a state table. We illustrate one such technique which, in general, leads to a simplification by inspection.

Example 4-6.   Simplify the state-table description given by Table 4-6.

A careful examination of this table reveals that whether the sequential circuit is in state *b* or state *c* the next state and output are identical for both possible inputs. Thus, as far as the output is concerned, the operation of the circuit would be the same if the circuit is in either state. Or, in other words, states *b* and *c* are *equivalent*. Similarly, we observe that states *a* and *d* are equivalent. We can thus eliminate the third and fourth rows in Table 4-6 and replace in the second row *d* by *a* and *c* by *b*. The resultant reduced state table is as shown in Table 4-8.

TABLE 4-8
Reduced State-Table Description of the
Sequential Circuit of Example 4-3

Present State	$X_n = 0$		$X_n = 1$	
	Next State	W_n	Next State	W_n
a	*a*	0	*b*	1
b	*a*	0	*b*	0

The corresponding state diagram is sketched in Fig. 4-26.

As a second example consider the simplification of the state table of Table 4-7.

Example 4-7.   Minimize the number of states necessary to design the sequential circuit of Example 4-4.

If we examine the original state table given by Table 4-7, we note that the entries for fourth row are identical to the entries for sixth row, that is, whether the machine is in state *d* or state *f*, the next state is state *a* and output is 0 regardless of the truth value of the input. Thus, as far as the output is concerned, the operation of the machine would be same for either state, that is, the two states *d* and *f* are equivalent. We thus eliminate the sixth row and replace the symbol *f* in the third row by *d* as indicated in Table 4-9. Similarly we find states *e* and *g* are the same as far as the next state and the output are concerned and replace the symbol *g* in the third row by *e* as shown in Table 4-9. The simplified state table obtained after eliminating states *f* and *g* is

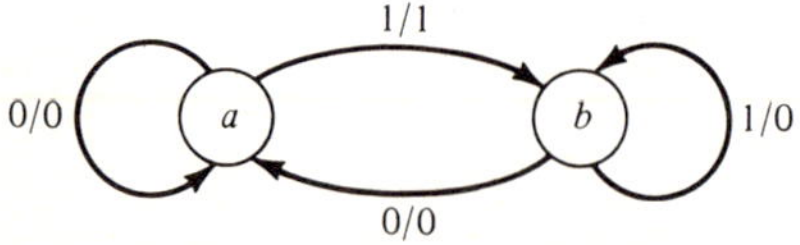

Figure 4-26   Simplified state diagram describing the sequential circuit of Example 4-5.

TABLE 4-9

Present State	$X_n = 0$			$X_n = 1$	
	Next State	Y_n		Next State	Y_n
a	b	0		c	0
b	d	0		e	0
c	~~f~~ d	0		~~g~~ e	0
d	a	0		a	0
e	a	0		a	1
~~f~~	~~a~~	~~0~~		~~a~~	~~0~~
~~g~~	~~a~~	~~0~~		~~a~~	~~1~~

shown in Table 4-10. We examine this table next and find that the entries in the second and third rows are now identical, indicating states b and c are the same. We can thus eliminate the third row and replace the symbol c in the first row by b. This leads to further simplification and leads to the state table of Table 4-11. The corresponding state diagram is sketched in Fig. 4-27. The simplification process has led to a sequential circuit which has now four states requiring two flip-flops to implement—a saving of one flip-flop.

TABLE 4-10

Present State	$X_n = 0$			$X_n = 1$	
	Next State	Y_n		Next State	Y_n
a	b	0		c	0
b	d	0		e	0
c	d	0		e	0
d	a	0		a	0
e	a	0		a	1

TABLE 4-11
Reduced State-Table Description of the Circuit of Example 4-4

Present State	$X_n = 0$			$X_n = 1$	
	Next State	Y_n		Next State	Y_n
a	b	0		b	0
b	d	0		e	0
d	a	0		a	0
e	a	0		a	1

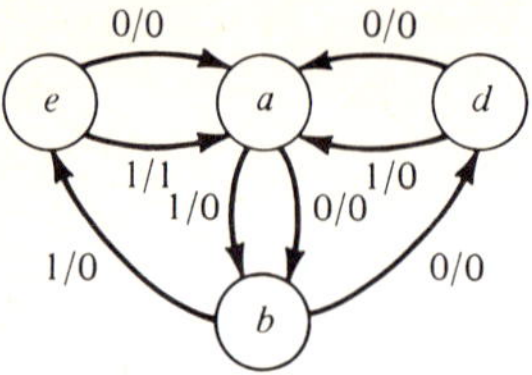

Figure 4-27   State diagram of the sequential circuit of Example 4-4.

The basic idea behind the state table simplification process is thus to identify groups of equivalent states and then to combine each group of equivalent states into a single state. Two or more states of a sequential circuit are *equivalent* if and only if the same output sequence is produced for every possible input sequence, irrespective of which of these equivalent states is used as the initial state when the input is applied. We have outlined above a simple method of state minimization procedure. Further simplification may be possible using more complex procedures. A detailed discussion on minimization is beyond the scope of this text and the interested reader is referred to texts dealing exclusively with digital system design.[1,3,4]

4-4   Synthesis of Sequential Circuits

The synthesis procedure to be outlined in this section assumes the existence of a state table description of the circuit. Hopefully the state table has been simplified so that a realization with fewer flip-flops can be obtained. Note that in some cases it may be easier to arrive at a state diagram description of the desired circuit. However, it is a simple exercise to obtain an equivalent state table description. The key idea in developing the synthesis procedure is to modify the state table by incorporating the flip-flop characteristics in the table. To this end it is convenient to describe each flip-flop by its excitation table.

Excitation Tables of Flip-Flops

The transition table of a flip-flop tells us precisely how the flip-flop is going to behave when inputs are applied. That is, knowing the present state and the present inputs, we know exactly the next state of the flip-flop. The excitation table of a flip-flop also contains the same information except here it tells us what should be the status of the inputs if we know ahead of time the next state of the flip-flop and its present state.

The excitation tables of the four flip-flops are given in Tables 4-12, 4-13, 4-14, and 4-15. These can be readily derived from their respective transition tables. In these tables, *d* denotes don't care condition.

TABLE 4-12
Excitation Table of an R–S Flip-Flop

Q_n	Q_{n+1}	R_n	S_n
0	0	d	0
0	1	0	1
1	0	1	0
1	1	0	d

TABLE 4-13
Excitation Table of a J–K Flip-Flop

Q_n	Q_{n+1}	J_n	K_n
0	0	0	d
0	1	1	d
1	0	d	1
1	1	d	0

TABLE 4-14
Excitation Table of a T Flip-Flop

Q_n	Q_{n+1}	T_n
0	0	0
0	1	1
1	0	1
1	1	0

TABLE 4-15
Excitation Table of a D Flip-Flop

Q_n	Q_{n+1}	D_n
0	0	0
0	1	1
1	0	0
1	1	1

Synthesis Examples

The basic idea behind the synthesis procedure is best illustrated with the aid of examples.

Example 4-8.   A 2-state single-input, single-output sequential circuit is described by the state table of Table 4-8. The input variable is X and the output variable is W. Realize the circuit using R–S flip-flops.

The first step in the synthesis procedure is to determine the number of flip-flops needed to implement the machine. Table 4-8 indicates that one flip-flop is adequate. However, the states a and b must be the complement of each other. Let us arbitrarily assign a to be the ZERO state and b to the ONE state of the R–S flip-flop.

With this new notation a modified table of the circuit is then formed. This modified table has additional initially empty columns, for the control inputs R_n and S_n which are required in order to make the indicated transition from Q_n to Q_{n+1}. These are filled by examining the excitation table of the R S flip-flop given by Table 4-12. The complete modified state table is indicated in Table 4-16.

From the modified table, expressions for the control inputs are developed next. To this end, Karnaugh maps may be used to simplify the expressions. The Karnaugh maps for R_n, S_n, and W_n are given in Fig. 4-28. These maps lead to the switching function expressions:

$$R_n = \bar{X}_n, \quad S_n = X_n, \quad W_n = \bar{Q}_n X_n \qquad (4\text{-}16)$$

TABLE 4-16
Modified State Table of the 2-State Sequential Circuit

		$X_n = 0$			$X_n = 1$			
Q_n	Q_{n+1}	W_n	R_n	S_n	Q_{n+1}	W_n	R_n	S_n
0	0	0	d	0	1	1	0	1
1	0	0	1	0	1	0	0	d

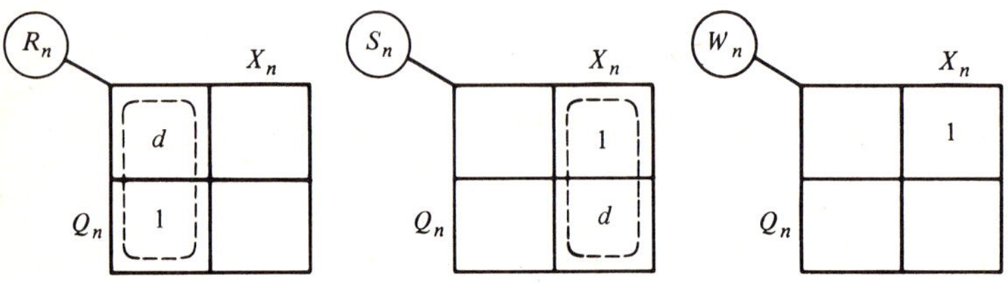

Figure 4-28 Karnaugh map representations of the control inputs and the output.

The final circuit diagram is obtained by implementing these switching functions as shown in Fig. 4-29.

Example 4-9. Realize, using J–K flip-flops, the single-input, single-output, 4-state sequential circuit described by the state table of Table 4-11.

It is seen from this table that two flip-flops are required for the realization. Let us denote the two flip-flops as flip-flop A and flip-flop B. The state of the sequential circuit is then given as $A_n B_n$ where A_n and B_n are the present states of flip-flops A and B, respectively. We arbitrarily make the following state assignments:

$$\begin{aligned}
\text{State } a: \quad & A_n = 0, \quad B_n = 0 \\
\text{State } b: \quad & A_n = 1, \quad B_n = 1 \\
\text{State } d: \quad & A_n = 0, \quad B_n = 1 \\
\text{State } e: \quad & A_n = 1, \quad B_n = 0
\end{aligned} \qquad (4\text{-}17)$$

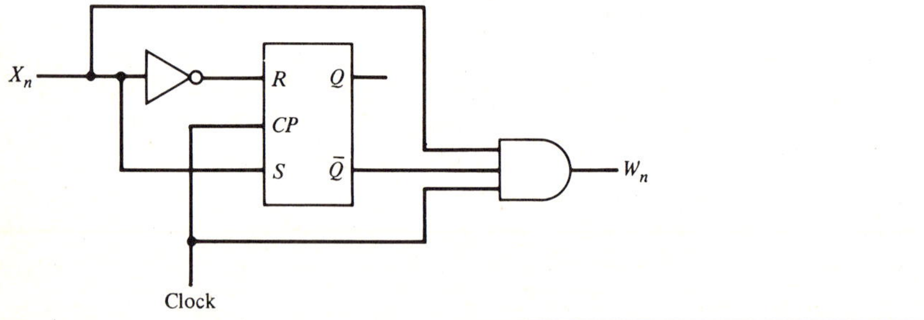

Figure 4-29 Implementation of the sequential circuit of Example 4-8.

TABLE 4-17
Modified State Table of the Sequential Circuit of Example 4-4

		$X_n = 0$							$X_n = 1$						
A_n	B_n	A_{n+1}	B_{n+1}	J_n^A	K_n^A	J_n^B	K_n^B	Y_n	A_{n+1}	B_{n+1}	J_n^A	K_n^A	J_n^B	K_n^B	Y_n
0	0	1	1	1	d	1	d	0	1	1	1	d	1	d	0
0	1	0	0	0	d	d	1	0	0	0	0	d	d	1	0
1	0	0	0	d	1	0	d	0	0	0	d	1	0	d	1
1	1	0	1	d	1	d	0	0	1	0	d	0	d	1	0

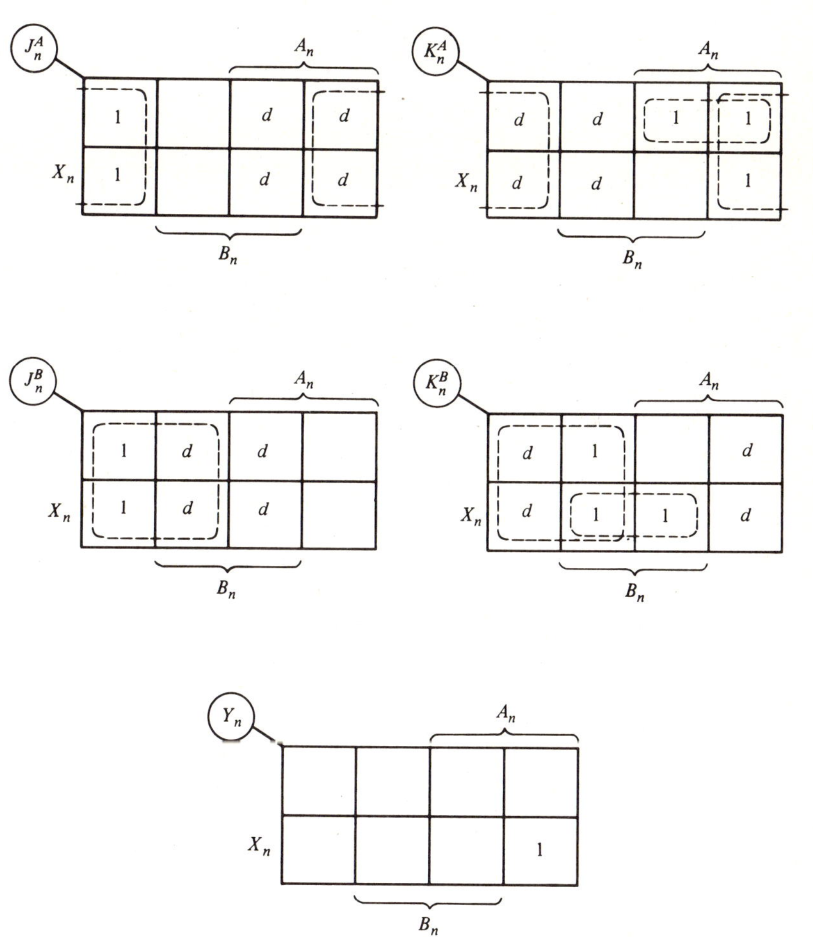

Figure 4-30 Karnaugh maps for the control inputs and the output.

Making use of the excitation table of the J–K flip-flop as given by Table 4-13 and the state assignment of Eq. (4-17) in Table 4-11, we arrive at the modified state table of our sequential circuit. This is shown in Table 4-17.

The Karnaugh map representations of the four control inputs and the output are shown in Fig. 4-30 from which we arrive at the desired switching function expressions:

$$J_n^A = \overline{B}_n, \qquad K_n^A = \overline{B}_n + A_n \overline{X}_n$$
$$J_n^B = \overline{A}_n, \qquad K_n^B = \overline{A}_n + B_n X_n \qquad (4\text{-}18)$$
$$Y_n = A_n \overline{B}_n X_n$$

Implementing the switching functions given above and connecting the two J–K flip-flops, we arrive at the desired realization of the complete sequential circuit as shown in Fig. 4-31.

The synthesis method can be used in principle to design complex special purpose sequential circuits such as arithmetic units, code converters, and so on. However, in many cases, a design can be arrived at directly. In the remaining part of this chapter, we describe the design of a number of widely used sequential circuits. A number of the sequential circuits to be described are available in IC

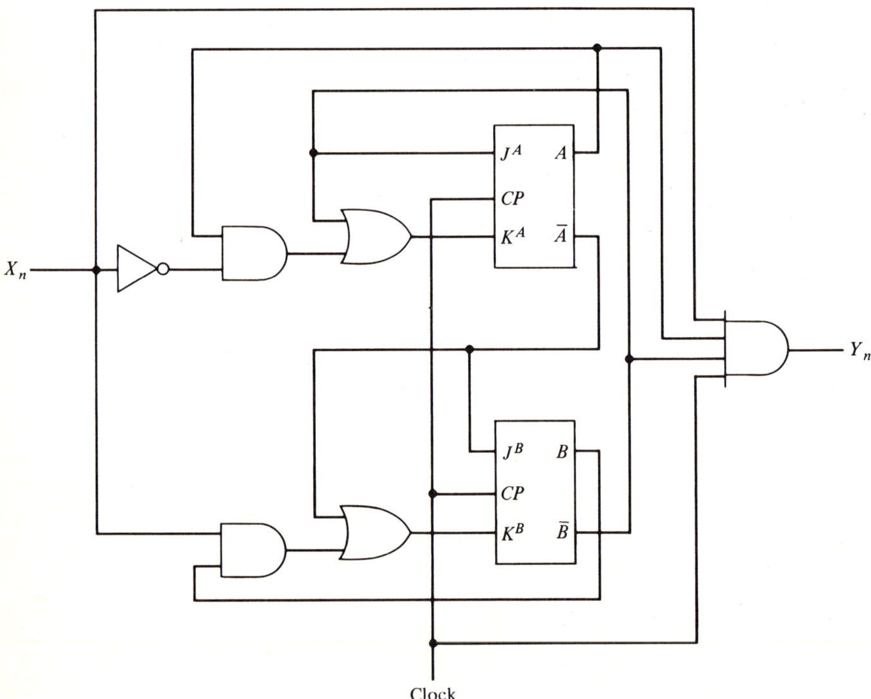

Figure 4-31   A realization of the sequential circuit described by Table 4-11.

form, usually as MSI and LSI packages. These sequential circuits and the combinational circuits described in the previous chapter are used as basic building blocks in the design of more complex digital systems.

4-5 Registers

Registers are used to temporarily store binary information and often are part of digital systems implementing computational algorithms. Depending on the type of application, they can contain from one to a number of flip-flops. There are basically two types of registers: storage register and shift register. We describe these next.

Storage Register

The purpose of the storage register is to store each bit of a binary data in individual storage devices and make it available for later processing. All bits are stored simultaneously by gating with clock pulse. As a result, clocked flip-flops are used primarily as the storage devices. Figure 4-32(a) shows a 4-bit storage register which uses clocked D flip-flops.

Here all the input bits A_1, A_2, A_3, and A_4 are entered at once (in parallel form) when the clock is HIGH. The state of each D flip-flop then assumes the same value as that of its input after some finite time delay when the data stored in the register will be $Q_1 = A_1$, $Q_2 = A_2$, $Q_3 = A_3$, and $Q_4 = A_4$. Stored data can be read out in parallel form. After the clock goes LOW, any further changes in the input data will not affect the status of the flip-flops.

The storage register of the type shown is often used as a *buffer register* for temporary storing of data being transferred between two asynchronous digital logic circuits that may be operating at different speeds. A typical IC buffer register contains two independent 5-bit registers with separate clock inputs.

A variation of the storage register available in IC form includes gating circuitry at the inputs of the flip-flops which allows the storage of two different data by appropriately selecting the logic level of a data selector input line. A possible implementation of such a 4-bit data selector/storage register unit is shown in Fig. 4-32(b). Here, if the data select input S is a 1, the word $A_1 A_2 A_3 A_4$ is stored in the register; and if S is a 0, the word $B_1 B_2 B_3 B_4$ is stored instead.

Shift Register

A very important type of register is the shift register which includes appropriate interconnection between the flip-flops to allow the transferring of data between adjacent flip-flops, resulting in storage of input binary data available in serial form.

Figure 4-33 shows a 4-bit shift register in its simplest form. The flip-flops are of the master-slave type. Data is read into the register through the leftmost flip-flop serially and can be read out serially from the output of the rightmost flip-flop. Each flip-flop is also called a *stage* of the register. Before the data is read in, the flip-flops can be reset to 0 by setting the *clear* input to HIGH.

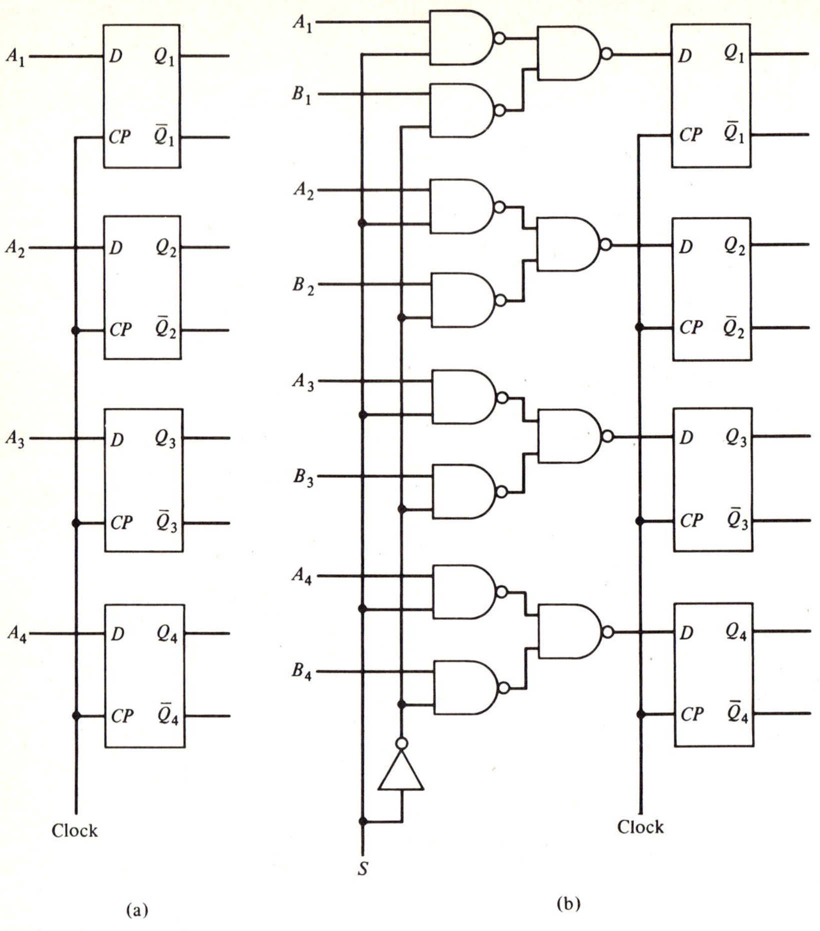

Figure 4-32   Storage register.

To understand the operation of the shift register, assume that the clock pulses appear at times $t_1, t_2, t_3, \ldots$. Before the arrival of first clock pulse at time t_{1-}, assume also that all stages have been cleared to 0 by a common reset (R_D) pulse. Let the input data be a 4-bit word $WXYZ$ appearing serially with Z as the first bit. At time t_1, the input D_0 to the first stage is Z, whereas the inputs $D_1, D_2,$ and D_3 to the other three stages are $Q_0, Q_1,$ and Q_2, respectively. All flip-flops are at logical ZERO. After the first clock pulse arrives, the state of each flip-flop changes to that of its input. After all flip-flops have settled to their new state at time t_{1+}, $Q_0 = Z, Q_1 = 0, Q_2 = 0,$ and $Q_3 = 0$. This process continues with the appearance of subsequent clock pulses. The operation of the 4-bit shift register is summarized in Table 4-18. Note that after four clock pulses, the 4-bit input word is stored in the register. Also, the stored data can be read out serially at

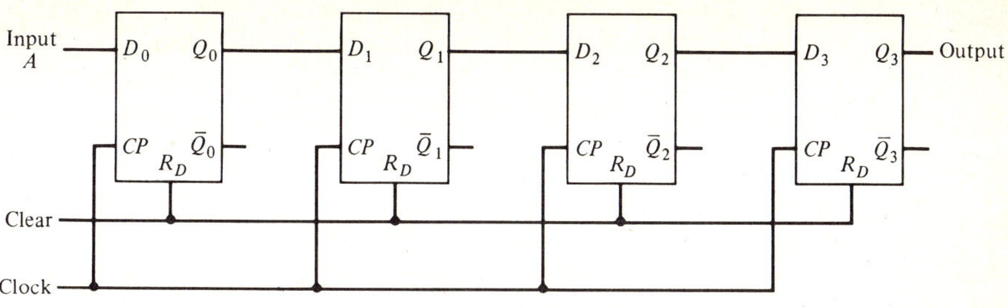

Figure 4-33 A 4-bit left-to-right shift register.

the output of the last stage by clocking with additional clock pulses. By construction, the data in the register moves only in one direction (left-to-right) and this type of register is thus called a *unidirectional shift register.*

Data entry and data exit in the shift register of Fig. 4-33 are in serial mode. Data can also be read out in parallel by making the outputs of each flip-flop externally available. If the data is entered serially, then to read out the input word in parallel, it would be necessary to wait for four clock periods. Using additional logic circuits or using flip-flops with both direct set and direct reset input terminals, data can also be entered in parallel. Figure 4-34 shows the logic diagram of Fairchild 9300 4-bit shift register in which data entry and data exit can both be either in serial or in parallel mode. When the parallel enable input ($\overline{PE}$) is LOW, the circuit appears essentially as four flip-flops with common clocks and, as a result, data can be entered in parallel synchronously with the clock through the parallel input terminals P_0, P_1, P_2, and P_3. When $\overline{PE}$ input is HIGH, the circuit operates as a regular left-to-right shift register with a serial input $J + \overline{K}$. Connecting together the J and $\overline{K}$ inputs results in a D-type input. Data can be read out serially via the Q_3 ($\overline{Q}_3$) output or in parallel, via the Q_0, Q_1, Q_2, and Q_3 outputs. When the master reset ($\overline{MR}$) input is pulled LOW, all stages are reset to ZERO.

TABLE 4-18
Shift Register Operation Table

Time	State of Flip-Flops			
	Q_0	Q_1	Q_2	Q_3
t_{1-}	0	0	0	0
t_{1+}	Z	0	0	0
t_{2+}	Y	Z	0	0
t_{3+}	X	Y	Z	0
t_{4+}	W	X	Y	Z
t_{5+}	0	W	X	Y
t_{6+}	0	0	W	X
t_{7+}	0	0	0	W

Leads		Loading
$\overline{PE}$	Parallel enable (active low) input	2.3 UL
P_0, P_1, P_2, P_3	Parallel inputs	1 UL
J	First stage J (active high) input	1 UL
$\overline{K}$	First stage K (active low) input	1 UL
CP	Clock (active high going edge) input	2 UL
$\overline{MR}$	Master reset (active low) input	1 UL
Q_0, Q_1, Q_2, Q_3	Parallel outputs	6 UL
$\overline{Q}_3$	Complementary last stage output	8 UL

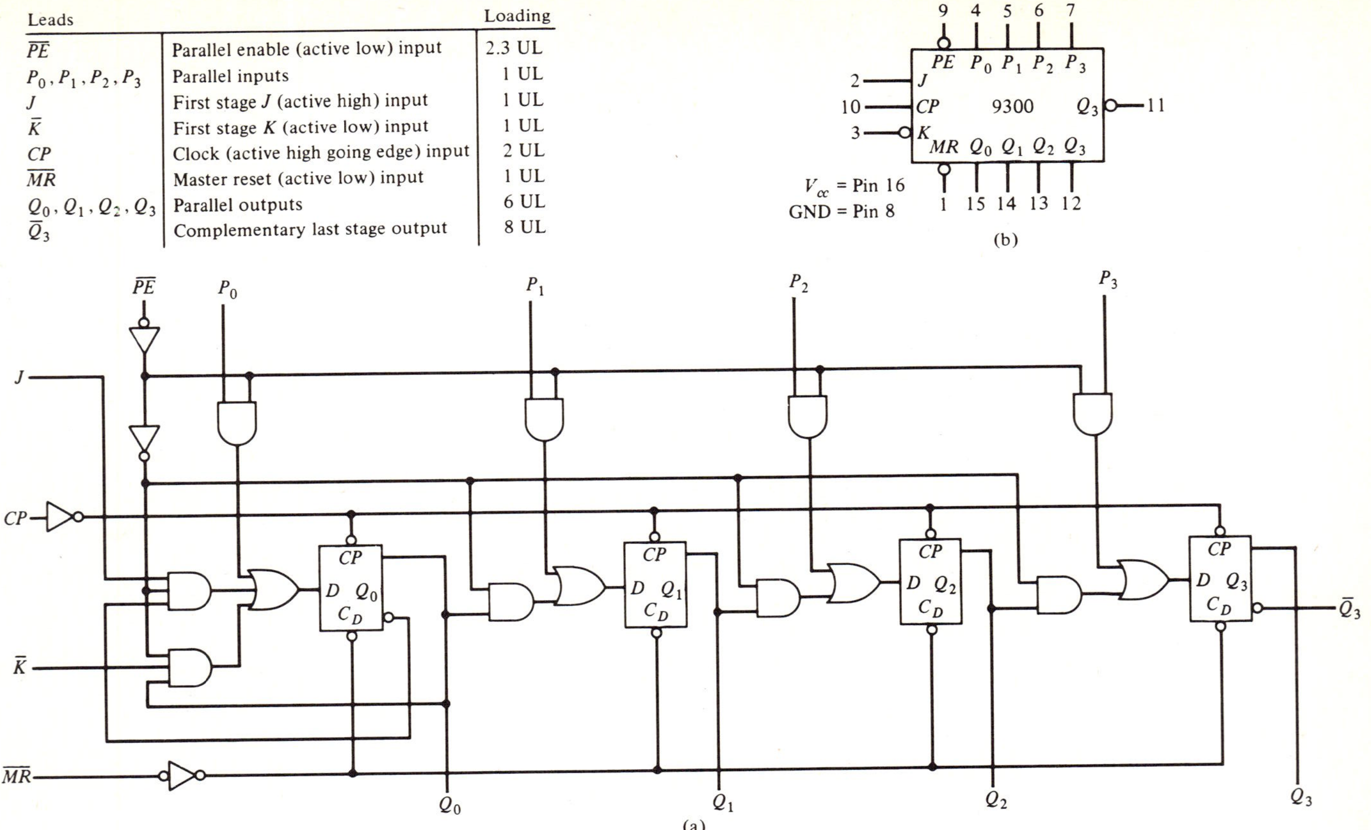

Figure 4-34 Type 9300 4-bit shift register: (a) logic diagram, (b) logic symbol. (Courtesy Fairchild Camera and Instrument Corp.)

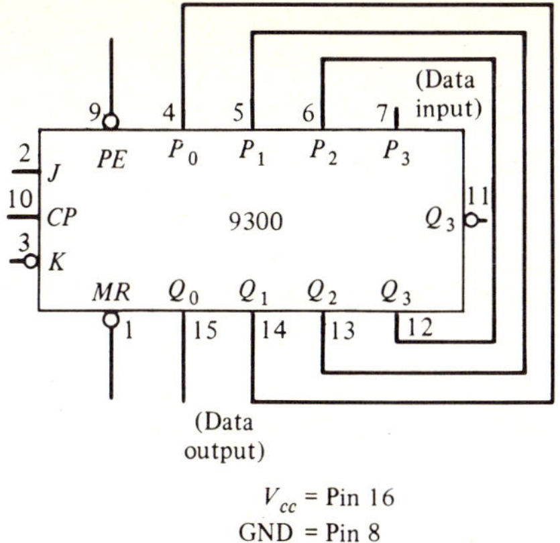

Figure 4-35 Type 9300 4-bit shift register externally connected to work as right-to-left shift register.

The 9300 shift register is basically a left-to-right unidirectional unit. However, it can be externally connected to work as a bidirectional unit with both left-to-right and right-to-left shift capabilities. If we connect the Q_3 output terminal to the P_2 input terminal, Q_2 to P_1 and Q_1 to P_0, as shown in Fig. 4-35, then with $\overline{PE}$ input LOW, the data will shift right-to-left synchronously with each clock pulse. In this arrangement data cannot be read in parallel but data can be read serially through the P_3 input terminal, and shifted left.

Shift registers are used to convert serial data to parallel form and vice versa, and to transfer serial data between two digital systems operating at different speeds. They are used to delay binary data in digital filter design. The shift register is a basic unit in a serial multiplier.

If the data stored in a shift register is read out serially, it is no longer available for further use. To circumvent this problem, we can connect the output of the last stage to the serial input terminal and form what is known as a *circulating register* which finds applications in some digital computers.

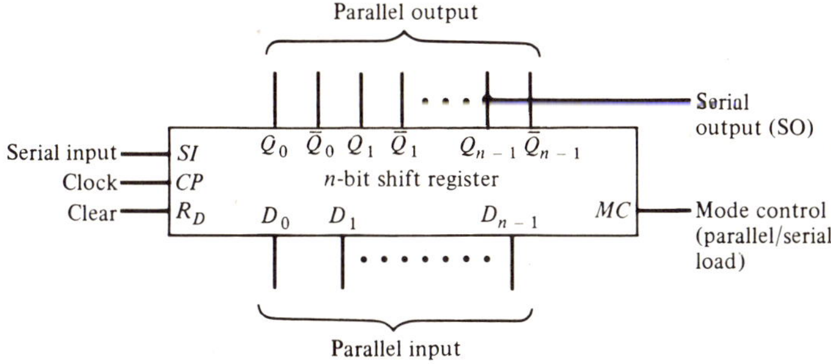

Figure 4-36 Logic symbol of a *n*-bit shift register.

There are various other applications of a shift register such as sequence generators and detectors, counters, and so on. In the next section we consider the latter application.

The logic symbol of an n-bit left-to-right shift register with serial/parallel data entry and exit is sketched in Fig. 4-36. The status of the mode control input determines whether the shift register is to be loaded in serial or parallel mode.

4-6   Counters

The counter is a special type of sequential circuit that is used to count pulses, usually clock pulses. In addition to simple counting applications, it is used for operation sequencing, frequency division, analog-to-digital conversion, and various other applications. There are several different types of counters, some of which are described in this section. A variety of such counters are available in IC form.

Ripple Counter

The ripple counter operates in an asynchronous mode and is the simplest of all counters. It can be designed using a chain of T flip-flops whose T inputs are permanently set to 1. The clock pulses are fed into the clock input of the first flip-flop. The clock input of each subsequent flip-flop in the chain is connected to the output of the previous flip-flop. For example, Fig. 4-37(a) shows a 4-bit binary ripple counter. The operation of the counter will be clear from its state diagram sketched in Fig. 4-37(b). Before the counting begins, all flip-flops are in ZERO state. By examining the output of the flip-flops after the flip-flops have changed states one can easily determine the number of pulses counted up to that time. Note that the counting is done in natural binary code with the count increasing with every clock pulse. Such a counter is called an *up* counter. After counting 16 pulses, the counter resets to zero. That is, after the fifteenth pulse the counter reads 1111_2 whose decimal equivalent is 15. The sixteenth pulse causes a reading of 0000_2.

A *down* counter counts by decreasing the stored count number with each clock pulse. The circuit of Fig. 4-37(a) can be used as a *down* counter by observing the count at the complemented output terminals of each flip-flop, $\bar{Q}_0$, $\bar{Q}_1$, $\bar{Q}_2$, and $\bar{Q}_3$.

A ripple counter being asynchronous in its operation is limited in its speed of operation. The maximum time, which is 4τ sec if τ is the propagation delay of each flip-flop, is taken when the counter in going from the 1111 state to the 0000 state.

Figure 4-37(c) shows the waveforms of each flip-flop output along with that of the input clock. Note that the output Q_0 of the first flip-flop produces one pulse for every two clock pulses. Hence the frequency of the output pulses appearing here is half of that of the clock frequency. In a similar manner, the frequency is divided by 4 at the output of the second flip-flop, divided by 8 at the output of the third flip-flop, and divided by 16 at the output of the fourth flip-flop. As a consequence, a counter can also be used as a *frequency divider*.

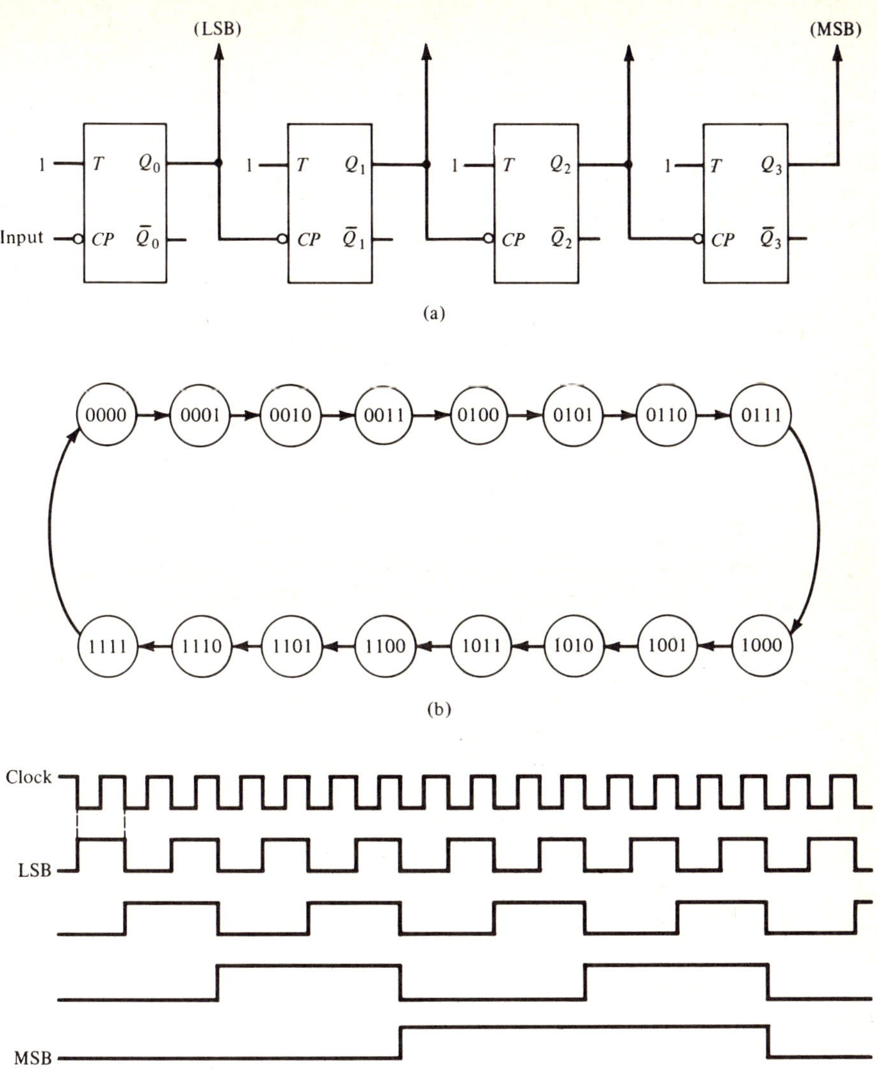

Figure 4-37 A 4-bit binary ripple counter and its state diagram.

An n-bit binary counter can divide by 2^k, where k is an integer in the range $1 \leq k \leq n$. However, with additional logic circuits it can be converted to divide by any integer m where

$$2^{n-1} < m < 2^n \tag{4-19}$$

First, the number of flip-flops is determined from Eq. (4-19). Then the flip-flops are connected as a binary ripple counter. The binary representation of m

is determined next. The output of all flip-flops for which $Q = 1$ at the count m are connected to a NAND gate. The output of this NAND gate is finally connected to the direct reset terminals of all flip-flops. We have assumed here that a ZERO at the direct reset input clears the flip-flop.

Example 4-10.   Design a decade (divide by 10) ripple counter using 8-4-2-1 binary coding of the counter.

Here $m = 10$ and hence, from Eq. (4-19), we observe that n must be 4. The binary representation of m is 1010. Since the first flip-flop in a ripple counter represents the LSB and the last flip-flop represents the MSB, it follows that after ten pulses have been counted the counter state is given by $Q_0 = 0, Q_1 = 1,$ $Q_2 = 0,$ and $Q_3 = 1$. We connect Q_1 and Q_3 to the inputs of a NAND gate whose output is then fed into the direct reset inputs of all flip-flops. The complete design is shown in Fig. 4-38.

It should be noted that if the propagation delays (from the direct reset input to the flip-flop output) of the flip-flops are unequal, then the reset pulse may not stay at ZERO long enough to clear all the flip-flops. A recommended method to eliminate this problem is to use a latch in the resetting circuit part, as shown in Fig. 4-39.

Synchronous Counter

In the synchronous counter, all flip-flops are controlled by the same clock pulse and as a result the counter can operate at a much faster speed. Design of this type of counter to count in powers of 2 is quite simple. However, for other cases, the design is a little more complicated and the use of the synthesis technique outlined in Section 4-4 is then recommended.

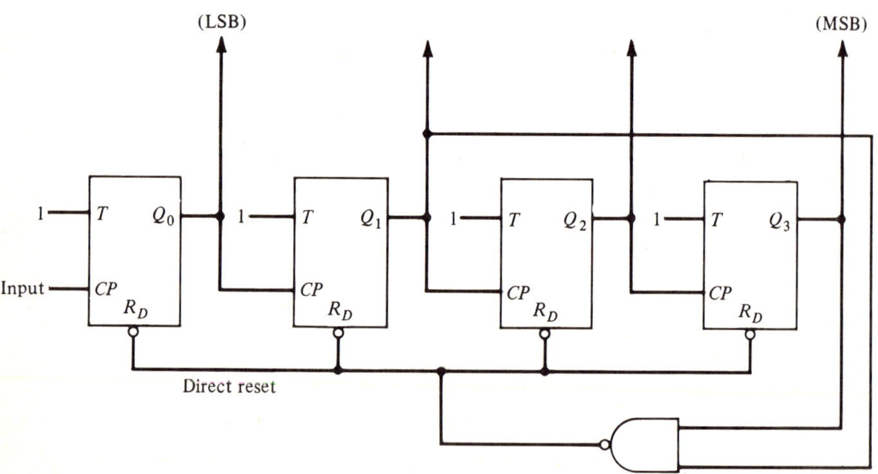

Figure 4-38   Divide-by-10 ripple counter.

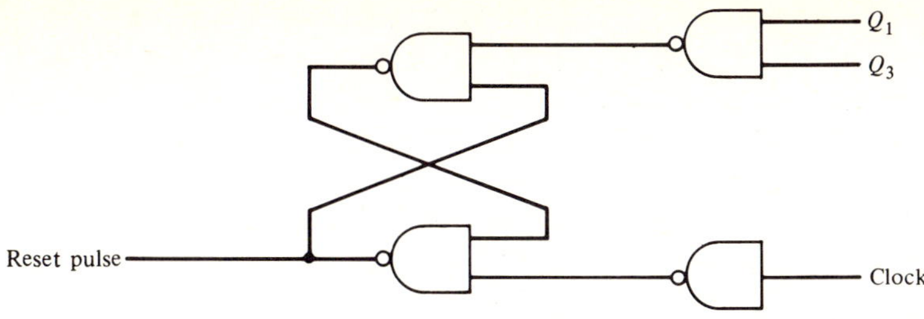

Figure 4-39 Modified resetting circuit for the divide-by-10 ripple counter of Fig. 4-38.

Figure 4-40 shows two different implementations of a 4-bit binary synchronous counter. In each circuit, at least one flip-flop changes state with each clock pulse. The change of state of the ith flip-flop is determined by the status of the carry signal C_i at its T input which in turn is dependent upon the states of the preceding flip-flops.

The basic difference between the two synchronous counters of Fig. 4-40 is the way their carry signals are generated. The circuit at the top is a synchronous counter with *parallel carry* (also known as *look-ahead carry*). Here the maximum delay between change of counter states is the sum of propagation delays of a flip-flop and an AND gate. However, as the number of stages increases, the fan-in requirements of the AND gates and the fan-out requirements of flip-flops increase. These difficulties are avoided in the circuit shown at the bottom which shows a synchronous counter with ripple carry. However, the second circuit has a lower speed of operation compared to the first. The design of an n-bit synchronous binary counter with parallel or ripple carry should be self-evident.

We next consider the design of a synchronous divide-by-m counter where m cannot be expressed as a power of 2. The design procedure is best illustrated with the aid of an example.

Example 4-11. Design a synchronous divide-by-3 counter.

The first step in the design procedure is to determine the number of flip-flops needed to implement the counter. This number is determined from the inequality constraint given by Eq. (4-19). For our problem $m = 3$. Hence from Eq. (4-19) we observe that the total number of flip-flops needed, n, is 2. Let us select T flip-flops for the design. We designate the first flip-flop as A and the second one as B.

Now, a sequential circuit containing two flip-flops has four states. Of these, three will be used to operate the counter. We can choose any three out of the four states as the *regular* states. Depending on which states are chosen as the regular states and the fourth unused state, many different state diagrams can be formulated to describe the divide-by-3 counter. Figure 4-41 shows two such state diagrams. In Fig. 4-41(a), the three regular states are "00," "01," and "10," whereas state "11" is an *isolated* state. For proper

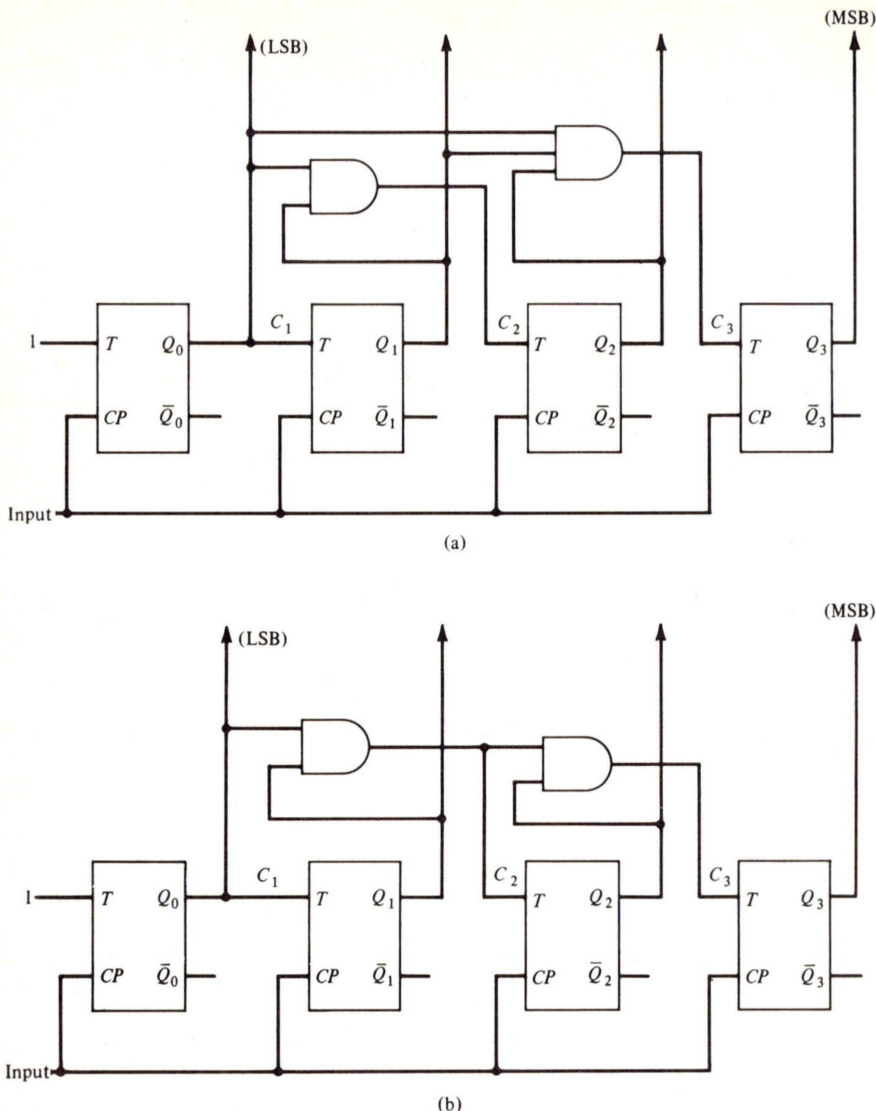

Figure 4-40   4-bit synchronous binary counter with (a) parallel carry and (b) ripple-carry.

operation of the corresponding counter, it is necessary to start the counter with one of the regular states as the initial state.

If the machine starts in state " 11," then it remains locked in that state forever. In Fig. 4-41(b) the regular states are also " 00," " 01," and " 10." However, here the state " 11 " is a *transient* state which has an exit to a regular state. Its corresponding relization as a result does not require any

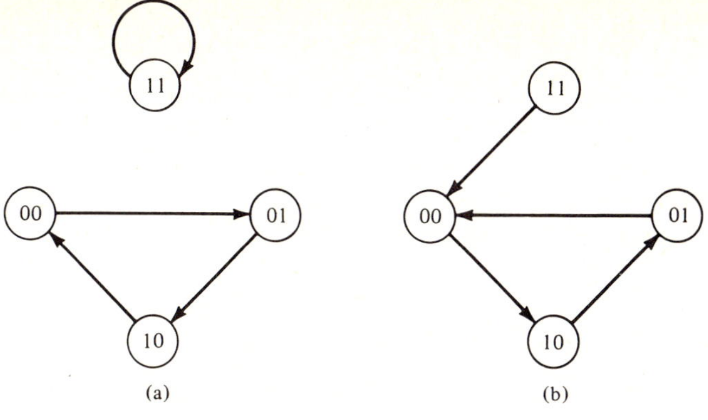

Figure 4-41   Two different state diagrams of a divide-by-3 counter.

initialization and is therefore *self-starting*. Observe that the count sequence depicted in Fig. 4-41(a) is that of an *up* counter and that shown in Fig. 4-41(b) represents a *down* counter.

For illustrative purpose, let us consider the design of the divide-by-3 *down* counter. From the excitation table of the T flip-flop (Table 4-14) and the state diagram of Fig. 4-41(b), we arrive at the modified state table shown in Table 4-19. Here we have assumed that A is the MSB and B is the LSB. The expressions for the control variables are obtained from this state table as

$$T_n^A = A_n + \bar{B}_n, \qquad T_n^B = A_n + B_n \qquad (4\text{-}20)$$

Implementing Eq. (4-20) and making the necessary connections to the two T flip-flops, we finally arrive at the complete logic diagram of the synchronous divide-by-3 *down* counter, as shown in Fig. 4-42.

Shift Register Counters

With additional external connections, a shift register can also be used for counting applications. The simplest form of such a counter is obtained by feeding

TABLE 4-19
Modified State Table of a Divide-by-3 Down Counter

A_n	B_n	A_{n+1}	B_{n+1}	T_n^A	T_n^B
0	0	1	0	1	0
0	1	0	0	0	1
1	0	0	1	1	1
1	1	0	0	1	1

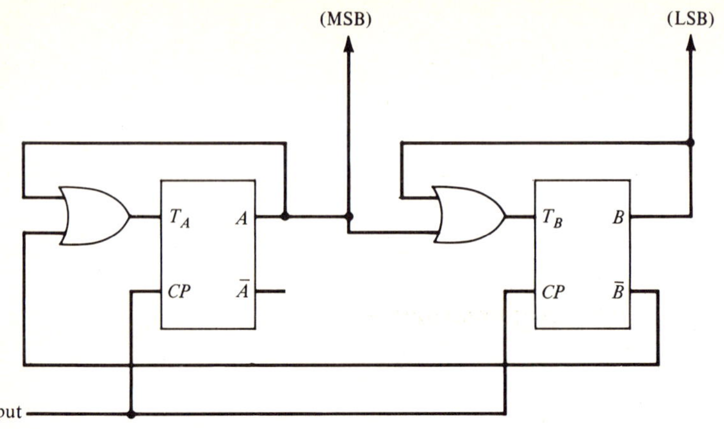

Figure 4-42   A synchronous divide-by-3 *down* counter.

back the output (uncomplemented) of the last stage to the input of the first stage. This type of counter is more commonly known as a *ring counter*. If n is the number of stages in the flip-flop, then the number of valid states of a ring counter is also n, although there is a total of 2^n states. If all of the flip-flops except one in the shift register are set to ZERO, then a ONE will circulate around the counter synchronized with the clock. Hence the outputs of each flip-flop can be checked directly to determine the number of clock pulses counted without any additional decoding logic circuits, which are necessary in the other types of counters discussed earlier.

Figure 4-43(a) shows a simple 4-bit ring counter and its state diagram. Note that to get into the counting sequence where a ONE circulates around, it would be necessary to apply a separate initial pulse to preset the first stage to a 1 and clear the other three stages when the counter is started. Two other counting sequences also exist for this 4-bit ring counter, as well as four more states which will not serve as a 4-bit counter. A self-starting version of a ring counter requires additional logic circuits as shown in Fig. 4-43(b).

The number of states in a shift register counter can be increased to $2n$ by feeding back the complemented output of the last stage to the input. In this version, it is known as a *Johnson* (also *twisted-ring*) counter. A regular 4-bit Johnson counter and its state diagram is indicated in Fig. 4-44. In this particular counter, there are two counting sequences of period 8. In general, a regular Johnson counter has one valid counting sequence and the counter has to be preset to one of the valid states before starting. With additional logic circuits, self-starting Johnson counters can be constructed. Johnson counters with odd cycle lengths can also be implemented with complex feedback arrangement. Although the Johnson counters make more efficient use of the flip-flops, they require special decoding circuits to decode each valid state onto a separate line.

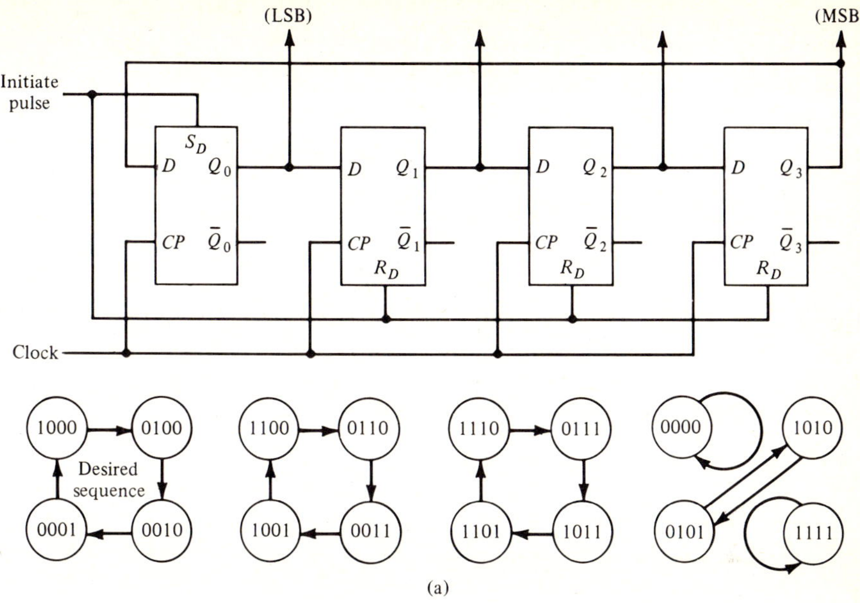

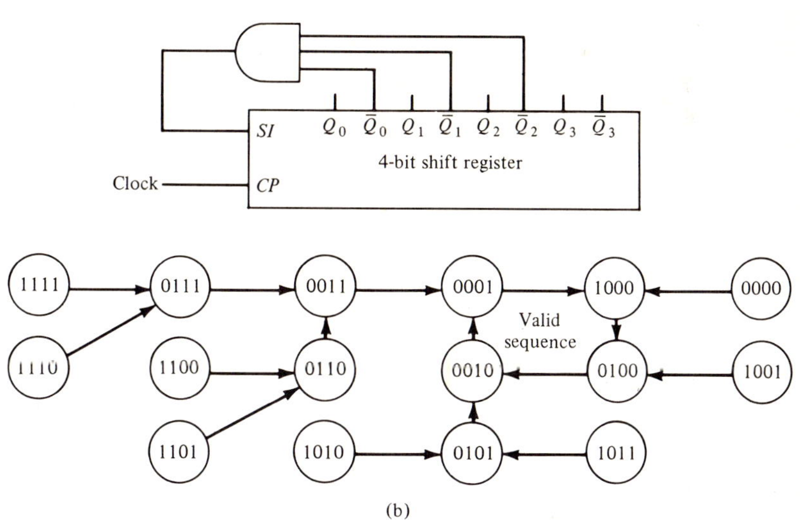

Figure 4-43   4-bit ring counter: (a) nonself-starting, (b) self-starting.

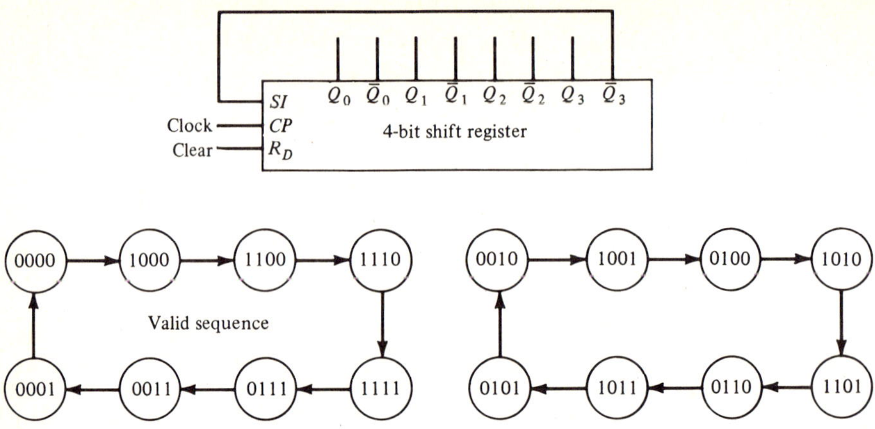

Figure 4-44 A 4-bit Johnson counter.

4-7 Arithmetic Circuits

Many of the arithmetic operations discussed in Section 3-9 can also be performed sequentially if the data is available in serial form. Even though the sequential circuit implementations are in general much simpler than their corresponding combinational logic implementations, their speeds of operation are relatively slower. In this section we describe a number of useful sequential arithmetic circuits.

Serial Comparator

The magnitudes of two n-bit binary numbers A and B can be compared with the aid of the simple circuit of Fig. 4-45. The comparator employs two J–K flip-flops which could be either the edge-triggered type or the master-slave type. Initially, the flip-flops are cleared to ZERO which results in $f_{(A=B)}$ being set to 1. The two input numbers are compared serially bit by bit with the MSBs entering first and the LSBs last. The output line remains at 1 as long as $A_i = B_i$. As soon as the two input bits become unequal, one of the flip-flops is set to a 1. If $A_i > B_i$, Q_1 is set to 1. On the other hand, if $A_i < B_i$, Q_2 is set to 1. As soon as the output of one of the flip-flops is set to 1, its complemented output keeps the other flip-flop on 0 state for the remaining input bits. Thus, after n bits have been examined, the states of the three outputs will indicate whether one number is equal to, greater than, or less than the other number. Note that in some applications, the two numbers to be compared may be stored in shift registers.

Serial Adder

Addition of two binary numbers appearing serially one bit at a time with LSBs first can be accomplished with a single full adder and a D flip-flop which stores the resulting carry bit. The addend and the augend are usually stored in shift

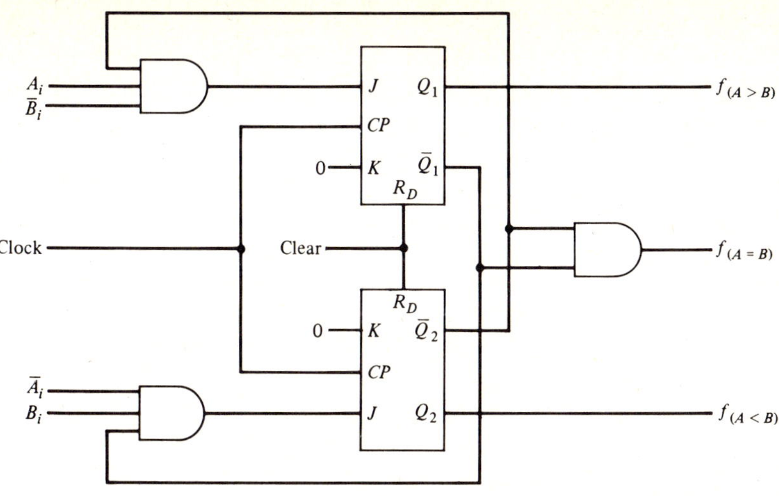

Figure 4-45 A serial comparator circuit for data appearing with MSB first.

registers, and one of these registers or a separate one can be employed to store the sum. If one of the numbers is repeatedly added, its corresponding shift register can be made to be the circulating type. A schematic of the serial adder is sketched in Fig. 4-46. If the augend and the addend are available in parallel form, then they can be loaded into their respective shift registers through the parallel input lines.

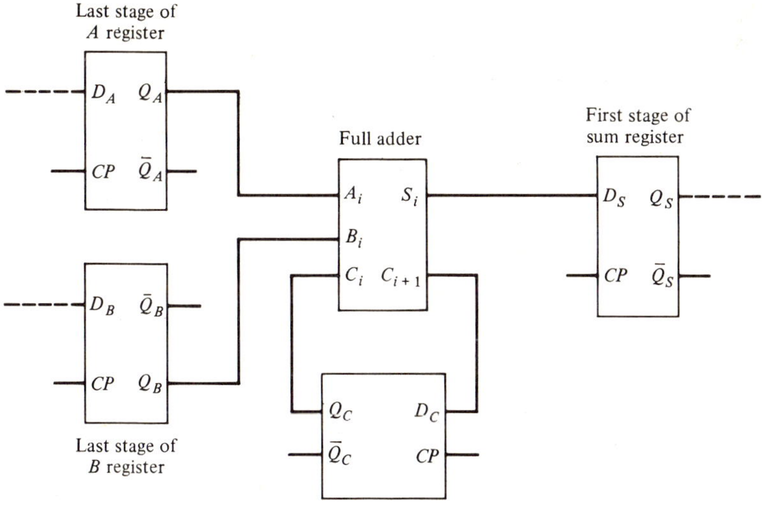

Figure 4-46 Serial adder.

The serial adder can also be used for binary subtraction if negative numbers are represented in 2's complement form. The simple flip-flop circuit of Fig. 4-47(a) can be used to obtain the 2's complement of a binary number. The input must appear with LSB first and the flip-flop must be reset to ZERO before the input appears. The flip-flop is of the master-slave type. The corresponding state diagram is sketched in Fig. 4-47(b).

Serial Multiplier

The implementation of the multiplication of two binary numbers in a serial fashion, as explained in Section 2-7, is considered next. For simplicity, the design of a 3-bit × 3-bit multiplier is discussed. Let the multiplicand be denoted as $A_2 A_1 A_0$ and the multiplier as $B_2 B_1 B_0$ with A_0 and B_0 being the LSBs. Their product, in general, is a 6-bit number to be denoted as $P_5 P_4 P_3 P_2 P_1 P_0$. From our discussion on the design of the parallel multiplier in Section 3-12, we can write the product as

$$P_5 P_4 P_3 P_2 P_1 P_0 = (A_2 A_1 A_0) \times B_0$$
$$+ (A_2 A_1 A_0) \times B_1 \times 2^1 + (A_2 A_1 A_0) \times B_2 \times 2^2$$

$$\underline{\qquad\text{binary addition}\qquad}$$

In the serial implementation of the binary multiplication process, the multiplicand is multiplied by B_0 and stored. If B_0 is a 0, the product is all zeros and if B_0 is a 1, the product is simply the multiplicand itself. Next, the multiplicand is multiplied by B_1 and shifted to the left by one bit position. The shifted partial product is then added to the previous stored partial product and the result is stored again in the same register. Note that there would be no change in the intermediate sum of two partial products if the first partial product is shifted to the right by one bit position, and, then, to which is added the second partial product. In the final step, the partial product $(A_2 A_1 A_0) B_2$ is formed and added to the sum of the previous two partial products which has been shifted to the right by one position. The circuit arrangement, which can repeatedly add

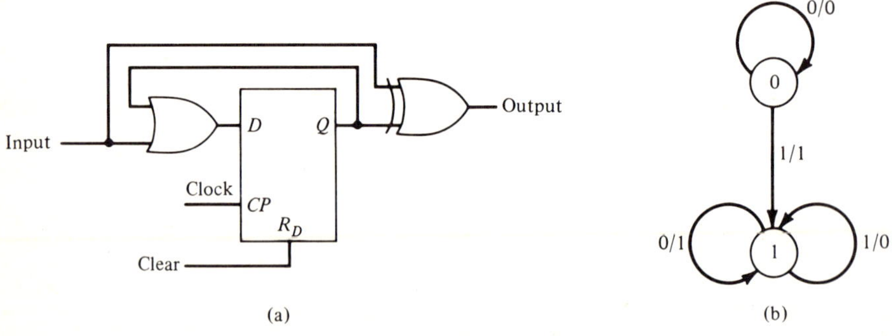

Figure 4-47   Serial generation of TWO's complement.

numbers and store the partial sum after each addition to be used for the next addition, is known as an *accumulator*.

A schematic of the 3-bit × 3-bit serial multiplier arrangement is shown in Fig. 4-48.[5] Here the 6-bit storage register block is of the type shown in Fig. 4-32(a) and is initially reset to zero. The design of a parallel adder has been discussed earlier in Section 3-9. Note also that the contents of the 5-bit shift register SR_2 at the start of the multiplication process is $0\,0\,A_2\,A_1\,A_0$. The serial-out terminal of SR_1 has been labeled SO. The output of the top AND gate is used as a clock for the storage register. If the rightmost bit of the 3-bit shift register SR_1 is a 1, the contents of the 5-bit shift register SR_2 is added with that of the storage register, and the result is stored back in the storage register. If the rightmost bit is a 0, the contents of the storage register does not change. Each

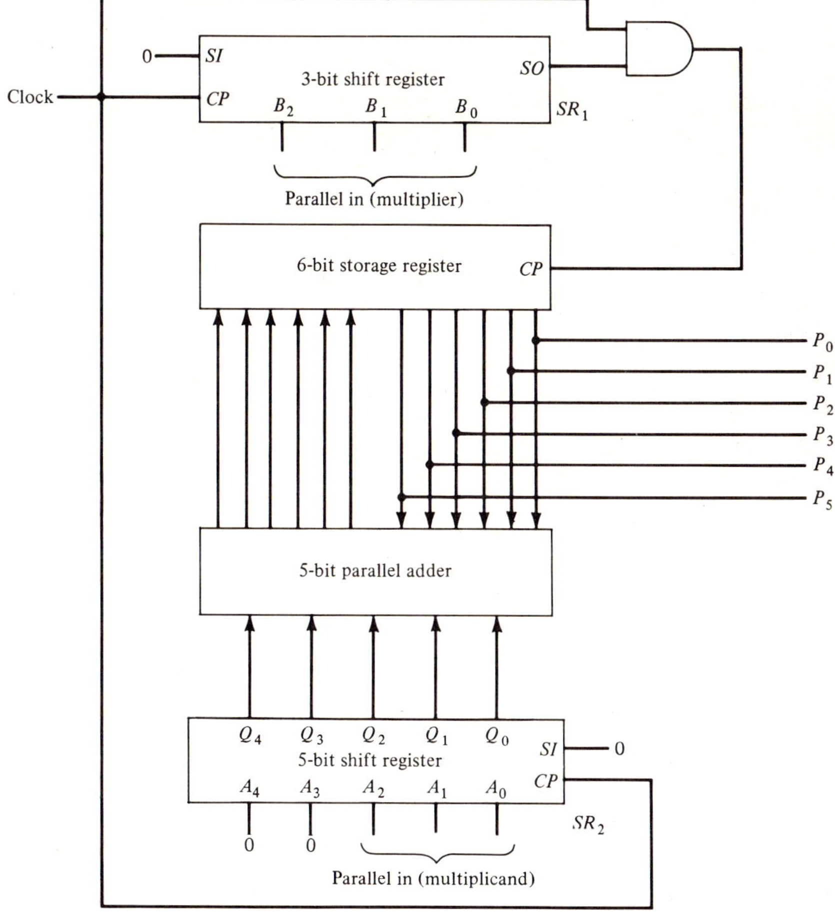

Figure 4-48 A 3-bit × 3-bit serial multiplier.

clock pulse shifts the contents of both the shift registers to the right by one bit position. The following example illustrates the operation of the multiplier unit for a typical case.

Example 4-12.   Describe the basic steps involved in the multiplication of two binary numbers, 110 and 101, using the serial multiplier of Fig. 4-48.

Let the multiplier $B_2 B_1 B_0$ be 101 and the multiplicand $A_2 A_1 A_0$ be 110. Initially, the contents of the storage register is 000000 and that of SR_2 is 00110. The serial-out SO is same as B_0.

Since initially $B_0 = SO = 1$, the first clock pulse causes the contents of SR_2 to be added to that of the storage register (accumulator) and the result $P_5 P_4 P_3 P_2 P_1 P_0 = 000110$ is stored back in the accumulator. The first clock pulse also shifts SR_1 to the right and SR_2 to the left by one bit position. Hence the contents of SR_1 is now 010 and that of SR_2 is 01100.

When the second clock pulse arrives, SO being a 0, there is no change in the contents of the accumulator. However, the second clock pulse causes SR_1 and SR_2 to shift, resulting in 001 in SR_1 and 11000 in SR_2.

Now SO is 1. As a result, the third clock pulse causes the contents of SR_2, which is 11000, to be added to 000110 stored in the accumulator. This results in $P_5 P_4 P_3 P_2 P_1 P_0 = 011110$ which is stored back. The contents of accumulator is now the final product.

4-8   Code Converters

Serial conversion of binary data from one code to another is also economical provided the relatively larger value of conversion time is acceptable. To illustrate the basic ideas involved only two serial converters are described here.

Serial Gray-to-Binary Converter

Recall from our discussion of combinational (parallel) Gray-to-binary converter if G_i represents the ith bit of a number in Gray code and B_i represents the same for the converted number in binary code, then the relations between the two bits are given by

$$B_n = G_n$$
$$B_j = B_{j+1} \oplus G_j, \qquad j = 0, 1, \ldots, n-1 \tag{4-21}$$

where B_n and G_n are the MSBs. Equation (4-21) is simply implemented by a T flip-flop as shown in Fig. 4-49. The input is fed serially with MSB first and the output appears serially with MSB first. It is also assumed that the flip-flop is cleared to 0 before the conversion process begins. Note that the finite propagation delay of the flip-flop should be taken into account before the binary number is observed at the output of the flip-flop.

The above idea can be extended to design a converter for the conversion of parallel input data. The Gray-coded input is read into a shift register in parallel

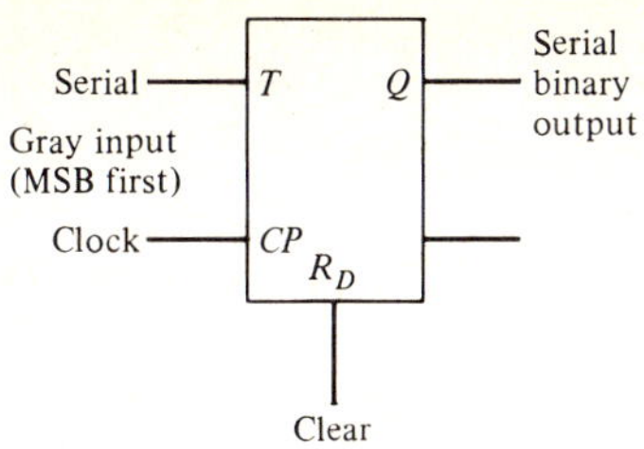

Figure 4-49   A serial Gray-to-binary code converter.

which is then shifted serially into a T flip-flop. The output of the flip-flop gives serially the converted data in binary form. By feeding the output back to the serial input of the shift register, the converted data can be stored temporarily and then read out in parallel. Such a scheme is illustrated in Fig. 4-50 for the conversion of a 4-bit number.

The circuit of Fig. 4-49 can also be used for *serial parity checking*. After all the bits of the data word and the parity bit have been examined, the output of the flip-flop is a 1 for odd-parity checking and a 0 for even-parity checking.

For conversion of a binary number to a Gray-coded number, recall that the relations between the individual bits are given as

$$G_n = B_n$$
$$G_j = B_j \oplus B_{j+1}, \qquad j = 0, 1, \ldots, n - 1 \tag{4-22}$$

A serial implementation of the above equations is sketched in Fig. 4-51 where it has been assumed that the binary data appear with MSB first. The flip-flop is

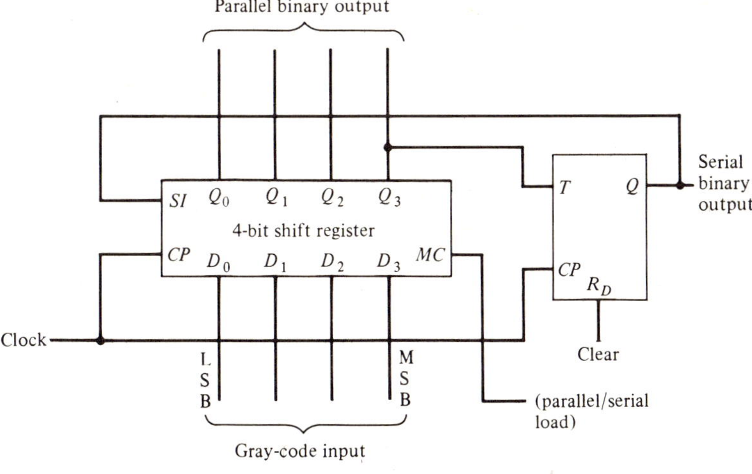

Figure 4-50   A parallel Gray-to-binary code converter.

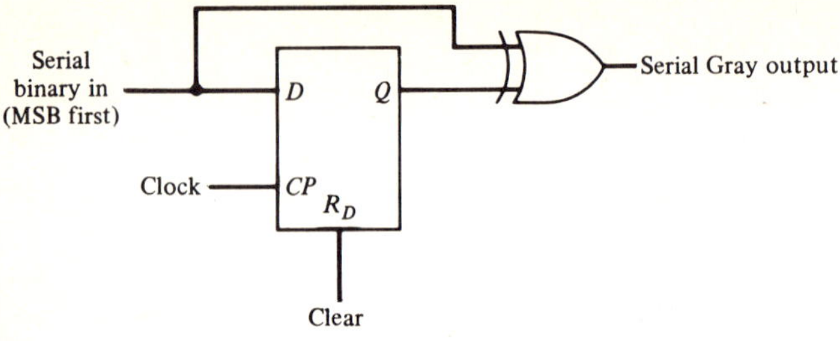

Figure 4-51 A serial binary-to-Gray code converter.

cleared to 0 before the input data is read in. The propagation delay of the flip-flop must be taken into account before the output is observed.

Sequential BCD-to-Binary Converter

Conversion of a BCD number into its binary equivalent can be achieved rather easily using a count-comparison–type approach. The basic idea behind this approach is illustrated in Fig. 4-52 in block diagram form. The operation of the converter is as follows. First, the BCD number is read into a BCD *down* counter and the binary *up* counter is reset to zero. Then, with the application of a " start " signal, clock pulses are applied simultaneously to both counters. Each clock pulse increments the *up* counter and decrements the *down* counter. When the *down* counter contents become zero, the output of the NOR gate (or some other suitable all-zero detector circuit) terminates the clock pulses. The binary equivalent of the BCD number is now given by the contents of the *up* counter.

If we choose the *up* counter to be a BCD *up* counter, and the *down* counter

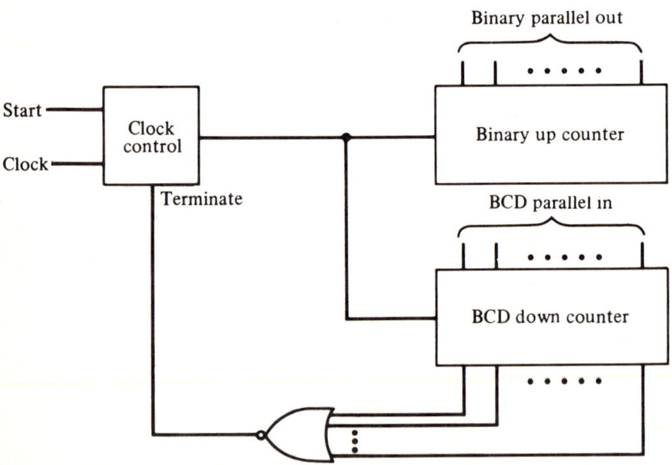

Figure 4-52 Sequential BCD-to-binary converter.

as a binary *down* counter, the arrangement shown in Fig. 4-52 can then be used to convert a binary number into a BCD number. In spite of its simplicity, the count-comparison method is quite slow.

4-9 Sequence Generators

We described earlier two types of shift register based counters: the ring counter and the Johnson counter. Consider the self-starting 4-bit ring counter of Fig. 4-43(b). Once the counter is cycling through the states of the desired sequence, the bit pattern observed at the output of any flip-flop is of the form

$$\ldots 0\,0\,0\,1\,0\,0\,0\,1\,0\,0\,0\,1 \ldots$$

Note that the output is a periodic (repetitive) sequence where the basic period (cycle) of the sequence is 0001 and is of length 4. An n-bit ring counter thus produces a periodic sequence whose basic period is of length n. The basic period is increased to a length $2n$ in a Johnson counter without increasing the number of stages in the shift register.

Further increase in the period length is obtained by using a modulo-2 (i.e., exclusive-OR) feedback. By suitably selecting the feedback logic circuit, the period can be made equal to $2^n - 1$, which is the maximum length obtainable with an n-stage shift register. This type of sequence generator is more commonly called a *maximum-length-sequence* (MLS) generator.

The general form of an MLS generator using an n-bit shift register is shown in Fig. 4-53. In this figure, α_i is a binary constant. If α_i is 1, there is a connection from the *i*th flip-flop output A_{m-i} to its respective exclusive-OR gate and if α_i is a 0, there is no connection. The operation of the MLS generator is described by the linear recurrence relation obtained by expressing the output A_m of the leftmost exclusive-OR gate as

$$A_m = \alpha_1 A_{m-1} \oplus \alpha_2 A_{m-2} \oplus \cdots \oplus \alpha_{n-1} A_{m-n+1} \oplus \alpha_n A_{m-n} \qquad (4\text{-}23)$$

A_m is the generated state to be assumed by the first flip-flop after the arrival of the next clock.

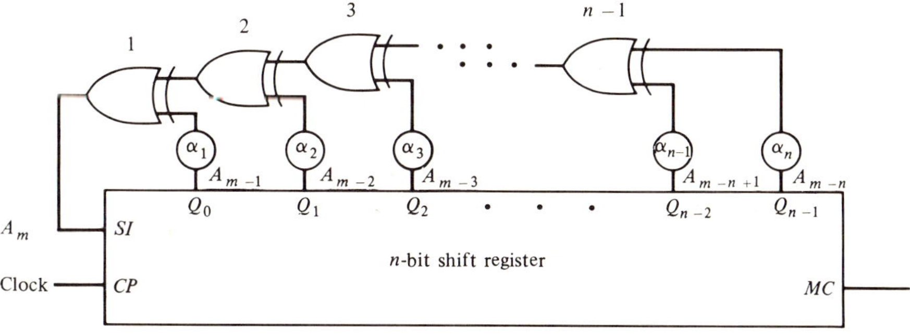

Figure 4-53 General form of an n-bit MLS generator.

**TABLE 4-20
Feedback Connections for
MLS Generators[9]**

n	Nonzero α_i Values
3	$\alpha_2 = \alpha_3 = 1$
4	$\alpha_3 = \alpha_4 = 1$
5	$\alpha_3 = \alpha_5 = 1$
6	$\alpha_5 = \alpha_6 = 1$
7	$\alpha_6 = \alpha_7 = 1$
8	$\alpha_4 = \alpha_5 = \alpha_6 = \alpha_8 = 1$
9	$\alpha_5 = \alpha_9 = 1$
10	$\alpha_7 = \alpha_{10} = 1$

The generation of a maximum-length sequence is usually achieved with only a few of the α_i's being 1. The derivation of the values of α_i are beyond the scope of this text but Table 4-20 lists typical unity value α_i's needed to generate maximum-length sequences for various numbers of stages in the shift register. It should be evident from Eq. (4-23) that if all flip-flops in the shift register are in the ZERO states, then $A_m = 0$ always. As a result, the MLS generator would continue to remain in the all-ZERO state. Consequently some non-zero values of A_j must be initially inserted into the shift register. Any pattern of 1 to n non-zero values will do since they constitute a valid state and the MLS generator goes through every state, including the one inserted.

Example 4-13. Using Table 4-20, design a 4-bit MLS generator and analyze it.

From Table 4-20, for $n = 4$, we note that $\alpha_3 = 1$ and $\alpha_4 = 1$. This also implies that $\alpha_1 = 0$ and $\alpha_2 = 0$. Substituting these values in Eq. (4-23) we arrive at

$$A_m = 0 \oplus 0 \oplus A_{m-3} \oplus A_{m-4} = A_{m-3} \oplus A_{m-4} \qquad (4\text{-}24)$$

The corresponding realization is sketched in Fig. 4-54(a) which is not self-starting. With the aid of a shift register with parallel-loading capability, it is simple to add a zero-suppression arrangement as indicated in Fig. 4-54(b).

Analysis of a MLS generator is achieved by assuming it to be in a certain state and then systematically following through every state transition as dictated by Eq. (4-23). Let us assume that the initial state of the sequential circuit is given by

$$A_{m-1} = A_{m-2} = A_{m-3} = A_{m-4} = 1 \qquad (4\text{-}25)$$

The initial state is recorded in the first row of the state table of Table 4-21. Using Eq. (4-25) in Eq. (4-24), we observe that the next state of the first flip-flop is given by

$$A_m = 1 \oplus 1 = 0 \qquad (4\text{-}26)$$

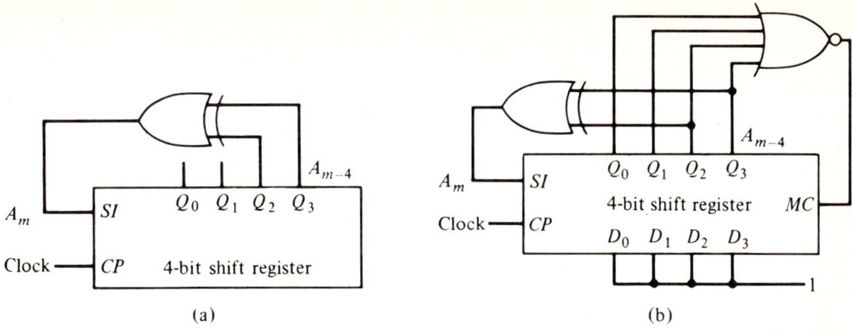

Figure 4-54 4-bit MLS generators: (a) nonself-starting, (b) self-starting.

which is then recorded as the second entry in the first row as shown. As the clock pulse arrives, all the A_j values move to the right by one bit position, as shown in Table 4-21 with the aid of arrows. Using the new values of A_{m-3} and A_{m-4}, A_m is computed from Eq. (4-24) and inserted in its position in the second column. This process is continued until the states of A_m, A_{m-1}, A_{m-2}, A_{m-3}, A_{m-4} become identical to that at the beginning for $n = 1$. The complete state table is given in Table 4-21. If we observe A_m as the output; the sequence generated is seen to be periodic with a period of length $2^4 - 1 = 15$.

Using different types of feedback connections it is possible to generate *nonminimum length sequence* (NMLS) whose length is less than $2^n - 1$. Both

TABLE 4-21
State Table of a 4-Bit MLS Generator

State Number n	A_m	A_{m-1}	A_{m-2}	A_{m-3}	A_{m-4}
1	0	1	1	1	1
2	0	0	1	1	1
3	0	0	0	1	1
4	1	0	0	0	1
5	0	1	0	0	0
6	0	0	1	0	0
7	1	0	0	1	0
8	1	1	0	0	1
9	0	1	1	0	0
10	1	0	1	1	0
11	0	1	0	1	1
12	1	0	1	0	1
13	1	1	0	1	0
14	1	1	1	0	1
15	1	1	1	1	0
16 → 1	0	1	1	1	1

MLS and NMLS generators are also called *pseudo-random sequence* (PRS) generators, which find application in improving the signal-to-noise ratio in digital communications and related areas. Another interesting application of these types of sequence generators is in the coding and decoding of binary characters. Messages are often transmitted in coded form to ensure secrecy.

The coding of a *message* sequence S can be achieved through modulo-2 addition of the message with the output of a PRS generator, generating an encoded sequence C. The original message S is recovered at the receiving end through another modulo-2 addition of the encoded message with the output of an identical PRS generator. It is necessary that the PRS generators at the transmitting and the receiving end work synchronously with each starting at the same initial states. Let S_m and C_m denote the mth bits of the message and its coded version. Then it is seen that the relation between these is given by

$$C_m = A_m \oplus S_m \tag{4-27}$$

If we add A_m and C_m with modulo-2, then the decoded message is

$$D_m = C_m \oplus A_m = (A_m \oplus S_m) \oplus A_m = S_m \tag{4-28}$$

as expected.

To illustrate the coding and decoding process, consider the use of a self-starting 5-bit MLS generator coding messages made up of 3-bit characters as indicated below:

$$a - 0\,0\,0; \quad b - 0\,0\,1; \quad c - 0\,1\,0; \quad d - 0\,1\,1$$
$$e - 1\,0\,0; \quad f - 1\,0\,1; \quad g - 1\,1\,0; \quad h - 1\,1\,1 \tag{4-29}$$

A 5-bit MLS generator designed using Table 4-20 generates a sequence A_m according to the relation

$$A_m = A_{m-3} \oplus A_{m-5} \tag{4-30}$$

Figure 4-55 illustrates the encoding and decoding of a message sequence formed from Eq. (4-30) for a typical sequence generated by the MLS generator.

Pseudorandom sequence A_m: 000 110 111 010 100 001 001 011

Message S_m:	010	101	110	001	100	010	001	111
	c	f	g	b	e	c	b	h
Coded message C_m:	010	011	001	011	000	011	000	100
	c	d	b	d	a	d	a	e
Decoded message D_m:	010	101	110	001	100	010	001	111
	c	f	g	b	e	c	b	h

Figure 4-55   Typical sequences in the encoding and decoding of messages.

4-10 Special-Purpose Circuits

The registers, counters, arithmetic circuits, and code converters are commonly used in digital system design. Some typical examples of each of these circuits have been described in the last four sections. Of these, the registers and the counters find more frequent applications and, as a result, a large variety of these two types of sequential circuits have been made available in IC forms, primarily in MSI packages.

There are some special-purpose sequential circuits that often find applications in the design and testing of digital systems. We describe a number of these circuits in this and the following section.

Contact-Bounce Eliminator

For testing digital circuits, mechanical switches are usually used to connect either the supply voltage or the ground to provide a ONE or a ZERO to certain input terminals. However, because of unavoidable " bouncing " on contact, a series of 1's and 0's may be applied to these terminals instead of a single 1 or a single 0. This problem can be avoided with the circuit arrangement of Fig. 4-56 which employs a single-pole, double-throw switch, a NAND latch, and two resistors. The resistor values shown are for TTL gates but a similar arrangement can be had with other logic families. The switch is a push-type switch, and its normal resting position is terminal ① as shown. Contact with terminal ② is made by pushing the lever toward the terminal and, upon release, the switch returns automatically to position ①.

In the resting position with the switch connected to terminal ①, the variable $\bar{S}$ is at 0 V, that is, at logical-0 level, and the variable $\bar{R}$ is at 5 V, that is, at logical-1 level. Thus, in accordance with the truth table of the R–S latch as given by Table 4-1, Q is set to 1 and $\bar{Q}$ is at 0. When the switch is pushed to make contact with terminal ②, $\bar{S}$ goes to 1 as soon as the switch lever is disconnected from terminal ①. Since both $\bar{S}$ and $\bar{R}$ are 1, no change in the status of Q and $\bar{Q}$ occurs. Upon first contact of the switch lever with terminal ②, $\bar{R}$ is set to 0, which resets $\bar{Q}$ to 1 and Q to 0. If the contact at terminal ② bounces with the switch lever being physically disconnected from terminal ②, but not remaking contact with terminal ①, $\bar{R}$ goes to 1 and with both $\bar{R}$ and $\bar{S}$ at 1, Q remains at 0 and $\bar{Q}$ remains at 1. As soon as the switch is released, it makes contact with terminal

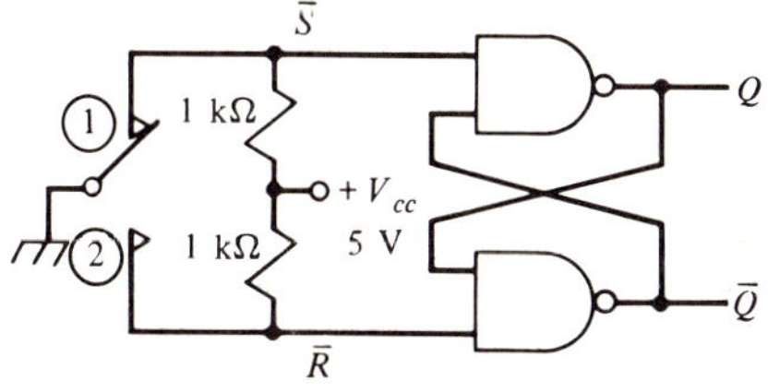

Figure 4-56　Contact-bounce eliminator.

①, setting Q to 1 and $\bar{Q}$ to 0. If the contact bounces at terminal ① without making physical contact with terminal ②, Q continues to remain at 1 and $\bar{Q}$ at 0. Thus this circuit can be used to provide a single 1 pulse at the $\bar{Q}$ terminal, and a single 0 pulse at the Q terminal by pressing the switch once.

Data Synchronizer

The input data to a synchronous sequential digital circuit may sometimes be asynchronous. Thus it is necessary to synchronize the input with respect to the clock pulses before it is processed by the sequential circuit. This can be achieved simply with either an edge-triggered or a master-slave J–K flip-flop connected as a D flip-flop. Figure 4-57 shows the input and output waveforms along with that of the clock in the case of a master-slave and a negative-edge-triggered J–K flip-flop. Note that the edge-triggered flip-flop can synchronize data of width less than that of the clock pulse provided the data pulse is present for the flip-flop set-up time at the activating edge of the clock pulse. On the other hand, if a master-slave type flip-flop is used, since both edges of the clock pulse are used here, the width of the data pulse should be at least greater than the width of the clock pulse.

Clock-Burst Generator[9]

The circuit of Fig. 4-58 produces a train of pulses as long as the J input is at 1. Assuming the flip-flop to be a negative-edge-triggered J–K flip-flop, we observe that if the J input is at 1, Q is set to 1 at the first negative-edge transition of the

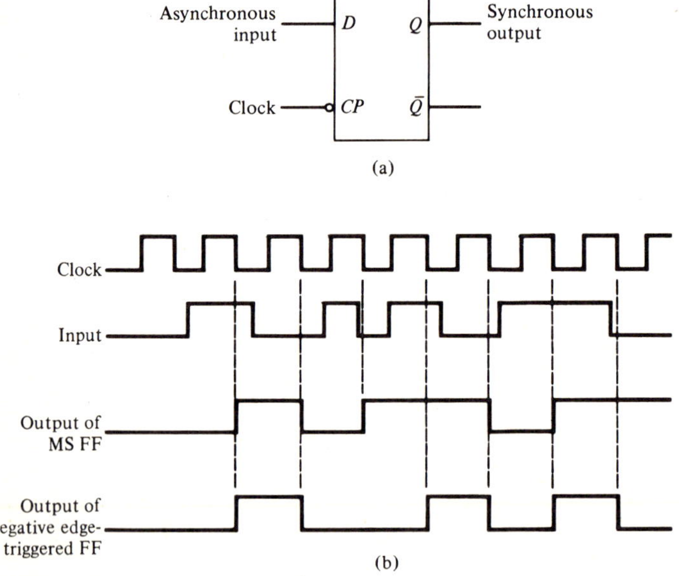

Figure 4-57   Synchronization of asynchronous input.

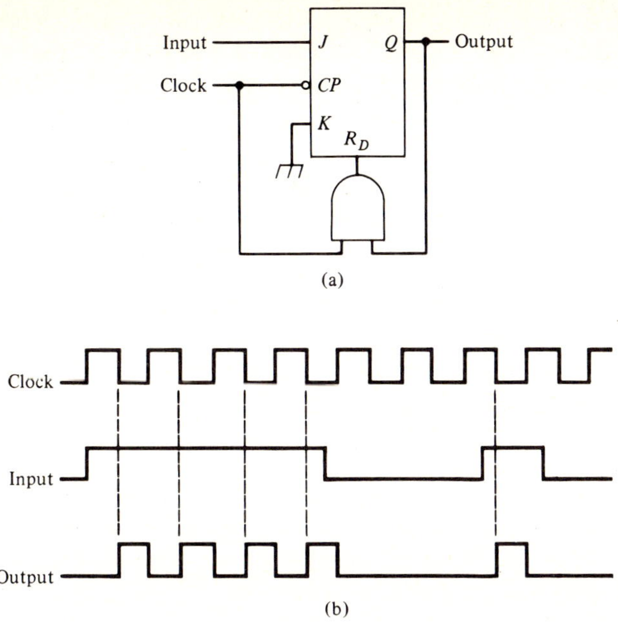

Figure 4-58 Clock-burst generator.

clock. Q remains at 1 during the period the clock is at the logical-0 level. When the clock goes to the logical-1 level, the asynchronous direct reset input is set to a 1, which clears the flip-flop and Q is set to 0. Q is set to 1 again at the next negative edge transition of the flip-flop. Thus Q output switches back and forth between the 1 and 0 levels synchronously with the clock as long as J is held at 1. Typical waveforms are also shown in Fig. 4-58.

Gated Single-Pulse Selector[9]

For testing purposes, it is useful to have a circuit that generates a single pulse on command from a gating signal. The circuit of Fig. 4-59 can be used for this

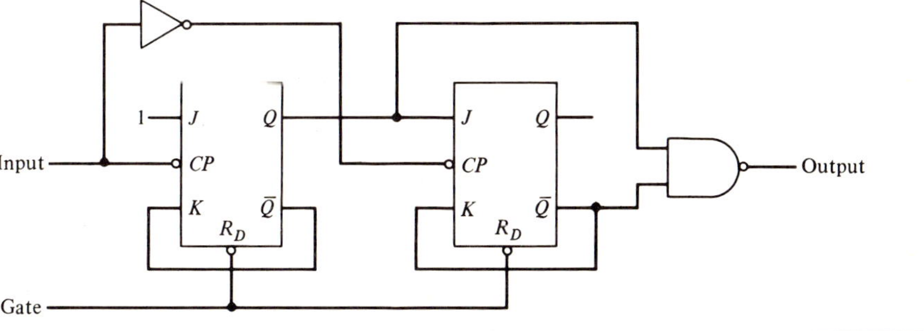

Figure 4-59 Single-pulse selector.

purpose. When the gating signal is LOW, both flip-flops are cleared to 0. The first complete pulse appearing at the input after the gate goes HIGH appears at the output. Any subsequent input pulses that occur while the gate remains at the HIGH level are suppressed.

4-11   Timing Circuits

There is another type of special-purpose circuit that often finds applications in the design of digital systems in solving various types of timing problems. These circuits are generally labeled as timing circuits. Three most important types of timing circuits are the *Schmitt trigger* and the *monostable* and *astable multivibrators*. We discuss these circuits in this section.

Schmitt Trigger

Often, data sources are basically slow in changing, that is, they generate data with slowly rising and falling edges. This type of input data may cause difficulty in the operation of a digital system for one or more of the following reasons: (i) Rise-time sensitive circuits in the system, if present, may not function properly; (ii) logic gates may be biased too long in the " active " region ; and (iii) prediction of propagation delays may not be possible and certain logical operations may be unintentionally performed while the input data are changing. To circumvent these problems, the Schmitt trigger is used to " sharpen up " the rising and falling edges of data waveforms.

The symbolic representation of the Schmitt trigger and its transfer characteristic is depicted in Fig. 4-60. As can be seen from this figure, the output of the circuit can be at one of two possible states: V_{O-} and V_{O+}. If the input voltage V_i is less than V_{T-}, the output is V_{O-}. Output remains at V_{O-} if the input voltage is increased slowly until $V_i = V_{T+}$, when the output voltage makes a sudden transition to V_{O+}. Any further increase in V_i has no additional effect on the output. If V_i is now decreased, the output continues to remain at V_{O+} until $V_i = V_{T-}$, when it makes a sudden transition to V_{O-} and continues to remain

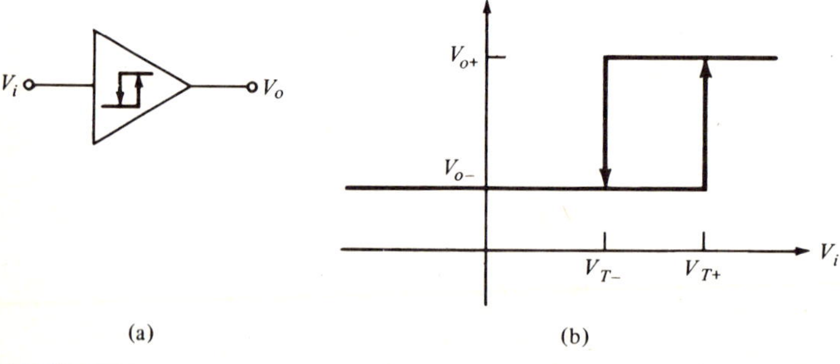

(a)   (b)

Figure 4-60   (a) Symbolic representation and (b) input–output transfer characteristic of a Schmitt trigger.

at this state for any further decrease in the input voltage. Because of the two different threshold levels for the input voltage at which the output transition occurs, depending on whether the input is increasing or decreasing, the Schmitt trigger is said to exhibit hysteresis of value $V_{T+} - V_{T-}$.

The presence of the hysteresis eliminates the effect of any superimposed noise on the input data waveform.[5] This is illustrated in Fig. 4-61 which shows a typical input waveform, and the output waveforms of a Schmitt trigger inverter along with that of a regular inverter. Note from Figs. 4-61(c) and (d), the

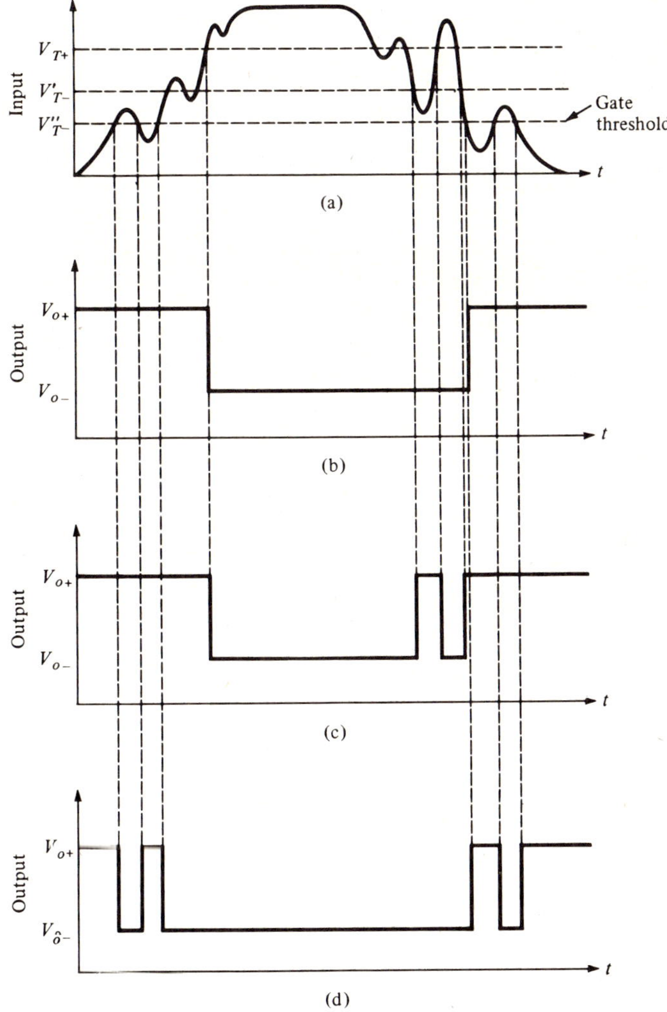

Figure 4-61 (a) Input waveform, (b) output of Schmitt trigger inverter with large hysteresis $V_{T+} - V''_{T-}$, (c) output of Schmitt trigger inverter with a smaller hysteresis $V_{T+} - V'_{T-}$, (d) output of an (ideal) NOT gate.

noisy input waveform, if fed directly to a logic gate or to a Schmitt trigger with small hysteresis, can cause multiple triggering. On the other hand, if the hysteresis is large, the output of a Schmitt trigger is a " clean " square pulse as shown in Fig. 4-61(b).

IC Schmitt triggers are also used as inverters or multiple input logic gates to enable additional logical operations to be performed along with the " squaring up " of input pulses. Examples of such gates are the Texas Instruments SN54/7413 dual 4-input NAND Schmitt trigger and the Texas Instruments SN54/7414 hex Schmitt trigger INVERTER packages. These gates are compatible with TTL or DTL logic circuits. They require a supply voltage of $+5$ V. The positive-going and negative-going threshold voltages are typically $V_{T+} = 1.7$ V and $V_{T-} = 0.9$ V, resulting in a hysteresis of 0.8 V. Typical values of the high-level and low-level output voltages are $V_{O+} = 3.4$ V and $V_{O-} = 0.2$ V, respectively. These gates have *totem-pole* output stages to provide low output resistance for both output states. The low output resistance characteristic allows fast charging and discharging of output capacitances. In using these packages, spare Schmitt triggers can be used as standard INVERTERs or NAND gates, as appropriate.

Monostable and Astable Multivibrators

All flip-flops, some of which were described earlier, are bistable circuits, that is, they have two stable states. A flip-flop can remain in one stable state as long as there is no change in the status of the input lines. Transition to the other stable state is affected with prescribed change in the status of input voltage levels.

The *monostable multivibrator*, more commonly called a *one-shot*, has also two states. However, one state is stable and the other is semistable or quasistable. Normally the one-shot remains in the stable state. When appropriately triggered, it temporarily moves to the semistable state and automatically returns to the stable state after a predetermined time period. The one-shot can be used as a time delay to regenerate pulses.

In contrast to the bistable flip-flops and monostable circuits, there are two-state circuits with both states being semistable. These latter circuits are known as *astable multivibrators* and are used as square-wave generators for clocking and other purposes.

There are two types of monostable multivibrators: *nonretriggerable* and *retriggerable*. In the former type, if the one-shot is triggered by two successive pulses of separation t less than the width t_w of the output pulse generated by a single trigger, the second trigger has no effect on the output and is thus ignored by the one-shot. In the latter type, under the same condition as above, the second input pulse essentially retriggers the one-shot, generating an output pulse of width approximately equal to $t + t_w$, neglecting the propagation delay of the one-shot. By applying a continuous series of triggers, the output of the one-shot can be maintained at the quasistable state as long as it is necessary or, in other words, produce a pulse with 100 percent duty cycle. A retriggerable one-shot can be used to construct an astable multivibrator.

Both astable and monostable multivibrators can be designed with logic gates

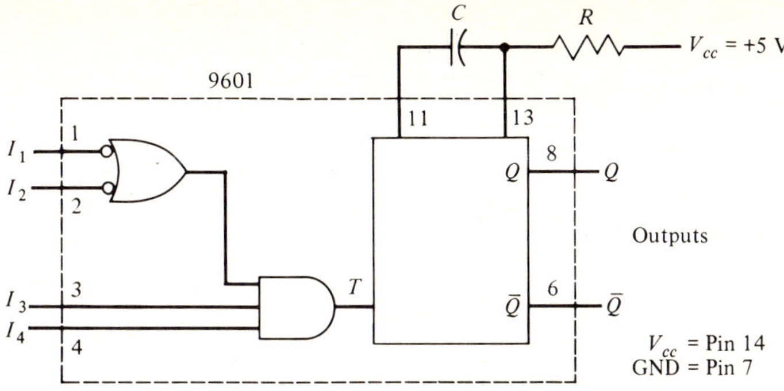

Figure 4-62 Logic diagram of 9601 retriggerable one-shot along with external R–C network connection. (Courtesy Fairchild Camera and Instruments Corp.)

along with additional resistor-capacitor networks which control the timing properties of these circuits (Problems 4-46, 4-47, and 4-48). However, it is usually more convenient to use IC one-shots in system design because of their wide availability and low cost.

Examples of nonretriggerable one-shots are the Fairchild 9603 and 74121 TTL circuits, whereas examples of the retriggerable type are the Fairchild 9600, 9601, 9602, and 74123 TTL circuits. These circuits are designed with multiple-input logic circuits at the input to provide additional flexibility in the triggering. In all of these one-shots, the output pulse width is controlled by an external resistor and an external capacitor. Figure 4-62 shows the logic diagram of the 9601 unit along with the connection of the resistor and the capacitor. The operation of this one-shot is outlined in Table 4-22. Note that the trigger input T to the one-shot is given by

$$T = (\overline{I_1 I_2})I_3 I_4 \qquad (4\text{-}31)$$

where I_1, I_2, I_3, and I_4 are the four possible external trigger inputs. As can be seen from Table 4-22, the triggering or retriggering occurs only when the trigger

TABLE 4-22
9601 Triggering Scheme[a]

I_1	I_2	I_3	I_4	T	Operation
$1 \to 0$	1	1	1	$0 \to 1$	Trigger
1	$1 \to 0$	1	1	$0 \to 1$	Trigger
0	d	$0 \to 1$	1	$0 \to 1$	Trigger
d	0	$0 \to 1$	1	$0 \to 1$	Trigger
0	d	1	$0 \to 1$	$0 \to 1$	Trigger
d[b]	0	1	$0 \to 1$	$0 \to 1$	Trigger

[a] Courtesy Fairchild Camera and Instruments Corp.
[b] d: don't care.

input T goes from a logical 0 to a logical 1. The combinational circuit at the input of the one-shot allows triggering on either LOW-to-HIGH or HIGH-to-LOW transitions on the input waveform.

The width t_w of the output pulse for a single trigger input depends on the resistor-capacitor network connected to terminals 11 and 13, and is given by

$$t_w = 0.32\,RC\left(1 + \frac{700}{R}\right) \tag{4-32}$$

If the resistor value R is in kiloohms and the capacitor value C is in picofarads, the width t_w will be in nanoseconds. The above formula is accurate within ± 10 percent and is valid for $C > 1000$ pF. Adjustable resistors and/or capacitors should be used to trim output pulses to exact, desired width. For circuits requiring capacitor values less than 1000 pF, nomograms provided in the manufacturer's data sheet should be consulted to determine appropriate resistor values for desired output pulse widths.

The 9601 can be connected as shown in Fig. 4-63 to work as a nonretriggerable one-shot with negative-edge triggering. The unused inputs I_3 and I_4 can be left unconnected, but this reduces the noise immunity of the circuit. It is therefore advisable to tie these inputs to the power supply ($+V_{cc}$) through a 1-kΩ resistor.

The 9601 can be configured readily to work as a Schmitt trigger. The pertinent connection is sketched in Fig. 4-64. The diode at the input ensures the triggering by positive signals only. The input signal should have peak amplitude of at least 2.5 V and a current in the range of 100 μA–1 mA. Another application of the 9601 is in designing an astable multivibrator as indicated in Fig. 4-65.

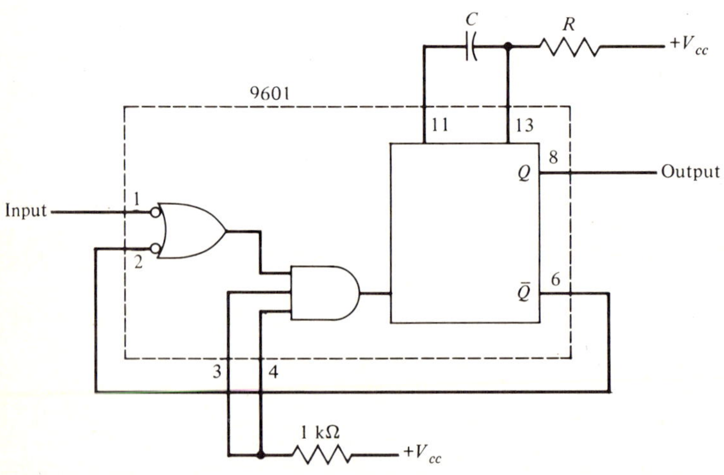

Figure 4-63   Realization of a nonretriggerable one-shot using the 9601.

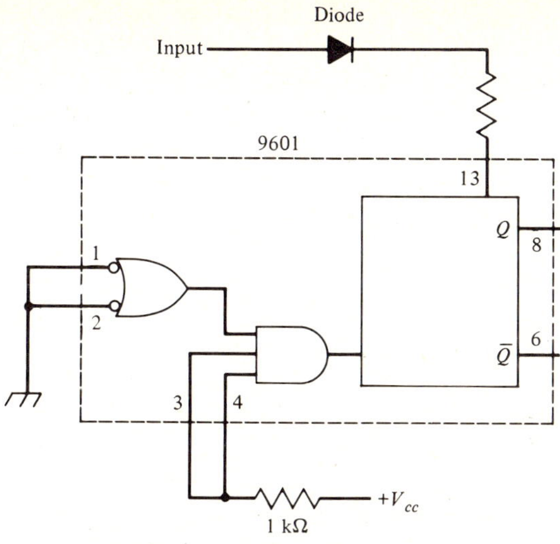

Figure 4-64 Schmitt trigger operation using a 9601.

Various other applications of the 9601 such as programmable delay line, pulse-duration detector, frequency discriminator, double-pulse detector, malfunction indicator, gated clock generator, and so on, are described in the Fairchild TTL Applications Handbook.[12]

The monostable and astable multivibrators can also be constructed using an analog comparator. These circuits are described in Section 7-6.

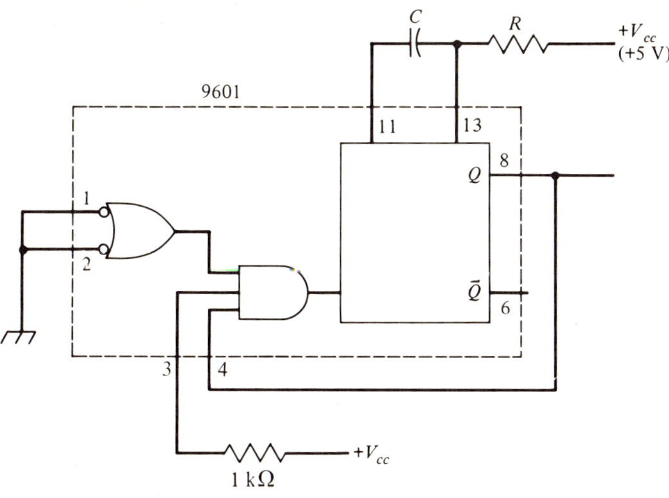

Figure 4-65 Astable multivibrator design using a 9601.

4-12   Summary

This chapter outlines the techniques for analyzing and synthesizing sequential circuits. Primarily sequential circuits of the synchronous type are considered. A sequential circuit can be considered as an interconnection of a combinational circuit and storage devices (Fig. 4-1). The widely used storage unit is a flip-flop. The four basic flip-flops are described in Section 4-1 along with their properties, logic gate implementations, and interrelationships. A systematic procedure for analyzing a sequential circuit is outlined in Section 4-2. Analysis leads to a tabular description (state table) and a graphical description (state diagram) of the sequential circuit. Section 4-3 considers the development of a state table (or state diagram) characterizing the sequential circuit from a given problem statement and delineates a method to obtain an equivalent state table description with fewer states. A method to synthesize a synchronous sequential circuit from its state table description is discussed in the following section. This is achieved by modifying the state table of the sequential circuit to include the required truth values of the controlling inputs of the flip-flops being used.

Several widely used sequential circuits are described in the following five sections. The data in digital systems are often stored temporarily in registers for later use. Two types of registers are discussed in Section 4-5. A very useful type of sequential circuits is the counter which is the subject of discussion of Section 4-6. The design of three types of counters is treated here. Section 4-7 considers the design of serial sequential arithmetic circuits such as comparator, adder, and multiplier. The following section is concerned with the design of serial code converters. The generation and detection of periodic sequences is outlined in Section 4-9. Section 4-10 describes a number of simple useful special purpose circuits that are often needed in the design and testing of complex digital systems. Finally the last section outlines several typical timing circuits.

References

1. D. L. Dietmeyer, *Logic Design of Digital Systems*, Allyn and Bacon, Boston, Mass., 1971.
2. W. E. Wickes, *Logic Design with Integrated Circuits*, John Wiley & Sons, New York, 1971.
3. F. J. Hill and G. R. Peterson, *Introduction to Switching Theory and Logical Design*, John Wiley & Sons, New York, 1968.
4. H. Torng, *Introduction to the Logical Design of Switching Circuits*, Addison-Wesley, Reading, Mass., 1964.
5. A. Barna and D. I. Porat, *Integrated Circuits in Digital Electronics*, Wiley-Interscience, New York, 1973.
6. V. T. Rhyne, *Fundamentals of Digital System Design*, Prentice-Hall, Englewood Cliffs, N.J., 1973.
7. G. K. Kostopoulos, *Digital Engineering*, Wiley-Interscience, New York, 1975.
8. T. R. Blakeslee, *Digital Design with Standard MSI and LSI*, Wiley-Interscience, New York, 1975.

9. R. L. Morris and J. R. Miller, Eds., *Designing with TTL Integrated Circuits*, McGraw-Hill Book Co., New York, 1971.
10. J. D. Lenk, *Handbook of Logic Circuits*, Reston Publishing Co., Reston, Va., 1972.
11. H. V. Malmstadt and C. G. Enke, *Digital Electronics for Scientists*, W. A. Benjamin, New York, 1969.
12. P. Alfke and I. Larsen, Eds., *The TTL Applications Handbook*, Fairchild Semiconductor, Mountain View, Calif., 1973.
13. M. Schwartz and L .Shaw, *Signal Processing: Discrete Spectral Analysis, Detection, and Estimation*, McGraw-Hill Book Co., New York, 1975.

Problems

4-1[13] A simple way to eliminate most of the effects of the disturbance superimposed on a signal is by filtering the noise mixed signal with the aid of an averaging filter described by

$$y_n = \tfrac{1}{2} \cdot [x_n + x_{n-1}]$$

where x_n and y_n denote the input and output of the filter. Calculate the output of the averaging filter if the signal s_n and the disturbance d_n are given by

$$s_n = n(0.95)^n, \qquad d_n = (-0.95)^n$$

4-2 Show that an R–S flip-flop is characterized by Eq. (4-5).

4-3 Analyze the all-NAND realization of an R–S latch given in Fig. 4-4(b).

4-4 Modify the R–S latch of Fig. 4-4(b) to a clocked R–S flip-flop.

4-5 Verify the J–K flip-flop characterization given by Eq. (4-6).

4-6 Verify the T flip-flop characterization given by Eq. (4-7).

4-7 Verify the flip-flop conversions shown in Figs. 4-13, 4-15, and 4-16.

4-8 Show that the circuits in Fig. 4-66 behave as T flip-flops.

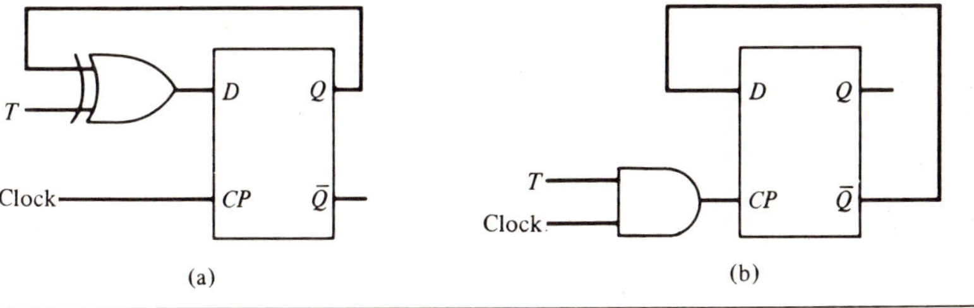

Figure 4-66 Circuits for realizing T flip-flops.

4-9 A sequential comparator circuit containing several flip-flops is used to compare two binary numbers A and B. At the end of the comparison cycle, the normal output terminal (Q) of one flip-flop, if at logical-1 level, indicates that $A > B$. If the complement output terminal ($\bar{Q}$) of the same flip-flop is at logical-1 level at the end of the comparison cycle, what can we say about the two binary numbers?

4-10 Analyze the sequential circuit of Fig. 4-67(a) and develop its state table and state diagram.

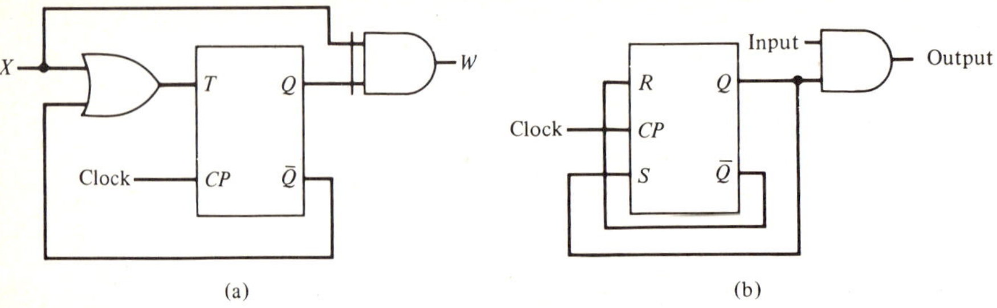

(a)   (b)

Figure 4-67

4-11 Derive the state table and the state diagram of the sequential circuit of Fig. 4-67(b).

4-12 Verify the state transitions and the output for the sequential circuit of Fig. 4-21 as given by the last three rows of Table 4-5.

4-13 Analyze the sequential circuit of Fig. 4-68 and derive its state table and state diagram.

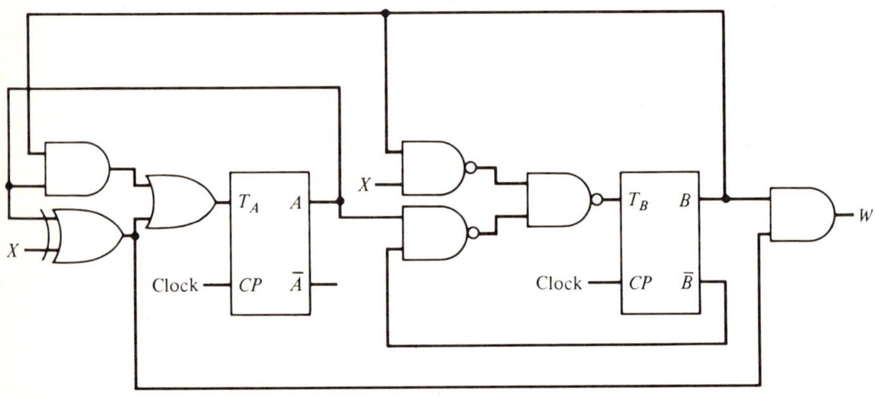

Figure 4-68

4-14 A synchronous sequential circuit is described by the state diagram of Fig. 4-69. Determine its operation. Simplify the state diagram.

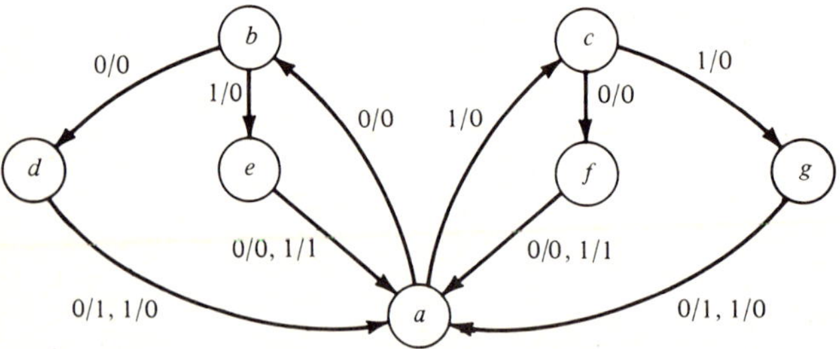

Figure 4-69

4-15 A synchronous sequential circuit is characterized by the state table of Table 4-23. Derive the corresponding state diagram. Simplify the state diagram.

TABLE 4-23

Present State	Input = 0		Input = 1	
	Next State	Output	Next State	Output
a	c	0	g	0
b	e	0	b	0
c	f	0	h	0
d	d	0	a	0
e	f	0	h	0
f	d	0	a	0
g	e	0	b	1
h	c	0	g	0

4-16 Using the formal synthesis method of Section 4-4 realize a T flip-flop using a D flip-flop and logic gates.

4-17 Realize a J–K flip-flop using a T flip-flop and logic gates. Use the synthesis method of Section 4-4.

4-18 Realize a J–K flip-flop using a D flip-flop and logic gates.

4-19 Determine the number of flip-flops needed to implement the sequential circuit of Fig. 4-69.

4-20 Determine the number of flip-flops needed to implement the sequential circuit described by the state table of Table 4-23.

4-21 An alternate realization of the circuit of Example 4-8 is obtained by assigning state a to be the ONE state and state b to be the ZERO state of the R–S flip-flop. Develop the alternate realization.

4-22 Obtain another realization of the circuit of Example 4-9 using a D flip-flop.

4-23 Repeat the design problem of Example 4-8 using T flip-flops.

4-24 A synchronous sequential circuit is to be designed to examine a 4-bit data serially. The circuit produces an output of 1 after all 4 bits have been examined and if the number of 1's in the data is greater than two. The output is 0 otherwise. After all 4 bits have been tested, the circuit resets itself to the initial state and is ready to examine another 4-bit data. Develop the state diagram of the circuit and simplify the state diagram.

4-25 Implement the sequential circuit of Problem 4-24 using J–K flip-flops.

4-26 Implement the sequential circuit characterized by the state diagram of Fig. 4-70 using **(a)** R–S flip-flops, and **(b)** D flip-flops.

4-27 Repeat Problem 4-26 with a different state assignment scheme.

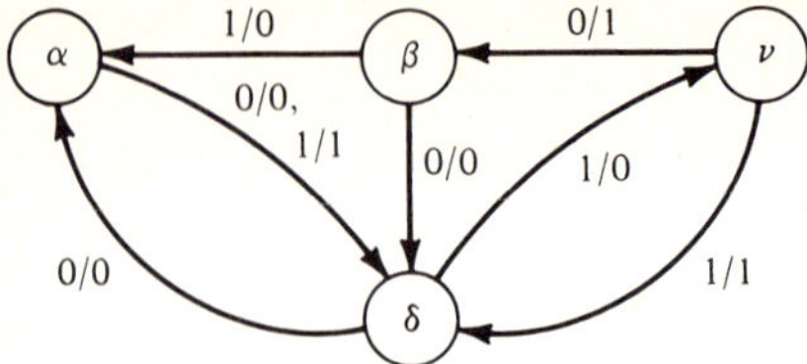

Figure 4-70

4-28 A synchronous sequential circuit is to be designed to examine a binary data train. The circuit produces a 1 if three consecutive bits are 1; otherwise the output is 0. Design the circuit with at most two J–K flip-flops.

4-29 Repeat Problem 4-28 with a different state assignment scheme.

4-30 A three-state sequential circuit described by the state diagram of Fig. 4-71 requires two flip-flops for its realization. However, the total number of states generated by two flip-flops is four. Since the fourth state connection has not been specified, there are two ways the original state diagram can be modified to incorporate the fourth state: either the fourth state can be left as an isolated state or it can be treated as a transient state. Develop realizations using T flip-flops for both modifications. In the second modification assume that for both values of the input bit, the circuit moves from the fourth state to state b.

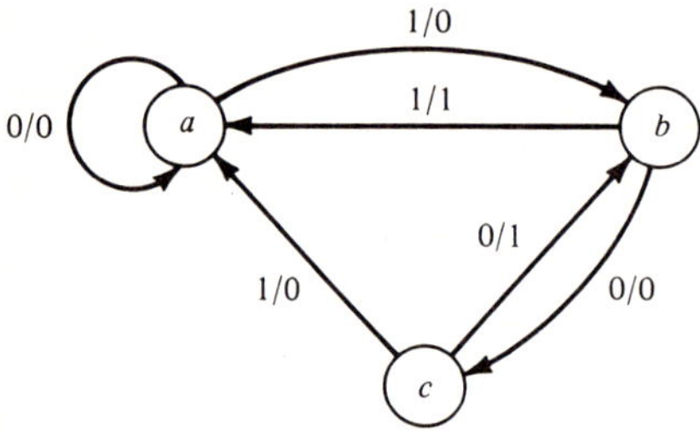

Figure 4-71

4-31 Design a 4-bit unidirectional (left-to-right) serial-in, serial-out shift register using J–K flip-flops.

4-32 Design a divide-by-8 ripple counter.

4-33 Design a divide-by-12 ripple counter.

4-34 Design a divide-by-9 ripple counter.

4-35 Design a 6-bit (divide-by-64) synchronous counter with look-ahead carry.

4-36 Design a 6-bit synchronous counter with ripple carry.

4-37 Synthesize a synchronous divide-by-3 *up* counter using T flip-flops.

4-38 Synthesize using J–K flip-flops a synchronous divide-by-4 counter to count in the sequence shown in Fig. 4-72.

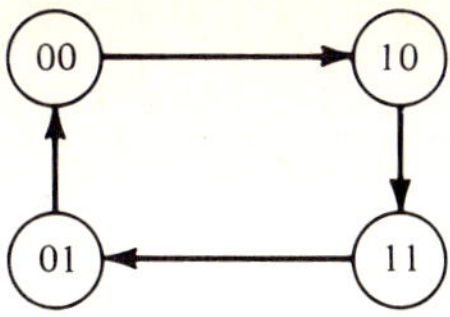

Figure 4-72

4-39 Synthesize a synchronous divide-by-7 counter using *J–K* flip-flops.

4-40 The circuit schematic of a synchronous ripple carry *up/down* counter is shown in Fig. 4-73. It operates as an *up* counter if the mode-control (MC) input is at logical-1 level and operates as a *down* counter if the MC input is at logical-0 level. Analyze the operation of this dual-purpose counter.

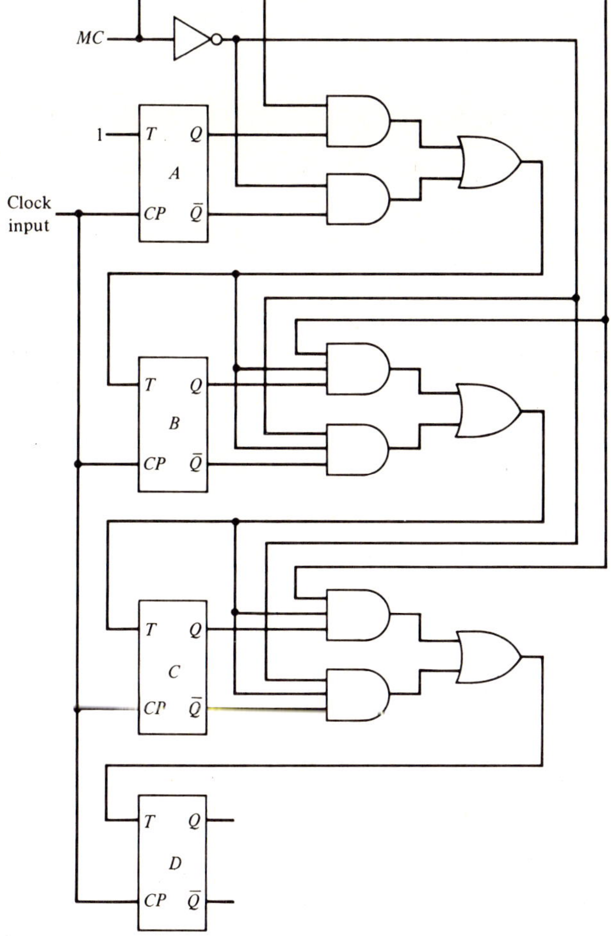

Figure 4-73

4-41 Figure 4.74 shows the schematic and the pin connection of Texas Instruments SN54/7492 IC counter. Show that if the output A (pin 12) is connected to the input BC (pin 1), and input count pulses are applied to input A (pin 14), then output A (pin 12), output C (pin 9), and output D (pin 8) provide a divide-by-2, a divide-by-6, and a divide-by-12 counting, respectively.

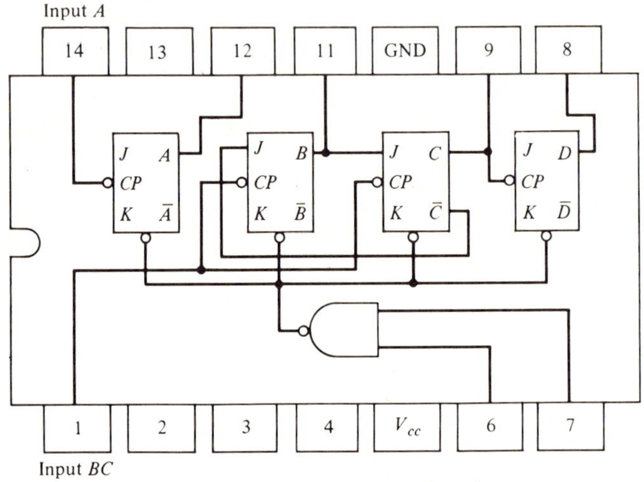

Figure 4-74　Logic diagram of SN54/7492 IC counter.

4-42 Figure 4-75 shows three other counter connections of the SN54/7492 package. Verify the operations of these circuits.

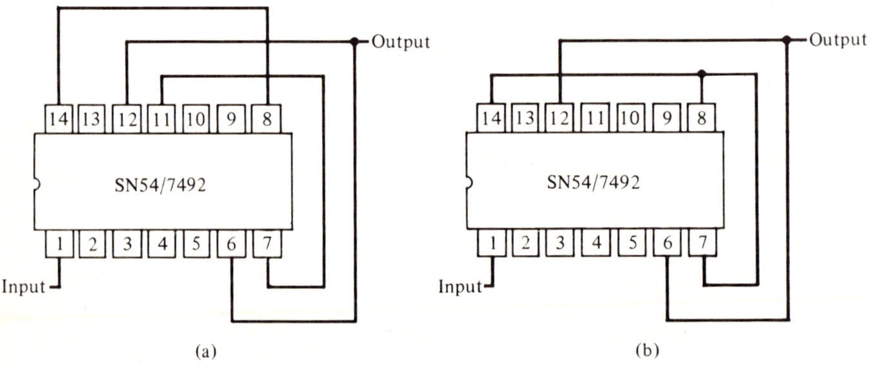

Figure 4-75　(a) Divide-by-7 ripple counter, (b) divide-by-9 ripple counter, and (c) divide-by-11 ripple counter.

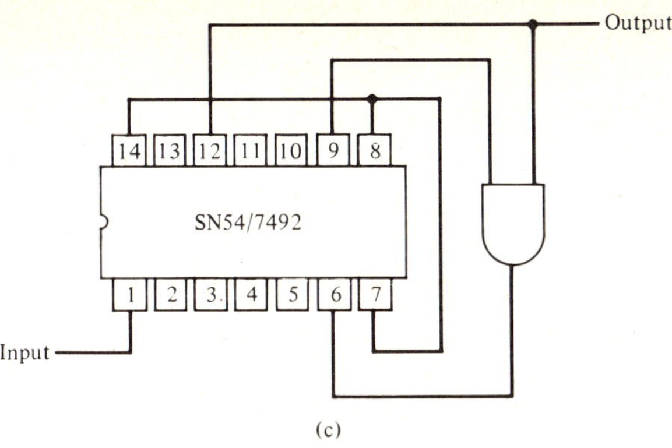

Figure 4-75 (continued).

4-43 A sequential comparator to compare two binary numbers appearing with LSBs first is shown in Fig. 4-76. Before the numbers appear, the flip-flops are cleared to ZERO using the asynchronous clear inputs. Show that the output terminal at logical-1 level after all bits have been examined indicate the result of the comparison.

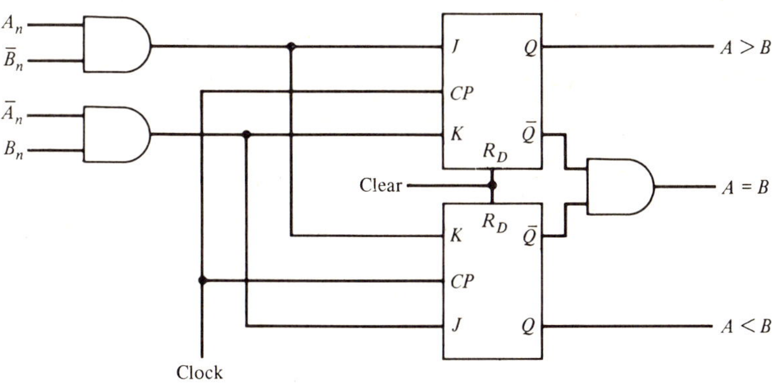

Figure 4-76 A sequential comparator for data appearing with LSB first.

4-44 Design a 6-bit MLS generator.

4-45 Design an 8-bit MLS generator.

4-46 Analyze the monostable circuit of Fig. 4-77 and sketch its input and output waveforms.

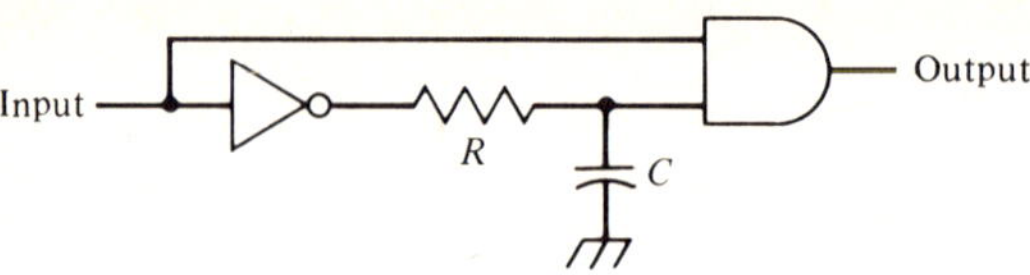

Figure 4-77

4-47 Another realization of a monostable circuit is shown in Fig. 4-78. Analyze the operation of this circuit.

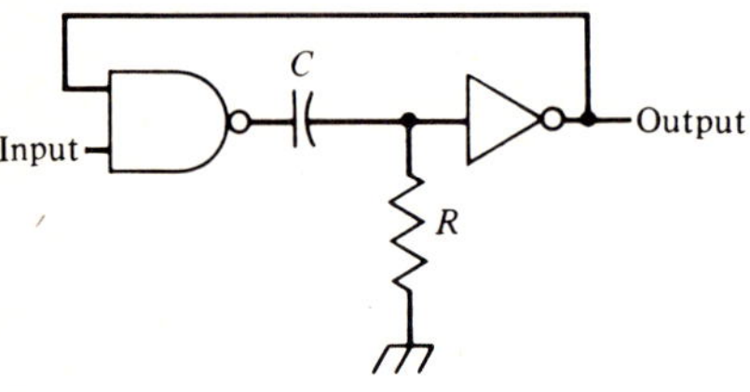

Figure 4-78

4-48 Analyze the operation of the astable multivibrator circuit of Fig. 4-79.

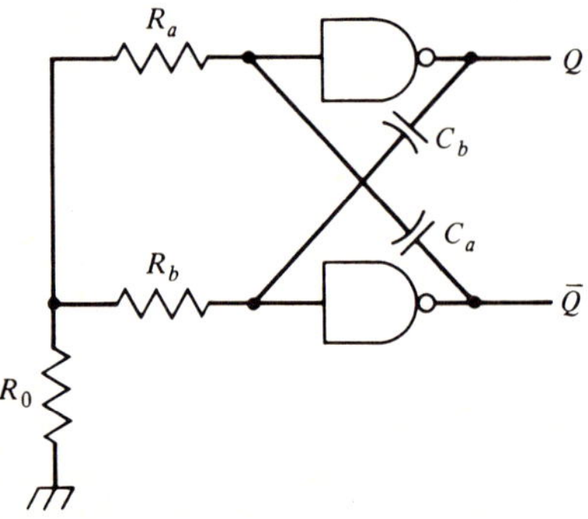

Figure 4-79

CHAPTER 5

Amplifier Fundamentals

In the last three chapters we were concerned with the analysis and design of digital electronic circuits and systems. In this type of circuit, signal variables are discrete functions of time, that is, they are defined only at certain instants of time. Moreover, the amplitudes of the signals in the digital circuits considered can take one of two possible "values." As indicated in Chapter 1, there is another class of electronic circuits, more commonly known as analog circuits, where the signal variables are continuous functions of time and their amplitudes in theory can take any real value. In this chapter and the following two chapters we discuss this latter type of circuit. In Chapter 8 we consider the problem of designing hybrid electronic circuits which work as an interface between analog and digital electronic circuits.

Conceptually the *amplifier* is the simplest of all analog signal-processing devices. Its purpose is to develop an output signal which is an amplified replica of an input signal. Amplifiers are thus primarily used to amplify weak signals so that they can be processed without significant interference from noises generated at intermediate points in the system. With the aid of additional linear and nonlinear circuit elements, amplifiers can also be used to perform many other types of linear and nonlinear signal-processing operations such as modulation, detection, signal generation, and so on. Undoubtedly, it is the most versatile and widely used analog device, finding applications in telephone communication systems, hi-fi systems, television systems, radios, computers, radar, and in many other industrial, consumer, and military electronic systems.

There are many different types of amplifiers, with most types being available commercially in modular form. A large variety of the amplifier modules are fabricated either in monolithic or in hybrid integrated circuit form. However, certain types are available only as discrete circuit modules.

A major part of this book is devoted to amplifiers and their use in linear and nonlinear electronic systems. In this chapter we outline the basic concepts and the general characteristics of the amplifier along with a few simple applications. Chapter 6 is concerned with a very special type of amplifier, the so-called operational amplifier and its applications to linear circuit design. Nonlinear applications of amplifiers are discussed in Chapter 7.

Our first objective in this chapter is to define the ideal amplifier. Later on, we point out the characteristics of a practical amplifier. We then evaluate the performance of a practical amplifier and compare it with that of an ideal amplifier. How "good" a practical amplifier is depends on how closely its performance in a system resembles that of the idealized device.

5-1   Ideal Amplifiers

In its basic form, the amplifier is a 4-terminal network as shown in Fig. 5-1. The signal to be amplified is connected to one pair of terminals, called the *input port*, and the amplified version appears between the other pair of terminals, called the *output port*. Or, in other words, the amplifier is used as a 2-port network. If the input and the output ports of the device have a common terminal, then it is called a *3-terminal amplifier*, otherwise, it is a *4-terminal amplifier*. The 3-terminal amplifier is most commonly encountered in practice with the common terminal connected to the system ground.

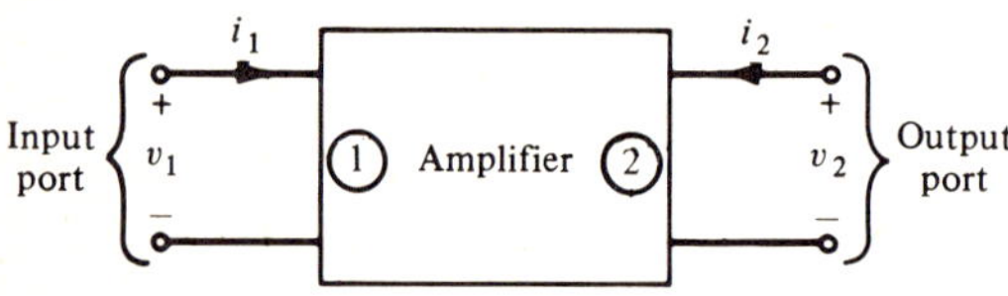

Figure 5-1   2-port representation of an amplifier.

In a typical system application, the amplifier is connected between two other 2-ports as shown in Fig. 5-2(a). For analysis purposes, the circuit on the left connected to the input side of the amplifier can be replaced by an equivalent circuit which is a nonideal voltage source v_s with internal resistance R_s.* Likewise, the circuit connected at the output can be replaced by a resistance R_L (usually known as the *load* resistance).* A simplified representation of the complete system is thus as shown in Fig. 5-2(b).

As an example of the type of application illustrated in Fig. 5-2, consider the hi-fi system of Fig. 5-3 for playing a phonograph record. The essential compo-

---

* For simplicity we are, for the present, assuming that the system has no reactive elements.

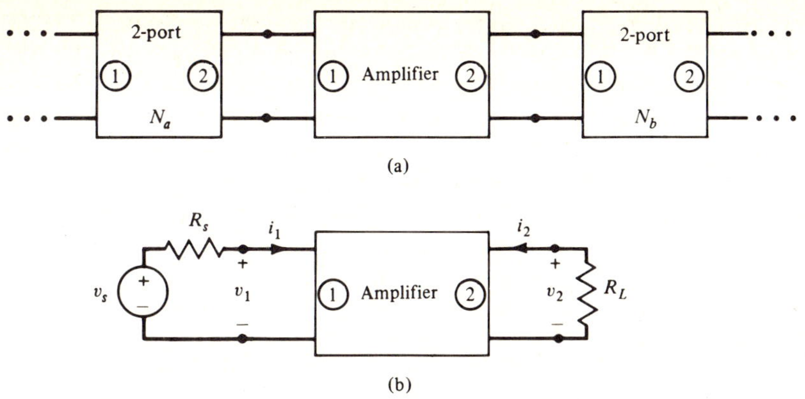

Figure 5-2   A typical system application of an amplifier and its simplified model.

nents of the hi-fi system, from an electrical point of view, are the cartridge, the amplifier, and the speaker.

The cartridge in the tone arm of the record player converts the information stored in the record to an electrical signal. In one form, it contains a rigid metallic beam attached to a diamond needle at one end and a coil at the other end. The coil in turn is suspended between the poles of a magnet. The movement of the needle in the groove of the record causes the coil of wire to move in the fixed magnetic field developing a voltage across the coil.

In the speaker, a coil of wire connected to a diaphragm sits in a magnetic field. A current flowing through the coil in a magnetic field causes it to move which then results in the movement of the diaphragm. The movement of the diaphragm generates air-pressure waves in accordance with the variation of the current in the coil of the speaker.

Now the output voltage of a magnetic cartridge is typically 1–10 mV, and is not strong enough to cause an appreciable movement of the diaphragm. Typically, an 8-Ω speaker may require 10 W of peak power. Since the power dissipated in an R-Ω resistor with V V across it is given by V^2/R W, the above speaker would require a peak voltage of close to 9 V. As a result, the output signal of the cartridge needs to be amplified approximately 900 times. This is provided by the amplifier. Not only must the amplifier provide a voltage amplification, it must also be able to deliver sufficient current to drive the load.

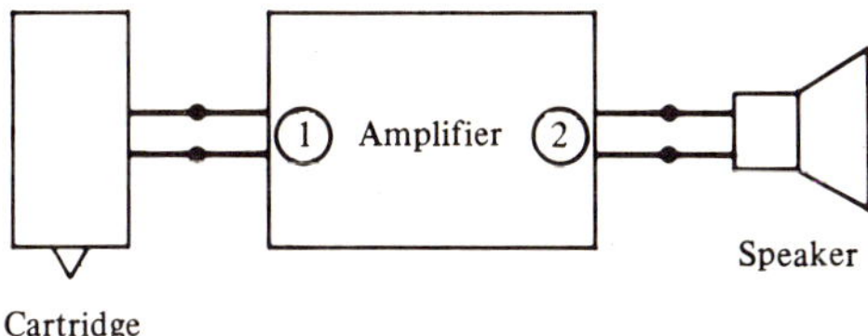

Figure 5-3   A simplified representation of a hi-fi system.

The above discussion on hi-fi amplifiers, strictly speaking, provides a very simplistic description of the system. In practice, the phonograph amplifier needs to provide a playback equalization also to compensate for the deliberately introduced frequency distortion in the phonograph recording and to reduce effects of noise and low inertia of the groove-cutting stylus. A detailed discussion of hi-fi amplifiers is beyond the scope of this book.

We now introduce the concept of an ideal amplifier. There are basically two types of ideal amplifiers: the voltage amplifier and the current amplifier.

Ideal Voltage Amplifier

An ideal voltage amplifier produces an output voltage which at every instant is proportional to the input voltage. The model of this device is a voltage-controlled voltage source as shown in Fig. 5-4(a). From the circuit representation of the ideal voltage amplifier terminated by a load at the output port and excited by a source at the input port as depicted in Fig. 5-4(b), the input–output relations characterizing the ideal amplifier are seen to be

$$v_2 = \mu v_1$$

$$i_1 = 0$$

(5-1)

The constant of proportionality μ is called the *voltage-amplification factor* (or simply the *voltage gain*) which, for an ideal device, is real and constant.

In addition to the voltage gain, a voltage amplifier is also characterized by its *input resistance* R_{in} and its *output resistance* R_{out}, defined as follows:

$$R_{in} = v_1/i_1 \tag{5-2}$$

$$R_{out} = v_2/i_2\big|_{v_1 = 0} \tag{5-3}$$

It follows from the definition (and also from Fig. 5-4) that the input resistance of an ideal voltage amplifier is infinite and the output resistance is zero.

As emphasized by the model, the ideal voltage amplifier is a unilateral device, that is, the electrical conditions at the input are independent of those at the output. The input terminals are said to be *isolated* from the output terminals. In many applications, the isolation property plays an important role.

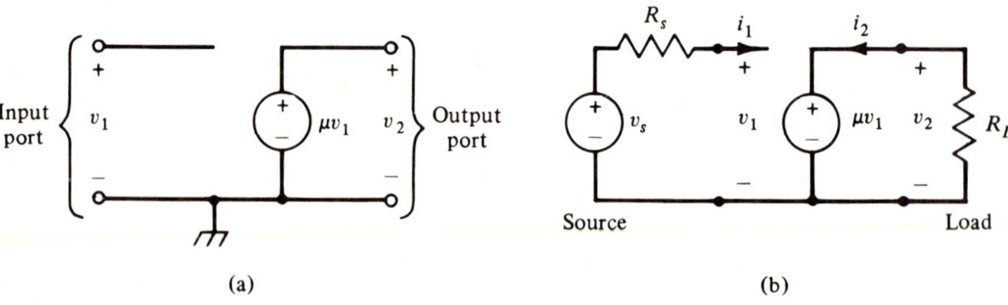

Figure 5-4 Models of (a) a 3-terminal ideal voltage amplifier, and (b) a terminated ideal voltage amplifier.

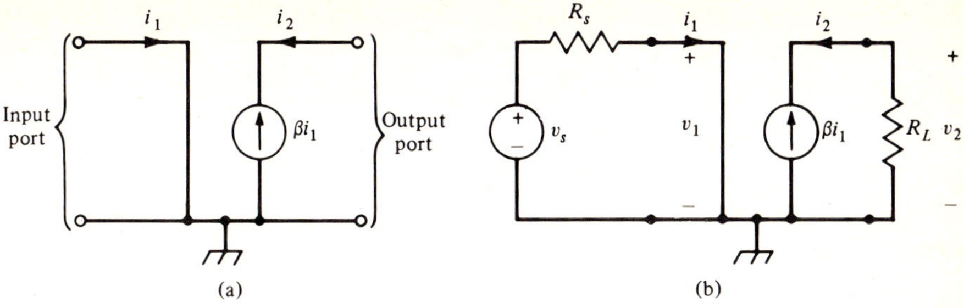

Figure 5-5 Models of (a) an ideal current amplifier, and (b) a terminated ideal current amplifier.

Ideal Current Amplifier

The second type of ideal amplifier is the current amplifier for which the output current is proportional to the input current. The model of this device is a current-controlled current source, as indicated by Fig. 5-5(a). The circuit representation of the ideal current amplifier terminated by a load and excited by a source is sketched in Fig. 5-5(b). From this figure we observe that the characterizing input–output equations of the ideal current amplifier are

$$i_2 = -\beta i_1$$
$$v_1 = 0$$

(5-4)

The constant of proportionality β is known as the *current-amplification factor* (or simply the *current gain*) and is real and constant. An ideal current amplifier, which is also a unilateral device with input terminals electrically isolated from the output terminals, is characterized in addition by a zero input resistance and infinite output resistance.

5-2 Voltage, Current, and Power Gains

Let us now examine the circuit representation of the terminated ideal voltage amplifier as shown in Fig. 5-4(b). We observe that an ideal voltage amplifier for a fixed input voltage v_1 can supply any amount of current to the load (the value of the current being dependent on the value of the load resistance) while maintaining a fixed voltage $v_2 = \mu v_1$ across it. Furthermore the current i_1 drawn by the amplifier input is zero. This implies that an ideal voltage amplifier can deliver any amount of output power to the load while receiving no input power from the source. Likewise, from the circuit representation of the ideal current amplifier sketched in Fig. 5-5(b), we similarly note that an ideal current amplifier for a fixed input current i_1 can develop any amount of voltage across the load, depending on the value of the load resistance, while maintaining a constant current i_2 through the load. Since the input voltage v_1 is zero, an ideal current amplifier can also supply any amount of power to the load while receiving no power from the source. In a practical amplifier, however, the input and the output powers are finite and limited by a number of factors, including the finite input and output resistances of the amplifier. Thus it is quite likely that in a practical voltage amplifier, for which the output voltage is greater than the input voltage, the output power delivered may be less than the input power received. Often, an amplifier such as the one used in the hi-fi system may be required to supply more power to the load than it receives, in addition to providing voltage or current amplification. Such an amplifier is known as a *power amplifier*.*

A practical amplifier is thus characterized by three types of gain: voltage gain, current gain, and power gain. With reference to Fig. 5-2, these gains are defined next.

The ratio of the output voltage v_2 across the load to the input voltage v_1 is the *voltage gain A_v* of the amplifier, that is,

$$A_v = v_2/v_1 \tag{5-5}$$

It is seen from Fig. 5-4(b) that for an ideal voltage amplifier, the voltage gain A_v is simply μ and is independent of the source and load resistances.

The *current gain A_i* is defined as the ratio of the output current $-i_2$ going through the load to the input current i_1 going into the amplifier, that is

$$A_i = -i_2/i_1 \tag{5-6}$$

As seen from Fig. 5-5(b), the current gain A_i of an ideal current amplifier is β which is independent of R_s and R_L.

The ratio of the power P_o delivered to the load to the power P_i supplied to the amplifier is defined as the *power gain A_p*. Since

$$P_o = v_2(-i_2)$$
$$P_i = v_1(i_1) \tag{5-7}$$

we have

$$A_p = P_o/P_i = v_2(-i_2)/(v_1 \cdot i_1) = A_v \cdot A_i \tag{5-8}$$

Very high gain values are often encountered in the design of an electronic system. For example, the Fairchild monolithic operational amplifier μA741C has a voltage gain of about 100,000. In order to facilitate the handling of such large numbers and also small numbers, a logarithmic unit called decibel (dB) is used to express the gains. If A_p is the power gain, then its decibel equivalent $\mathcal{G}_p$ is defined as

$$\text{Power gain in decibels} = \mathcal{G}_p \triangleq 10 \log_{10} A_p \tag{5-9}$$

---

* It should be noted that the principle of conservation of energy is not violated because the biasing dc supplies provide the required output power.

The voltage gain A_v and the current gain A_i are expressed in decibels, according to the following formulas:

$$\text{Voltage gain in decibels} = \mathcal{G}_v \triangleq 20 \log_{10} |A_v| \tag{5-10}$$

$$\text{Current gain in decibels} = \mathcal{G}_i \triangleq 20 \log_{10} |A_i| \tag{5-11}$$

In the case of cascaded amplifiers there is another advantage in using decibels to express the gains. Here, as we show later, the gain in decibels of the cascaded amplifier is the sum of the gains in decibels of the individual sections (provided the sections are isolated from each other).

It should be noted, however, that the term "power gain" has been used to denote both the numerical ratio as defined by Eq. (5-8) and the logarithm of the ratio as defined by Eq. (5-9). In order to avoid ambiguity, we shall always use the term "power gain" to indicate the numerical ratio, unless specifically stated otherwise. Similar statements apply to the other two gains.

Example 5-1. As an example of voltage-gain calculations, consider the circuit of Fig. 5-6. We wish to determine the voltage gain v_2/v_1.

Applying Kirchhoff's voltage law (KVL) at the input side of the circuit, we observe that the voltage across R_a is v_1. By applying KVL again, we observe next that the voltage across R_b is $v_2 - v_1$ with the polarity shown. The currents through these resistors are then

$$i_a = v_1/R_a$$

$$i_b = (v_2 - v_1)/R_b$$

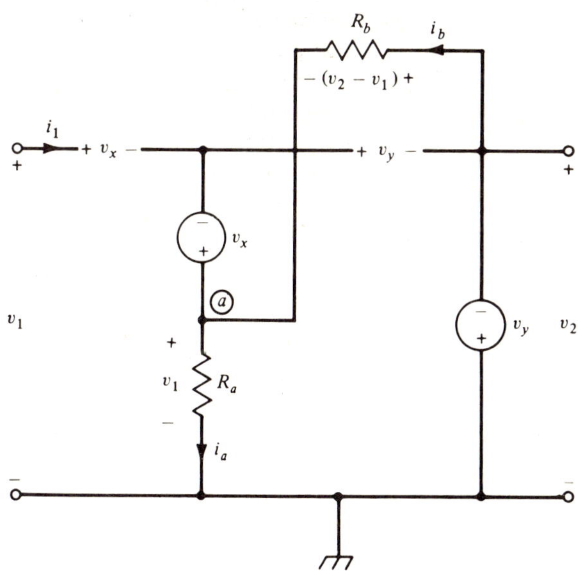

Figure 5-6

Now applying Kirchhoff's current law (KCL) at node $@$, we arrive at

$$(v_2 - v_1)/R_b = v_1/R_a$$

which when solved yields

$$A_v = v_2/v_1 = 1 + R_b/R_a \qquad (5\text{-}12)$$

Note that the input current i_1 is zero. Since the output voltage v_2 is developed across a voltage source, the circuit of Fig. 5-6 represents an ideal voltage amplifier. The two voltage-controlled voltage sources can be replaced by two unity-gain voltage amplifiers. As a result, this example also demonstrates the realization of a voltage amplifier with a gain greater than one using two unity-gain voltage amplifiers.

5-3   Voltage Amplifier

Out of the two basic amplifiers, the voltage amplifier is more frequently used in system design. Because of this, our discussion now and in the following chapters will concentrate mostly on the voltage amplifier. Nevertheless, in most cases, the results can be extended to current amplifiers as well.

Types of Voltage Amplifiers

The voltage amplifier of Fig. 5-4(a) is a 3-terminal amplifier, having one input port and one output port with a common ground. It is called a single-ended voltage amplifier and is often schematically represented by a triangle as shown in Fig. 5-7. The apex of the triangle is connected to the output terminal and the input terminal is connected to the base of the triangle. The common terminal between the input and the output is not explicitly shown, but both terminal voltages are measured with respect to that of the common terminal.

There are several other variations of the voltage amplifier that are used in system design. These amplifiers have either two input terminals and/or two output terminals. Figure 5-8(a) depicts the controlled-source model and the symbolic representation of a 2-input, single-output voltage amplifier. This amplifier responds to the difference of the input voltages and is also known as a *differential-input* or simply as a *difference* amplifier. Figure 5-8(b) shows the controlled-source model and the symbolic representation of a *differential-input, differential-output voltage amplifier* having two input and two output terminals. Finally, Fig. 5-8(c) shows a single-input, differential-output voltage amplifier, more commonly known as a *phase splitter*. In each case here, the common

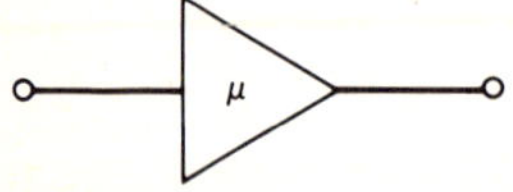

Figure 5-7   Symbolic representation of a single-ended voltage amplifier.

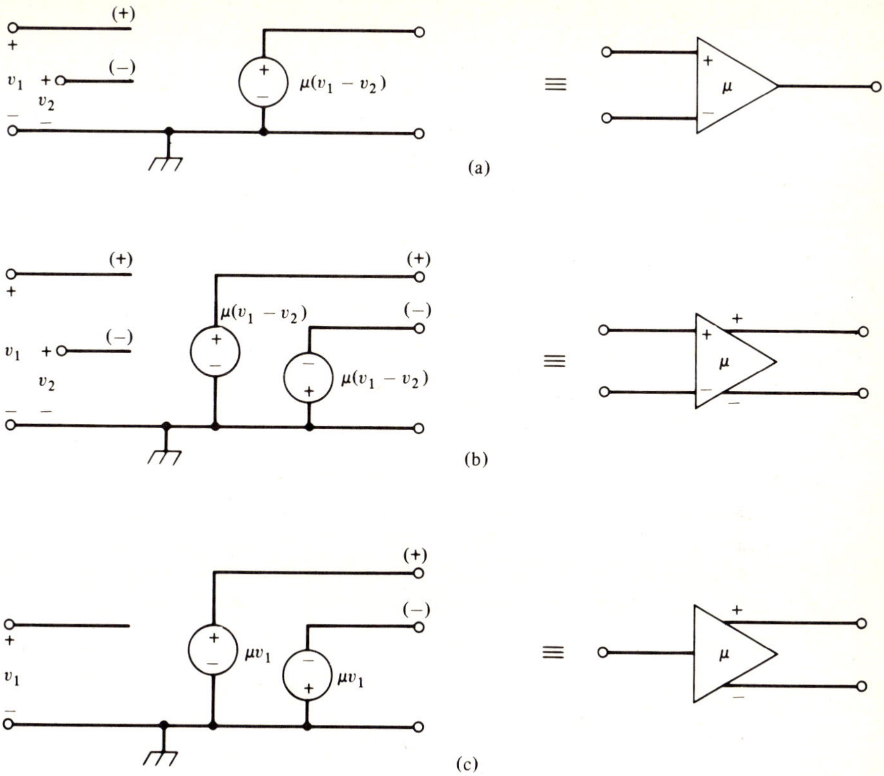

Figure 5-8 (a) Differential-input, single-ended output voltage amplifier, (b) differential-input, differential-output voltage amplifier, (c) phase splitter.

terminal between the input and the output ports is not explicitly shown in the schematic representation. Alternate symbolic representations of the voltage amplifiers of Fig. 5-8 are sketched in Fig. 5-9. Note that these symbols use the "circle" to indicate the inverting input and output terminals, commonly used in digital logic circuit symbols.

In the case of the single-ended voltage amplifier, the voltage gain μ can either be positive or negative. If $\mu > 0$, the polarity of the output voltage is the same

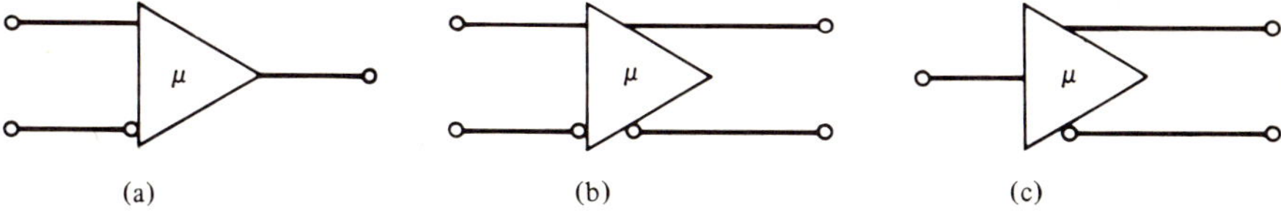

Figure 5-9 Alternate symbolic representations of (a) differential-input, single-ended output voltage amplifier, (b) differential-input, differential-output voltage amplifier, and (c) phase splitter.

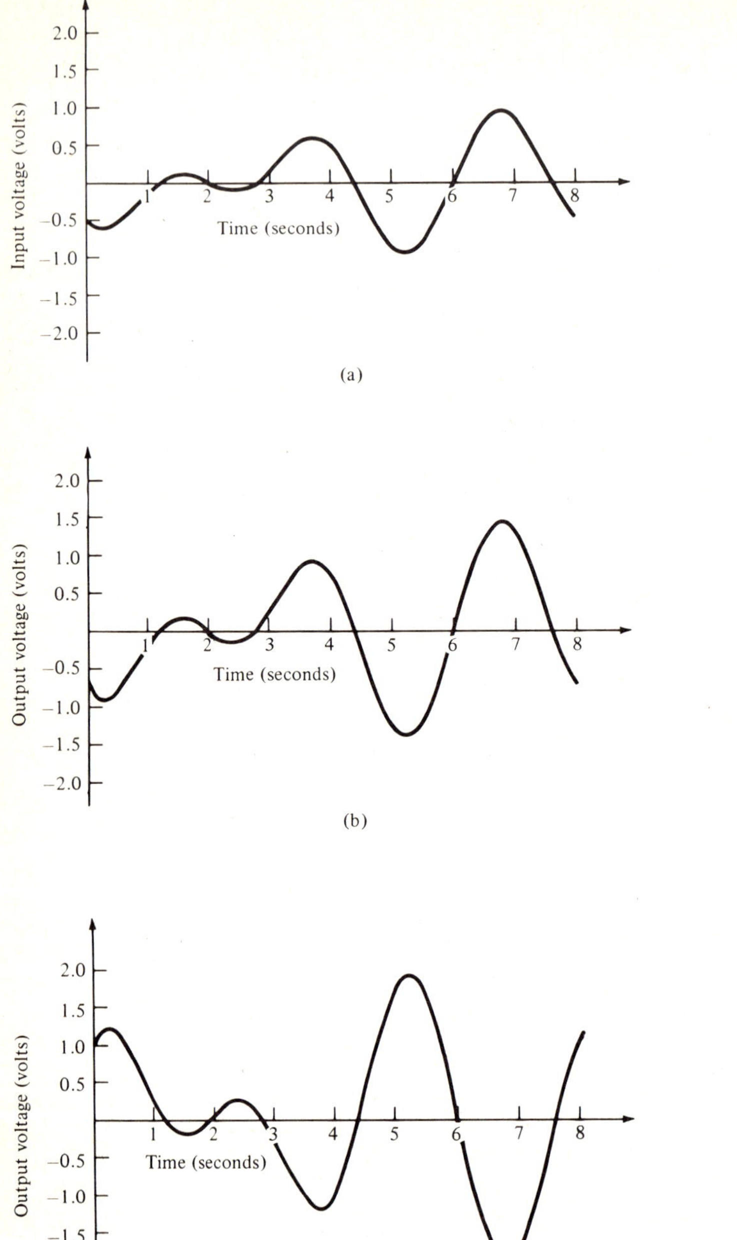

Figure 5-10 (a) Input waveform, (b) output from noninverting amplifier of gain 1.5, (c) output from inverting amplifier of gain 2.

as that of the input voltage and the amplifier is then called a *noninverting* voltage amplifier. If $\mu < 0$, the polarity of the output voltage is always opposite to that of the input voltage and the amplifier is called an *inverting* voltage amplifier. Thus, for sinusoidal input signals, the output voltage is *in phase* with the input voltage for a noninverting (positive-gain) voltage amplifier. For an inverting (negative-gain) voltage amplifier, the output voltage is 180° out of phase with the input voltage. When the gain of a noninverting amplifier is unity, the output voltage is an exact replica of the input voltage. This type of noninverting amplifier is known as a *voltage follower*. Figure 5-10 shows typical input–output waveforms for inverting and noninverting voltage amplifiers.

Transfer Characteristic

A graphical characterization of a voltage amplifier is by means of the input-output transfer characteristic as indicated in Fig. 5-11 for a noninverting unit where the output voltage v_2 has been plotted against the input voltage v_1. (In the case of a differential-input amplifier, the abscissa will represent the input difference voltage.) The slope of the voltage transfer characteristic is μ, which is constant for an ideal amplifier.

Cascaded Amplifiers

Often amplifiers are placed in a cascade as shown in Fig. 5-12. To understand the operation of the composite circuit, we note that

$$v_2 = \mu_1 v_1; \qquad v_3 = \mu_2 v_2$$

Hence

$$v_3 = \mu_1 \mu_2 v_1 \tag{5-13}$$

In other words, the two amplifiers in cascade behave like a single amplifier of voltage gain equal to $\mu_1\mu_2$. More than two amplifiers in cascade will lead to similar results. It is evident from Eq. (5-13) that the voltage gain in decibels of the cascaded amplifier will be the sum of the individual voltage gains in decibels. Thus, for example, if the first amplifier has a gain of 15 dB and the second has a gain of 20 dB, the overall gain will be 35 dB.

There are a number of applications of cascaded amplifiers. For example, if μ_2 and μ_1 are both negative, then their product is positive. This implies that we can design a noninverting voltage amplifier by cascading two inverting voltage

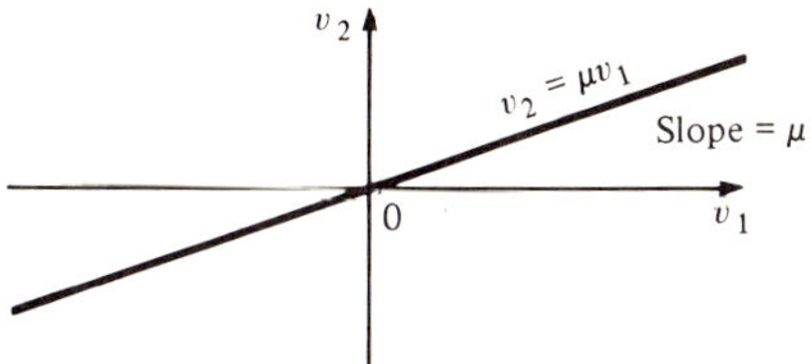

Figure 5-11 Transfer characteristic of an ideal noninverting amplifier.

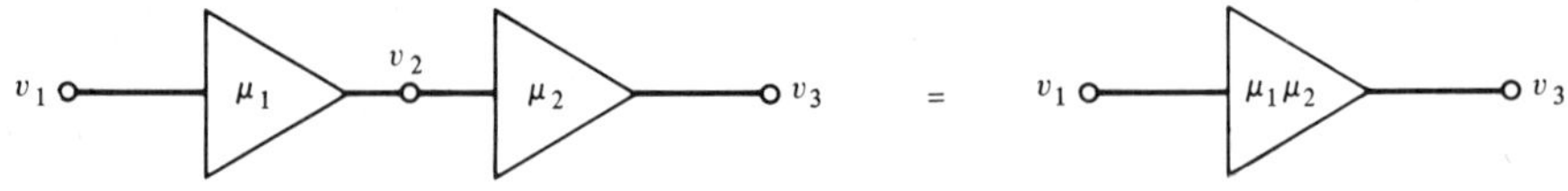

Figure 5-12   A cascaded amplifier and its equivalent.

amplifiers. Another application is in designing very high-gain amplifiers. For example, if $\mu_2 = \mu_1 = 10$, then their product is 100. Thus, by cascading two amplifiers of voltage gain 10, we can form a voltage amplifier of voltage gain 100. Theoretically at least, one can get almost any amount of gain by cascading a number of low-gain amplifiers.

The amplifier designed to provide both voltage and power gains such as the one used in a hi-fi system, is usually designed as a cascade of several amplifiers. Here the first stage, usually known as the *preamplifier* stage, provides the voltage amplification. The last stage is the power amplifier which provides the necessary power gain to drive the load.

5-4   Simple Applications

We indicated earlier that the amplifier is used to design many novel circuits and systems. Before we consider the practical amplifier and describe its properties, it would be profitable to point out several rather simple applications of this versatile device. Additional applications are described in later chapters.

Voltage Amplifier as a Buffer

A convenient way to design electronic systems is by cascading a series of 2-ports. If the 2-ports are noninteracting, then each 2-port can be designed and tested separately, making the design of the complete system much simpler. In some cases, the networks as designed may load or interact, leading to unsatisfactory operation of the complete system. One way to avoid the loading problem is to insert buffer or isolation amplifiers between the networks concerned. To illustrate the loading problem and its elimination with the aid of a buffer amplifier, consider the situation shown in Fig. 5-13(a) which shows a 10-V battery connected to a load resistance R_L. Because of the unavoidable internal resistance of the battery, the voltage developed across the load will not be 10 V but less than that. The actual value will depend on the value of the load resistance and also on the internal resistance of the battery. For example, if $R_L = 45 \ \Omega$ and $R_i = 5 \ \Omega$, the voltage v_2 developed will be 9 V. If $R_L = 15 \ \Omega$ instead, v_2 will be 7.5 V. In many cases, the exact value of the load resistance may not be known as it may vary with time or some other quantities. To avoid this type of loading problem (and uncertainty) we can insert a voltage amplifier between the source and the load as shown in Fig. 5-13(b). A commonly used buffer is the voltage follower which is available in monolithic IC form. As can be seen from Fig. 5-13(b), the voltage across the load is now maintained at 10 V, independent of the value of the load resistance.

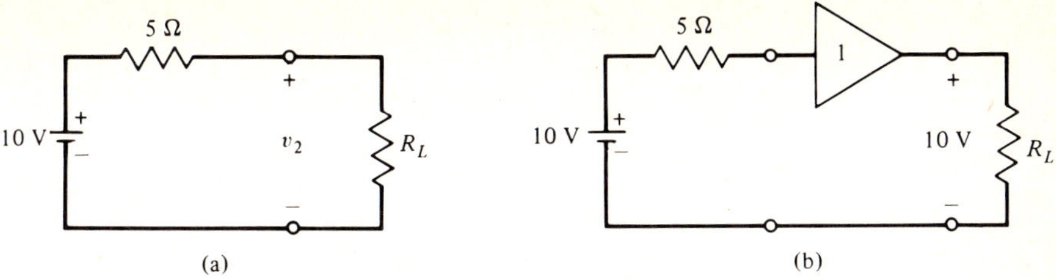

Figure 5-13   Illustration of the loading problem and its elimination using buffer amplifiers.

Simulation of an Inductor

Fabrication of circuits containing inductors in monolithic integrated circuit form has proved to be impossible. Even microminiaturization of an inductor comparable in size with available integrated circuits has not been very satisfactory. An alternate approach to solve this problem has been to develop inductorless circuits that simulate inductors. One such circuit arrangement[2] that uses a voltage follower is shown in Fig. 5-14(a). Applying KCL to node $\textcircled{a}$ we obtain first

$$\frac{v_1 - v_2}{R_2} = \frac{v_2}{R_1} + C\frac{d}{dt}(v_2 - v_1) \tag{5-14}$$

The input current i_1 flows through R_2 and hence

$$i_1 = \frac{v_1 - v_2}{R_2} \tag{5-15}$$

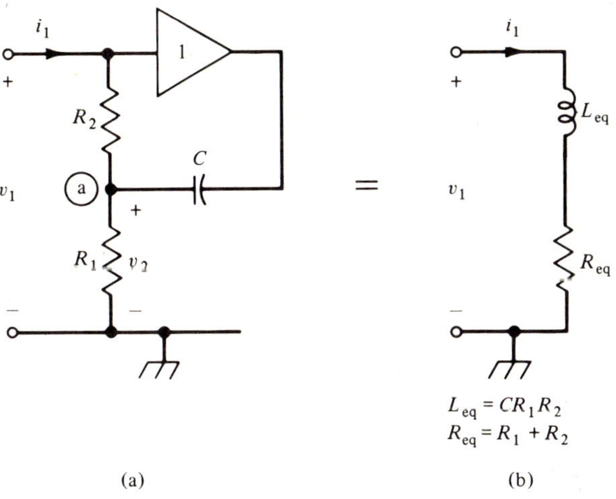

$$L_{eq} = CR_1 R_2$$
$$R_{eq} = R_1 + R_2$$

Figure 5-14   Simulation of an inductor: (a) circuit arrangement, and (b) equivalent model.

Eliminating v_2 from Eqs. (5-14) and (5-15) we arrive at

$$v_1 = (CR_1 R_2)\frac{di_1}{dt} + (R_1 + R_2)i_1 \qquad (5\text{-}16)$$

which indicates that the circuit of Fig. 5-14(a) behaves as a lossy inductor having an inductance value of $CR_1 R_2$ H and a resistance of $(R_1 + R_2)\,\Omega$ as shown in Fig. 5-14(b).

A Negative Resistance Circuit

Another application of a noninverting voltage amplifier is in the realization of a negative resistance as indicated in Fig. 5-15. To analyze this circuit we observe that the voltage across the resistor R is $(v_1 - \mu v_1)$ which causes a current $(1 - \mu)v_1/R$ to flow toward the output of the amplifier. Hence, applying KCL at the input of the amplifier, we arrive at

$$i_1 = \frac{(1 - \mu)v_1}{R}$$

As a result, the input impedance $v_1/i_1 = R_{\text{in}}$ of this circuit is

$$R_{\text{in}} = -R/(\mu - 1) \qquad (5\text{-}17)$$

which is negative for $\mu > 1$.

Capacitance Multiplier

Capacitors with large capacitance values tend to be bulky and are thus not attractive in circuit design. The circuit arrangement of Fig. 5-16(a), which employs a voltage follower, is suitable for converting a moderate-valued capacitor to larger values. To determine the input–output relations, we analyze the network. Applying KCL at nodes $@$ and $ⓑ$ we arrive at

$$i_1 = \frac{v_1 - v_2}{R_1} + \frac{v_1 - v_2}{R_2}$$

$$\frac{v_1 - v_2}{R_2} = C\frac{dv_2}{dt}$$

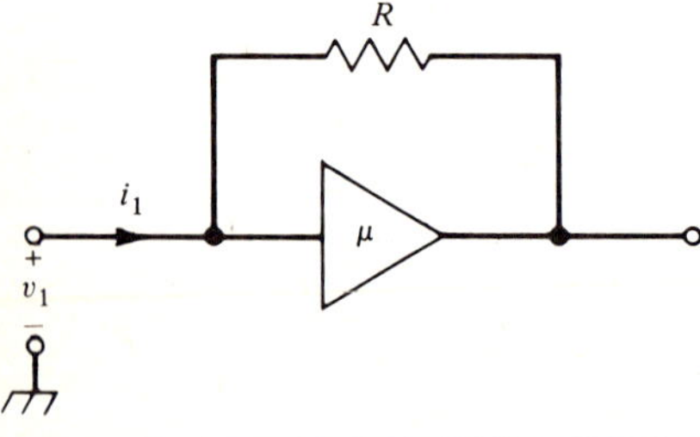

Figure 5-15 Simulation of a negative resistance.

Eliminating v_2 from the above equations we arrive at

$$v_1 = \frac{R_1}{C(R_1 + R_2)} \int_{-\infty}^{t} i_1 \, d\tau + \frac{R_1 R_2}{R_1 + R_2} i_1 \tag{5-18}$$

The above expression indicates that between the input terminal and ground, the circuit of Fig. 5-16(a) behaves as a lossy capacitor of capacitance value C_{eq} in series with a resistance R_{eq} as shown in Fig. 5-16(b). The values of these elements for $R_2 \gg R_1$ are

$$C_{eq} = \frac{C(R_1 + R_2)}{R_1} \simeq \frac{CR_2}{R_1}$$

$$R_{eq} = \frac{R_1 R_2}{R_1 + R_2} \simeq R_1$$

Typical values of the resistances are $R_1 = 1\ \text{k}\Omega$ and $R_2 = 10\ \text{M}\Omega$.

Voltage Follower as a Driver

A voltage follower capable of providing a large-valued output current is often used to *drive* a pulse transformer or a long transmission line. This type of voltage follower is usually called a *current driver* or a *current amplifier*, the latter name being a misnomer. The circuit arrangements for these two applications are sketched in Fig. 5-17.[13]

In Fig. 5-17(a) the pulse transformer is characterized by the following input–output relations:

$$v_1 = nv_2$$
$$i_2 = -ni_1$$

where n is the *turns ratio*. If the network connected at the output of the transformer presents an input impedance R_L, then the impedance seen by the output

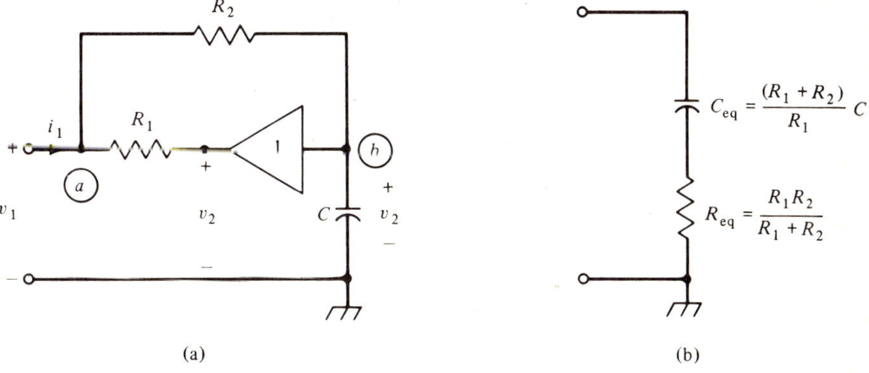

(a)

(b)

Figure 5-16   A capacitance multiplier: (a) circuit schematic, and (b) equivalent circuit.

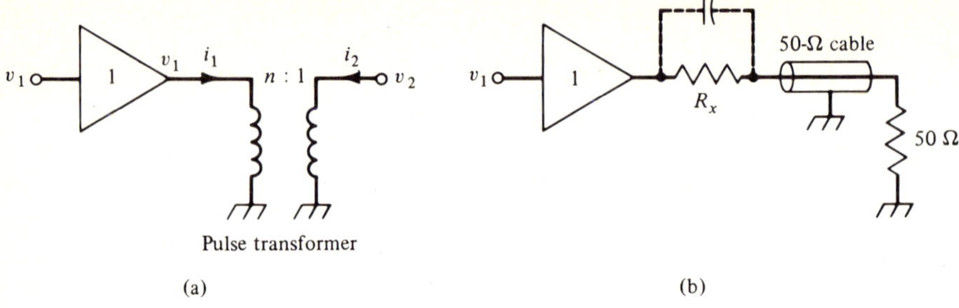

Figure 5-17   (a) Voltage follower as a driver for a pulse-transformer, and (b) voltage follower as a transmission-line driver.

of the voltage follower is $n^2 R_L$. Since $v_2 = v_1/n$, by choosing the turns ratio appropriately, the pulse transformer can be used to change the amplitude and the impedance level of the input pulse. If $n < 0$, then the polarity of the pulse can also be reversed. In addition, by choosing a pulse transformer with a center-tapped winding at the output, both positive and negative pulses can be produced simultaneously.

In the transmission-line driver arrangement of Fig. 5-17(b), the resistor R_x is used to *match* the output impedance of the practical amplifier to the load and the coaxial cable. Thus, if the output impedance of the amplifier is 10 Ω, the value of R_x should be 40 Ω. The matching is necessary to minimize reflections. A capacitor is often connected across R_x to adjust the time-domain response of the output signal.

5-5   Characteristics of a Practical Amplifier

So far we have assumed the amplifier to be ideal, characterized by a single parameter relating the output signal to the input signal. An electronic amplifier circuit in practice is a nonideal device. We shall now describe the sources of nonidealness in a practical amplifier and point out their effects on the performance of the device.

Voltage and Current Limitations

A practical amplifier has certain voltage and current limitations both at the input and at the output which, if exceeded, cause the amplifier to operate in a nonlinear mode leading to waveform distortion and may in some cases damage the amplifier permanently. It is extremely important that these limitations be strictly observed. Depending on the type of amplifier and its intended application, these limitations are specified in different ways. Some commonly specified limitations are as follows:

Output Voltage Limit.   It is the maximum peak-to-peak output voltage swing referred to zero that can be observed without introducing nonlinear distortion in the waveform. This limit is also called the *maximum peak-to-peak output voltage* or *full-scale* output.

Input Voltage Range.   It is the range of voltages at the input terminals within which the amplifier operates as specified. For differential-input amplifiers, there are two types of input voltage range. The maximum difference voltage that can be applied across the two input terminals is its *differential-input voltage limit.* The maximum peak voltage that can be applied to either terminal with respect to ground is the *common-mode input voltage limit* or simply the *input voltage limit.*

Output Current Limit.   It is the peak-to-peak output current that the amplifier can deliver to the load without introducing distortion. *Rated output* is defined as the maximum values of output voltage and current swings that can be simultaneously maintained.

It is recommended that for a specific module, the manufacturer's specification sheet be consulted initially to determine applicable limitations.

Input and Output Resistances

The input and output resistances of a practical amplifier are finite and nonzero. The nonideal models of the two types of amplifiers showing the input and the output resistances are sketched in Fig. 5-18. In this figure R_i represents the input resistance and R_o is the output resistance which take different values in the two types of amplifiers.

The input resistance of a practical voltage amplifier is usually very large and the output resistance is small. For example, the National Semiconductor monolithic voltage follower LM310 has an input resistance of about 10^{12} Ω and an output resistance of less than 1 Ω. On the other hand, the input resistance of a practical current amplifier is very small, of the order of a few ohms, and its output impedance is very large, usually greater than 10 kΩ.

It should be noted that the simplified nonideal models shown in Fig. 5-18 still assume that the amplifiers are unilateral, that is, the electrical conditions at the input are not affected by those at the output.

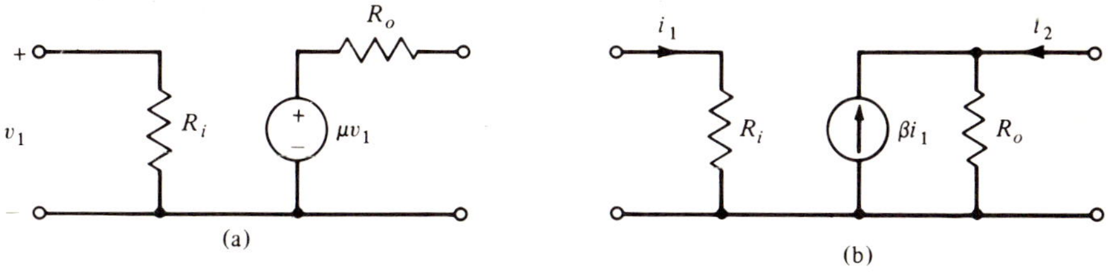

Figure 5-18   Models of practical amplifiers: (a) voltage amplifier, and (b) current amplifier.

Frequency Dependence of Amplifier Parameters

The parameters of a practical amplifier in general are not real and constants, but functions of frequency. The frequency dependence is a result of delay of charge carriers in the transistors, parasitic capacitances, and coupling and bypass capacitors in the amplifier circuit. For low-frequency applications, these parameters can be considered real in most cases. On the other hand, for high-frequency applications it is important to know the exact dependence of the parameters on the frequency. These are normally supplied graphically by the manufacturers.

Due to the frequency dependence of the amplifier parameters, the voltage and current gains of an amplifier are, in general, functions of the complex frequency variable s, and are thus complex numbers for a specified value of the frequency. If the input voltage is of the form $\mathbf{V}_1 e^{st}$, then the output voltage of the amplifier is of the form $\mathbf{V}_2 e^{st}$. The voltage gain, to be denoted as $A_v(s)$, now is defined as

$$A_v(s) = \mathbf{V}_2/\mathbf{V}_1 \qquad (5\text{-}19)$$

For sinusoidal signal frequencies, it is convenient to express the gain in terms of its magnitude and phase. Thus, for $s = j\omega$, we write

$$A_v(j\omega) = |A_v(j\omega)| \cdot e^{j\phi(\omega)} \qquad (5\text{-}20)$$

where $|A_v(j\omega)|$ is the *magnitude* of the voltage gain and $\phi(\omega)$ is its *phase*. Both magnitude and phase functions are again in practice functions of frequency. A plot of the magnitude and phase of the amplifier gain as a function of frequency is known as the *frequency response* of the amplifier. Figure 5-19 shows the frequency response of a typical voltage amplifier.

The current gain of a practical amplifier in general will also be a function of the complex frequency variable s and will be denoted as $A_i(s)$. For sinusoidal

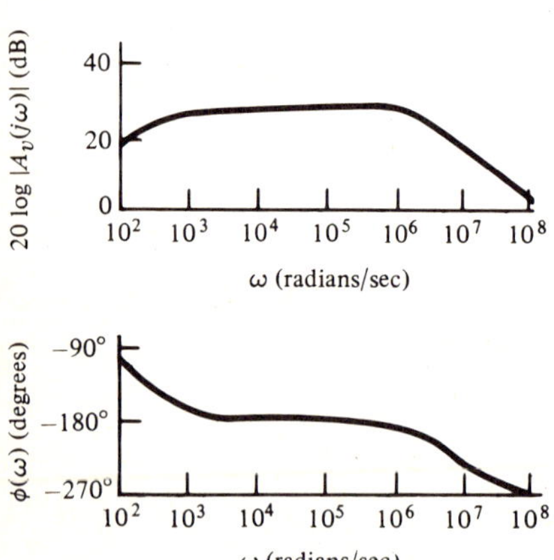

Figure 5-19   Frequency response of a typical voltage amplifier.

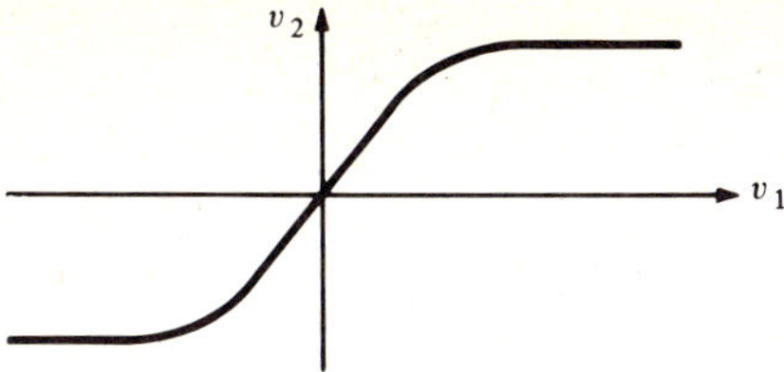

Figure 5-20 Nonlinear transfer characteristic of a practical amplifier.

applications, it is again convenient to express the current gain in terms of its magnitude and phase:

$$A_i(j\omega) = |A_i(j\omega)| \cdot e^{j\psi(\omega)} \tag{5-21}$$

where $|A_i(j\omega)|$ is the magnitude of the current gain and $\psi(\omega)$ is its phase.

Because of the frequency dependence of the gain of a practical amplifier, if the amplifier is excited by an input signal composed of a number of different frequency components, each component may be amplified by different amounts, causing the output waveform to be a distorted version of the input waveform. This type of distortion is called *frequency distortion* and is examined at length in the next section.

Frequency dependence of the amplifier parameters also affects the amplifier's behavior in the time domain. This aspect is discussed in Section 5-7.

Nonlinear Gain Characteristic

Another source of nonidealness in a practical amplifier is the nonlinear relation between the input and output signals. A typical input–output characteristic of a practical voltage amplifier appears as shown in Fig. 5-20, instead of the ideal straight-line characteristic of Fig. 5-11. From this curve we observe that for signals with small amplitude variations, the characteristic is essentially linear and the amplifier can be considered to be operating as a linear device. For signals with large voltage or current swings, the nonlinear characteristic cannot be ignored because different amplitude input signals are amplified by different amounts leading to distortion in the output waveform. This type of distortion is called *nonlinear distortion* (see Problems 5-33 and 5-34).

5-6   Frequency Distortion

Let us now investigate in detail the effect of the frequency response of the amplifier gain on its steady-state response. To illustrate the effect we consider here a voltage amplifier with a voltage gain $A_v(s)$. For sinusoidal steady-state analysis, it is more convenient to express the complex voltage gain in terms of its magnitude and phase, as indicated by Eq. (5-20). Now if the input voltage $v_1(t)$ is a sum of two cosinusoidal functions, say

$$v_1(t) = \alpha \cos \omega_a t + \beta \cos \omega_b t \tag{5-22}$$

then the steady-state response of the amplifier will be

$$v_2(t) = \alpha |A_v(j\omega_a)| \cos \{\omega_a t + \phi(\omega_a)\} + \beta |A_v(j\omega_b)| \cos \{\omega_b t + \phi(\omega_b)\} \quad (5\text{-}23)$$

It follows from the above that the waveform of $v_2(t)$ will be an amplified replica of $v_1(t)$ for all possible values of input frequencies if the magnitude of the gain satisfies the condition

$$|A_v(j\omega)| = K \qquad (5\text{-}24)$$

and the phase satisfies the condition

$$\phi(\omega) = m\pi - \omega\tau \qquad (5\text{-}25)$$

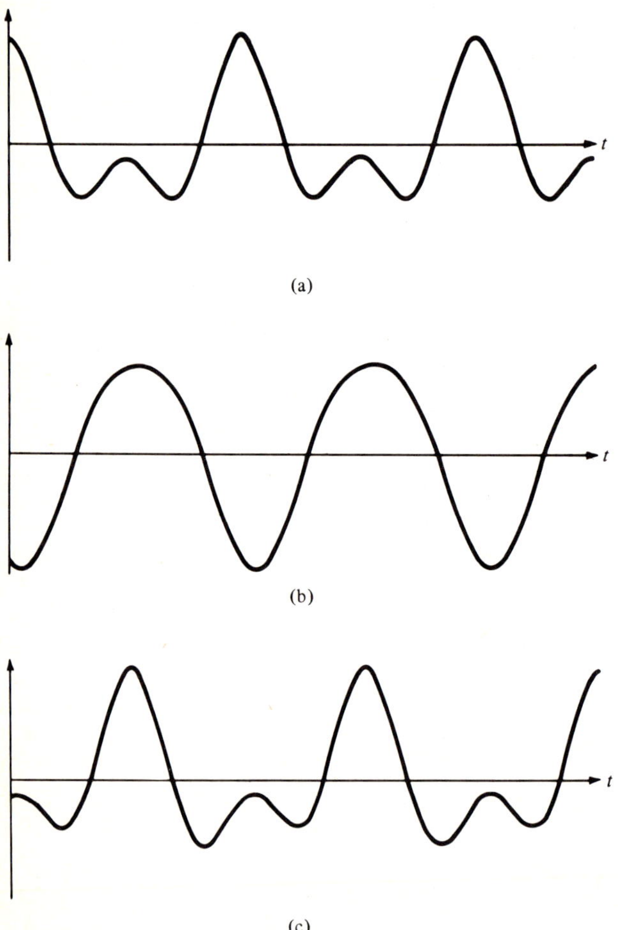

Figure 5-21 Typical waveforms illustrating amplitude and phase distortion: (a) input signal, (b) amplitude distorted output, and (c) phase distorted output.

where K and τ are positive constants and m is an integer. Substituting Eqs. (5-24) and (5-25) in Eq. (5-23) we arrive at

$$v_2(t) = K[\alpha \cos(\omega_a t - \omega_a \tau + m\pi) + \beta \cos(\omega_b t - \omega_b \tau + m\pi)]$$
$$= \pm K v_1(t - \tau) \tag{5-26}$$

We can now generalize the above results. In order that the waveform of the amplifier output be an amplified replica of the input waveform, the following two conditions must be met over the frequency range of interest:

1. the magnitude of the amplifier gain be a constant independent of frequency, and
2. the phase shift varies linearly with frequency as given by Eq. (5-25).

The frequency range over which the above two conditions are approximately satisfied is roughly the *bandwidth* or the useful band of the amplifier. If the above two conditions are not met, then the amplifier introduces *frequency distortion* in the output response. If condition (1) is not met, then the amplifier response has *amplitude distortion*. Condition (2), if not met, gives rise to *phase distortion*. Figure 5-21 illustrates the effect of these two types of distortion on the output waveform.

5-7 Transient Response

As indicated earlier, the unavoidable frequency dependence of amplifier parameters affects its time-domain response. For an amplifier intended for processing pulses, the time-domain response of the amplifier is of interest. Since a square pulse can be generated by taking the difference of a step function and its delayed version, the response of the amplifier to a step input also provides the circuit designer with enough information regarding the pulse response of the amplifier. The amplitude of the input step is chosen to yield a full-scale output. The response of the amplifier to such a step is also known as its *transient response*. A typical step response for an input step applied at time $t = t_0$ is sketched in Fig. 5-22. The response is characterized by several parameters.

Overshoot. The oscillations in the step-response waveform around the final steady-state value is in essence the overshoot. An estimate of the *overshoot* is given by the difference between the maximum value of the step response V_m and the steady-state value V_f and is specified as a percent of the steady-state value. Thus,

$$\text{Overshoot} = \left| \frac{V_m - V_f}{V_f} \right| \times 100 \text{ percent} \tag{5-27}$$

The maximum error in the transient response following the initial rise (or fall) is identified by the amount of overshoot.

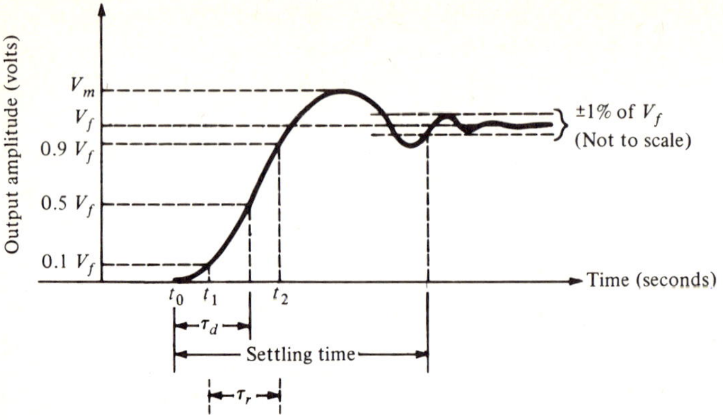

Figure 5-22   Typical response of an amplifier for a step input applied at time t_0.

Rise Time.   The time taken by the output to reach the final value from the time the unit step is applied at the input of the amplifier is roughly the rise time. There are actually a number of definitions of rise time. The most convenient and widely used definition is 10–90 percent rise time. Refer to Fig. 5-22. If at t_1, $v_2(t_1) = 0.1V_f$ and at time t_2, $v_2(t_2) = 0.9V_f$, where V_f is the final value of $v_2(t)$, then the *rise time* τ_r is given as

$$\tau_r = t_2 - t_1 \text{ sec} \qquad (5\text{-}28)$$

Delay Time.   The time taken by the response to reach 50 percent of its final value from the instant the step is applied is defined as the *delay time*, normally denoted by τ_d.

Settling Time.   In some cases, the time taken by the step response to settle to its final value is a useful measurement of the time-domain behavior of the amplifier. Strictly speaking, this time is infinite. A more convenient measure is the time taken by the step response to settle to within a specified percentage of the final rated value. This time period is defined as the *settling time*. Usually the specified tolerance is 0.01–0.1 percent. Figure 5-22 illustrates the concept of the settling time.

Slew Rate.   We have indicated before that a practical amplifier does not respond immediately to a step input. A similar situation arises when a large amplitude high-frequency square-wave signal is applied. The internal capacitances in the amplifier require some time to charge and discharge; consequently the output waveform is unable to follow the input. The maximum time rate at which output can change or *slew* when supplying the rated output is known as its *slew rate S* and is usually specified in volts/μsec. A simple way to measure the slew rate is to apply a high-frequency square wave whose amplitude is large enough to overdrive the output beyond its rated level in both directions. The slope of the output waveform between the two clipped levels determines the slew rate. If the slew rate for a positive-going edge is different from that of the negative-going edge, then the lower slew rate value is specified. Because of slew rate limitations, a very large amplitude high-frequency input signal will appear distorted at the output. A simple inequality relating the slew rate to the frequency and amplitude of the output can be derived. For a sine wave

$$v_2(t) = V_m \sin 2\pi f_o t$$

the maximum time rate of change occurs at points where $v_2(t) = 0$. It is easy to show that

$$\left. \frac{\Delta v_2}{\Delta t} \right|_{\max} = 2\pi f_o V_m \qquad (5\text{-}29)$$

and hence the slew rate must satisfy the condition

$$S \geq 2\pi f_o V_m \qquad (5\text{-}30)$$

Note that if the peak amplitude V_m is in volts and the frequency is in megahertz, then the slew rate will be in volts/μsec.

> ***Example 5-2.***   An IC amplifier is characterized by a slew rate of 5 V/μsec. Calculate the highest signal frequency that will result in an undistorted output of peak amplitude of (a) 4 V and (b) 1 V.
>
> From Eq. (5-30)
>
> $$f_o \leq \frac{S}{2\pi V_m}$$
>
> For part (a)
>
> $$\left. f_o \right|_{\max} = \frac{5}{2\pi \times 4} = 198.9 \text{ kHz}$$
>
> For part (b)
>
> $$\left. f_o \right|_{\max} = \frac{5}{2\pi \times 1} = 795.8 \text{ kHz}$$

This example points out that the waveform distortion due to slew rate limitation can be eliminated either by decreasing the signal frequency or by decreasing the input signal amplitude or both.

5-8   Performance of the Terminated Amplifier

In the previous several sections we pointed out a number of sources of non-idealness in a practical amplifier and outlined in general terms how some of the nonideal characteristics affect the performance of the amplifier. One of the

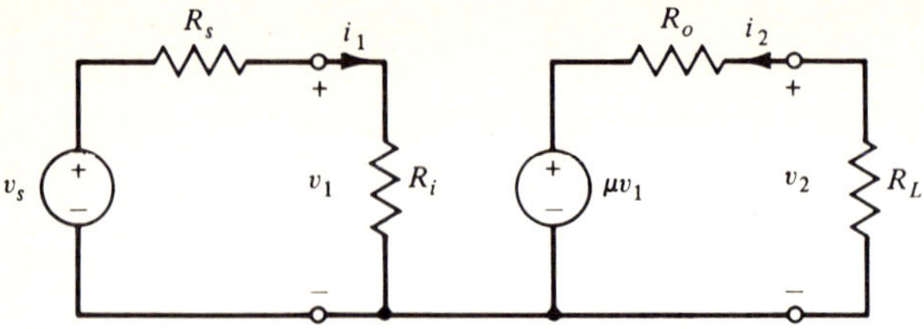

Figure 5-23   A terminated nonideal voltage amplifier.

amplifier characteristics mentioned was the finite nonzero values of the input and output resistances. It was further mentioned that due to the presence of the input and output resistances, the constant of proportionality relating the output signal to the input signal was dependent on the terminating resistances seen by the amplifier input and output terminals.

Let us now examine this problem more precisely. Figure 5-2(b) shows the terminated practical voltage amplifier. The input is connected to a voltage source v_s with an internal resistance R_s, and the output is terminated by a load resistance R_L. Replacing the amplifier by its nonideal model of Fig. 5-18(a), we arrive at the circuit equivalent shown in Fig. 5-23. The performance of the terminated amplifier is characterized by a number of system parameters which are evaluated next.

Voltage Gain A_v.   The ratio of the output voltage to the input voltage is the voltage gain. From Fig. 5-23, by the voltage-divider relation

$$v_2 = \frac{R_L}{R_o + R_L} \cdot \mu v_1$$

Hence

$$A_v = \frac{v_2}{v_1} = \frac{\mu R_L}{R_o + R_L} \tag{5-31}$$

Current Gain A_i.   The current gain of the terminated amplifier is the ratio of the output current to the input current. From Fig. 5-23,

$$i_1 = v_1/R_i$$

$$i_2 = -v_2/R_L$$

Thus

$$A_i = \frac{-i_2}{i_1} = \frac{R_i}{R_L}\left(\frac{v_2}{v_1}\right) = \frac{\mu R_i}{R_o + R_L} \tag{5-32}$$

Power Gain A_p.   The ratio of output power P_o to the input power P_i is defined as the power gain. Substituting Eqs. (5-31) and (5-32) in Eq. (5-8) we obtain

$$A_p = \frac{\mu R_L}{R_o + R_L} \cdot \frac{\mu R_i}{R_o + R_L} = \frac{\mu^2 R_i R_L}{(R_o + R_L)^2} \tag{5-33}$$

which can also be expressed as

$$A_p = \frac{R_i}{R_L}\left(\frac{\mu R_L}{R_o + R_L}\right)^2 = \left(\frac{R_i}{R_L}\right)(A_v)^2 \tag{5-34}$$

or as

$$A_p = \frac{R_L}{R_i}\left(\frac{\mu R_i}{R_o + R_L}\right)^2 = \left(\frac{R_L}{R_i}\right)(A_i)^2 \tag{5-35}$$

If we express the power gain in decibels, then from Eqs. (5-34) and (5-35) we arrive at

$$10 \log_{10} A_p = 20 \log_{10} |A_v| + 10 \log_{10}\left(\frac{R_i}{R_L}\right)$$

$$= 20 \log_{10} |A_i| + 10 \log_{10}\left(\frac{R_L}{R_i}\right) \tag{5-36}$$

In many communication circuits, $R_i = R_L$. Under this condition, as indicated by Eq. (5-36), the power gain, voltage gain, and current gain in decibels are all equal.

Input and Output Resistances. From Fig. 5-23 it follows that the input resistance R_{in} of the terminated amplifier is simply R_i and the output resistance R_{out} is equal to R_o. That is,

$$R_{\text{in}} = \frac{v_1}{i_1} = R_i \tag{5-37}$$

$$R_{\text{out}} = \frac{v_2}{i_2}\bigg|_{v_s = 0} = R_o \tag{5-38}$$

System Voltage Gain K_v. In some applications, the ratio of the output voltage v_2 to the source voltage v_s is of interest. This ratio defines the overall system voltage gain and is also called the composite voltage gain. From Fig. 5-23,

$$v_1 = \frac{R_i}{R_s + R_i} \cdot v_s \tag{5-39}$$

From Eqs. (5-31) and (5-39) we readily obtain

$$K_v = \frac{v_2}{v_s} = \mu\left(\frac{R_i}{R_s + R_i}\right) \cdot \left(\frac{R_L}{R_o + R_L}\right) \tag{5-40}$$

Let us now examine the expression for the system voltage gain as given by Eq. (5-40). Since $(R_o + R_L) \geq R_L$ and $(R_i + R_s) \geq R_i$, the system voltage gain K_v is always less than or equal to μ, the voltage-amplification factor of the amplifier. In other words, the system voltage gain can never exceed μ assuming of course that all of the pertinent resistances are positive. To maximize the value of K_v, the terms inside the parentheses in Eq. (5-40) should be maximized. Maximum value of $R_i/(R_i + R_s)$ is equal to one if either R_i is infinite or R_s is zero. Similarly, maximum value of $R_L/(R_L + R_o)$ is one if either R_L is infinity or R_o is zero. Normally, the circuit designer does not have too much choice in specifying the values of R_s and R_L. Hence, for an optimum result, the amplifier should be designed such that $R_i \gg R_s$ and $R_o \ll R_L$. For fixed R_L, R_o, and μ, K_v is determined by the ratio R_i/R_s. Thus, if R_s is changed from a value of 10 Ω to 1 kΩ, then to keep K_v invariant, the input resistance of the amplifier should increase 100 times. Similarly, for fixed R_i, R_s, and μ, K_v is invariant if the ratio R_L/R_o remains constant.

Example 5-3. Consider an amplifier having an input impedance R_i equal to 1 MΩ, an output impedance R_o of value 100 Ω, and a gain μ of value 10. The amplifier is terminated by a load resistance R_L equal to 600 Ω. Let us calculate the various gains.

The voltage gain is

$$A_v = \frac{\mu R_L}{R_o + R_L} = \frac{6000}{700} = 8.57$$

In decibels,

$$\mathcal{G}_v = 20 \log_{10} |A_v| = 20 \log_{10} (8.57) = 18.66 \text{ dB}$$

The current gain for this amplifier is

$$A_i = \frac{R_i}{R_L} A_v = \frac{10^6}{600} \cdot \frac{60}{7} = 14{,}286$$

which in decibels is

$$\mathcal{G}_i = 20 \log_{10} |A_i| = 20 \log_{10} (14{,}286) = 83.10 \text{ dB}$$

The power gain in decibels is therefore, from Eq. (5-8),

$$\mathcal{G}_p = \tfrac{1}{2}(\mathcal{G}_v + \mathcal{G}_i) = \tfrac{1}{2}(83.10 + 18.66) = 50.879 \text{ dB}$$

The device is thus a better current amplifier than a voltage amplifier.

Usually an IC amplifier is designed to work under wide variations in terminal conditions to increase its applicability in a large number of system design problems. The following example illustrates the development of design criteria for such amplifiers.

Example 5-4. An IC amplifier is to be designed to work with sources whose internal resistances are in the range 10 Ω to 1 kΩ and whose load resistances have values in the range 10–100 kΩ. It is required that the system voltage gain does not change by more than 10 percent for any combination of input and load resistances in this range.

For fixed R_i and R_o, the maximum value of K_v occurs for $R_s = 10$ and $R_L = 10^5$, and the minimum value of K_v is at $R_s = 10^3$ and $R_L = 10^4$. So, we require that the worst-case inequality,

$$\mu \cdot \frac{R_i}{R_i + 1000} \cdot \frac{10^4}{10^4 + R_o} \geq (0.9)\mu \cdot \frac{R_i}{R_i + 10} \cdot \frac{10^5}{10^5 + R_o}$$

be satisfied. The above expression simplifies (assuming the worst case) to

$$(R_i + 10)(10^5 + R_o) = 9(R_i + 10^3)(10^4 + R_o)$$

Since there are two unknowns and one equation, there is no unique answer. One way to solve this is to assume a value for one variable and solve for the other. Thus, if we set $R_o = 100$ and solve for R_i, we obtain

$$R_i = \frac{898,990}{92} = 9771.6 \ \Omega$$

This implies that if the amplifier is designed with an output impedance less than or equal to 100 Ω and an input impedance greater than or equal to 9772 Ω, then the composite voltage gain will not change more than 10 percent from its maximum value.

5-9   Frequency Limitations

We indicated before that the effect of the frequency dependence of the amplifier parameters is to make the voltage gains and other system parameters functions of frequency which causes the frequency distortion of signals. It was shown also that to minimize the frequency distortion, the magnitude of the amplifier gain must be constant and the associated phase shift must be a linear function of frequency over the frequency range of interest. We have shown how the frequency response plots provide qualitative information on the steady-state behavior of the amplifier for sinusoidal excitations. The time-domain behavior of the amplifier can also be determined from its frequency response.

In this section we examine in some detail a very simple type of gain function that is characterized by a single pole. There are a number of reasons for considering these types of gains. First, many practical amplifiers are specifically designed to have single-pole gain representation. These devices are usually the very high-gain dc amplifiers which are invariably used with negative feedback from the output to the input and the single-pole model ensures the stability of the feedback amplifier. Second, the relation between the time-domain behavior and the frequency response can be easily derived for such amplifiers. Finally, a number of characteristics of more complex gain functions can be obtained by straightforward extension of the results of the single-pole case.

A Single-Pole Gain Model

The gain of a voltage amplifier represented by a single-pole model can be written as

$$A_v(s) = \frac{V_2}{V_1} = \frac{K_o \omega_o}{s + \omega_o} \tag{5-41}$$

where $\omega_o > 0$. The corresponding magnitude and phase functions are

$$|A_v(j\omega)| = \frac{K_o \omega_o}{\sqrt{\omega^2 + \omega_o^2}} \tag{5-42}$$

$$\phi(\omega) = -\tan^{-1}(\omega/\omega_o) \tag{5-43}$$

A typical plot of the above magnitude and phase functions is given in Fig. 5-24. It is seen that both the magnitude and phase are monotonically decreasing functions of ω. At $\omega = 0$, the magnitude of the gain is a maximum and is equal to

$$|A_v(j0)| = K_o \tag{5-44}$$

K_o is thus the *dc gain*. At $\omega = \omega_o$,

$$|A_v(j\omega_o)| = \frac{K_o \omega_o}{\sqrt{2\omega_o^2}} = \frac{K_o}{\sqrt{2}} = 0.707|A_v(j0)| \tag{5-45}$$

That is, at $\omega = \omega_o$, the magnitude is 0.707 times the dc gain. In decibels,

$$20 \log_{10} |A_v(j\omega_o)| = 20 \log_{10} |A_v(j0)| + 20 \log_{10} (1/\sqrt{2})$$

$$\cong 20 \log_{10} |A_v(j0)| - 3 \text{ dB} \tag{5-46}$$

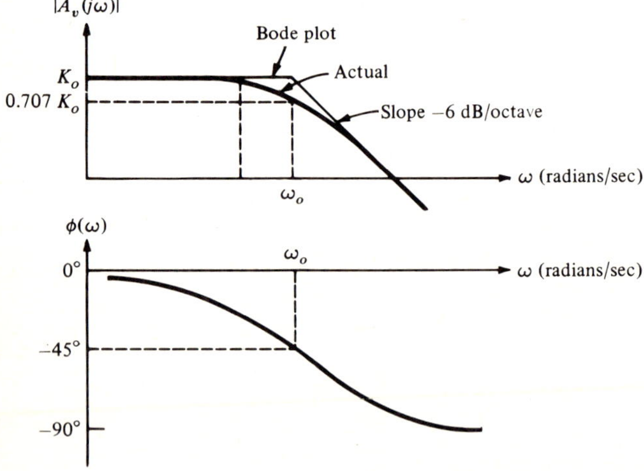

Figure 5-24   The frequency response of a single-pole amplifier.

Hence at $\omega = \omega_o$, the gain is 3 dB less than that at dc. ω_o is called the *3-dB cut-off angular frequency* in radians/sec.

Example 5-5. Consider the amplifier circuit of Fig. 5-25 where the capacitance C_o represents the total parasitic capacitances across the output port. Determine the 3-dB cut-off frequency of the amplifier for the following parameter values:

$$R_i = 5 \text{ k}\Omega, \quad R_o = 100 \ \Omega, \quad R_L = 1 \text{ k}\Omega, \quad \mu = 2, \quad C_o = 10 \text{ pF}$$

The complex voltage gain of the amplifier is given as

$$A_v(s) = \frac{V_2}{V_1} = \frac{\mu R_L}{(C_o R_o R_L)s + (R_o + R_L)} \tag{5-47}$$

Note that if the parasitic capacitance C_o is not present, that is, $C_o = 0$, the above expression reduces to that given by Eq. (5-31) as expected. Substituting the circuit parameter values in Eq. (5-47) we obtain

$$\omega_o = \frac{R_o + R_L}{C_o R_o R_L} = \frac{1100}{(10 \times 10^{-12})(100)(1000)} = \frac{1100}{10^{-6}} = 1100 \times 10^6$$

Therefore the 3-dB cut-off frequency in Hertz is

$$f_o = \frac{1100}{2\pi} \text{ MHz} = 175 \text{ MHz}$$

Bode Plot

A convenient way to visualize the frequency dependence of the amplifier gain is with the aid of the Bode plot which we now discuss. Rewriting Eq. (5-42) in decibels we obtain

$$\mathcal{G}(\omega) = 20 \log_{10} |A_v(j\omega)| = 20 \log_{10} \frac{K_o}{\sqrt{1 + (\omega/\omega_o)^2}}$$

$$= 20 \log_{10} K_o - 10 \log_{10} \{1 + (\omega/\omega_o)^2\}$$

$$= \mathcal{G}_o - 10 \log_{10} \{1 + (\omega/\omega_o)^2\} \tag{5-48}$$

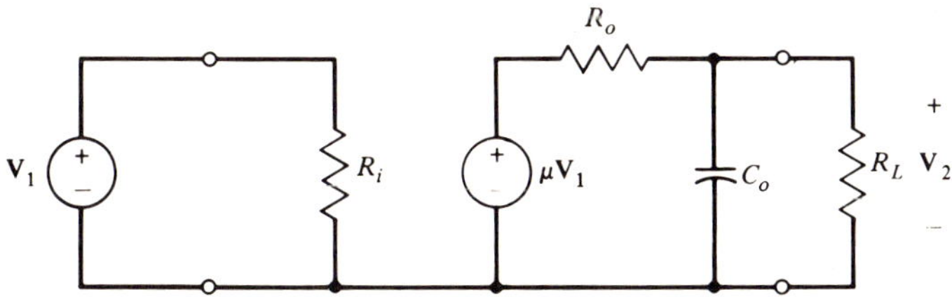

Figure 5-25 A high-frequency model of a dc amplifier.

For small values of ω such that $\omega \ll \omega_o$, we can write

$$\mathcal{G}(\omega) = \mathcal{G}_1(\omega) \cong g_o - 10 \log_{10}(1) = g_o \text{ dB} \tag{5-49}$$

For $\omega \gg \omega_o$, on the other hand,

$$\mathcal{G}(\omega) = \mathcal{G}_2(\omega) \cong g_o - 10 \log_{10}(\omega/\omega_o)^2 = g_o - 20 \log_{10}(\omega/\omega_o) \text{ dB} \tag{5-50}$$

Let us choose two frequencies ω_1 and $\omega_2 = 2\omega_1$ where both ω_1 and ω_2 are much larger than ω_o. At these frequencies

$$\mathcal{G}_2(\omega_1) = g_o - 20 \log_{10}(\omega_1/\omega_o) \text{ dB} \tag{5-51}$$

and

$$\begin{aligned}
\mathcal{G}_2(\omega_2) &= g_o - 20 \log_{10}(2\omega_1/\omega_o) \\
&= g_o - 20 \log_{10}(2) - 20 \log_{10}(\omega_1/\omega_o) \\
&\cong {} + \mathcal{G}_2(\omega_1) - 6 \text{ dB}
\end{aligned} \tag{5-52}$$

Thus at $\omega = 2\omega_1$, the gain is 6 dB less than that at $\omega = \omega_1$ or, in other words, the gain decreases at the rate of 6 dB/octave at frequencies far above ω_o. So, if we plot the two asymptotes to the gain curve, one will be a straight line parallel to the ω-axis having zero slope for $\omega < \omega_o$ and the other will be a straight line having a 6 dB/octave slope for $\omega > \omega_o$ (see Fig. 5-24). These two asymptotes meet at $\omega = \omega_o$ because at $\omega = \omega_o$, $\mathcal{G}_1(\omega_o) = \mathcal{G}_2(\omega_o)$. ω_o is consequently called the *break point* or *break frequency*.

It is clear from the above discussion that the gain plot in decibels can be reasonably well represented by its straight-line approximation which is also known as the *Bode plot*. From the Bode plot we can conclude that the gain is fairly constant for $\omega < \omega_o$ and all signals of frequencies less than ω_o are amplified equally. On the other hand, the gain starts decreasing rapidly for $\omega > \omega_o$ and input signals of frequencies greater than ω_o are attenuated. For $\omega \gg \omega_o$, the input signal barely passes through the amplifier. So, we can state that the amplifier acts like a low-pass filter passing frequencies below ω_o and stopping (or attenuating) frequencies above ω_o. ω_o is thus called the *cut-off frequency* and the *bandwidth* of the amplifier is from dc to ω_o, that is, ω_o radians/sec.

Let us now examine the phase characteristic. Expanding Eq. (5-43) in a power series, we obtain

$$\phi(\omega) = -(\omega/\omega_o) + \tfrac{1}{3}(\omega/\omega_o)^3 - \tfrac{1}{5}(\omega/\omega_o)^5 + \cdots \tag{5-53}$$

For very low values of ω such that $\omega \ll \omega_o$, we can approximately write

$$\phi(\omega) \cong -\omega/\omega_o \tag{5-54}$$

Thus the phase is linear only at frequencies much smaller than the cut-off. Since the magnitude is also constant at these frequencies, only signals of very low frequencies will be faithfully reproduced at the output.

5-10 Classification of Amplifiers

A practical amplifier has a frequency-dependent gain characteristic that is approximately constant over a certain finite frequency band but drops outside the band. A typical gain characteristic is shown in Fig. 5-26. The frequency range over which the gain is almost constant is the bandwidth of the amplifier and in this range the characteristic of the amplifier approaches that of its idealized counterpart. This useful frequency range of the practical amplifier is often used to classify the device.

Amplifiers are broadly classified as *dc amplifiers* and *ac amplifiers*. The dc amplifier is capable of amplifying dc signals (direct currents or direct voltages) in addition to alternating signals. Here the gain stays constant from dc to some frequency f_H and decreases above f_H. Such amplifiers are used in computers, dc voltmeters, oscilloscope deflection amplifiers, and more generally in instrumentation-type applications. In the case of an ac amplifier, the gain is constant from a frequency f_L to frequency f_H and the gain drops for frequencies below f_L and above f_H.

The ac amplifiers are further classified according to their intended applications. For an *audio-frequency amplifier* or simply an *audio amplifier*, the useful frequency range is from 15 Hz to 20 kHz which is the normal audible frequency range. Such amplifiers are used mainly to amplify voice frequencies and are found in hi-fi systems, radios, and so on. The *radio-frequency* (r-f) *amplifier* is normally used in the frequency range 10 kHz–100,000 MHz. There are two types of r-f amplifiers. The *narrow-band* r-f amplifier is used for voice frequency modulated radio frequencies and the *broadband* r-f amplifier is used for television video signals or short-pulse modulated signals.

The *video amplifier* is a wide-band amplifier having a bandwidth from 60 Hz to about 4 MHz. It was originally developed to handle picture signals of TV systems which gave rise to the name *video amplifier*. The amplifiers used in radar systems to amplify short pulses are also called video amplifiers because they involve the same frequency range.

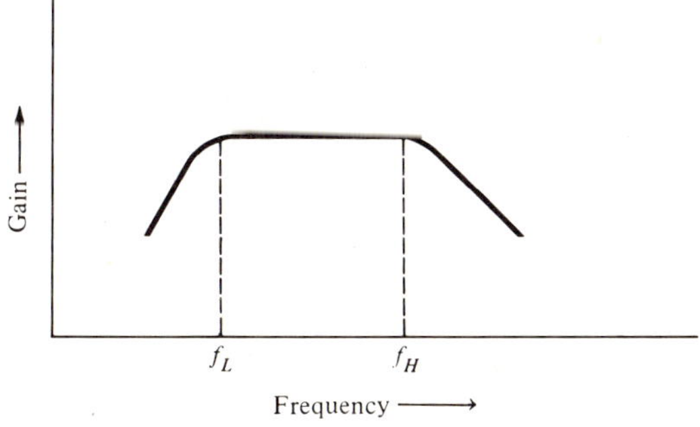

Figure 5-26 A typical amplifier gain characteristic.

The *intermediate frequency* (i-f) amplifier is used primarily in superheterodyne systems. The television i-f amplifier is a tuned amplifier that passes signal components in the frequency range 42.17–45.75 MHz. The i-f amplifier used in a superheterodyne radio has a selective frequency response centered at 455 kHz with a bandwidth of 5–10 kHz.

The amplifiers used in analog computers are very high-gain dc amplifiers and are more commonly known as *operational amplifiers.* These amplifiers are discussed in detail in the following chapter.

5-11   Typical IC Amplifiers

There is a large variety of IC linear amplifiers available from various manufacturers for applications as audio, video, r-f, and i-f amplifiers, and other nonlinear applications at corresponding frequency ranges. A number of IC amplifiers have additional features such as automatic gain control (AGC). A detailed discussion of all such IC amplifiers is beyond the scope of this book ; however, we provide a rather brief discussion of a sample of these IC packages along with some typical applications which hopefully will provide the reader with some idea regarding the versatility and potential of linear IC amplifiers.

Most electronic circuits, including amplifiers, require external dc supplies or batteries for biasing the transistors and supplying power. Thus, in addition to the input and output terminals, an IC amplifier package includes terminals for connecting the bias supplies and the system ground. In some packages, additional terminals may be provided for external adjustment of the amplifier characteristic. Even though an amplifier can be constructed in principle using a single transistor, most IC amplifiers are rather complex circuits containing many active and passive components. However, as explained in Chapter 1, their costs are still comparable to that of a discrete transistor.

Audio Amplifier

IC amplifiers designed specifically for audio-frequency applications are in general low-power circuits capable of delivering 1–2 W of power to a load. As a result, these modules are used primarily as preamplifiers, servo drivers, and microphone amplifiers. They may be used as audio power amplifiers in applications where low audio output is acceptable, such as in car radios, pocket radios, and so on. In most applications, these ICs require appropriate external heat sink.

An example of an IC audio-amplifier package is the National Semiconductor LM377 module.[11] It is a dual audio-amplifier package with each amplifier in the package being capable of delivering 2 W into an 8-Ω load. It features on-chip frequency compensation, thermal shut-down protection, fast turn-on and turn-off, output current limiting, and a 5–20 MHz gain-bandwidth product. This unit has been designed for applications as the audio power output stage for phonographs, tape recorders, or radios. Other applications include servo amplifiers, oscillators, and so on. It can operate over a power supply range of

10–35 V. Figure 5-27 shows the circuit schematic for using this IC as a stereo amplifier in the inverting mode. If adequate heat-sink capacity is not provided to the package, limits on device dissipation on signal peaks is automatically provided by the internal thermal limit circuitry. This, however, results in severe distortion in signal waveform which is similar to that obtained by peak clipping. The 300-μF capacitor is used to couple the ac part of the amplifier output to the speaker while blocking the flow of any dc. The input signal is similarly coupled to the input of the amplifier through the 0.47-μF capacitor. Additional applications of this IC package can be found in Reference 11.

Other examples of monolithic audio amplifiers are RCA CA3007, Motorola 1554, and National Semiconductor LM378 and LM379 modules.

Video Amplifier

A differential-input voltage amplifier with high open-loop bandwidth is usually classified as a video amplifier. For a typical IC video amplifier, the 3-dB bandwidth may be around 100 MHz with a dc gain of about 40 dB. It is usually used in the open-loop mode as its excessive phase-shift prevents the use of connection of external circuitry from the output to the inverting input terminal to control gain. Furthermore, in general, a video amplifier is characterized with a rather low output voltage swing which is typically a few volts at high frequencies. The

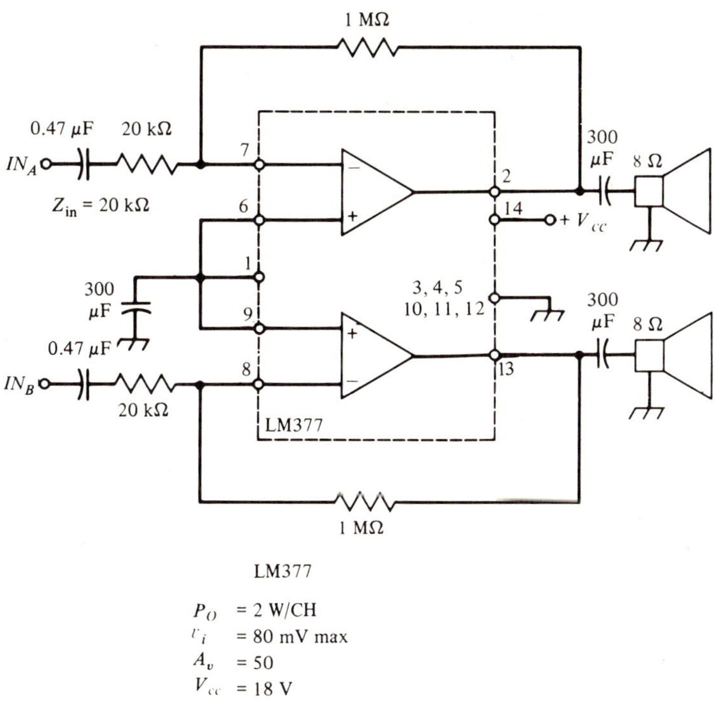

Figure 5-27   Inverting stereo amplifier. (Courtesy National Semiconductors Corp.)

video amplifier is used as a wide-band line amplifier, pulse amplifier, oscillator, phase shifter, fast Schmitt trigger, and tuned amplifier. Some IC video amplifiers are designed to operate with AGC and some have gated inputs allowing selection of a particular input by an appropriate gating signal. The gated video amplifier can also be used as a video switch, amplitude modulator, balanced modulator, pulse-width modulator, gated oscillator, and gated astable multivibrator.

An example of an IC video amplifier is the Motorola MC1552G package.[10] It provides a variety of gain options which may be used to establish a broad range of gain levels from unity to about 3000 with the aid of an external resistor. Depending on the gain level selected, the bandwidth of the amplifier can range from 5 to about 60 MHz. Figure 5-28 shows the circuit schematic of a pulse

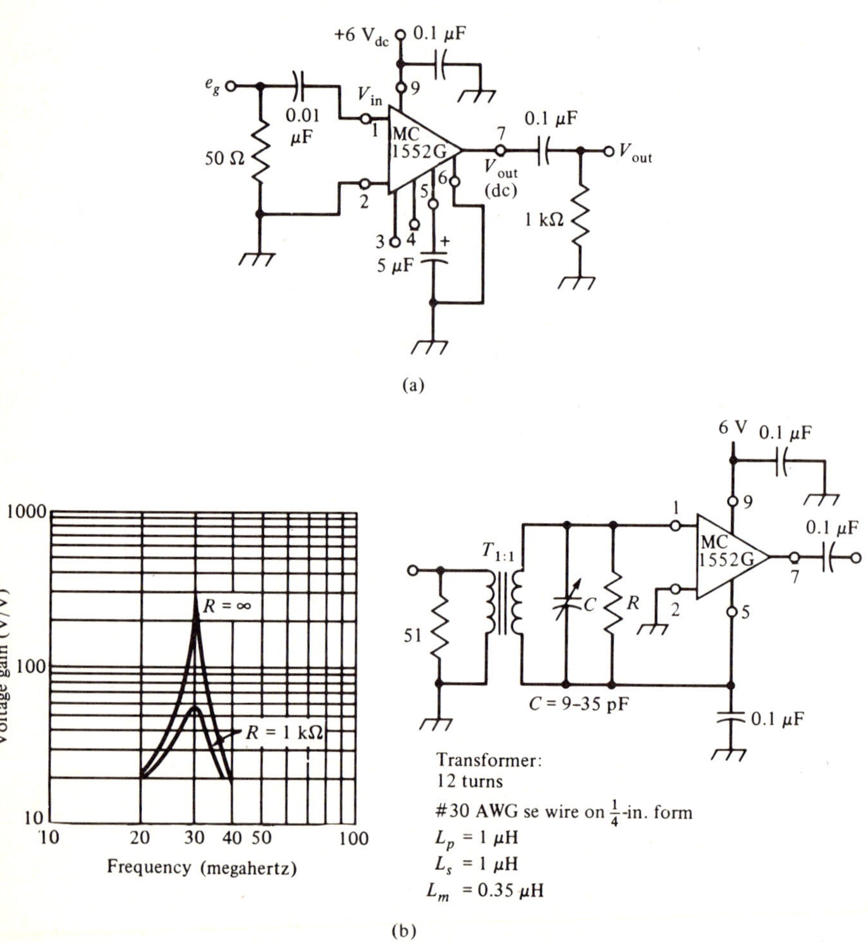

Figure 5-28   (a) Circuit schematic of a pulse amplifier using MC1552G video amplifier, and (b) circuit schematic and frequency response of a 30-MHz tuned amplifier using MC1552G IC video amplifier. (Courtesy Motorola Semiconductor Products Inc.)

amplifier and a tuned amplifier designed using the MC1552G. In using this amplifier, a high-frequency capacitor (0.1 μF) should be used to bypass the power supply as shown in Fig. 5-28. Additional layouts and other recommendations for proper use of this IC will be found in Reference 10.

Other examples of IC video amplifiers are the RCA CA3001 and Texas Instruments SN7511 modules. An example of a gated video amplifier is the Motorola MC1545 unit.

Operational Amplifiers

The operational amplifier is also a differential-input voltage amplifier. However, in contrast to the video amplifier it has a much larger dc gain and a smaller bandwidth. Typical dc gain is around 100 dB with a 3-dB bandwidth of 100 kHz. It is designed primarily for use with feedback path from the output to the inverting input terminal which allows the control of its gain characteristic. It is the most widely used linear IC with numerous applications. A detailed discussion of this versatile IC along with a variety of useful applications is provided in Chapter 6.

Voltage Follower

As indicated earlier, a unity-gain noninverting voltage amplifier is more commonly called a voltage follower. It is primarily used as a buffer for transducers or other high-impedance sources. It also finds applications in the design of frequency-selective networks without inductors, and sample-and-hold circuits (Section 8-2). As we shall see in the next chapter, a voltage follower can be readily designed by connecting the output of an IC operational amplifier directly to its inverting input terminal. However, a voltage follower built using a general-purpose IC operational amplifier is not suitable in many applications because of its low slew rate and high input current. Monolithic voltage followers overcoming these limitations are currently available. An example of such an IC is the National Semiconductor LM102 module. It has a voltage gain of 0.9995 and a bandwidth of 10 MHz. The input resistance is $10^{12}\,\Omega$; the output resistance is about 1 Ω; the input capacitance is 3 pF; the input current is 3 nA; and the slew rate is 10 V/μsec. Some applications of the voltage follower are outlined in the following chapter.

5-12   Summary

The ideal voltage and current amplifiers are introduced in Section 5-1. Practical amplifiers are characterized by three types of gain: the voltage gain, the current gain, and the power gain. These gains are defined in Section 5-2 along with a logarithmic unit of the gain, called the decibel. The voltage amplifier, more frequently used in system design, is considered again in Section 5-3. It outlines the different variations of the basic voltage amplifier. Some simple applications of the voltage amplifier are described in Section 5-4. Section 5-5 considers the nonideal properties of a practical amplifier and also describes its important characteristics. The effect of the frequency dependence of the gain of the amplifier for sinusoidal steady-state applications and for time-domain applications are treated in Sections 5-6 and 5-7, respectively. In many applications, an amplifier is used with input and output terminations. Performances of such a terminated amplifier are evaluated in Section 5-8. Section 5-9 considers in detail the single-pole gain model of a practical amplifier. It also outlines a simple approximate method (Bode plot) for plotting the gain characteristic of a nonideal amplifier. Section 5-10 introduces the various names that are used to describe amplifiers depending on their frequency range of applications. Finally, Section 5-11 provides a brief discussion on some commonly available IC amplifiers.

References

1. E. J. Angelo, Jr., *Electronics: BJTs, FETs, and Microcircuits*, McGraw-Hill Book Co., New York, 1969.
2. A. J. Prestcott, "Loss Compensated Active Gyrator Using Differential-Input Operational Amplifiers," *Electronics Letters*, Vol. 2, July 1966, pp. 283–284.
3. J. Millman and C. C. Halkias, *Integrated Electronics: Analog and Digital Circuits and Systems*, McGraw-Hill Book Co., New York, 1972.
4. V. H. Grinich and H. G. Jackson, *Introduction to Integrated Circuits*, McGraw-Hill Book Co., New York, 1975.
5. S. D. Senturia and B. O. Wedlock, *Electronic Circuits and Applications*, John Wiley & Sons, New York, 1975.
6. P. M. Chirlian, *Electronic Circuits: Physical Principles, Analysis and Design*, McGraw-Hill Book Co., New York, 1971.
7. M. S. Ghausi, *Electronic Circuits*, Van Nostrand Reinhold, New York, 1971.
8. D. J. Comer, *Modern Electronic Circuit Design*, Addison-Wesley, Reading, Mass., 1976.
9. R. J. Smith, *Circuits, Devices, and Systems*, John Wiley & Sons, New York, 1976.
10. Staff, "A Wide-band Monolithic Video Amplifier," Application Note AN-404, Motorola Semiconductor Products Inc., Phoenix, Ariz.
11. J. Sherwin, "LM377, LM378, and LM379 Dual Two, Four, and Six Watt Power Amplifiers," Application Note AN-125, National Semiconductor Corp., Santa Clara, Calif., January 1975.
12. Staff, "A Fast Integrated Voltage Follower with Low Input Current," Application Note AN-5, National Semiconductor Corp., Santa Clara, Calif., May 1968.
13. Staff, "Application of the NH0002 Current Amplifier," Application Note AN-13, National Semiconductor Corp., Santa Clara, Calif., September 1968.

Problems

5-1 The models of a *transconductance amplifier* and a *transresistance amplifier* are shown in Fig. 5-29 which are seen to be a voltage-controlled current source and a current-controlled voltage source, respectively. Implement a voltage amplifier and a current amplifier utilizing these two latter types of amplifiers.

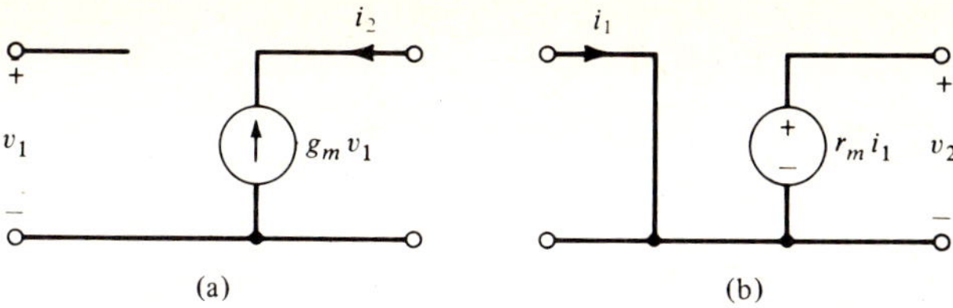

Figure 5-29 Models of (a) a transconductance amplifier, and (b) a transresistance amplifier.

5-2 Develop the circuit schematics for designing a transconductance amplifier and a transresistance amplifier using ideal voltage and current amplifiers and resistances.

5-3 Calculate the voltage gain v_2/v_1 of the circuits of Fig. 5-30.

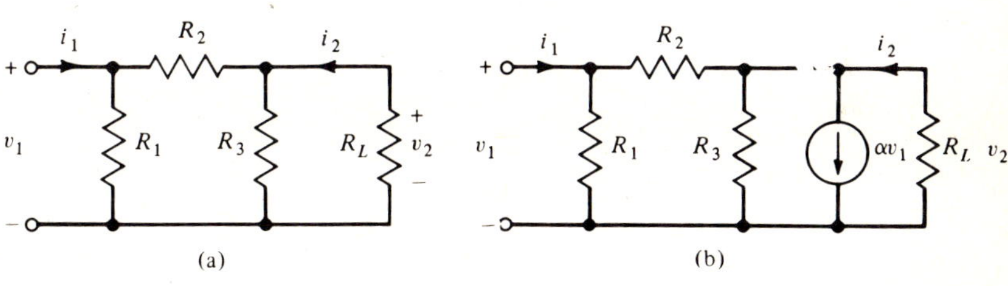

Figure 5-30

5-4 Calculate the current gain $-i_2/i_1$ of the circuits of Fig. 5-30.

5-5 Determine the voltage gain v_2/v_1 of the negative resistance amplifier of Fig. 5-31. For what values of the load resistance R_L will the output be in phase with the input and for what values of R_L will the output be 180° out of phase with the input?

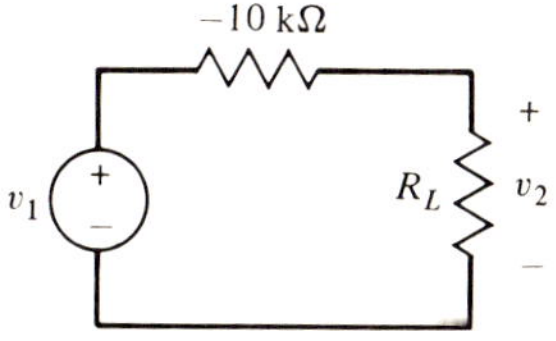

Figure 5-31

5-6 Show that the circuit of Fig. 5-32 realizes an ideal current amplifier of the type of Fig. 5-5(a). Determine the expression for the current gain β. Show that if the current gains α_a and α_b of the constituent current amplifiers are each positive and less than 1, the current gain of the composite circuit is still positive and less than 1 but greater than α_a and α_b.

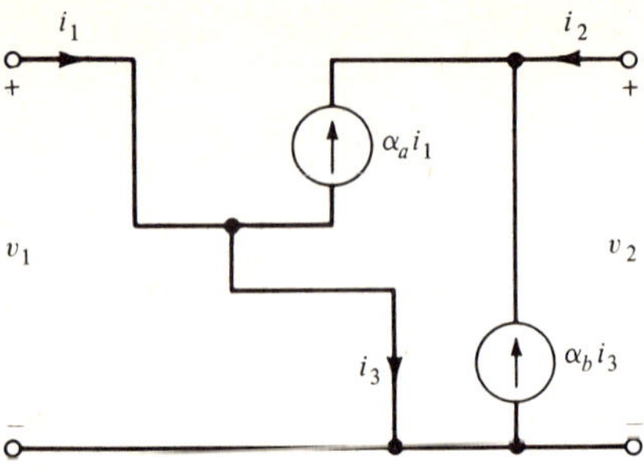

Figure 5-32

5-7 Show that the circuit of Fig. 5-33 is equivalent to an ideal voltage amplifier with v_2 being dependent on v_1. Determine the expression for the voltage gain of the amplifier.

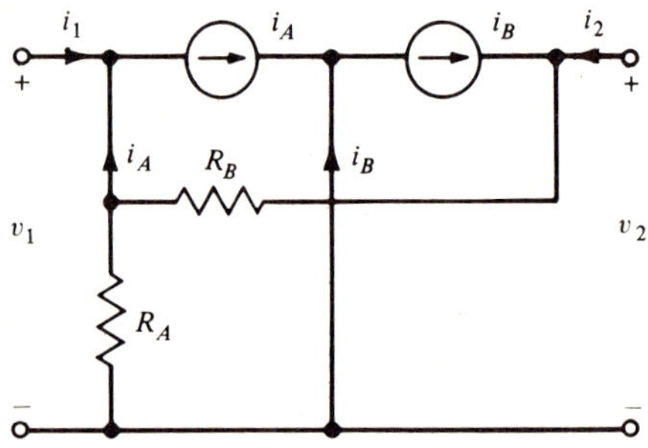

Figure 5-33

5-8 If the resistors of the circuits of Fig. 5-30 have a tolerance of ± 5 percent, what is the worst-case deviation of the voltage gains in each case.

5-9 Determine graphically the waveform of the output voltage of an ideal voltage amplifier of voltage gain 3 if the input voltage is 2-V $p\text{--}p$ sine wave.

5-10 Develop the formulas to calculate the decibel equivalents of the following voltage gains: (a) nA_v and (b) $(A_v)^n$, where n is a positive real number.

5-11 We wish to design a voltage amplifier with at least 80-dB gain by cascading three low-gain voltage amplifiers having gains μ_1, μ_2, and μ_3. If $\mu_1 = 15$ and $\mu_2 = 40$, what should be the value of μ_3?

5-12 A high-gain voltage amplifier is to be designed using three identical inverting voltage amplifiers. If the required overall gain is 90 dB, what should be the gain of each individual amplifier in decibels and as a numerical ratio?

5-13 Determine the voltage gains in decibels of the circuit of Fig. 5-31 for the following values of the load resistance R_L: (a) 5 kΩ and (b) 15 kΩ.

5-14 Two identical noninverting voltage amplifiers having a gain of 40 dB each are connected as shown in Fig. 5-34 with the aid of a summing amplifier. Determine the overall gain v_2/v_1 of the system in decibels.

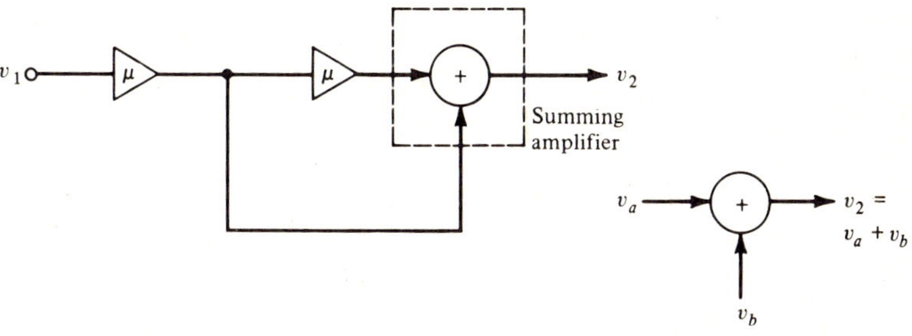

Figure 5-34

5-15 Determine the values of the resistors R_A and R_B if the desired voltage gain of the amplifier of Fig. 5-33 is 25 dB.

5-16 Determine the input resistance v_1/i_1 of the circuit of Fig. 5-35. For what values of the gain μ will it be negative?

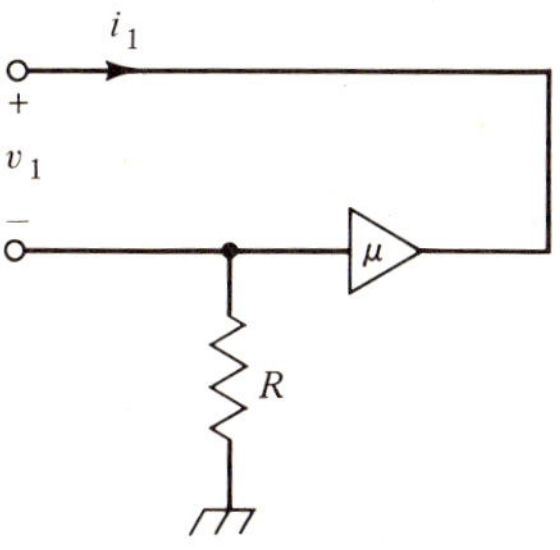

Figure 5-35

5-17 Show that the input resistance R_{in} of the circuit of Fig. 5-15 is that given by Eq. (5-17).

5-18 Show that the input capacitance C_{in} of the circuit of Fig. 5-36 is given as $(A + 1)C$.

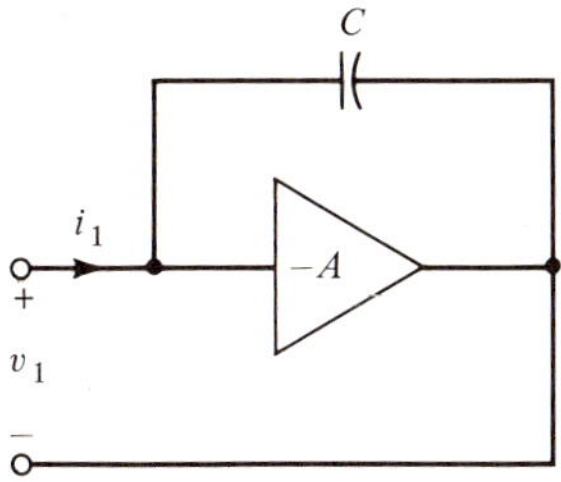

Figure 5-36

5-19 Determine suitable values of the passive components for the circuit of Fig. 5-14 to simulate an inductance of value 0.5 H and having a Q value of 10 at 1 kHz. (*Note:* The Q of a lossy inductor is defined as $Q = \omega L/R$, where L is the inductance value in henries and R is the effective series resistance in ohms.)

5-20 An inductance simulation circuit can be designed by connecting two transconductance amplifiers in parallel in a manner shown in Fig. 5-37. Determine the expression for the equivalent inductance L_{eq}.

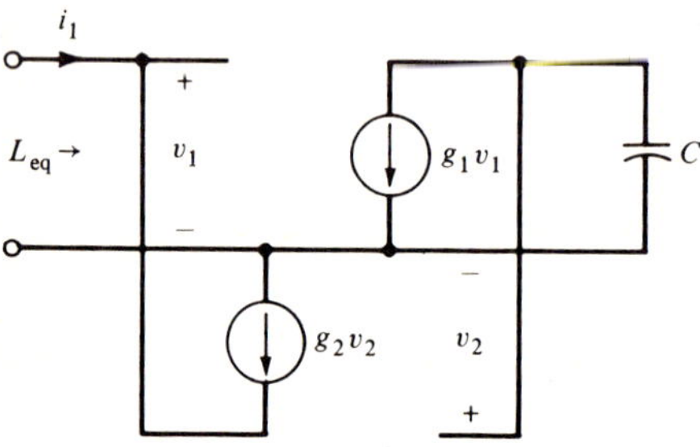

Figure 5-37

5-21 Show that the 2-port network of Fig. 5-38 will simulate an inductor if either port is terminated by a capacitor.

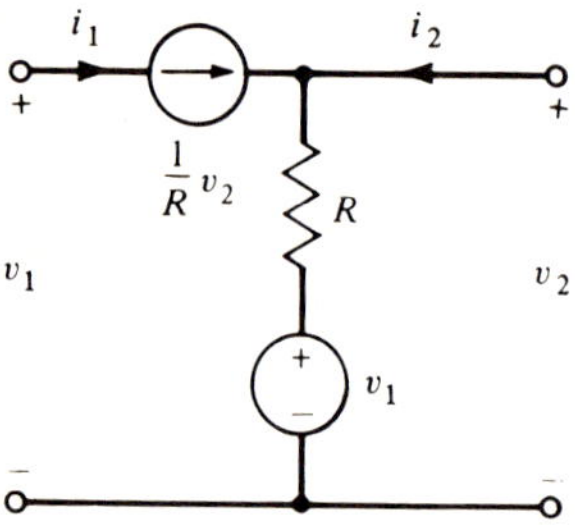

Figure 5-38

5-22 An amplifier with an input resistance of 2 MΩ and terminated by a load resistance of 1 kΩ has a current gain A_i of 150. What is its power gain in decibels?

5-23 A more accurate model of a practical voltage amplifier is shown in Fig. 5-39, where the parameters z_{11}, z_{12}, z_{21}, and z_{22} are known as the open-circuit impedance parameters which can be measured as follows:

$$z_{11} = \left.\frac{v_1}{i_1}\right|_{i_2=0}, \quad z_{12} = \left.\frac{v_1}{i_2}\right|_{i_1=0}, \quad z_{21} = \left.\frac{v_2}{i_1}\right|_{i_2=0}, \quad z_{22} = \left.\frac{v_2}{i_2}\right|_{i_1=0}$$

If the amplifier is excited by a voltage source v_s with internal resistance R_s Ω and terminated by a load resistance R_L Ω, determine the expressions for the voltage gain v_2/v_1, the current gain $-i_2/i_1$, the input impedance v_1/i_1, and the output impedance $v_2/i_2|_{r_s=0}$ of the terminated amplifier.

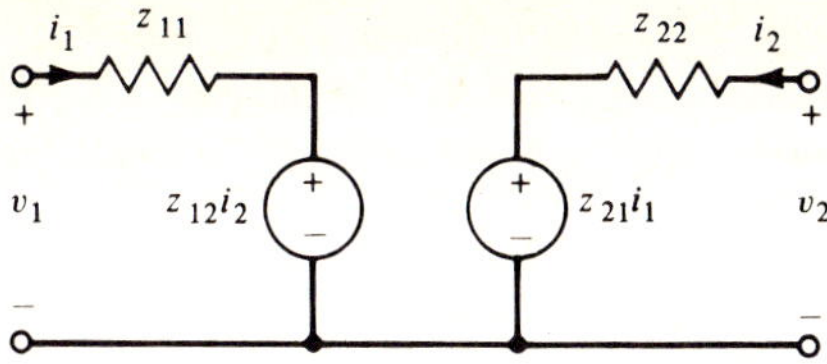

Figure 5-39

5-24 (a) Determine the expression for the voltage gain v_o/v_s of the three-stage amplifier circuit shown in Fig. 5-40.

(b) Given: $R_1 = 1.5\,\text{k}\Omega$, $R_s = 600\,\Omega$, $R_4 = 100\,\Omega$, $R_5 = 5\,\text{k}\Omega$. If R_1 changes from 1.5 to 4.5 kΩ, and R_4 changes from 100 to 10 Ω, what should be the new values of R_s and R_5 so that the voltage gain v_o/v_s remains invariant?

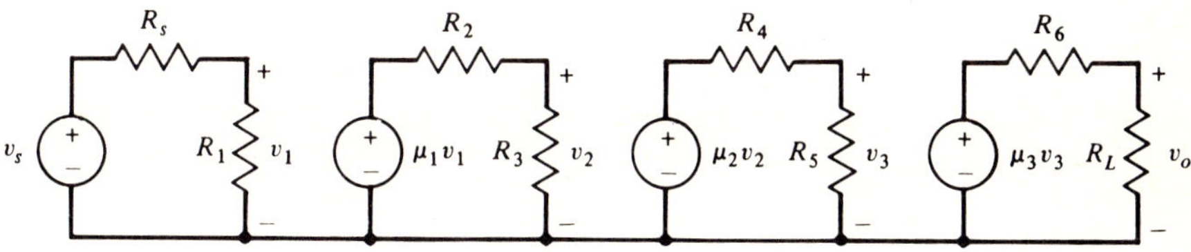

Figure 5-40

5-25 Determine the minimum value of the input resistance and the maximum value of the output resistance of an IC amplifier so that the system voltage gain does not change more than 5 percent. The source resistance is in the range 50 Ω–5 kΩ and the load resistance is in the range 20 kΩ–1 MΩ.

5-26 An amplifier is designed with the following parameter values: $\mu = 2$, $R_L = 5\,\text{k}\Omega$, and $R_o = 100\,\Omega$. If μ can change by ± 5 percent and R_L by ± 20 percent, what is the worst-case deviation of the voltage gain A_v?

5-27 Consider the cascaded amplifier of Fig. 5-41 for which the gain μ of each individual amplifier is 20 dB.

(a) Find the overall voltage gain v_4/v_1 in decibels.

(b) To what value should the load resistance R_L be adjusted to realize an overall voltage gain v_4/v_1 of 30 dB?

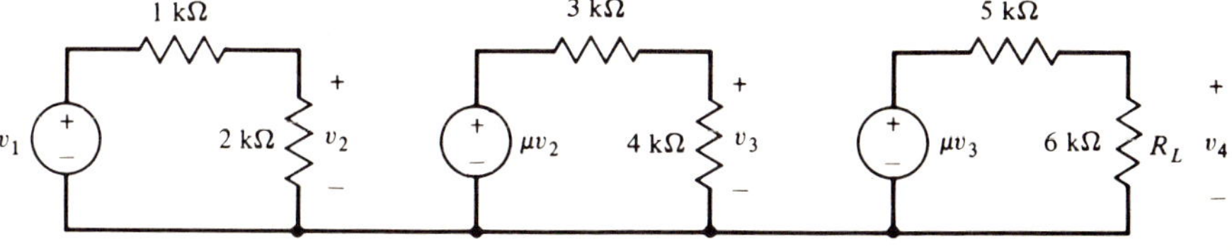

Figure 5-41

5-28 Consider a cascade of n identical amplifiers characterized by the single-pole model of Eq. (5-41). Determine the expression for the magnitude of the overall gain and its dc value. If ω_x is the new 3-dB cut-off frequency of the cascaded amplifier circuit, show that it is given as

$$\omega_x = \omega_o\sqrt{2^{1/n} - 1}$$

Is the new bandwidth greater or less than that of an individual amplifier?

5-29 A multistage amplifier is to be designed by cascading three identical amplifiers. The desired 3-dB bandwidth of the multistage amplifier is 10 MHz. What is the 3-dB bandwidth of each individual amplifier if each amplifier is assumed to be represented by a single-pole model?

5-30 The step response of an IC amplifier is shown in Fig. 5-42. Determine its **(a)** overshoot, **(b)** rise time, **(c)** delay time, and **(d)** settling time (to within 5 percent of its final value).

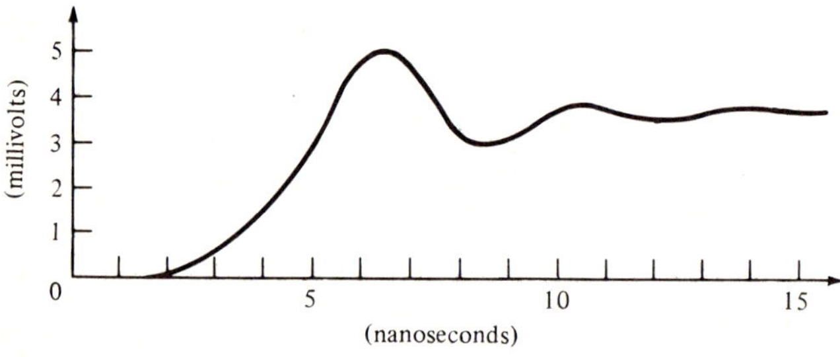

Figure 5-42

5-31 To measure the slew rate S of an IC amplifier, a high-frequency square wave was applied at the input. The plot of the output as measured is shown in Fig. 5-43.
(a) What is the slew rate of this amplifier in volts/μsec?
(b) Calculate the highest signal frequency that will result in an undistorted output of peak amplitude of 5 V.

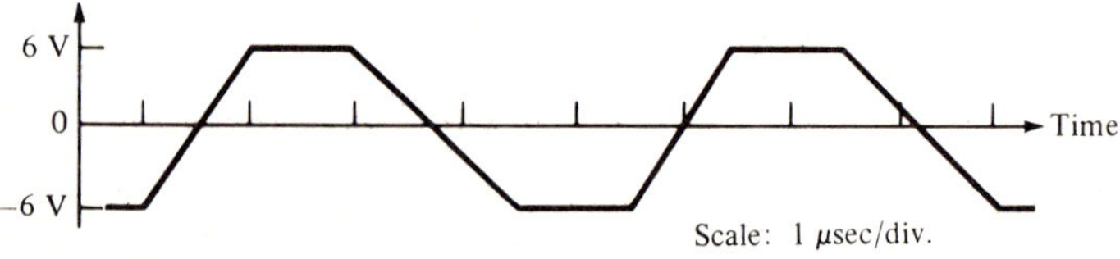

Figure 5-43

5-32 Determine the largest undistorted output voltage obtainable from an IC amplifier at **(a)** 100 Hz, **(b)** 5 kHz, and **(c)** 2 MHz if the slew rate of the amplifier is 20 V/μsec.

5-33 A voltage amplifier with a nonlinear gain characteristic has an input–output relation given by

$$v_2(t) = \alpha_1 v_1(t) + \alpha_2 v_1^2(t) + \alpha_3 v_1^3(t)$$

where $v_1(t)$ is the input and $v_2(t)$ is the output. If $v_1(t) = V_1 \cos \omega t$, show that the output $v_2(t)$ is of the form $v_2(t) = B_0 + B_1 \cos \omega t + B_2 \cos 2\omega t + B_3 \cos 3\omega t$ indicating that the output not only contains a *fundamental* component of same frequency as the input, but also *harmonic* components having frequencies that are integral multiples of the fundamental frequency. Determine B_0, B_1, B_2, and B_3. The distortion of the output waveform due to the nonlinear gain characteristic is known as *harmonic distortion*.

5-34 The nonideal amplifier of Problem 5-33 is excited by an input $v_1(t) = V_1 \cos \omega_1 t + V_2 \cos \omega_2 t$. Determine the expression for the output $v_2(t)$ and show that it contains not only the original frequency components and their harmonics but also components having frequencies that are sums and differences of the input signal frequencies and their harmonics. This type of distortion is known as *intermodulation distortion*.

5-35 A voltage amplifier has a nonlinear transfer characteristic as described by the following equation:

$$v_2(t) = 5v_1(t) + v_1^2(t) + 0.1v_1^3(t)$$

where $v_1(t)$ is the input and $v_2(t)$ is the output. If $v_1(t) = V_m \sin \omega t$, determine $v_2(t)$ and then calculate the harmonic distortions due to the second and third harmonics and the total harmonic distortion. For what value of the input signal amplitude V_m can we regard the amplifier as a linear device?

5-36 Determine the two frequencies above and below ω_o in the magnitude plot of Fig. 5-24 where the difference between the Bode plot and the actual value of the gain is 1 dB.

CHAPTER 6

Operational Amplifiers

Thus far we have discussed in general terms the properties of an amplifier and its performance in a system. In this chapter we apply these concepts in designing practical circuits using a very specific commercially available amplifier module —the operational amplifier. Even though this highly popular module has properties similar to those discussed in the previous chapter, it does have some additional unique characteristics that merit special discussion.

The operational amplifier is a very high-gain dc differential-input voltage amplifier. It is usually used with a large amount of negative feedback which makes the functional performance of the amplifier with feedback insensitive to changes in the internal parameters of the operational amplifier with time and temperature. Also, the effect of circuit loading is substantially reduced. The behavior of the operational amplifier with feedback, then, depends only on the feedback network which can be precisely controlled by the circuit designer. Originally, the operational amplifier was used primarily in analog computers to perform mathematical operations such as addition, subtraction, multiplication, integration, and differentiation. This gave rise to the name operational amplifier for this type of amplifier. In 1965, monolithic integrated circuit operational amplifiers were introduced as off-the-shelf items and are presently being marketed by a number of manufacturers in various forms. In addition, they are also available in hybrid IC form. Besides their smaller sizes, modern operational amplifiers are considerably cheaper and more reliable than their earlier versions. As a result, in recent years they have found extensive use in signal processing and conditioning, instrumentation, control system design, nonlinear function generation, regulators, and so on. Undoubtedly the monolithic operational amplifier is the most widely used analog IC module.

We first define the ideal operational amplifier and outline a number of simple linear applications. Characteristics of a practical operational amplifier are then described along with their effect on its performance. Applications of these devices in analog computation and filtering are next considered. Nonlinear applications of the operational amplifier are discussed in Chapter 7.

6-1   The Ideal Model

The operational amplifier is a 2-input, single-output voltage amplifier with "infinite" voltage gain, whose idealized control source model is as shown in Fig. 6-1. From the figure we observe that the idealized model is defined by

$$v_o = \mu(v_{i2} - v_{i1}), \qquad \mu \to \infty \tag{6-1}$$

$$i_{i1} = i_{i2} = 0 \tag{6-2}$$

Equation (6-1) indicates that the output voltage v_o responds to the difference of voltages at the input terminals and hence the amplifier is known as a *differential-input operational amplifier*. Note that the output polarity is the same as that of the signal at terminal ② and is the opposite to that of the input signal at terminal ①. Terminal ② is thus called the *noninverting input terminal* and terminal ① is called the *inverting input terminal*. Since an operational amplifier is invariably used in *closed-loop* form with the output connected to the input terminals through other circuit elements, the gain μ in Eq. (6-1) is called the *open-loop* gain to distinguish it from the gain of the circuit containing the operational amplifier.

An alternate way of writing Eq. (6-1) is by defining the input voltages as

$$v_{i1} = v_{cm} - (v_e/2)$$
$$v_{i2} = v_{cm} + (v_e/2) \tag{6-3}$$

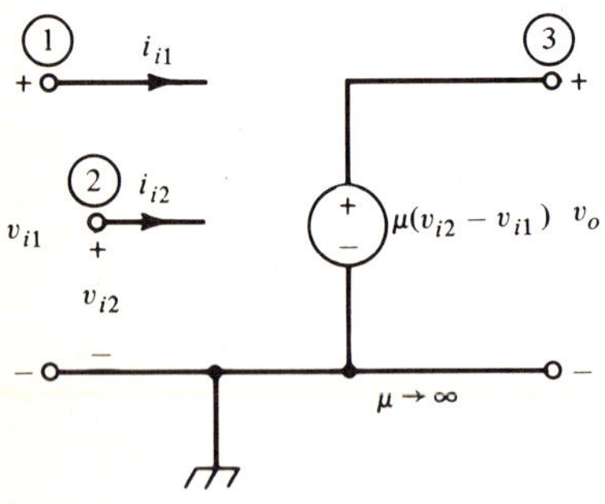

Figure 6-1   Controlled-source model of an ideal operational amplifier.

which imply

$$v_{cm} = \tfrac{1}{2}(v_{i1} + v_{i2}) \tag{6-4}$$

$$v_e = v_{i2} - v_{i1} \tag{6-5}$$

From Eq. (6-4) it follows that v_{cm} can be considered as the signal common to both inputs of the operational amplifier and, as a result, is called the *common-mode input voltage*. Likewise, Eq. (6-5) indicates that v_e is the *differential-input voltage* across the two input terminals. We can thus express the output voltage v_o of the operational amplifier in terms of v_{cm} and v_e as

$$v_o = \mu_c \cdot v_{cm} + \mu_d \cdot v_e \tag{6-6}$$

where

$$\mu_c = \left. \frac{v_o}{v_{cm}} \right|_{v_e = 0} \tag{6-7a}$$

is the *common-mode gain* of the operational amplifier and

$$\mu_d = \left. \frac{v_o}{v_e} \right|_{v_{cm} = 0} \tag{6-7b}$$

is the *differential gain* of the amplifier. It is easy to show that these gains can be determined readily as follows:

$$\mu_c = \left. \frac{v_o}{v_{i2}} \right|_{v_{i1} = v_{i2}} \tag{6-8a}$$

$$\mu_d = \left. \frac{v_o}{2v_{i2}} \right|_{v_{i2} = -v_{i1}} \tag{6-8b}$$

From Eq. (6-1) we can conclude that for an ideal operational amplifier, the common-mode gain μ_c is zero and differential gain μ_d is simply μ.

The commonly used symbolic representation of the differential-input operational amplifier is as shown in Fig. 6-2. Here the inverting input terminal is marked by a minus sign $(-)$ and the noninverting input terminal by a plus sign $(+)$ inside the amplifier symbol. It should be noted that, just as in the case of finite gain voltage amplifiers, the common ground between the input and output terminals is not specifically shown in the symbolic representation.

It follows from the definition that the output resistance of the ideal operational amplifier is zero. Likewise, the input resistance seen between the two input terminals (called the *differential-input resistance*) and the input resistance seen

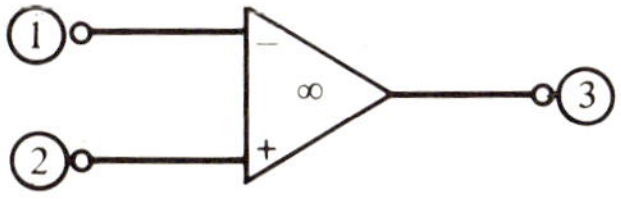

Figure 6-2 Symbolic representation of an operational amplifier.

between each input terminal and the ground (called the *common-mode input resistance*) are infinite. Equation (6-1) also implies that if $v_{i1} = v_{i2}$, then $v_o = 0$ or, in other words, the ideal operational amplifier has a zero voltage *offset*. The input currents are also zero even with excitations present and, ideally, the device has thus no current *offset*. In addition, the voltage gain μ is real and constant, implying that the idealized device has an infinite bandwidth.

Some additional properties of the ideal operational amplifier can be easily derived by analyzing a simple circuit as shown in Fig. 6-3(a) whose equivalent circuit is shown in Fig. 6-3(b). We first observe that the voltage v_f developed across R_2 is given as

$$v_f = \frac{R_2}{R_1 + R_2} \cdot \mu v_e \tag{6-9}$$

obtained using the voltage-divider relation. Applying KVL at the input side we obtain

$$v_1 = v_e + v_f \tag{6-10}$$

Substituting Eq. (6-9) in Eq. (6-10), we arrive at

$$v_e = \frac{R_1 + R_2}{R_1 + R_2 + \mu R_2} \cdot v_1 \tag{6-11}$$

Since $v_2 = \mu v_e$, we then obtain from above

$$v_2 = \frac{\mu(R_1 + R_2)}{R_1 + R_2 + \mu R_2} \cdot v_1 \tag{6-12}$$

Letting $\mu \to \infty$ in Eq. (6-12) we derive the desired relation

$$\frac{v_2}{v_1} = 1 + \frac{R_1}{R_2} \tag{6-13}$$

Since the input current supplied by the source v_1 is zero, the input resistance of the circuit of Fig. 6-3(a) is infinite. The output is developed across a controlled voltage source. As a result, the output resistance is zero. In other words, the circuit shown behaves as a finite gain ideal noninverting voltage amplifier whose gain depends only on the ratio of the two external resistors, R_1 and R_2. The circuit is also unilateral.

The infinite gain feature makes the analysis of circuits containing operational amplifiers rather straightforward. To illustrate this, we first observe that the differential-input voltage v_e across the input terminals of the operational amplifier is given by Eq. (6-11). As $\mu \to \infty$, we note that

$$v_e \to 0 \tag{6-14}$$

or, in other words, the two input terminals of the operational amplifier are at the same potential. Conditions (6-14) and (6-2) are the keys to the simplified analysis procedure for operational amplifier networks. We now apply these

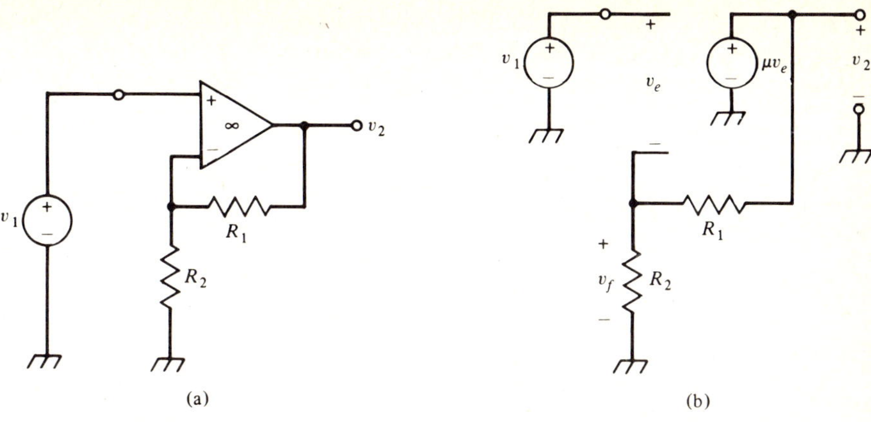

Figure 6-3 Noninverting voltage amplifier: (a) circuit connection and (b) equivalent circuit.

conditions to the analysis of the circuit of Fig. 6-3(a). Observe that condition (6-14) implies that the voltage at the inverting input terminal is also v_1. The current through the resistor R_2 flowing toward the ground is then v_1/R_2. The current flowing through the resistor R_1 toward the amplifier inverting input terminal is $(v_2 - v_1)/R_1$. Now by assumption (6-2) no current flows into the amplifier through any of its input terminals. Hence we have, by KCL,

$$\frac{v_2 - v_1}{R_1} = \frac{v_1}{R_2}$$

or

$$v_2 = \left(1 + \frac{R_1}{R_2}\right)v_1 \qquad (6\text{-}15)$$

which is same as Eq. (6-13).

Condition (6-14) holds provided there is a signal path from the output terminal to the inverting input terminal through the external network. This ensures what is known as a negative feedback situation (see Section 6-4).

6-2   Simple Applications

Before we get involved with the characteristics of a practical operational amplifier and their effect on the performance of circuits containing these modules, we outline here the design of a number of circuits for performing some simple operations. In spite of their simplicity, these circuits are frequently encountered in system design. Additional applications are described in later parts of this chapter and in the following chapters.

Finite-Gain Voltage Amplifiers

We have already shown that the circuit of Fig. 6-3(a) behaves like a *positive-gain voltage amplifier*. The following example illustrates a simple design problem using this circuit.

Example 6-1. Realize a noninverting voltage amplifier of 30-dB voltage gain.

From Eq. (6-13), the gain of the pertinent circuit of Fig. 6-3(a) in decibels is $20 \log_{10} (1 + R_1/R_2)$. Therefore,

$$20 \log_{10} \left(1 + \frac{R_1}{R_2} \right) = 30$$

This implies

$$1 + \frac{R_1}{R_2} = 31.62$$

or

$$\frac{R_1}{R_2} = 30.62$$

Typical values of resistances in designing practical circuits using operational amplifiers are in the range of 1–200 kΩ. Thus we can choose one of the resistances in the circuit of Fig. 6-3(a) to be a value within this range and determine the value of the other resistance such that the above ratio is maintained. For example, we can choose $R_2 = 1$ kΩ. This leads to $R_1 = 30.62$ kΩ.

The realization of a *voltage follower* follows readily from Fig. 6-3(a). If we make $R_1 = 0$, that is, replace it by a short circuit, then Eq. (6-15) reduces to

$$v_2 = \frac{0 + R_2}{R_2} v_1 = v_1 \tag{6-16}$$

implying that with $R_1 = 0$, the circuit of Fig. 6-3(a) will be a voltage follower for any arbitrary value of R_2. A convenient value is to make $R_2 \to \infty$ as indicated by Fig. 6-4.

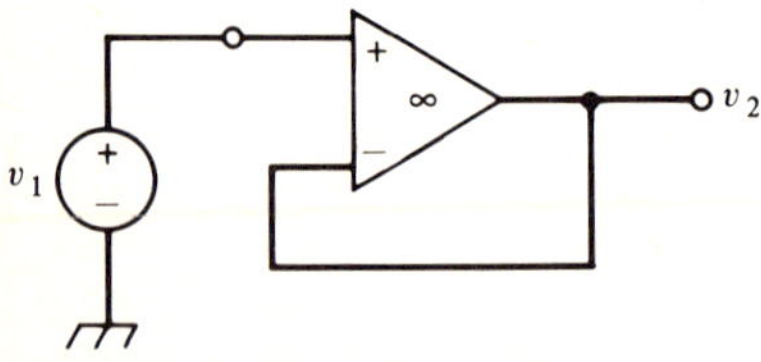

Figure 6-4 A voltage follower.

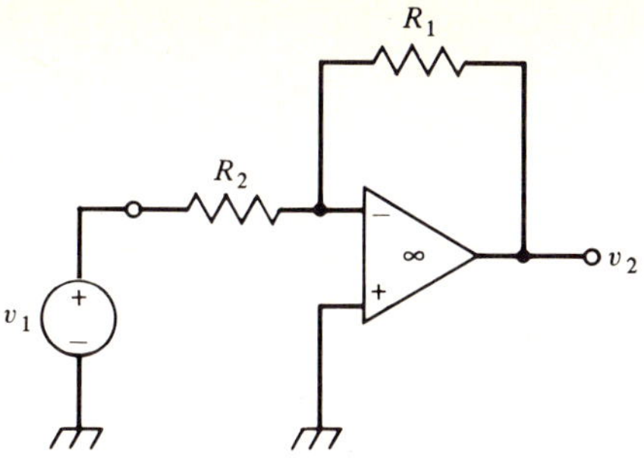

Figure 6-5 An inverting voltage amplifier.

The realization of a negative-gain voltage amplifier is shown in Fig. 6-5. Let us determine the relation between v_2 and v_1 of this circuit. Because of condition (6-14), the inverting input terminal will be at ground potential (zero voltage) as the noninverting input terminal is grounded. Note that the inverting input terminal is not physically connected to ground but simply acts as if grounded due to the active intervention of the operational amplifier and the surrounding circuitry. This is called *virtual ground*. Applying KCL at the inverting input terminal we obtain

$$\frac{v_1}{R_2} + \frac{v_2}{R_1} = 0$$

since condition (6-2) also holds. Therefore,

$$v_2 = -(R_1/R_2) \cdot v_1 \tag{6-17}$$

Since the inverting input terminal is at virtual ground, the input impedance of this amplifier is essentially R_2. However, the output impedance is zero if the operational amplifier is ideal.

The circuits of Figs. 6-3(a), 6-4, and 6-5 are used widely in electronic system design. For example, the noninverting amplifier configuration of Fig. 6-3(a) is used for electrometer applications such as pH meters. The unity gain configuration will be found in many sample-and-hold circuits (Section 8-2). It is also the basis of isolation amplifiers for medical instrumentation. However, in this case, to provide total isolation between the input and output signal ground circuits, battery power supply is employed along with transformer coupling. All of these amplifier configurations have found extensive applications in inductorless filtering, a brief discussion of which appears in Section 6-8.

We consider next a very typical design problem.

Example 6-2. An analog front end to an instrumentation system is to be designed. It is required to work with three full-scale ranges of input signals: ± 1 V, ± 10 V, and ± 100 V. For each input range, the output is required to

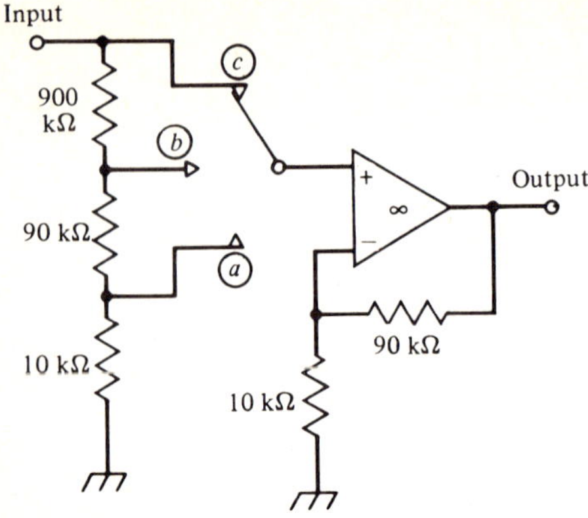

Figure 6-6 An analog front-end to an instrumentation system.

be ± 10 V full scale. The input resistance of the front end should be 1 MΩ and it must provide a buffered output.

Note that the problem reduces to the design of a voltage amplifier with an input impedance of 1 MΩ and having an adjustable gain of 10, 1, and 0.1 for the three different input ranges. One possible design is sketched in Fig. 6-6 which consists of a resistive potential divider with an adjustable tap connected to a noninverting voltage amplifier. The three gain settings are done by a manual switch. When the switch is connected to contacts $\textcircled{a}$, $\textcircled{b}$, and $\textcircled{c}$, respectively, the voltage gain of the potential divider takes the values 0.01, 0.1, and 1.0, respectively. The gain of the voltage amplifier is 10. The sum of the three resistances in the potential divider circuit is 1 MΩ—the desired input resistance.

Charge Amplifier

A number of transducers such as crystal phono pick-up microphones, capacitance alarms, and piezoelectric accelerometers have capacitive internal impedances. A circuit suitable for amplifying signals from such capacitive sources is sketched in Fig. 6-7 and is known as a charge amplifier. Note that this circuit is similar to the inverting amplifier of Fig. 6-5. Applying KCL to the inverting input terminal (and ignoring the resistance R_1 for the present) we arrive at

$$C_2 \frac{dv_1}{dt} + C_1 \frac{dv_2}{dt} = 0$$

If the initial voltages across the two capacitors are zero, then the above reduces to

$$\frac{v_2}{v_1} = -\frac{C_2}{C_1} \qquad (6\text{-}18)$$

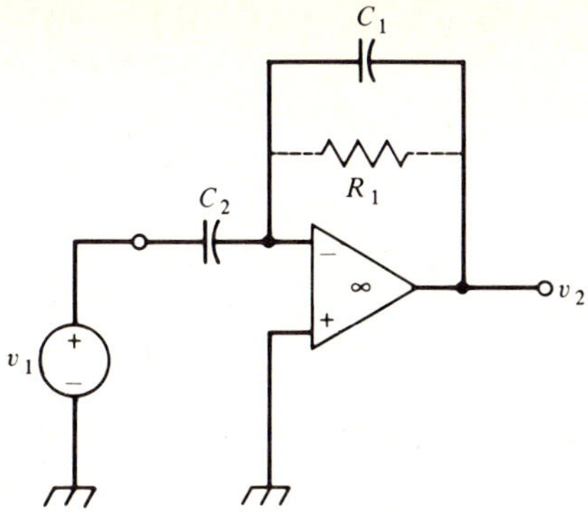

Figure 6-7   A charge amplifier.

   To provide a path for the dc bias current, a very high-valued resistance is connected across C_1 as shown by the dotted lines in the figure. Typical value of this resistance is of the order of 10,000 MΩ.

Transresistance Amplifier

The *transresistance amplifier* is a current-controlled voltage source. One realization of this circuit is shown in Fig. 6-8(a), for which

$$v_2 = -R_1 i_1 \tag{6-19}$$

Since the inverting input terminal is at virtual ground, the input resistance is zero. The output resistance is also seen to be zero. Note that the operation of the inverting voltage amplifier of Fig. 6-5 is essentially similar to that of Fig. 6-8(a) if the voltage source in series with R_2 is considered as a current source $i_1 = v_1/R_2$ in shunt with a resistance R_2.

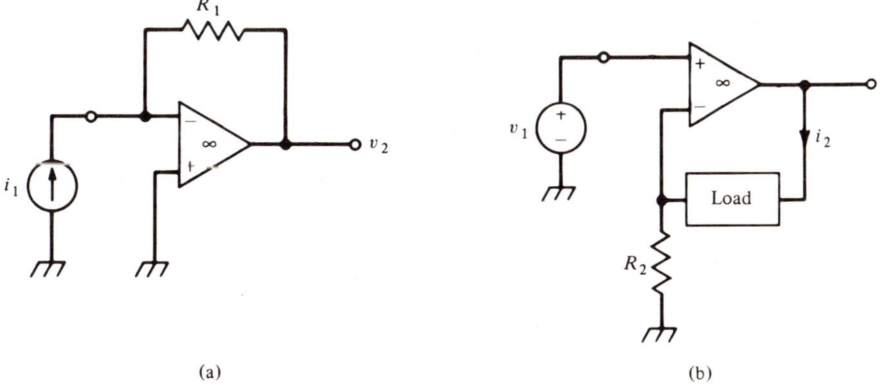

(a)

(b)

Figure 6-8   (a) A transresistance amplifier and (b) a transconductance amplifier.

The circuit of Fig. 6-8(a) can be used to convert the current output of many ICs to a voltage signal. It also finds applications for measuring currents below 1 pA as in gas chromatographs, ionization gauges, and photomultipliers.

Transconductance Amplifier

This type of basic amplifier is a voltage-controlled current source and is also called a *voltage-to-current converter*. A routine application of this circuit is as a current source. If a dc current source is desired, then the input voltage can be supplied by a battery or a dc power supply. It is often used as a line driver for signal transmission over long lines by converting the voltage signal to a current.

A very simple implementation of this amplifier is sketched in Fig. 6-8(b). It is seen that the voltage across R_2 is v_1 which causes a current v_1/R_2 to flow toward ground. This current also flows through the load. Hence,

$$i_2 = v_1/R_2 \tag{6-20}$$

The input and output resistances are infinite if the operational amplifier is ideal. Note, however, that the load here is floating, that is, neither of its terminals is grounded.

For a grounded load, the circuit of Fig. 6-9 can be used. Denoting the voltages at the input and the output terminals of the operational amplifier as v_2 and v_o, respectively, we write the KCL at the two input terminals as

$$\frac{v_1 - v_2}{R_1} = \frac{v_2 - v_o}{R_2}$$

$$\frac{v_o - v_2}{R_4} = \frac{v_2}{R_3} + i_2$$

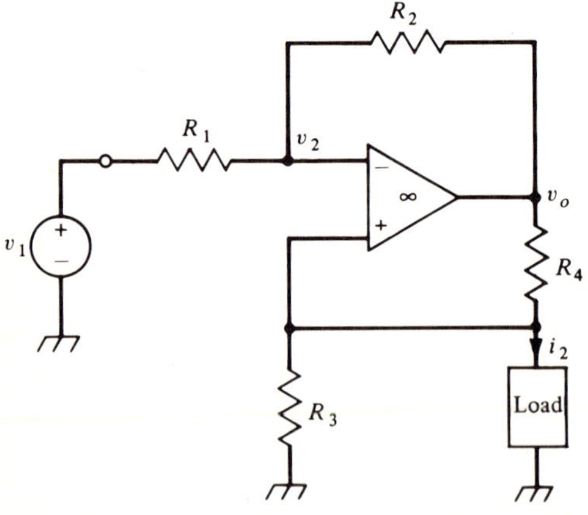

Figure 6-9 A transconductance amplifier for grounded load.

Eliminating v_o from above equations we arrive at

$$\frac{R_2}{R_1}v_1 + R_4 i_2 + \left(\frac{R_4}{R_3} - \frac{R_2}{R_1}\right)v_2 = 0 \tag{6-21}$$

If the resistors are chosen so that

$$R_1 R_4 = R_2 R_3$$

then we obtain from Eq. (6-21)

$$i_2 = -\frac{R_2}{R_1 R_4}v_1 = -\frac{v_1}{R_3} \tag{6-22}$$

Current Amplifier

One design of a *current amplifier* (current-controlled current source) for a floating load is indicated in Fig. 6-10. Analysis yields for this circuit

$$i_2 = -\left[1 + \frac{R_1}{R_2}\right]i_1 \tag{6-23}$$

It can also be designed by cascading a transresistance amplifier and a transconductance amplifier. This realization is left as an exercise (Problem 6-10).

Reference Voltage Source

The circuits of Figs. 6-3(a) and 6-5 are also suitable as reference voltage sources if the input voltage signal source v_1 is replaced by a battery or other type of dc voltage sources. Both circuits offer very low output resistances. These circuits are also capable of supplying large amounts of load currents if the operational amplifier used has the required output current capability. In the noninverting configuration of Fig. 6-3(a) the current from the input dc source will be negligible.

The input reference voltage can also be obtained with the aid of a zener diode from the power supply used for the operational amplifier.

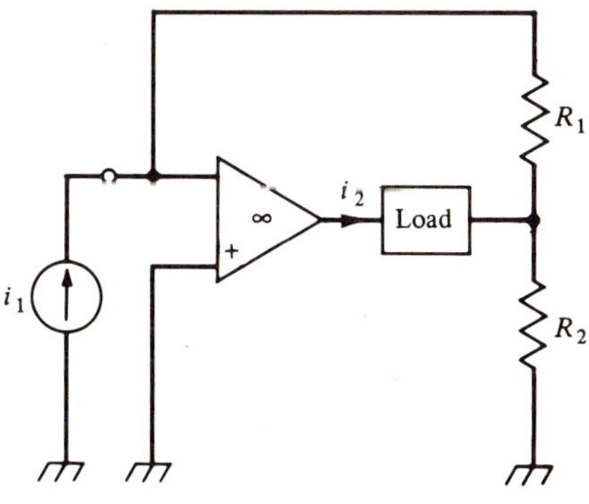

Figure 6-10   A current amplifier.

Summing Amplifiers

Often it is necessary to sum the voltages of a number of sources with appropriate scaling. This can be achieved by using a summing amplifier. Figure 6-11(a) shows a summing amplifier with negative gain. Applying KCL at the inverting input terminal we obtain

$$\frac{v_1}{R_1} + \frac{v_2}{R_2} + \frac{v_3}{R_3} + \frac{v_o}{R_0} = 0$$

Thus,

$$v_o = -\frac{R_0}{R_1} v_1 - \frac{R_0}{R_2} v_2 - \frac{R_0}{R_3} v_3 \tag{6-24}$$

A summing amplifier with positive gain is shown in Fig. 6-11(b). Because of negative feedback, the two input terminals of the operational amplifier are at the same potential v_x with respect to ground. Applying KCL at the noninverting input terminal we first obtain

$$\frac{v_1 - v_x}{R_1} + \frac{v_2 - v_x}{R_2} + \frac{v_3 - v_x}{R_3} = \frac{v_x}{R_0} \tag{6-25}$$

Next, applying KCL at the inverting input terminal we obtain

$$\frac{v_o - v_x}{R_b} = \frac{v_x}{R_a} \tag{6-26}$$

Eliminating the variable v_x from Eqs. (6-26) and (6-25) we arrive at the expression for the output voltage in terms of the input voltages as

$$v_o = \frac{(R_b + R_a)/R_a}{\dfrac{1}{R_1} + \dfrac{1}{R_2} + \dfrac{1}{R_3} + \dfrac{1}{R_0}} \left[\frac{v_1}{R_1} + \frac{v_2}{R_2} + \frac{v_3}{R_3} \right] \tag{6-27}$$

Extension of the summing amplifier circuits of Fig. 6-11 to more than three sources is straightforward. The summing amplifier is a very useful circuit. It is widely used in analog computation, digital-to-analog conversion, and many other applications.

Example 6-3. Design a summing amplifier to form the sum

$$v_o = 8v_1 + 4v_2 + 2v_3$$

where v_1, v_2, and v_3 are the input voltages.

Comparing above with Eq. (6-27) we set $R_1 = \frac{1}{8}$, $R_2 = \frac{1}{4}$, and $R_3 = \frac{1}{2}$. Then Eq. (6-27) reduces to

$$v_o = \frac{(R_b + R_a)/R_a}{14 + (1/R_0)} (8v_1 + 4v_2 + 2v_3)$$

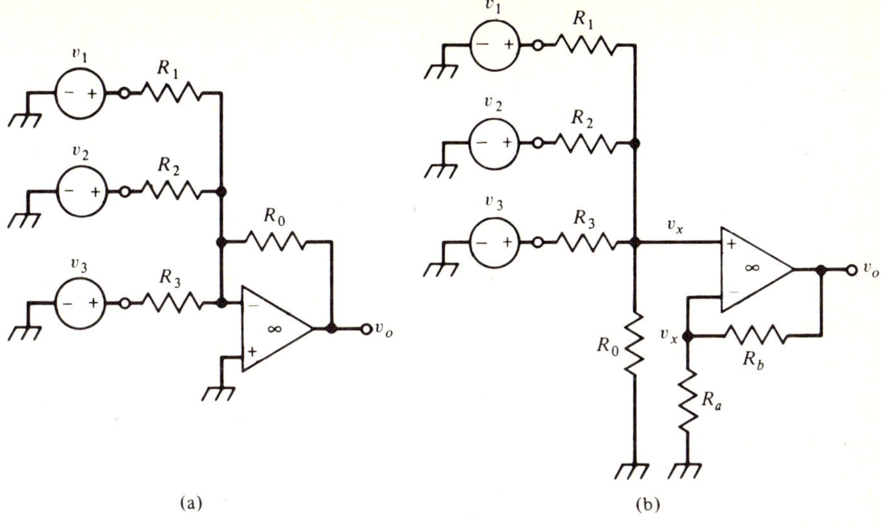

(a)   (b)

Figure 6-11 (a) Summing amplifier with negative gain, and (b) summing amplifier with positive gain.

We thus set

$$\frac{(R_b + R_a)/R_a}{14 + (1/R_0)} = 1$$

If we arbitrarily choose $R_0 = \frac{1}{2}$, then above equation implies $R_b/R_a = 15$. One possible solution to these two resistor values would be $R_b = 1$ and $R_a = 1/15$. We next scale the resistor values to bring them into a more practical range. Multiplying all the resistors by 15×10^3 we obtain one set of possible resistor values as

$$R_a = 1 \text{ k}\Omega, \qquad R_b = 15 \text{ k}\Omega$$
$$R_0 = 7.5 \text{ k}\Omega, \qquad R_3 = 7.5 \text{ k}\Omega$$
$$R_2 = 3.75 \text{ k}\Omega, \qquad R_1 = 1.875 \text{ k}\Omega$$

Difference Amplifier

In many applications, a circuit producing a difference of two voltages with appropriate scaling is needed. Such circuits are known as *difference amplifiers.* In its most commonly used form it is a dc amplifier amplifying the difference of two signals. It is also known as a differential amplifier, instrumentation amplifier, error amplifier, and bridge amplifier. Because of their widespread use, monolithic differential amplifiers are being made available commercially. Here we describe a number of realizations of these amplifiers using operational amplifiers.

A popular circuit for the design of a difference amplifier is shown in Fig. 6-12. Because of the connection from the output to the inverting input terminal, the

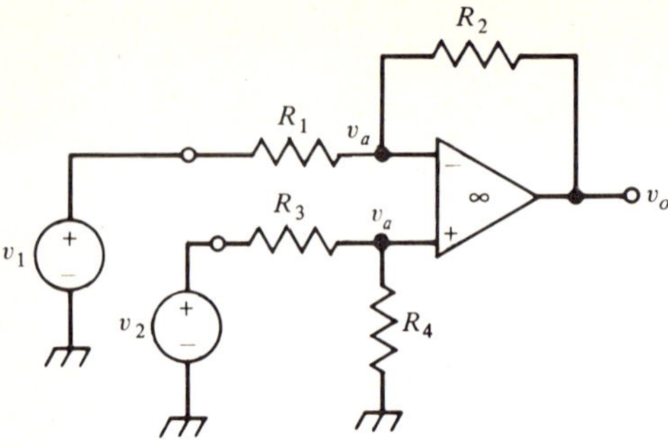

Figure 6-12   A difference amplifier.

two input terminals are at the same potential, which we denote as v_a. Now summing the currents at the two input nodes, we get

$$\frac{v_1 - v_a}{R_1} = \frac{v_a - v_o}{R_2}; \qquad \frac{v_2 - v_a}{R_3} = \frac{v_a}{R_4}$$

which yields the desired result

$$v_o = K_2 v_2 - K_1 v_1 \qquad (6\text{-}28)$$

where

$$K_1 = R_2/R_1 \qquad (6\text{-}29a)$$

$$K_2 = \frac{(R_2 + R_1)R_4}{(R_3 + R_4)R_1} \qquad (6\text{-}29b)$$

Next let us determine the restrictions on the scaling coefficients K_1 and K_2 in Eq. (6-28). Combining the two equations of Eq. (6-29) we arrive at

$$\frac{R_3}{R_4} = \frac{K_1 + 1}{K_2} - 1 \qquad (6\text{-}30)$$

Since K_1 and K_2 are required to be positive, R_3/R_4 will be positive if

$$K_1 + 1 > K_2 \qquad (6\text{-}31)$$

which is the desired realizability condition. In the following example we consider the design of a specific type of difference amplifier.

Example 6-4.   Develop a circuit to implement the difference

$$v_o = K(v_2 - v_1) \qquad (6\text{-}32)$$

Here $K_1 = K_2 = K$. With these values it is seen from Eq. (6-31) that the realizability condition is satisfied. From Eq. (6-30) the design equation is

$$R_2/R_1 = R_4/R_3 = K \qquad (6\text{-}33)$$

which can be used to design the difference amplifier for a specified value of K. For example, if $K = 5$, we can choose $R_1 = R_3 = 20\text{ k}\Omega$ and $R_2 = R_4 = 100\text{ k}\Omega$ to implement the difference amplifier. The gain K in Eq. (6-32) is more commonly known as the *differential-mode gain* [see Eq. (6-7b)].

The difference amplifier with $K_1 = K_2$, as discussed in the above example, is used frequently to measure the difference between two almost equal voltages. How accurately this difference can be measured depends on the *common-mode gain* of the difference amplifier which we define next. If we set $v_1 = v_2$ in Fig. 6-12, the common-mode gain A_{vcm} is defined as [see Eq. (6-7a)]

$$A_{vcm} = \left.\frac{v_o}{v_1}\right|_{v_1 = v_2} \tag{6-34}$$

It follows from Eq. (6-32) that if the operational amplifier in Fig. 6-12 is ideal and if all the resistors are matched according to Eq. (6-33), then the common-mode gain is zero. Or, in other words, the difference amplifier provides an infinite *common-mode rejection*. As a result, in theory, the circuit of Fig. 6-12 can be used to measure the difference of two arbitrarily close voltages. However, in practice, due to the nonideal characteristics of a practical operational amplifier, the common-mode rejection is finite. Later (in Section 6-6) we discuss the effect of finite common-mode rejection. It should be noted that under the matched condition given by Eq. (6-33), the input impedance seen across the two input terminals is $2R_1$.

Another typical application of the above difference amplifier is in processing low-level outputs of a "floating" transducer. Usually in such applications, the output of the transducer needs to be amplified first by a preamplifier before it can be processed. In many cases, the preamplifier cannot be placed close to the transducer, necessitating the use of a long cable to carry the transducer output to the amplifier. To prevent contamination of the signal by the extraneous noise such as that induced by ac power lines, the cable is shielded with the shield connected to the ground. However, some amount of noise may still appear as a common-mode signal at each input of the difference amplifier. To minimize the effect of this noise, a difference amplifier with very high common-mode rejection is used as the preamplifier.

If $K_1 \neq K_2$, the common-mode gain of the circuit of Fig. 6-12 is no longer zero but equal to $K_2 - K_1$. If finite common-mode rejection is not a problem, then this circuit can be used to provide the difference of two voltages with unequal scaling. However, in this case, condition (6-31) must be met to ensure realization with positive resistors.

One drawback of the circuit of Fig. 6-12 as a differential amplifier (i.e., $K_1 = K_2 = K$) is that to change the differential-mode gain K, two resistors have to be adjusted while maintaining relation (6-33) (to ensure zero common-mode gain). This may prove difficult in practice. A modification of this basic circuit shown in Fig. 6-13 circumvents this problem.[15] Here gain can be varied

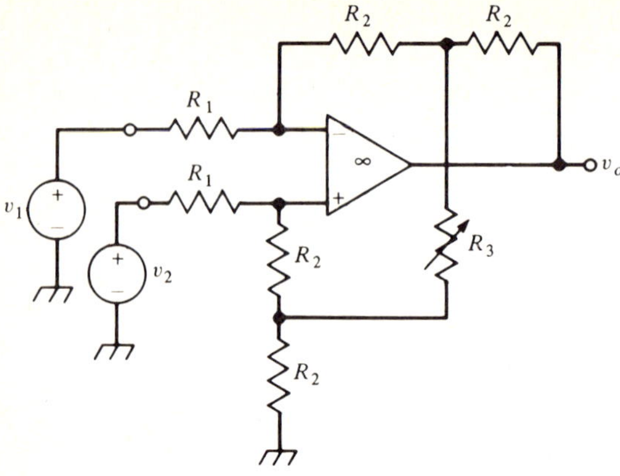

Figure 6-13 A differential amplifier with adjustable gain.

by adjusting a single resistor R_3. For this circuit it can be shown that

$$v_o = 2\left(1 + \frac{1}{\alpha}\right)\frac{R_2}{R_1}(v_2 - v_1) \tag{6-35}$$

where we have used $R_3 = \alpha R_2$.

Another realization of the differential amplifier can be obtained with a 2-input summing amplifier and an inverting amplifier (Problem 6-13).

All of the differential amplifier circuits considered so far have low-input impedances and thus may load the input sources in some cases. The circuit of Fig. 6-12 can be changed to a high-input impedance differential amplifier by buffering the input v_1 through a noninverting amplifier of the form of Fig. 6-3(a) and feeding v_2 directly to the noninverting input terminal without the potential divider. The complete design is left as an exercise (Problem 6-14).

Bridge Amplifiers

Some applications require the measurement of the change in the resistance value of a transducer. If the resistance value depends on a physical parameter, then the change of resistance value provides a measure of the change in the physical parameter. An example of such a transducer is the strain gauge which is made of a metal wire or a semiconductor structure cemented to an insulating base. Lengthening of the resistor caused by tension increases the resistor value and shortening of the resistor by compression decreases the resistance value. If the strain gauge is cemented to a physical structure such as a beam of a bridge, monitoring of the change in resistor value provides a measure of the stress of the physical structure caused by heat, wind, or other effects.[15]

One way to measure the change in resistance value is to connect the transducer in a bridge circuit as shown in Fig. 6-14. Here the resistor R_t represents the resistance of a transducer whose change is being monitored. The resistors

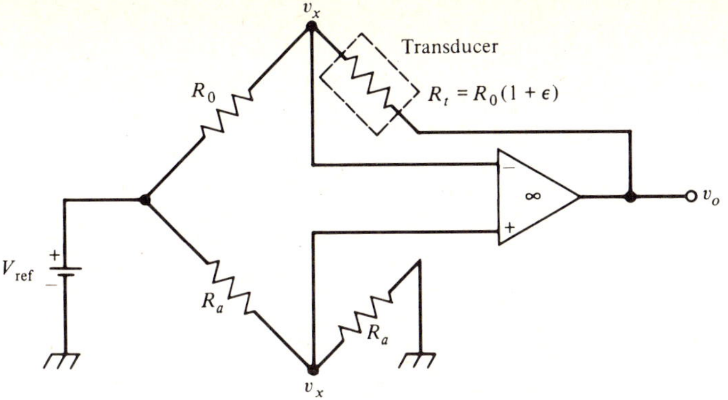

Figure 6-14 A bridge amplifier.

R_0 and R_a are precision resistors. An analysis of this circuit yields the expression
for the output voltage as

$$v_o = \frac{V_{ref}}{2}\left(1 - \frac{R_t}{R_0}\right) \tag{6-36}$$

Normally, the bridge is balanced with R_t nominally at a value R_0. This results
in v_o being zero. If R_t changes to a new value $R_t = R_0(1 + \varepsilon)$, the output voltage
becomes

$$v_o = -(V_{ref}/2) \cdot \varepsilon$$

and is thus linearly related to the change.

An alternate realization of the bridge amplifier makes use of the differential
amplifier of Example 6-4 and is sketched in Fig. 6-58 of Problem 6-18. However,
this circuit develops an output voltage approximately linearly related to ε for
$|\varepsilon| \ll 1$.

Differential-Input, Differential-Output Voltage Amplifier

With the aid of two differential-input operational amplifiers it is possible to
design a differential-input, differential-output voltage amplifier. The pertinent
circuit[14] arrangement is sketched in Fig. 6-15.

If we denote the input terminal voltages with respect to ground as v_{1A} and
v_{1B}, and the output terminal voltages with respect to ground as v_{2A} and v_{2B},
then it follows from the figure that

$$v_1 = v_{1A} - v_{1B}$$
$$v_2 = v_{2A} - v_{2B} \tag{6-37}$$

The circuit of Fig. 6-15 can be recognized as two noninverting amplifiers connected in a back-to-back fashion. Hence from Eq. (6-15) we can write

$$v_{2A} = \left(1 + \frac{R_1}{R_2}\right)v_{1A}, \qquad v_{2B} = \left(1 + \frac{R_3}{R_4}\right)v_{1B} \tag{6-38}$$

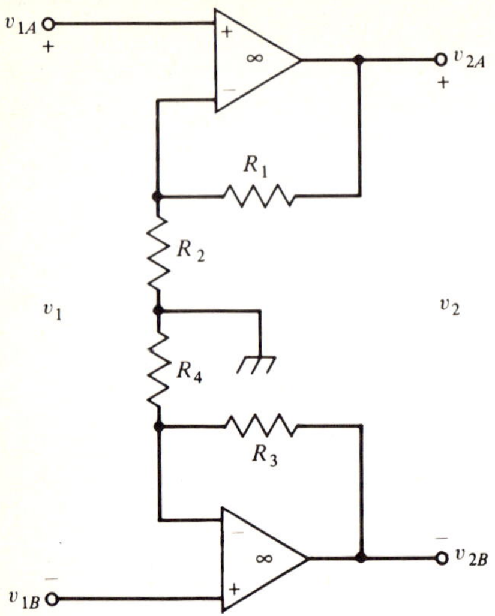

Figure 6-15   A differential-input differential-output amplifier.

If we set $R_1/R_2 = R_3/R_4$ then take the difference of the above two equations, and make use of Eq. (6-37), we arrive at the desired result:

$$v_2 = \left(1 + \frac{R_1}{R_2}\right)v_1 \tag{6-39}$$

Thus the differential-mode gain is $(1 + R_1/R_2)$

To determine the common-mode gain, we set $v_{1A} = v_{1B}$, and obtain from Eqs. (6-37) and (6-38),

$$A_{vcm} = \left(1 + \frac{R_1}{R_2}\right) - \left(1 + \frac{R_3}{R_4}\right) \tag{6-40}$$

which is zero when $R_1/R_2 = R_3/R_4$.

A variation of this circuit is found in Problem 6-12.

6-3   Characteristics of a Practical Operational Amplifier

Practical operational amplifier modules are available in three different forms—discrete, hybrid IC, and monolithic IC. These modules usually contain a single amplifier circuit. Recently monolithic IC modules containing two or more separate operational amplifiers have appeared in the market. As the name implies, a *dual operational amplifier* contains two such devices, a *triple operational amplifier* has three amplifier circuits, and so on. Several varieties of standard packages are used to package these amplifiers and some typical ones were described in Section 1-6.

A number of basically different approaches are followed in designing a practical operational amplifier with each design offering some advantages over the others with regard to the performance of the amplifier in a system. These approaches are usually used to classify the practical operational amplifiers[11] and are as follows: (1) general-purpose bipolar transistor types, (2) FET-input types, (3) varactor-bridge input types, and (4) chopper-stabilized types. We point out later the basic characteristics of these various types of operational amplifiers. However, a detailed discussion on them is beyond the scope of this book.

No matter how the practical operational amplifier is designed and manufactured, its characteristics are different from those of the ideal model discussed in Section 6-1. However, for linear applications an operational amplifier is always used with a path from the output to the inverting input terminal which improves its functional performance; and with careful design the overall behavior of a circuit containing a practical operational amplifier approaches that obtainable with the idealized model.

In addition to the characteristics of the nonideal amplifier discussed earlier in Section 5-5, a practical operational amplifier exhibits some additional features. In this section we outline some of the important characteristics of these versatile devices. Effect of these limitations are evaluated in later sections. A thorough understanding of these practical limitations is necessary to make the best use of the devices.

A Typical IC Module and Its Characteristics

Some of the external features of the operational amplifier module are best described in terms of a typical package. Just like other electronic circuit modules, an operational amplifier package has a number of external terminals in addition to the input and output terminals. Usually it has one or two terminals to connect dc power supplies to provide the necessary power and to bias the transistors properly for linear operation. A terminal is provided for connecting the system to ground. Some modules may have terminals for connecting RC compensating networks to modify the frequency response of the open-loop gain. Terminals are also included for nulling the offset voltage and current externally.

A highly popular monolithic module is the Type 741 operational amplifier. It is manufactured by a number of IC companies. The schematic diagram of the Fairchild μA741 is shown in Fig. 6-16(a). This device is constructed on a single silicon chip and is available in five different packages: 8-lead TO-5, 10-lead flatpack, 14-lead cavity DIP, and 8-lead and 14-lead molded DIPs. Figure 6-16(b) shows the lead connection diagram for the 10-lead flat package.

Both positive and negative dc supply voltages are usually used for this amplifier. The supply voltages are typically ± 15 V. It is possible to operate this device using a single power supply. In most cases, the 741 (741C) is a direct plug-in replacement for the 709C, 748, LM201, and MC1439.

The parameters characterizing this operational amplifier are shown in Table 6-1 along with the absolute maximum ratings.

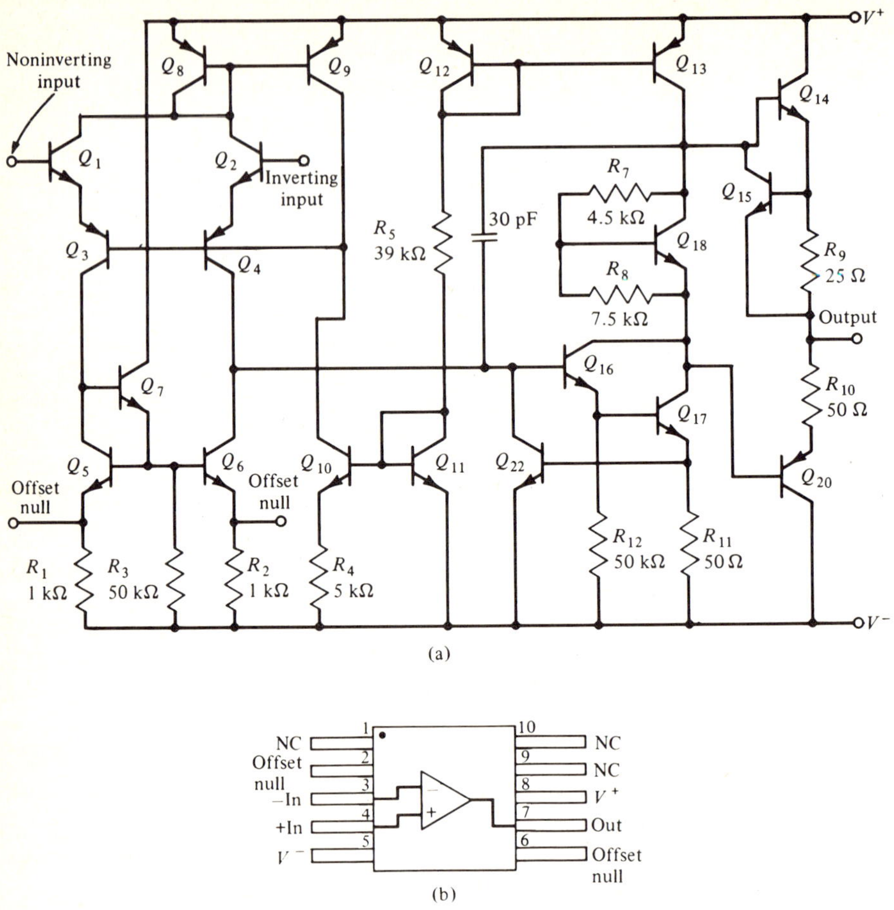

Figure 6-16 μA741 operational amplifier: (a) circuit diagram, and (b) pin connection for the 10-lead flatpak. (Courtesy Fairchild Camera and Instruments Corp.)

Some of the pertinent parameters characterizing an operational amplifier module are elaborated next.

Input and Output Resistances

The input and output resistances of a practical operational amplifier are finite and nonzero. Since there are two input terminals, the input resistance can be measured in three different ways. The input resistance seen between each input terminal and the ground is called the *common-mode input resistance*, R_{ic}. Because of matched input stages, the two common-mode resistances are almost equal and usually their values are very high. The input resistance seen between the two input terminals is called the *differential-input resistance* R_{id} whose value is also high but less than that of the common-mode input resistances. The differential-input impedance of the μA741 is typically 2 MΩ.

TABLE 6-1[a]

μ**A741 : Electrical Characteristics ($V_S = \pm 15V$, $T_A = 25°C$ unless otherwise specified)**

Parameters		Conditions	Min	Typ	Max	Units
Input Offset Voltage		$R_S \leq 10 \text{ k}\Omega$		1.0	5.0	mV
Input Offset Current				20	200	nA
Input Bias Current				80	500	nA
Input Resistance			0.3	2.0		MΩ
Input Capacitance				1.4		pF
Offset Voltage Adjustment Range				± 15		mV
Large Signal Voltage Gain		$R_L \geq 2 \text{ k}\Omega$, $V_{\text{out}} = \pm 10 \text{ V}$	50,000	200,000		
Output Resistance				75		Ω
Output Short-Circuit Current				25		mA
Supply Current				1.7	2.8	mA
Power Consumption				50	85	mW
Transient Response (Unity Gain)	Rise time	$V_{\text{in}} = 20 \text{ mV}$, $R_L = 2 \text{ k}\Omega$, $C_L \leq 100 \text{ pF}$		0.3		μsec
	Overshoot			5.0		%
Slew Rate		$R_L \geq 2 \text{ k}\Omega$		0.5		V/μsec
The following specifications apply for $-55°C \leq T_A \leq +125°C$:						
Input Offset Voltage		$R_S \leq 10 \text{ k}\Omega$		1.0	6.0	mV
Input Offset Current		$T_A = +125°C$		7.0	200	nA
		$T_A = -55°C$		85	500	nA
Input Bias Current		$T_A = +125°C$		0.03	0.5	μA
		$T_A = -55°C$		0.3	1.5	μA
Input Voltage Range			± 12	± 13		V
Common-Mode Rejection Ratio		$R_S \leq 10 \text{ k}\Omega$	70	90		dB
Supply Voltage Rejection Ratio		$R_S \leq 10 \text{ k}\Omega$		30	150	μV/V
Large-Signal Voltage Gain		$R_L \geq 2 \text{ k}\Omega$, $V_{\text{out}} = \pm 10 \text{ V}$	25,000			
Output Voltage Swing		$R_L \geq 10 \text{ k}\Omega$	± 12	± 14		V
		$R_L \geq 2 \text{ k}\Omega$	± 10	± 13		V
Supply Current		$T_A = +125°C$		1.5	2.5	mA
		$T_A = -55°C$		2.0	3.3	mA
Power Consumption		$T_A = +125°C$		45	75	mW
		$T_A = -55°C$		60	100	mW

[a] Courtesy Fairchild Camera and Instruments Corp.

Likewise, the *output resistance* R_o is the resistance seen between the output terminal and the ground with the two input terminals connected to the ground. Typically R_o has a very low value. In the case of μA741, it is of the order of 75 Ω.

The nonideal model of a practical operational amplifier taking into account these resistances is sketched in Fig. 6-17.

Output Voltage Limitations

Just like any other amplifier, the input–output relation of a practical operational amplifier is nonlinear. However, for small values of the input differential voltage, within about a millivolt range, the relation is linear. Outside this range, the output of the amplifier gets saturated, remaining at a constant value. The maximum output voltage that can be developed by the amplifier before going to saturation is given by its *output voltage swing*. For the μA741, the output voltage swing is about ± 14 V (with power supply voltages at ± 15 V and a load resistance R_L of 10 kΩ connected at the output). An IC operational amplifier can usually withstand saturated output conditions for a certain amount of time, but if this condition is maintained for too long, permanent damage may occur to the device. However, the 741 operational amplifier has been designed to provide overload protection both on the input and output.

Input Voltage Limitations

There are two types of input voltage limitations that must be observed. The maximum voltage that can be applied across the two input terminals without damaging the amplifier is its *differential-input voltage limit*. For the μA741 this limit is about ± 30 V.

If we set $v_{i1} = v_{i2}$, then the same voltage is applied to both the input terminals and the input difference voltage is zero. v_{i1} can be made physically equal to v_{i2} by shorting the two input terminals and connecting them to a single source. Alternately, when there is a path from the output to the inverting input terminal, the two input terminals are also at the same potential. The voltage common to

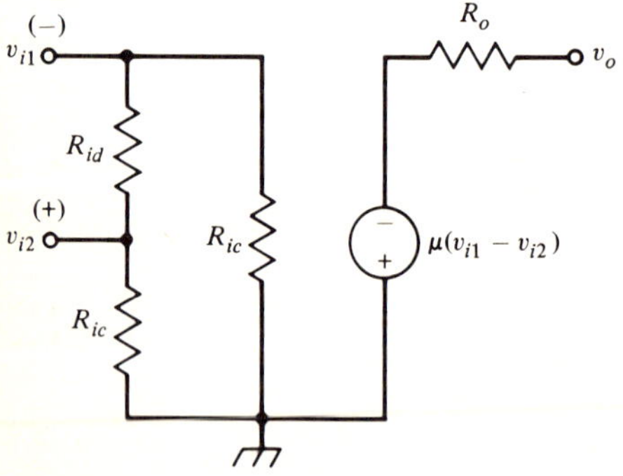

Figure 6-17 Model of a practical operational amplifier incorporating input and output resistances.

both input terminals is the *common-mode input voltage*. The maximum peak voltage that can be applied to either terminal without driving the amplifier into a nonlinear mode is its *common-mode input voltage limit* or simply the *input voltage limit*. Note that if a sine wave having a peak value greater than the input voltage limit is applied between the two inputs joined together and the ground, the output will be distorted. The input voltage range of the μA741 is about ± 15 V. If the supply voltage is less than ± 15 V, then the input voltage range should not exceed the supply voltage.

It should be noted that in the inverting amplifier configuration (Fig. 6-5) both the input terminals are at ground potential. Hence the common-mode input voltage range will never be exceeded. On the other hand, in the case of the noninverting amplifier configuration [Figs. 6-3(a) and 6-4], the common-mode input voltage limit must be observed. Some operational amplifiers, when used as voltage followers (Fig. 6-4), may exhibit *latch-up* if the common-mode voltage limits are exceeded. This is due to the saturation of the input transistor at the inverting input causing it to behave temporarily as a noninverting input and resulting in positive feedback. However, it should be noted that a number of amplifiers, such as the μA741, are designed specifically to avoid latch-up.

Offset

As pointed out earlier, for an ideal operational amplifier the output voltage should be zero if the input differential voltage is zero. In the case of a practical operational amplifier, output may not be identically zero in the absence of an input signal but may have a dc offset voltage, usually a small voltage. The offset voltage is caused by the inability to match components precisely in a practical circuit. More on offset will be discussed later in Section 6-5.

Frequency Response

As indicated before, the open-loop gain of a practical amplifier is not constant but is a monotonically decreasing function of frequency. Some monolithic operational amplifiers such as the 741 are internally compensated to provide a 6 dB/octave roll-off up to the *unity-gain crossover frequency*. The open-loop frequency response of the μA741 is shown in Fig. 6-18. Many other operational amplifiers, however, have an open-loop frequency response rolling off at 12 dB/octave or higher before the open-loop gain becomes unity. These units require external compensation to make overall open-loop gain to have a 6 dB/octave roll-off. It should be noted that because of the finite bandwidth, the assumption of very high gain is valid over a small frequency range. Even in this frequency range there is practically unlimited applications of these devices.

Transient Response and Slew-Rate Limitation

Since the internal and parasitic capacitances present at various places in the operational amplifier take a finite amount of time to charge and discharge, the practical amplifier will be unable to respond instantly to a step input or a large amplitude sinusoidal input signal. The step response, known more commonly

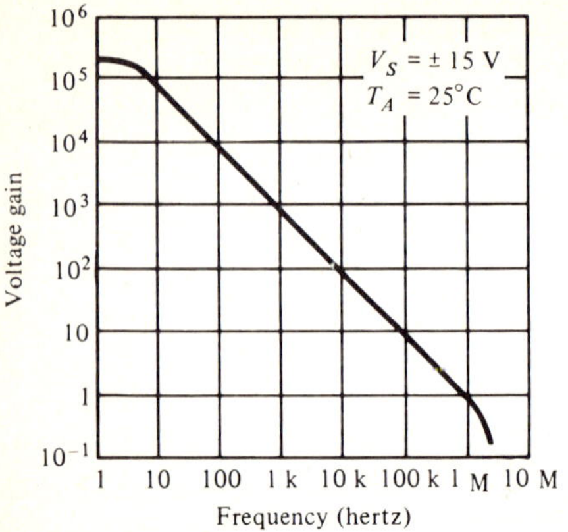

Figure 6-18   Open-loop frequency response of the μA741 operational amplifier. (Courtesy Fairchild Camera and Instruments Corp.)

as the transient response, is usually measured in the voltage follower configuration, and is typically as shown in Fig. 6-19. It is characterized by three parameters—*rise time*, *delay time*, and *overshoot*—defined earlier in Section 5-7.

The ability of the operational amplifier to follow a rapidly changing sinusoidal signal is described by its slew-rate limit which was also defined in Section 5-7. The slew-rate limit determines, for a prescribed frequency, the maximum peak amplitude of the input sinusoidal signal that would result in an undistorted output. A typical value of the slew rate for the μA741 is 0.5 V/μsec.

Common-Mode Rejection

An operational amplifier is required to amplify only the input difference signal and reject the common-mode input signal. In a practical amplifier the common-

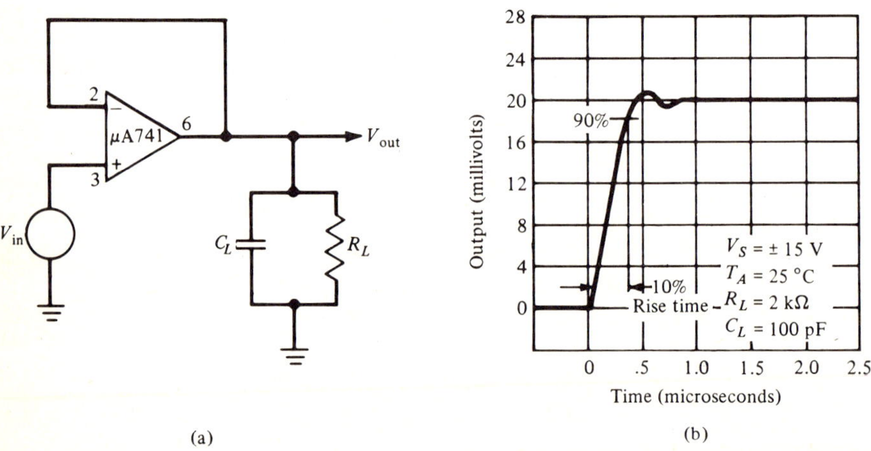

Figure 6-19   (a) Transient response test circuit for μA741. Pin connection shown is for TO-99 package, (b) transient response. (Courtesy Fairchild Camera and Instruments Corp.)

mode input signal is not completely rejected and causes an error voltage to appear at the output. The amount of rejection relative to the signal caused by the input difference signal is determined by the common-mode rejection ratio (CMRR). A good amplifier should have a very high CMRR. This aspect is discussed in detail in Section 6-6.

Supply Voltage Rejection

In some applications if the power supplies used to bias the monolithic operational amplifier are not well regulated, the power supply variation (ripple) may affect the performance of the operational amplifier. The effect of the ripple can be estimated from its *supply voltage rejection ratio* (SVRR) which is normally supplied by the manufacturer. SVRR is defined as the ratio of the change in input offset voltage to the change in supply voltage; this factor in essence determines how good the amplifier is in discriminating against power supply variations. Usually, a practical monolithic operational amplifier is characterized by a relatively low value of SVRR. For example, the SVRR for the Fairchild μA741 is 30 μV/V.

Let us now consider the effect of the nonideal characteristics on the performance of an operational amplifier in a closed-loop configuration. To this end, we shall analyze the performances of the finite-gain inverting and noninverting amplifier configurations, taking into account some of the major nonideal characteristics of a practical operational amplifier. However, our analysis procedures can be readily extended to other circuits as well.

6-4   Effects of Input, Output Resistances, and Finite Gain

The effect of finite gain and the finite, nonzero input and output resistances of a practical operational amplifier on the operation of a circuit containing these devices is best illustrated by analyzing typical circuit configurations as feedback amplifiers. These results can then be extrapolated to other circuits and, if necessary, exact results can be derived following similar analysis techniques. We consider first the noninverting amplifier circuit of Fig. 6-3(a) to illustrate the basic concepts of feedback and its properties. We shall represent the operational amplifier by the simplified nonideal model shown in Fig. 6-20.

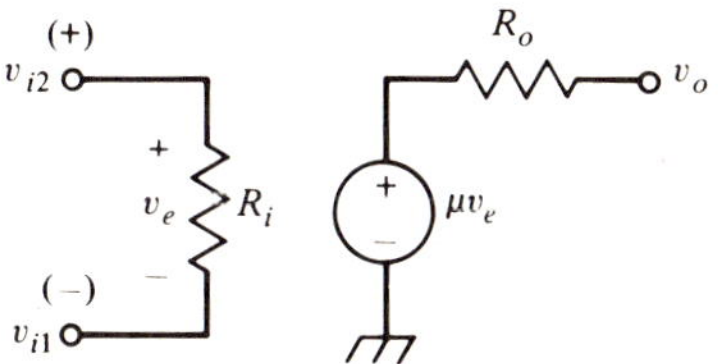

Figure 6-20   A simplified model of a nonideal operational amplifier.

Noninverting Voltage-Amplifier Configuration

The noninverting voltage amplifier of Fig. 6-3(a) with the nonideal model of the operational amplifier given by Fig. 6-20 is shown in Fig. 6-21 showing explicitly the feedback amplifier arrangement. A feedback voltage v_f, proportional to the output voltage (generated by the potential divider feedback network) is fed back to the input side. The input voltage v_e developed across the operational amplifier input terminals is given by the difference of the signal voltage v_1 and the feedback voltage v_f; that is,

$$v_e = v_1 - v_f \tag{6-41}$$

Since the feedback signal is proportional to the output *voltage* and is fed in *series* at the input side, the circuit of Fig. 6-21 is an example of *voltage-series* feedback amplifier configuration.

Closed-Loop Gain.   To simplify the analysis of this circuit and to show more clearly the effect of feedback, let us assume that the resistors R_1 and R_2 have been chosen such that the feedback network does not load the output of the amplifier, that is, the current flowing through R_1 can be neglected. From Fig. 6-21 we now have

$$v_2 = \mu v_e \tag{6-42}$$

and

$$v_f = \frac{R_2}{R_1 + R_2} \cdot v_2 = \beta v_2 \tag{6-43}$$

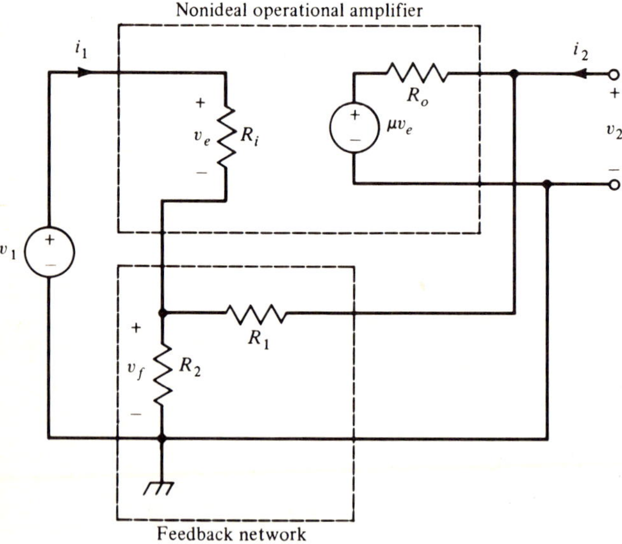

Figure 6-21   Feedback amplifier representation of the noninverting voltage amplifier.

where $\beta = R_2/(R_1 + R_2)$, known as the *feedback factor*, denotes the amount of output being fed back to the input. From Eqs. (6-41), (6-42), and (6-43) we obtain the closed-loop voltage gain G_v as

$$G_v = \frac{v_2}{v_1} = \frac{\mu}{1 + \mu\beta} \tag{6-44}$$

Note that if there is no feedback (that is, $\beta = 0$), the gain v_2/v_1 becomes μ, the voltage amplification factor of the operational amplifier. This is why μ is also called the *open-loop gain.*

In a practical amplifier μ is very large and for all practical purposes $\mu\beta \gg 1$. Then Eq. (6-44) reduces to

$$G_v \simeq \frac{1}{\beta} = \frac{R_1}{R_2} + 1 \tag{6-45}$$

which is identical to Eq. (6-15) derived in the case of an ideal operational amplifier. Equation (6-45) indicates that the closed-loop gain is pretty much independent of the open-loop gain provided $\mu\beta \gg 1$ and depends only on a resistor ratio which can be accurately maintained. Since $|1 + \mu\beta| > 1$, the type of feedback is known as *negative feedback.**

Input and Output Resistances.   To compute the input resistance R_{in} of the feedback amplifier of Fig. 6-21, we observe from the figure that

$$v_e = R_i i_1 \tag{6-46}$$

From Eqs. (6-46), (6-42), and (6-44) we arrive at

$$R_{\text{in}} = \frac{v_1}{i_1} = R_i(1 + \mu\beta) \tag{6-47}$$

Again, $(1 + \mu\beta)$ being much greater than 1, the effective input resistance of the noninverting amplifier is increased.

Next, to compute the output resistance R_{out}, we first set the signal voltage v_1 equal to zero. Then, from Eqs. (6-41) and (6-43), we derive

$$v_e = -v_f = -\beta v_2 \tag{6-48}$$

From the figure, we also observe that

$$v_2 = R_o i_2 + \mu v_e \tag{6-49}$$

Combining Eqs. (6-48) and (6-49) we arrive at the expression for the output resistance as

$$R_{\text{out}} = \frac{v_2}{i_2}\bigg|_{v_1=0} = \frac{R_o}{(1 + \mu\beta)} \tag{6-50}$$

Again, $(1 + \mu\beta)$ being much greater than 1, the effective output resistance is considerably lower than R_o.

---

* For *positive feedback* $|1 + \mu\beta| < 1$.

Example 6-5.   Evaluate the performance of a noninverting amplifier of gain 10 designed using an operational amplifier whose open-loop gain, and input and output resistances are given by

$$\mu = 5000 \qquad R_i = 25 \text{ k}\Omega \qquad R_o = 100 \,\Omega \qquad\qquad (6\text{-}51)$$

Observe from Eq. (6-45) a closed-loop gain of 10 is achieved with a $\beta = 0.1$ assuming the operational amplifier to be ideal. Hence, from Eqs. (6-44), (6-47), and (6.50), we obtain

$$G_v = \frac{5000}{1 + 500} = 9.98$$

$$R_{\text{in}} = 25 \times 10^3 \times (1 + 500) = 12.525 \text{ M}\Omega \qquad\qquad (6\text{-}52)$$

$$R_{\text{out}} = \frac{100}{1 + 500} = 0.1996 \,\Omega$$

It should be noted that in spite of a low open-loop gain, the closed-loop properties of the noninverting amplifier are quite close to that obtained with an ideal operational amplifier. Let us now develop an approximate expression to determine the fractional change in the closed-loop gain due to the finiteness of μ.

Gain Error.   Let us denote the ideal value of the closed-loop gain as G_{vi} which is equal to $1/\beta$. We can then rewrite Eq. (6-44) as

$$G_v = \left(\frac{1}{\beta}\right) \cdot \frac{\mu\beta}{1 + \mu\beta} = G_{vi}\left(\frac{\mu\beta}{1 + \mu\beta}\right) \qquad\qquad (6\text{-}53)$$

Since the closed-loop gain in the idealized case is G_{vi}, we consider $\mu\beta/(1 + \mu\beta)$ as the error factor. This factor can be expressed as

$$\frac{\mu\beta}{1 + \mu\beta} = \left[1 + \frac{1}{\mu\beta}\right]^{-1} \cong 1 - \frac{1}{\mu\beta}$$

as for most applications $\mu\beta \gg 1$. Using above in Eq. (6-53) we obtain

$$G_v = G_{vi} - \frac{G_{vi}}{\mu\beta}$$

Therefore the fractional change in the closed-loop gain due to a finite value of the open-loop gain μ is

$$\frac{\Delta G}{G_{vi}} = \frac{G_{vi} - G_v}{G_{vi}} \simeq \frac{1}{\mu\beta} \qquad\qquad (6\text{-}54)$$

Example 6-6.   For the noninverting voltage amplifier of Example 6-5, $\mu = 5000$ and $\beta = 0.1$. Hence the percentage error in the closed-loop gain is $100/(5000 \times 0.1) = 0.2$ percent.

The gain μ of monolithic operational amplifiers of the same type made in large volume quantities by batch processing would not be the same for each amplifier. Depending on the application we can determine the minimum gain necessary for a prescribed maximum error in closed-loop gain. The following example illustrates the approach.

Example 6-7.   A noninverting amplifier with a nominal value of closed-loop gain of 20 is to be designed. Determine the minimum value of the operational amplifier open-loop gain (to be designated as μ_m) so that the closed-loop gain does not change by more than 0.1 percent due to variation in μ.

From Eq. (6-54) with $\mu = \mu_m$, the greatest fractional change in the closed-loop gain is

$$\frac{\Delta G}{G_{vi}} = \frac{1}{\mu_m \beta}$$

For our problem

$$\frac{\Delta G}{G_{vi}} \leq 0.001$$

Since $\beta = 1/G_{vi} = 0.05$, we get from above two equations,

$$\frac{1}{\mu_m(0.05)} \leq 0.001$$

or

$$\mu_m \geq 20{,}000$$

Hence the minimum gain of the operational amplifier must be 20,000, that is, 86.02 dB.

Inverting Voltage Amplifier Configuration

As a second example to determine the effects of finite gain, and finite input and output resistances of a practical operational amplifier we consider next the inverting amplifier of Fig. 6-5 which has been redrawn in Fig. 6-22 to show clearly the feedback path. Note that in this circuit, the feedback signal generated is a current proportional to the output *voltage*. It is added to the current from the excitation and fed into the operational amplifier input. Since the output of the feedback network is in *shunt* with the amplifier input terminals, this circuit is an example of a *voltage-shunt feedback amplifier*.

Closed-Loop Gain.   To simplify the analysis, let us first assume that the input resistance R_i of the operational amplifier is very large and hence the current flowing through it can be neglected. Likewise the output resistance R_o is very small and the voltage developed across it can also be neglected. Under these

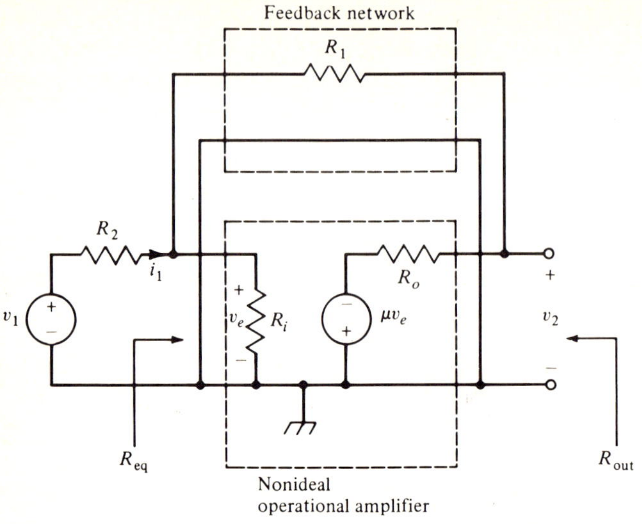

Figure 6-22   Feedback amplifier representation of the inverting voltage amplifier.

assumptions, then, we observe from the circuit diagram that we can write

$$v_2 = -\mu v_e \tag{6-55}$$

and

$$\frac{v_1 - v_e}{R_2} = \frac{v_e - (-\mu v_e)}{R_1} \tag{6-56}$$

Eliminating v_e from the above two equations, we arrive at the expression for the closed-loop gain as

$$G_v = \frac{v_2}{v_1} = \frac{-\mu R_1/(R_1 + R_2)}{1 + \mu R_2/(R_1 + R_2)} \tag{6-57}$$

If we denote

$$\beta = \frac{R_2}{R_1 + R_2} \tag{6-58}$$

then we can rewrite Eq. (6-57) as

$$G_v = \left(-\frac{R_1}{R_2}\right) \cdot \frac{\mu\beta}{1 + \mu\beta} \tag{6-59}$$

Note that for a practical operational amplifier, $\mu\beta \gg 1$, and as a result

$$G_v \simeq -\frac{R_1}{R_2} \tag{6-60}$$

which is seen to be the gain obtained in the case of an ideal operational amplifier. Thus, for most practical purposes, the closed-loop gain is independent of μ and depends only on the ratio of two resistors which can be accurately maintained. If we denote the ideal gain as $G_{vi} = -R_1/R_2$, then Eq. (6-59) reduces to

$$G_v = G_{vi} \cdot \frac{\mu\beta}{1 + \mu\beta} \tag{6-61}$$

which is identical in form to Eq. (6-53) derived for the noninverting amplifier. Hence, to compute the fractional change in the closed-loop gain due to finite μ, we can also use the expression given in Eq. (6-54).

Input and Output Resistances. Note that the feedback signal is proportional to the output voltage in both the noninverting and the inverting amplifier configurations. Hence the output resistance R_{out} of the inverting amplifier is of a form similar to that of the noninverting amplifier as given by Eq. (6-50) and is thus considerably smaller than R_o.

We observe next that the input resistance R_{in} is given by the sum of R_2 and the effective input resistance R_{eq} developed across the input terminals of the operational amplifier. To compute R_{eq}, we assume for simplicity that R_o can be neglected and redraw the circuit as shown in Fig. 6-23. By applying KCL at the input terminal we arrive at

$$i_1 = \frac{v_e}{R_i} + \frac{v_e - (-\mu v_e)}{R_1} = \left[\frac{1}{R_i} + \frac{1 + \mu}{R_1}\right]v_e \tag{6-62}$$

Hence,

$$R_{\text{eq}} = \frac{v_e}{i_1} = \left[\frac{1}{R_i} + \frac{1 + \mu}{R_1}\right]^{-1} \tag{6-63}$$

Thus R_{eq} is equivalent to a parallel combination of R_i and a resistance $R_1/(1 + \mu)$. The latter resistance is that developed across the input terminals due to the Miller effect (see Section 5-5). Note that for large values of μ and small

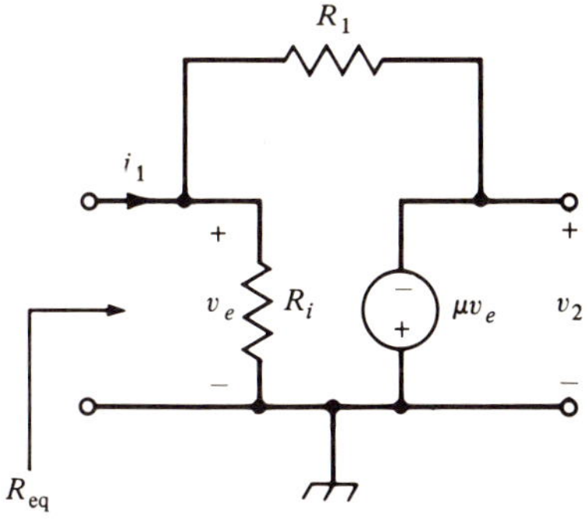

Figure 6-23 Input resistance calculation.

values of R_i, R_{eq} approaches zero. This is the property of shunt feedback connection at the input.

We thus have, for the complete circuit of Fig. 6-22, the expression for the input resistance as

$$R_{in} = R_2 + \frac{R_1 R_i}{R_1 + (1 + \mu)R_i} \tag{6-64}$$

which approaches R_2 for a practical operational amplifier.

Example 6-8.   Determine the actual closed-loop gain and the input resistance of an inverting amplifier of gain -10 designed using an operational amplifier characterized by the following parameters:

$$\mu = 5000 \qquad R_i = 25 \text{ k}\Omega \qquad R_o = 100 \text{ }\Omega$$

To obtain a closed-loop gain of nominal value -10, we can choose the two external resistors as $R_1 = 10$ kΩ and $R_2 = 1$ kΩ. Substituting the appropriate values in Eqs. (6.61) and (6.64) we arrive at

$$G_v = -9.978$$

$$R_{in} = 1001.9994 \text{ }\Omega$$

Note that the closed-loop gain is very close to its nominal value and the input resistance is almost equal to R_2.

6-5   Offset Current and Voltage Errors

From the definition of the ideal operational amplifier as given by Eq. (6-1), it is seen that the output voltage is exactly zero if the two inputs are shorted together. In practice, because of the unavoidable small mismatch existing between components in the integrated circuit implementation of the operational amplifier and other effects, there exists a finite dc voltage at the output when the inputs are shorted together. This offset voltage is directly proportional to the closed-loop gain of the circuit designed using the operational amplifier.

In addition, currents flowing into the two input terminals of a practical operational amplifier are not zero as postulated in Eq. (6-2) for the ideal model. Even with both input terminals grounded, some residual dc currents flow through the terminals to bias the transistors in the circuit. These input currents also produce a dc voltage across the input terminals because of voltage drops in the resistors connected to the input terminals; they also appear as a dc offset voltage at the output. The average of the two input currents is called the *input bias current* and their difference is called the *input offset current*.

Offset Error Analysis

The main sources of offset are shown in the modified equivalent circuit of the inverting type voltage amplifier as shown in Fig. 6-24. Here we have added an

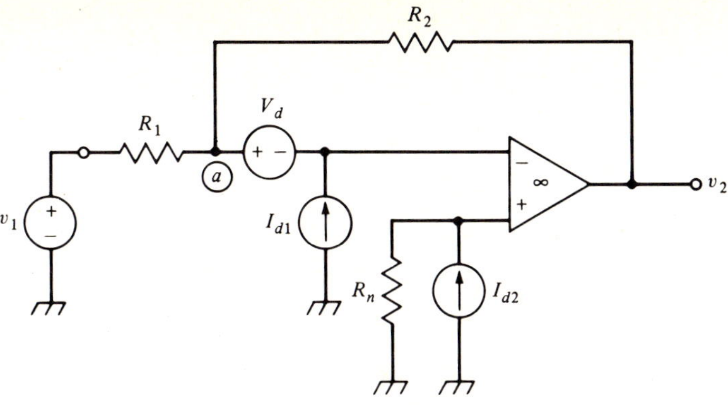

Figure 6-24   The model of an inverting amplifier including the sources of offset.

additional resistor R_n whose purpose will be explained later. The *input offset voltage* V_d is defined as the voltage that should be applied between the input terminals to make v_2 equal to zero. Typically the input offset voltage for the μA741 is 5 mV. The *input offset current* I_{os} is the difference of I_{d1} and I_{d2} that is flowing into the two inputs under the condition of zero-output voltage. The *input bias current* is given by $(I_{d1} + I_{d2})/2$ with output nulled. For the μA741, the typical value of the input offset current is 200 nA and that of the input bias current is 500 nA.

In evaluating the output offset voltage of a circuit containing operational amplifiers, all the three sources of offset must be considered along with the external circuitry. Let us now calculate the contribution $v_{2,os}$ to the output voltage due to these sources, which is the output voltage obtained with v_1 equal to zero. The voltage at the noninverting input terminal of the operational amplifier with respect to ground is seen to be $R_n I_{d2}$ which also appears at the inverting input terminal of the operational amplifier by virtue of the feedback path. Hence the voltage across the resistor R_1 is $V_d + R_n I_{d2}$ and that across the resistor R_2 is $V_d + R_n I_{d2} - v_{2,os}$. Now applying KCL at node $\textcircled{a}$ we arrive at

$$I_{d1} = \frac{V_d + R_n I_{d2}}{R_1} + \frac{V_d + R_n I_{d2} - v_{2,os}}{R_2}$$

which upon rearrangement yields

$$v_{2,os} = \left(1 + \frac{R_2}{R_1}\right)V_d + R_n\left(1 + \frac{R_2}{R_1}\right)I_{d2} - R_2 I_{d1} \tag{6-65}$$

If we choose

$$R_n = \frac{R_1 R_2}{R_1 + R_2} \tag{6-66}$$

then Eq. (6-65) reduces to

$$v_{2,os} = \left(1 + \frac{R_2}{R_1}\right)V_d + R_2(I_{d2} - I_{d1}) \tag{6-67}$$

In a practical amplifier, I_{d1} and I_{d2} are almost equal to each other, and track with changes in power supply voltages and temperature. Consequently the effect of current offset is almost cancelled with R_n chosen as expressed in Eq. (6-66).

The expression for the offset $v_{2,os}$ produced at the output as given by Eq. (6-67) also holds for the noninverting voltage-amplifier configuration shown in Fig. 6-25. The resistor R_n is again chosen as given by Eq. (6-66) to reduce the current offset.

It should be noted that the input-referred offset component $v_{1,os}$ is obtained by dividing $v_{2,os}$ by the closed-loop gain of the amplifier and consequently $v_{1,os}$ will be different in the two cases.

Offset Balancing Techniques

In most operational amplifiers, terminals are provided for external nulling of the remaining component of the offset. This nulling procedure, more commonly called *offset balancing*, is achieved by adding a small dc voltage at appropriate internal terminals of the operational amplifier such that the output dc voltage becomes zero. Figure 6-26 shows the recommended connection for offset balancing for the μA741. If such offset adjustment terminals are not available for internal adjustments, balancing can be achieved by adding an offset-cancelling voltage at the input terminal with the aid of the arrangements shown in Fig. 6-27.

Operational amplifiers working in the inverting mode can be balanced using the arrangement of Fig. 6-27(a).[22] The resistors R_a and R_b work as a potential divider, and R_c is a potentiometer. Typically R_a is 10–200 kΩ, R_b is 10–200 Ω, and R_c is 10–50 kΩ. For example, for $R_a = 100$ kΩ, $R_b = 50$ Ω, $V^+ = 10$ V, and $V^- = -10$ V, the dc voltage appearing at the noninverting terminal of the operational amplifier for offset adjustment is in the range ± 5 mV.

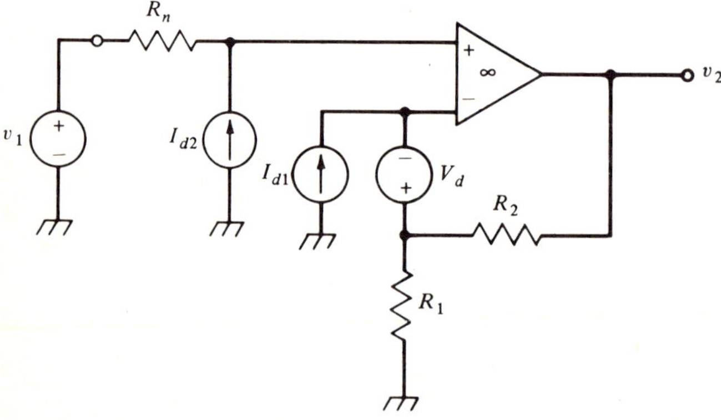

Figure 6-25   The model of a noninverting amplifier including the sources of offset.

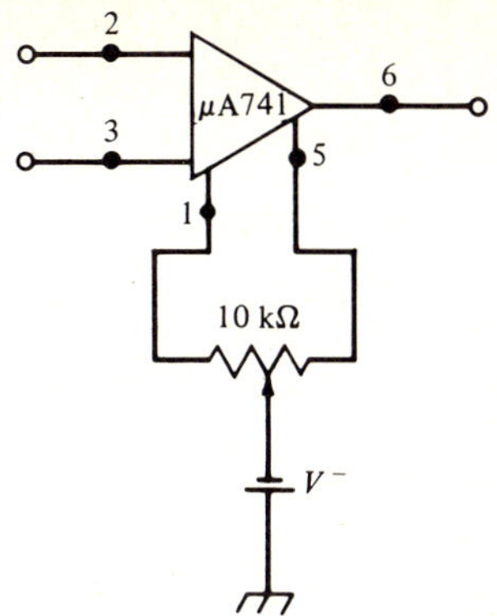

Figure 6-26 Offset balancing arrangement for the μA741. Pin connection shown is for TO-99 package. (Courtesy Fairchild Camera and Instruments Corp.)

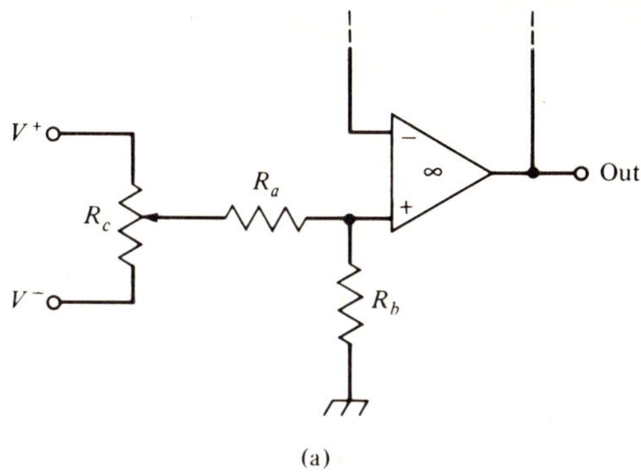

(a)

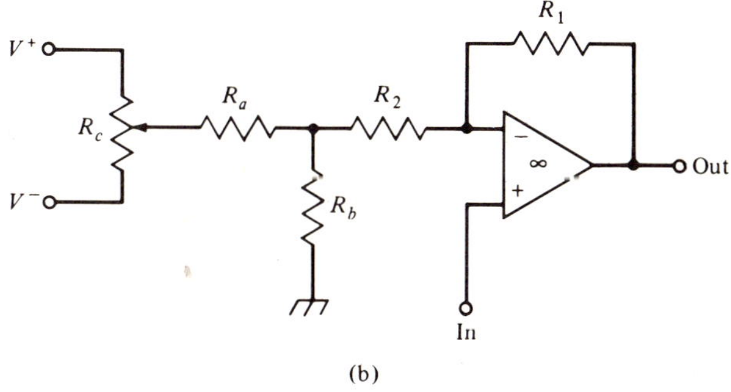

(b)

Figure 6-27 General offset balancing arrangement: (a) inverting amplifier case, and (b) noninverting amplifier case.

The circuit arrangement for balancing the noninverting amplifier configuration is as shown in Fig. 6-27(b).[22] The ranges of values of the resistors R_a and R_b, and the potentiometer R_c are the same as those for the earlier balancing circuit of Fig. 6-27(a). If $R_b \ll R_2$, then we can calculate readily the range of the dc voltage available at the inverting input terminal for offset adjustment using the same approximate analysis technique used earlier. Otherwise the effect of R_2 must be incorporated.

Offset Drift

The offset voltage and current parameters V_d, I_{d1}, and I_{d2} usually do not remain constant, but drift as a result of the changes in power-supply voltages and/or temperature. Thus, in spite of prior balancing, some dc offset voltage will always be present at the output of the operational amplifier. If we denote by ΔV_d and ΔI_{os} the changes in the offset voltage and current from their nominal values whose effects have been balanced out, the dc offset voltage at the output caused by these changes can be estimated from Eq. (6-67) to be given by

$$|\Delta v_{2,os}| \leq \left(1 + \frac{R_2}{R_1}\right)|\Delta V_d| + R_2|\Delta I_{os}|$$

$$= \left(1 + \frac{R_2}{R_1}\right)\{|\Delta V_d| + R_n|\Delta I_{os}|\} \tag{6-68}$$

Values of ΔV_d and ΔI_{os} are usually provided by the manufacturers.

6-6   Common-Mode Rejection

We consider now another source of nonidealness in a practical operational amplifier. As defined by Eq. (6-1), an ideal operational amplifier exhibits identical voltage gain μ from each of its input terminals to the output. In practice, due to the mismatch at the input stage, these two gains are not equal. As a result, Eq. (6-1) is in effect of the form

$$v_o = \mu_2 v_{i2} - \mu_1 v_{i1} \tag{6-69}$$

which can also be written as

$$v_o = (\mu_2 - \mu_1)v_{cm} + \tfrac{1}{2}(\mu_2 + \mu_1)v_e \tag{6-70}$$

where v_{cm} and v_e are the common-mode input voltage and the differential-input voltage as defined by Eqs. (6-4) and (6-5), respectively. Comparing Eq. (6-70) with Eq. (6-6), we observe that $\mu_2 - \mu_1$ is the common-mode gain μ_c and $(\mu_2 + \mu_1)/2$ is the differential gain μ_d as defined in Eq. (6-7). As a result, Eq. (6-70) reduces to

$$v_o = \mu_d\left[v_e + \frac{\mu_c}{\mu_d}v_{cm}\right] \tag{6-71}$$

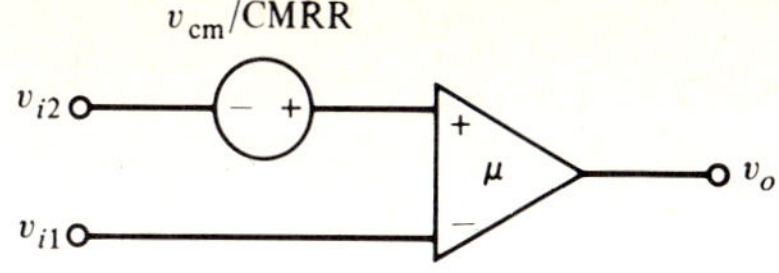

Figure 6-28 Model of an operational amplifier taking into account the common-mode rejection.

which indicates that the nonideal operational amplifier can be represented as shown in Fig. 6-28 where $(\mu_c/\mu_d)\cdot v_{cm}$ is the input offset signal producing the error in the output signal. The ratio of the common-mode signal v_{cm} and the above input offset signal is defined as the common-mode rejection ratio (CMRR), that is,

$$\text{CMRR} = \frac{v_{cm}}{\dfrac{\mu_c}{\mu_d}\cdot v_{cm}} = \frac{\mu_d}{\mu_c} \tag{6-72}$$

CMRR is thus seen to be the differential gain divided by the common-mode gain. It is generally specified in decibels. Now ideally μ_c is zero, that is, CMRR is infinite. Typical values of CMRR in a practical amplifier are from 70–100 dB. Note that the large signal voltage gain given in the specification sheet is essentially the differential gain μ_d.

Note that if $|v_{i2} - v_{i1}| \cong 0$, then for analysis purposes we can write

$$\frac{v_{i2} + v_{i1}}{2} \cong \frac{v_{i2} + v_{i2}}{2} = v_{i2}$$

and rewrite Eq. (6-71) as

$$v_o = \mu\left[(v_{i2} - v_{i1}) + \frac{v_{i2}}{\text{CMRR}}\right] \tag{6-73}$$

without any significant loss of accuracy. The following example illustrates an application of the above result.

Example 6-9. Determine the accuracy of the voltage follower of Fig. 6-4 by taking into account the finite CMRR of a practical operational amplifier.[1]

It follows from the figure that for the voltage follower $v_{i1} = v_o$ and $v_{i2} = v_1$. Substituting these in Eq. (6-73) we obtain

$$v_o = \mu\left[(v_1 - v_o) + \frac{v_1}{\text{CMRR}}\right]$$

which after rearrangement gives the gain of the voltage amplifier as

$$\frac{v_o}{v_1} = \frac{1 + (1/\text{CMRR})}{1 + (1/\mu)} \tag{6-74}$$

Let us calculate the dc accuracy of the voltage follower constructed using the μA741 for which typical value of μ is 200,000 and that of the CMRR is 90 dB which is equivalent to 31,623. Using these numbers in Eq. (6-74) we arrive at

$$\frac{v_o}{v_1} = \frac{1 \pm (1/31{,}623)}{1 + (1/2 \times 10^5)} = 1.00002661 \quad \text{or} \quad 0.999963378$$

where both the plus and the minus signs have been used for CMRR since its sign is not known a priori. The dc accuracy is therefore better than 0.0037 percent.

The CMRR is an important factor when the operational amplifier is used to measure accurately the difference of two almost equal signals. This is illustrated in the following example.

Example 6-10. Determine the accuracy of measuring the difference of two input dc signals of values 5 and 4.998 V, respectively, using the difference amplifier of Fig. 6-12 designed with all equal-valued resistors and an operational amplifier having a 80-dB CMRR.

If the operational amplifier has infinite CMRR, then the difference amplifier with all equal-valued resistors is characterized by the input–output relation

$$v_o = v_2 - v_1$$

Since v_2 is 5 V and v_1 is 4.998 V, v_o is ideally 2 mV.

For an operational amplifier with a finite CMRR, the output voltage of the difference amplifier can be expressed as $v_o + \Delta v_o$, where Δv_o is the error component. To compute the error component Δv_o we use the model of Fig. 6-28 with $v_{cm} \cong v_{i2}$ resulting in the equivalent circuit of Fig. 6-29(a). Now

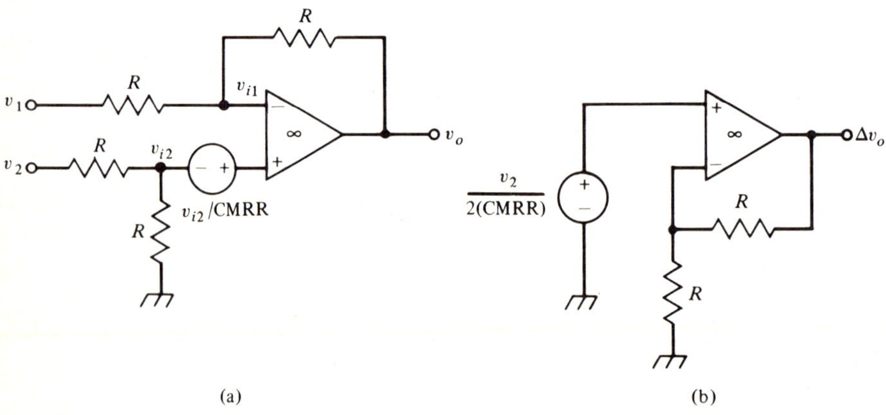

Figure 6-29   Calculation of the accuracy of the differential amplifier.

from the figure it is seen that

$$v_{i2} = \frac{R}{R + R} \cdot v_2 = \frac{v_2}{2}$$

In order to calculate the error component Δv_o we can use the superposition theorem and ground the two input terminals leading to the equivalent circuit of Fig. 6-29(b) which is recognized as the familiar noninverting amplifier configuration of Fig. 6-3(a) with a gain of 2. It follows then

$$\Delta v_o = 2 \times \frac{v_2}{2(\text{CMRR})} = \frac{v_2}{\text{CMRR}}$$

$$= \pm \frac{5}{10{,}000} = \pm 0.5 \text{ mV}$$

which is less than 2 mV. Thus, with this operational amplifier, the two input signals can be measured with $\Delta v_o / v_o \times 100$ percent $= 0.5/2 \times 100$ percent $= 25$ percent accuracy.

6-7   Analog Computation

In Section 6-2 a number of applications of the operational amplifier were outlined. Another very common application of this versatile device is as a basic component in an analog computer, which is considered in this section. An analog computer is employed to solve a differential equation by setting up an equivalent dynamic system characterized by an "identical" differential equation. Only the fundamental concepts of analog computation are illustrated here.

The four basic building blocks of an analog computer are the inverter, the summer, the integrator, and the summing integrator, as shown symbolically in Fig. 6-30.* As defined, the inverter is a negative-gain voltage amplifier and can thus be realized as indicated in Fig. 6-5. For scaling a signal by a gain of less than 1, a simple potentiometer can be used. The summer is a summing amplifier with negative gain and is thus realizable by the circuit of Fig. 6-11(a). The realizations of the integrator and the summing integrator are considered next.

Integrator

The pertinent circuit is sketched in Fig. 6-31(a). Summing the current at the inverting input terminal we obtain

$$\frac{v_1}{R} = -C \frac{dv_o}{dt} \tag{6-75}$$

---

* It should be noted that the commonly used symbols for these components in the analog computer literature do not have the circle at the output.

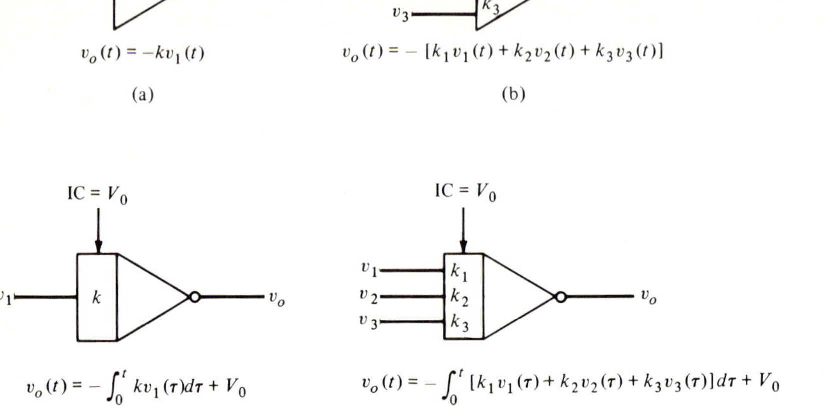

$$v_o(t) = -kv_1(t)$$

(a)

$$v_o(t) = - [k_1 v_1(t) + k_2 v_2(t) + k_3 v_3(t)]$$

(b)

$$v_o(t) = - \int_0^t kv_1(\tau)d\tau + V_0$$

(c)

$$v_o(t) = - \int_0^t [k_1 v_1(\tau) + k_2 v_2(\tau) + k_3 v_3(\tau)]d\tau + V_0$$

(d)

Figure 6-30 Basic building blocks of an analog computer: (a) inverter, (b) summer, (c) integrator, and (d) summing integrator.

which is equivalent to

$$v_o(t) = -\frac{1}{RC}\int_0^t v_1(\tau)d\tau + v_C(0) \tag{6-76}$$

where $v_C(0)$ is the initial voltage across the capacitor at the time of application of the input signal $v_1(t)$ at $t = 0$. If we set $RC = 1/k$, then

$$v_o(t) = -\int_0^t kv_1(\tau)d\tau + v_C(0) \tag{6-77}$$

Hence the output voltage $v_o(t)$ is the negative of the integral of the input voltage $v_1(t)$ plus an arbitrary constant of integration, the initial value of the output voltage, $v_o(0)$. Thus the circuit of Fig. 6-31(a) is an integrator. By choosing R and C values properly, this circuit can provide integration with a scale change.

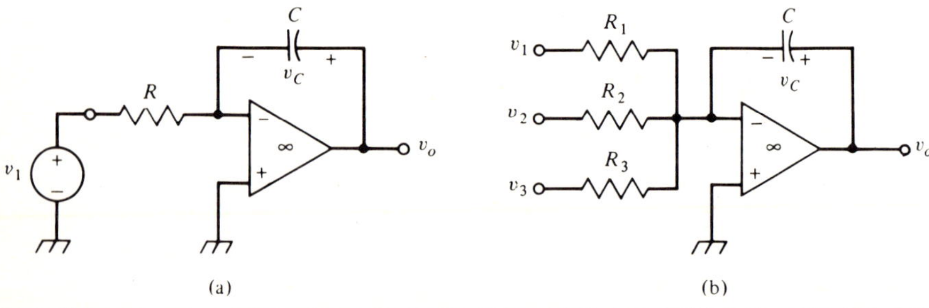

(a)

(b)

Figure 6-31 Realization of (a) an integrator, and (b) a summing integrator.

In practice a large-valued resistor R_s is connected in parallel with the capacitor C to provide dc stabilization of the integrator. This resistor in effect limits the low-frequency gain of the amplifier and as a result will minimize drift. A typical value of R_s is 10 times that of R. It can be shown that the input signal frequency must now be greater than $(1/2\pi R_s C)$ Hz so that the modified circuit will act as an integrator.

The realization of a summing integrator is readily obtained by connecting additional resistors to the inverting input of the operational amplifier in Fig. 6-31(a). For example, Fig. 6-31(b) shows the implementation of a 3-input summing integrator. Following a similar analysis as outlined above, we can derive the expression for the output voltage $v_o(t)$ as

$$v_o(t) = -\int_0^t \left[k_1 v_1(\tau) + k_2 v_2(\tau) + k_3 v_3(\tau) \right] d\tau + v_C(0) \tag{6-78}$$

where $k_1 = 1/R_1 C$, $k_2 = 1/R_2 C$, $k_3 = 1/R_3 C$, and $v_C(0)$ is the initial voltage across the capacitor.

It should be noted that all the four basic components of an analog computer use the operational amplifier in the inverting configuration and hence the common-mode input voltage limit need not be a problem.

Simulation of Differential Equations

Let us now illustrate, with the aid of an example, the simulation of a linear differential equation with constant coefficients on an analog computer.

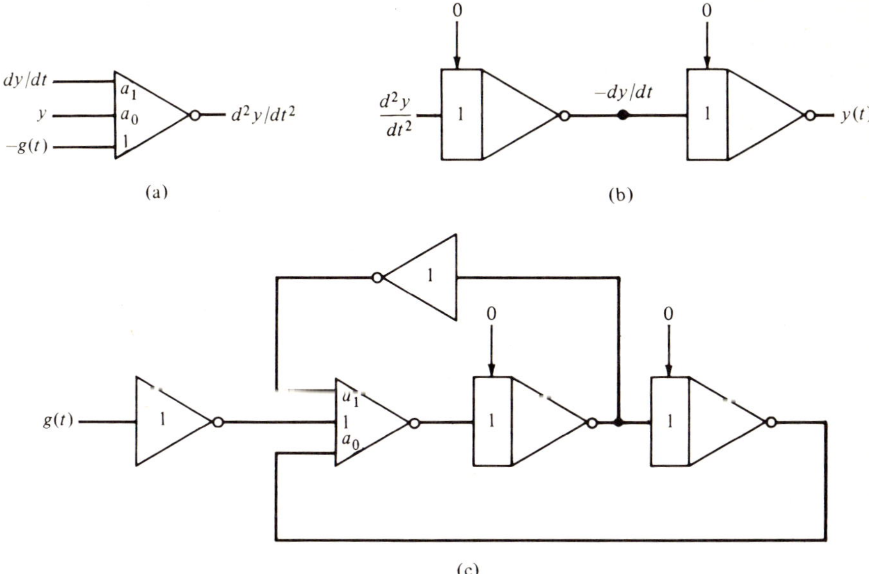

Figure 6-32 Development of the block-diagram representation of the analog computer simulation of a second-order differential equation.

Example 6-11. Consider the simulation of the second-order differential equation

$$\frac{d^2 y}{dt^2} + a_1 \frac{dy}{dt} + a_0 y = g(t) \tag{6-79}$$

where $y(t)$ is the dependent variable, the solution, and $g(t)$ is the forcing function, the "input."

We rewrite Eq. (6-79) as

$$\frac{d^2 y}{dt^2} = -a_1 \frac{dy}{dt} - a_0 y + g(t) \tag{6-80}$$

An implementation of Eq. (6-80) using a summer is as indicated in Fig. 6-32(a) in block-diagram form.

Now by integrating $d^2 y/dt^2$ twice we obtain $-dy/dt$ and y as shown in Fig. 6-32(b). Combining these two figures we arrive at the complete simulation diagram sketched in Fig. 6-32(c). The corresponding circuit realization is shown in Fig. 6-33. Note that this realization is for zero initial conditions, that is,

$$y(0) = 0; \qquad \frac{dy}{dt}(0) = 0 \tag{6-81}$$

Nonzero initial conditions can be handled easily.[23] Note that the resistor values $R_1, R_2, R_3, R_4,$ and R_5 can be chosen arbitrarily. A convenient choice is to make them all equal.

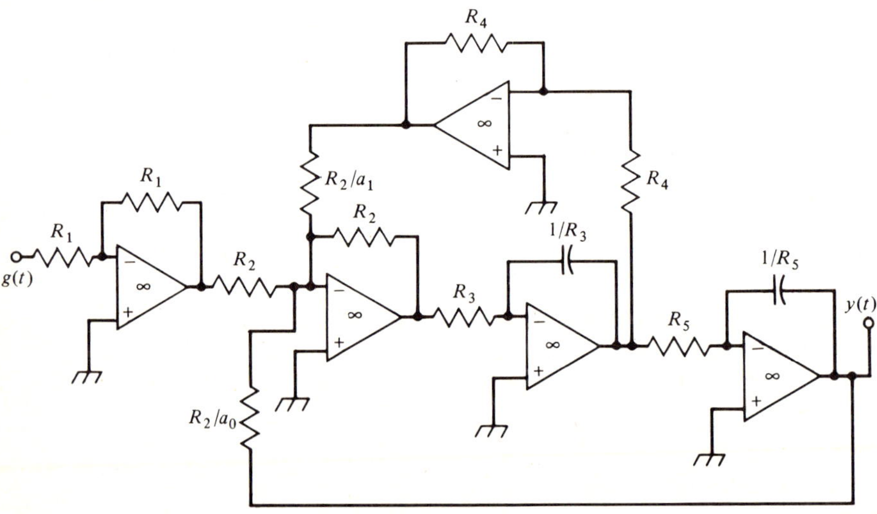

Figure 6-33   Analog computer simulation of a second-order differential equation. Element values are in ohms and farads.

6-8   Transfer Function Realization

The circuits described in Section 6-2 are characterized by transfer functions that are real and constant. As a result, they are in principle applicable for all signal frequencies from dc to some higher frequency limited by the bandwidth of the operational amplifier. For many applications, it is desirable to realize circuits with prescribed frequency response characteristics. We describe some of these circuits in this section. If the input signal to these circuits is of the form $\mathbf{V}_1 e^{st}$, then the output is of the form $\mathbf{V}_2 e^{st}$, where $\mathbf{V}_2$ and $\mathbf{V}_1$ are related via the transfer function $T(s)$ as

$$\mathbf{V}_2 = T(s) \cdot \mathbf{V}_1$$

These circuits are thus characterized by transfer functions that are functions of the complex frequency variable s.

AC Amplifiers

As mentioned in Section 5-10, ac amplifiers have a gain characteristic that is approximately constant between two frequencies ω_L and ω_H. The gain drops for frequencies below ω_L and above ω_H. Many IC operational amplifiers have very wide bandwidths and can also be used to design ac amplifiers. Both the noninverting amplifier of Fig. 6-3(a) and the inverting amplifier of Fig. 6-5 can be modified for this purpose. Figure 6-34(a) shows the inverting type ac amplifier. To determine its transfer function, we apply KCL in the s-domain at the inverting input terminal of the operational amplifier:

$$\frac{\mathbf{V}_1}{R_2 + (1/sC_2)} + \frac{\mathbf{V}_2}{R_1} = 0$$

From above we arrive at

$$T(s) = \frac{\mathbf{V}_2}{\mathbf{V}_1} = -\frac{R_1}{R_2} \cdot \frac{s}{s + (1/R_2 C_2)} \tag{6-82}$$

For sinusoidal input frequencies, the magnitude of the transfer function is of interest. Thus, for $s = j\omega$, the magnitude is given as

$$|T(j\omega)| = \frac{(R_1/R_2)\omega}{\sqrt{\omega^2 + \omega_L{}^2}} \tag{6-83}$$

where we have set

$$\omega_L = 1/R_2 C_2 \tag{6-84}$$

It is seen from Eq. (6-83) that for $\omega \gg \omega_L$, the magnitude approaches the constant value R_1/R_2 which is the midband gain of the amplifier. For $\omega < \omega_L$, the magnitude rolls off at 6 dB/octave and approaches zero value at dc. At $\omega = \omega_L$, the gain is 3 dB less than that at midband and ω_L is the lower 3-dB cut-off frequency. Because of the frequency dependence of the operational amplifier open-loop gain, the gain also drops at higher frequencies.

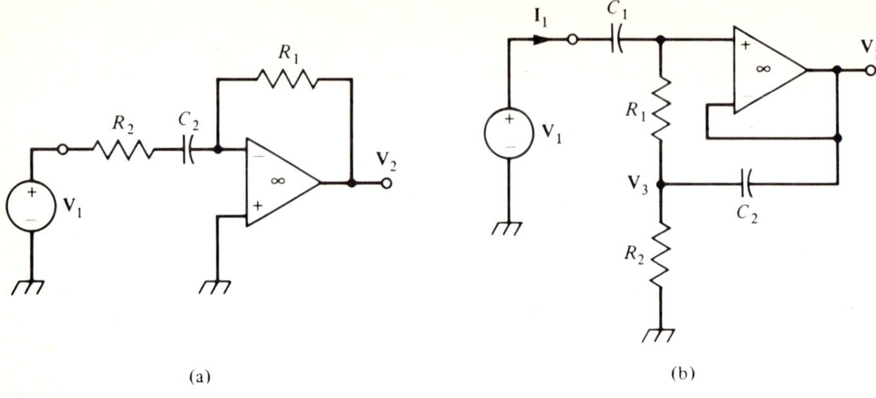

Figure 6-34 AC amplifiers.

A similar modification of the noninverting circuit of Fig. 6-3(a) to work as an ac amplifier is illustrated in Problem 6-36. Its analysis is left as an exercise. Both of these ac amplifiers have a very low input impedance at higher frequencies.

The ac amplifier circuit of Fig. 6-34(b) presents a high-input impedance by what is known as the "boot-strapping" technique.[24] The input impedance is given as

$$\frac{\mathbf{V}_1}{\mathbf{I}_1} = (R_1 + R_2) + sR_1R_2C_2 + \frac{1}{sC_1} \tag{6-85}$$

At low frequencies, the input impedance is essentially the impedance of the capacitor C_1 which is very high. At high frequencies, the magnitude of the input impedance approaches the value $\omega R_1 R_2 C_2$. However, at very high frequencies, the effective input impedance magnitude decreases due to the frequency dependence of the voltage follower gain.

Example 6-12.   Determine the magnitude of the input impedance of the ac amplifier of Fig. 6-34(b) at 100 Hz and at 1 kHz. The element values are $R_1 = R_2 = 100$ kΩ, $C_1 = 0.01$ μF, and $C_2 = 2$ μF.

With these values, it can be shown that the magnitude of the input impedance at both specified frequencies is approximately given by $2\pi f R_1 R_2 C_2$. Substituting these values the input impedance at 100 Hz is about 12.57 MΩ and at 1 kHz it is approximately 125.7 MΩ.

The transfer function $T(s) = \mathbf{V}_2/\mathbf{V}_1$ of the second ac amplifier can be readily determined by applying nodal analysis in the s-domain. Let us denote the voltage across the resistor R_2 as $\mathbf{V}_3$. Note also that due to the voltage follower configuration, the voltage at the noninverting input terminal of the operational amplifier is $\mathbf{V}_2$. Applying KCL to the two independent nodes we then arrive at

$$(\mathbf{V}_1 - \mathbf{V}_2)sC_1 = (\mathbf{V}_2 - \mathbf{V}_3)G_1$$

$$(\mathbf{V}_2 - \mathbf{V}_3)G_1 = (\mathbf{V}_3 - \mathbf{V}_2)sC_2 + \mathbf{V}_3G_2$$

where $G_1 = 1/R_1$ and $G_2 = 1/R_2$. Eliminating $\mathbf{V}_3$ from above we arrive at, after some algebra,

$$T(s) = \frac{\mathbf{V}_2}{\mathbf{V}_1} = \frac{(G_1 + G_2 + sC_2)sC_1}{C_1C_2 s^2 + C_1(G_1 + G_2)s + G_1G_2}$$

The magnitude of the transfer function for $s = j\omega$ is given as

$$|T(j\omega)| = \frac{\omega C_1[\omega^2 C_2{}^2 + (G_1 + G_2)^2]^{1/2}}{[(G_1G_2 - C_1C_2\omega^2)^2 + C_1{}^2(G_1 + G_2)^2\omega^2]^{1/2}}$$

The midband gain, obtained by letting ω become very large, is seen to be unity. The gain approaches zero value at dc. A plot of the magnitude function in the low- to medium-frequency range is sketched in Fig. 6-35 for the element values given in Example 6-12. As before, the gain also drops at higher frequencies because of the frequency dependence of open-loop gain of the operational amplifier.

We next consider more general types of networks characterized by complex transfer functions.

Active RC Filters

The term "filter" is normally used to designate a frequency-selective network with no or little attenuation of signals in a given frequency range but attenuating

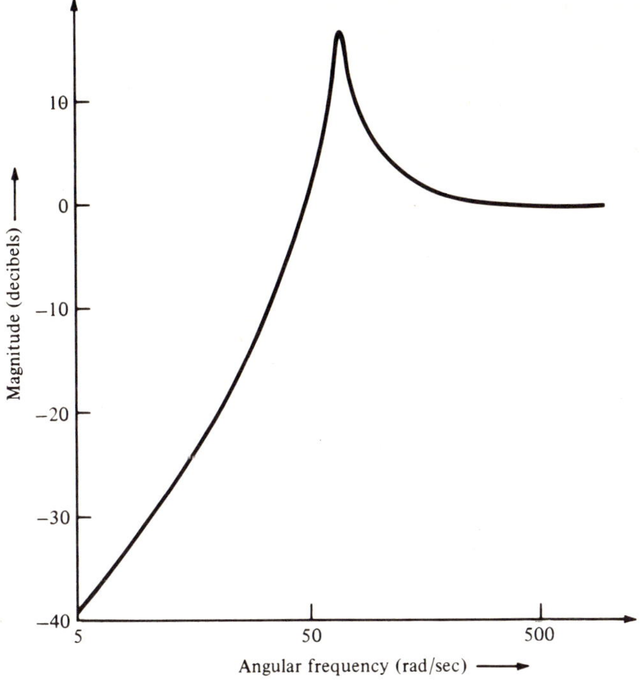

Figure 6-35 Magnitude response.

significantly signals outside this frequency range. To illustrate the concept of filtering, let the pertinent network be characterized by a stable transfer function $T(s)$ whose magnitude function is given by

$$|T(j\omega)| \cong K, \quad \text{a constant} \quad \text{for} \quad |\omega| \le \omega_1$$
$$\cong 0, \qquad\qquad\quad \text{for} \quad |\omega| > \omega_1 \qquad (6\text{-}86)$$

Let the input signal $x(t)$ be assumed to consist of two cosinusoidal signals of frequencies ω_a and ω_b:

$$x(t) = A_a \cos(\omega_a t + \phi_a) + A_b \cos(\omega_b t + \phi_b)$$

where

$$\omega_a \ll \omega_1 \ll \omega_b$$

By sinusoidal steady-state analysis, the output signal $y(t)$ of the network is given by

$$y(t) = A_a|T(j\omega_a)| \cos(\omega_a t + \phi_a + \beta_a)$$
$$+ A_b|T(j\omega_b)| \cos(\omega_b t + \phi_b + \beta_b) \qquad (6\text{-}87)$$

where $\beta_a = \arg T(j\omega_a)$ and $\beta_b = \arg T(j\omega_b)$. Because of condition (6-86), Eq. (6-87) reduces to

$$y(t) \cong A_a \cdot K \cos(\omega_a t + \phi_a + \beta_a)$$

Thus the network "passes" the low-frequency signal without changing its waveform, whereas it "stops" the high-frequency signal from appearing at the output. Or, in other words, the network behaves as a *low-pass filter* with a *cut-off frequency* ω_1 passing all signals within the passband $0 \le \omega \le \omega_1$ and stopping all signals with frequencies above ω_1. The idealized frequency response of the low-pass filter is sketched in Fig. 6-36(a). Two other basic filters are the high-pass and the bandpass filters. Figure 6-36(b) shows the characteristic of an ideal *high-pass filter* having a cut-off frequency ω_2 and the characteristic shown in Fig. 6-36(c) represents a *bandpass filter*, with cut-off frequencies ω_3 and ω_4.

The idealized characteristics are not realizable by a physical network and are thus approximated in practice. In recent years, operational amplifiers are

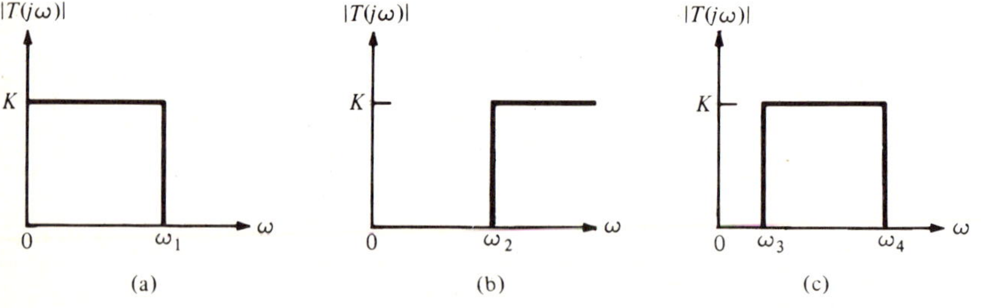

Figure 6-36   Basic filter transfer characteristics: (a) low-pass, (b) high-pass, and (c) bandpass.

being increasingly used in designing such special purpose networks without inductors. A convenient way of designing such filters is by cascading appropriate simpler filter sections characterized by second-order and first-order transfer functions. Since first-order sections can be easily realized using resistors and capacitors, the usual procedure has been to implement second-order active *RC* filter sections with operational amplifiers. We describe below three such basic sections. A number of other filter configurations are to be found in the problem section. It should be noted that several different types of second-order active filter configurations are commercially available in hybrid IC forms.

Bandpass Filter.   Consider the second-order transfer function

$$T_n(s) = \frac{b_1 s}{s^2 + b_1 s + b_0} \tag{6-88}$$

Its magnitude for $s = j\omega$ is given by

$$|T_n(j\omega)| = \frac{b_1 \omega}{[b_1^2 \omega^2 + (b_0 - \omega^2)^2]^{1/2}}$$

$$= \frac{1}{\left[1 + Q^2\left(\dfrac{\omega_n}{\omega} - \dfrac{\omega}{\omega_n}\right)^2\right]^{1/2}} \tag{6-89}$$

where we have used the notations

$$Q = \sqrt{b_0/b_1}$$
$$\omega_n = \sqrt{b_0} \tag{6-90}$$

It is seen from Eq. (6-89) that the magnitude goes to zero values at dc ($\omega = 0$) and at $\omega = \infty$. It reaches the maximum value of unity at $\omega = \omega_n$, called the *center frequency*. Plots of the magnitude as a function of ω/ω_n for several values of Q are sketched in Fig. 6-37. Such a magnitude characteristic is often used as an approximation to the bandpass filter characteristic of Fig. 6-36(c).

The passband of the magnitude characteristic of Fig. 6-37 is usually determined from the two frequencies ω_1 and ω_2 at which the magnitudes are equal to $1/\sqrt{2}$ times the maximum value. Since the maximum value occurs at ω_n and is of value unity, the two cut-off frequencies are determined from the equation

$$|T_n(j\omega)| = \frac{1}{\left[1 + Q^2\left(\dfrac{\omega_n}{\omega} - \dfrac{\omega}{\omega_n}\right)^2\right]^{1/2}} = \frac{|T_n(j\omega_n)|}{\sqrt{2}} = \frac{1}{\sqrt{2}}$$

which is equivalent to the equation

$$Q^2\left(\frac{\omega_n}{\omega} - \frac{\omega}{\omega_n}\right)^2 = 1$$

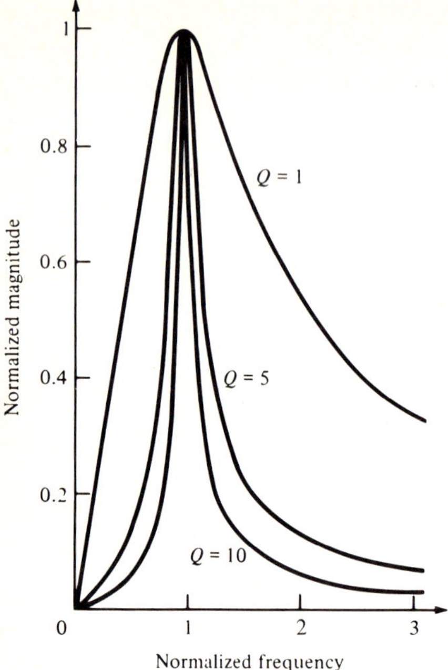

Figure 6-37   Frequency response of a second-order bandpass filter.

Solving the above for positive values of ω we arrive at

$$\frac{\omega}{\omega_n} = \sqrt{1 + \frac{1}{4Q^2}} \pm \frac{1}{2Q} \tag{6-91}$$

For $Q \gg 1$ (which is usually the case), we can approximate the above as

$$\frac{\omega}{\omega_n} \cong 1 \pm \frac{1}{2Q}$$

As a result, the two cut-off frequencies are

$$\omega_1 \cong \omega_n\left(1 - \frac{1}{2Q}\right)$$

$$\omega_2 \cong \omega_n\left(1 + \frac{1}{2Q}\right) \tag{6-92}$$

The bandwidth (i.e., the width of the passband) in radians/sec is

$$BW = \omega_2 - \omega_1 = \frac{\omega_n}{Q} \tag{6-93}$$

and is thus inversely proportional to Q. This means that the passband becomes narrower as Q increases, which is clearly evident from Fig. 6-37. The Q of the bandpass filter as a result provides a measure of the sharpness or *selectivity* of the bandpass characteristic.

It should be noted from our discussion in Section 5-2 that the magnitude in decibels is given by $20 \log_{10} |T_n(j\omega)|$. Since $|T_n(j\omega_n)| = 1$, the magnitude of the normalized bandpass transfer function of Eq. (6-88) at the center frequency ω_n is 0 dB. At the two cut-off frequencies, $|T_n(j\omega)| = 1/\sqrt{2}$ which in decibels is approximately -3 dB. As a result, the two cut-off frequencies ω_1 and ω_2 are normally called the 3-dB *cut-off frequencies*.

The general second-order bandpass transfer function (for which the maximum magnitude is not unity) is often expressed in terms of Q and ω_n as

$$T(s) = \frac{Hs}{s^2 + \dfrac{\omega_n}{Q} s + \omega_n^2} \tag{6-94}$$

A possible realization of above using an operational amplifier is sketched in Fig. 6-38. To determine its transfer function we make use of nodal analysis in the s-domain. Applying KCL to nodes $\textcircled{a}$ and $\textcircled{b}$ and observing $\mathbf{V}_2 = -k\mathbf{V}_b$ we arrive at

$$(\mathbf{V}_1 - \mathbf{V}_a)\frac{1}{R_1} = (\mathbf{V}_a - \mathbf{V}_b)sC_2 + (\mathbf{V}_a + k\mathbf{V}_b)sC_1$$

$$(\mathbf{V}_a - \mathbf{V}_b)sC_2 = \frac{\mathbf{V}_b}{R_2}$$

Eliminating $\mathbf{V}_a$ from above and replacing $\mathbf{V}_b$ with $-\mathbf{V}_2/k$, we arrive at the desired expression for the transfer function characterizing the filter as

$$\frac{\mathbf{V}_2}{\mathbf{V}_1} = T(s) = \frac{-\dfrac{k}{1+k} \cdot \dfrac{1}{R_1 C_1} s}{s^2 + \dfrac{C_1 R_1 + C_2 R_1 + C_2 R_2}{(1+k)C_1 C_2 R_1 R_2} s + \dfrac{1}{(1+k)C_1 C_2 R_1 R_2}} \tag{6-95}$$

Our next problem is to determine the design equations. To this end we compare the denominators of Eqs. (6-95) and (6-94) which yields

$$\frac{C_1 R_1 + C_2 R_1 + C_2 R_2}{(1+k)C_1 C_2 R_1 R_2} = \frac{\omega_n}{Q}$$

$$\frac{1}{(1+k)C_1 C_2 R_1 R_2} = \omega_n^2 \tag{6-96}$$

Note that there are five unknowns—the two resistor values, the two capacitor values, and the gain of the inverting amplifier. However, there are only two equations relating them. As a result, the solution is not unique. One way to

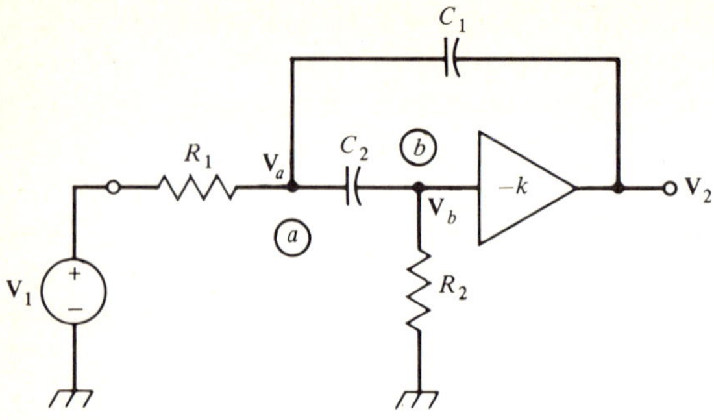

Figure 6-38   A second-order bandpass filter.

solve these equations is to select some parameters a priori, and then solve for the others. To this end, it is often convenient from the implementation point of view to make all resistors equal-valued, and all capacitors equal-valued. We thus set

$$R_1 = R_2 = 1$$
$$C_1 = C_2 = C$$

(6-97)

Then Eq. (6-96) reduces to

$$\frac{3}{(1 + k)C} = \frac{\omega_n}{Q}$$

$$\frac{1}{(1 + k)C^2} = \omega_n^2$$

which yields

$$C = 1/3Q\omega_n$$
$$k = 9Q^2 - 1$$

(6-98)

A typical design problem is considered in the following example.

Example 6-13.   Design a bandpass filter with a 3-dB passband of 40 Hz centered at 200 Hz.

From the specifications we note that

$$\omega_n = 2\pi \times 200$$

$$BW = \omega_n/Q = 2\pi \times 40$$

and as a result,

$$Q = 200/40 = 5$$

Substituting the above in Eq. (6-98) we obtain

$$C_1 = C_2 = \frac{1}{3 \times 5 \times 2\pi \times 200}$$

$$k = 9(5)^2 - 1 = 224$$

$$R_1 = R_2 = 1$$

To arrive at more suitable practical values for these elements, we raise the impedance level by 10^5 which yields

$$C_1 = C_2 = \frac{10^{-5}}{6000\pi} = 0.53 \text{ nF}$$

$$R_1 = R_2 = 100 \text{ k}\Omega$$

The inverting amplifier of Fig. 6-38 can be implemented using the circuit of Fig. 6-5. The final realization of the desired bandpass filter is thus as shown in Fig. 6-39.

Low-Pass Filter. The pertinent transfer function in this case is

$$T(s) = \frac{H\omega_n^2}{s^2 + \dfrac{\omega_n}{Q}s + \omega_n^2} \tag{6-99}$$

A plot of the magnitude of $|T(j\omega)|/H$ for a fixed ω_n and various values of Q is sketched in Fig. 6-40. Note that the response is essentially a low-pass characteristic for $Q < 1$. This is why the transfer functions of the form of Eq. (6-99) are known as low-pass transfer functions. For $Q \gg 1$, the response approaches that of a bandpass filter providing greater amplification of signals within a small band around ω_n. For very high Q values, ω_n is the frequency where peaking occurs.

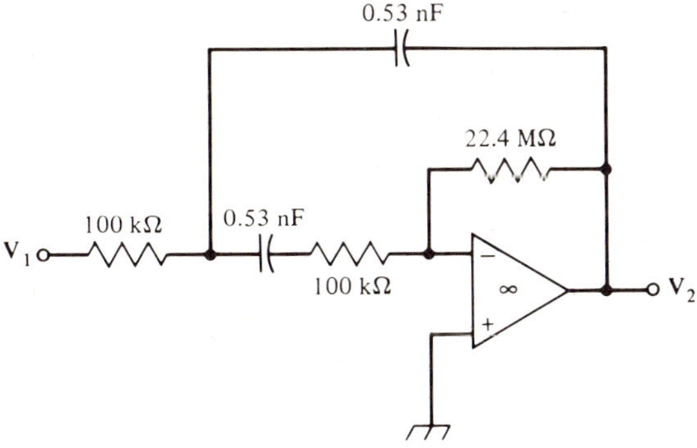

Figure 6-39 Bandpass filter of Example 6-13.

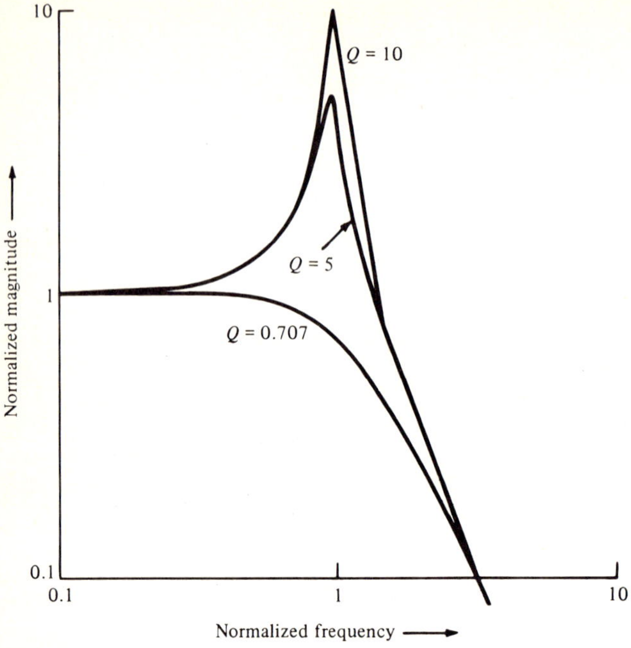

Figure 6-40   Low-pass filter frequency response.

One realization of Eq. (6-99) is sketched in Fig. 6-41. Nodal analysis of this circuit in the s-domain yields

$$\frac{1}{R_1}(\mathbf{V}_1 - \mathbf{V}_3) = \frac{1}{R_2}\left(\mathbf{V}_3 - \frac{\mathbf{V}_2}{\mu}\right) + sC_1(\mathbf{V}_3 - \mathbf{V}_2)$$

$$\frac{1}{R_2}\left(\mathbf{V}_3 - \frac{\mathbf{V}_2}{\mu}\right) = sC_2 \frac{\mathbf{V}_2}{\mu}$$

Eliminating $\mathbf{V}_3$ from above we arrive at the expression for the voltage-transfer ratio $T(s) = \mathbf{V}_2/\mathbf{V}_1$ characterizing the circuit which is given as

$$\frac{\mathbf{V}_2}{\mathbf{V}_1} = T(s) = \frac{\dfrac{\mu}{R_1 R_2 C_1 C_2}}{s^2 + \left(\dfrac{1-\mu}{R_2 C_2} + \dfrac{1}{R_1 C_1} + \dfrac{1}{R_2 C_1}\right)s + \dfrac{1}{R_1 R_2 C_1 C_2}} \qquad (6\text{-}100)$$

Comparing the denominators of Eqs. (6-99) and (6-100) we obtain

$$\frac{1-\mu}{R_2 C_2} + \frac{1}{R_1 C_1} + \frac{1}{R_2 C_1} = \frac{\omega_n}{Q}$$

$$\frac{1}{R_1 R_2 C_1 C_2} = \omega_n^2$$

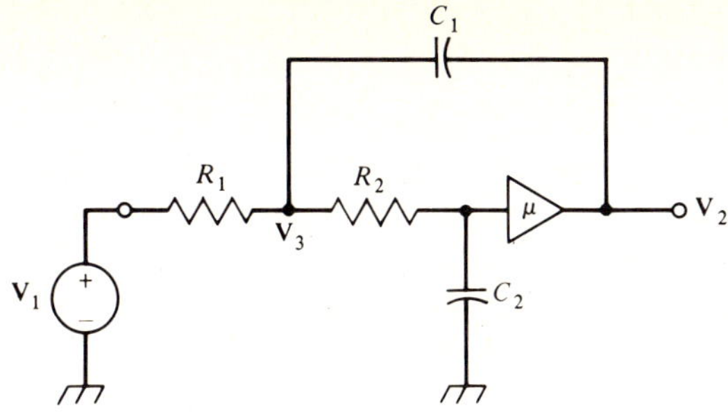

Figure 6-41 A second-order low-pass filter configuration.

which can be solved for the design equations. One possible solution of above is

$$R_1 = R_2 = R$$

$$C_1 = C_2 = 1/\omega_n R \qquad (6\text{-}101)$$

$$\mu = 3 - (1/Q)$$

The following example illustrates the design of a low-pass filter.

Example 6-14. Design an active filter to realize the second-order Butterworth transfer characteristic* having a 3-dB cut-off frequency of 1 kHz.
The desired transfer function is

$$T(s) = \frac{H\omega_n^{\,2}}{s^2 + \sqrt{2}\,\omega_n s + \omega_n^{\,2}} \qquad (6\text{-}102)$$

It is easy to show that at $\omega = \omega_n$, the magnitude characteristic is down 3 dB from its dc value. Therefore ω_n is the 3-dB cut-off frequency for this transfer function. We want

$$\omega_n = 2\pi \times 10^3, \qquad Q = 1/\sqrt{2} = 0.707$$

Using this information in Eq. (6-101) and setting $R = 10^5$ we obtain the element values as

$$C_1 = C_2 = 1.59 \text{ nF}, \quad R_1 = R_2 = 100 \text{ k}\Omega, \quad \mu = 3 - 1.414 = 1.586$$

The final realization is sketched in Fig. 6-42 where we have used the non-inverting voltage amplifier of Fig. 6-3(a).

---

* A commonly used approximation of the low-pass characteristic is the Butterworth approximation, details of which will be found in Reference 21.

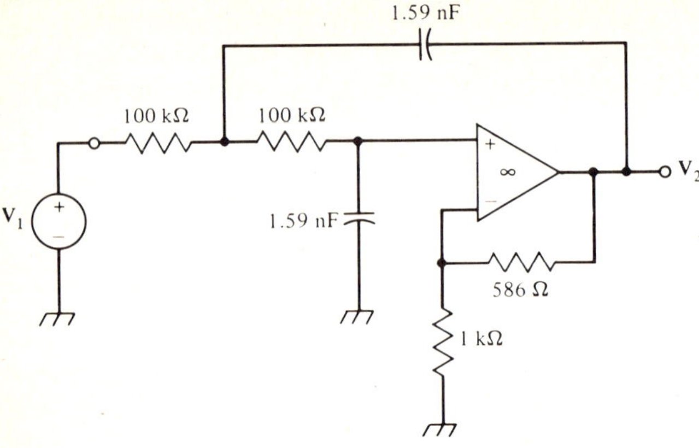

Figure 6-42 Realization of a second-order Butterworth low-pass filter of Example 6-14.

High-Pass Filter. A high-pass transfer function is of the form

$$T(s) = \frac{Hs^2}{s^2 + \dfrac{\omega_n}{Q} s + \omega_n^2} \qquad (6\text{-}103)$$

One realization of this transfer function is sketched in Fig. 6-43. Note that the noninverting amplifier used here can be implemented again using the circuit of Fig. 6-3(a).

Higher-Order Filter. As indicated earlier, it is usual practice to design a higher-order filter by cascading first- and/or second-order filter sections with proper isolation between sections. However, in each of the filter sections described above, the output of the operational amplifier is also taken as the output of the filter section. Because of the negative feedback situation, the output impedance of the operational amplifier in the closed-loop configuration (and hence

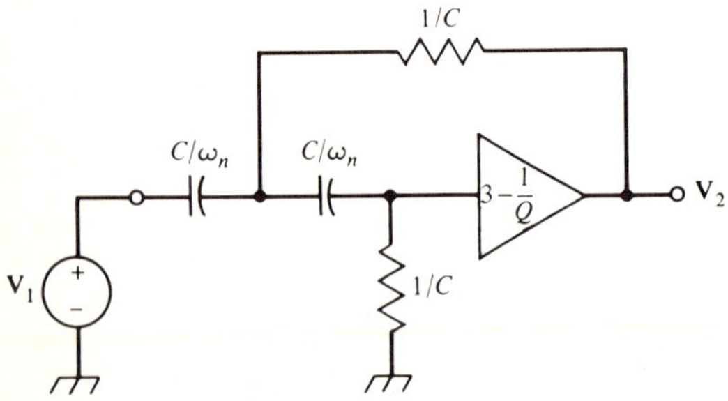

Figure 6-43 A second-order high-pass filter configuration. Element values are in ohms and farads.

the output impedance of the filter section) is extremely small. As a result, these structures can be cascaded without any additional buffers in between. In some other cases, as illustrated in the following example, such buffers may be needed.

Example 6-15.   Implement the third-order transfer function

$$T(s) = \frac{Hs^2}{(s + 2)(s^2 + 5s + 100)}$$

Note that there are two possible factorizations of the above transfer function:

$$T(s) = \frac{H_1}{s + 2} \cdot \frac{H_2 s^2}{s^2 + 5s + 100} \tag{6-104a}$$

$$= \frac{H_1 s}{s + 2} \cdot \frac{H_2 s}{s^2 + 5s + 100} \tag{6-104b}$$

The implementation of Eq. (6-104a) requires the cascading of a first-order low-pass section with a second-order high-pass section. On the other hand, the realization of Eq. (6-104b) requires the cascading of a first-order high-pass section with a second-order bandpass section. We make use of the decomposition of Eq. (6-104a) in our following discussion.

The implementation of the first-order low-pass section $H_1/(s + 2)$ is obtained with the aid of the RC structure of Fig. 6-44 which has a voltage transfer function

$$\frac{\mathbf{V}_2}{\mathbf{V}_1} = \frac{\dfrac{1}{RC}}{s + \dfrac{1}{RC}}$$

This implies that the RC product must equal $\frac{1}{2}$. Hence, if we choose $C = 1$ F, R becomes $(\frac{1}{2})$ Ω.

To implement the high-pass section, we use the circuit of Fig. 6-43. Comparing the second term on the right-hand side of Eq. (6-104a) with Eq. (6-103) we note that

$$\omega_n = 10$$
$$Q = 2$$

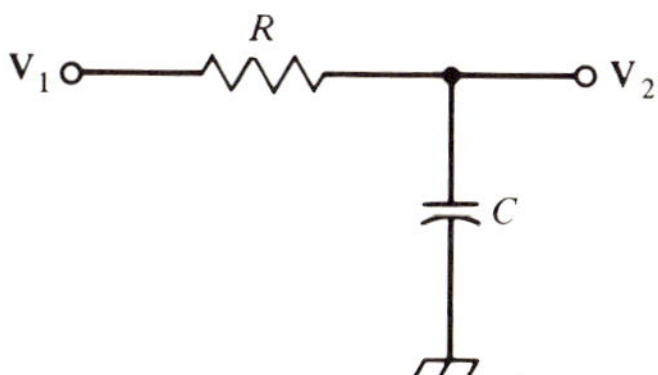

Figure 6-44   First-order low-pass section.

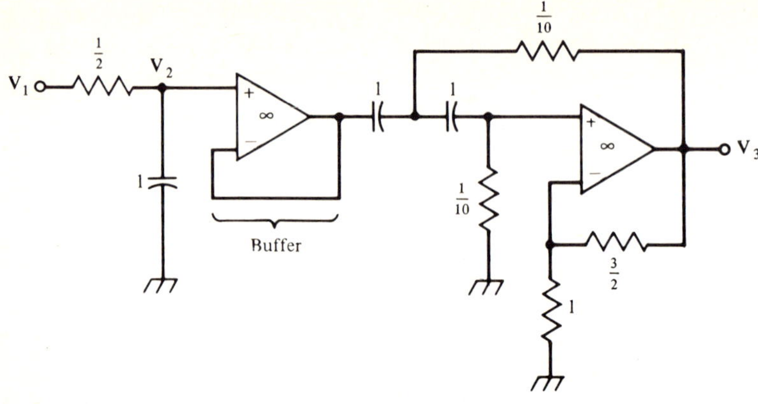

Figure 6-45 Complete realization of the third-order filter of Example 6-14. Element values are in ohms and farads.

If we now choose $C = 10$, the two resistors become equal to $(\frac{1}{10})$ Ω each and the two capacitors become equal to 1 F each. The gain of the noninverting amplifier is required to be 2.5.

The complete realization of the third-order filter is then as shown in Fig. 6-45.

6-9   Stability Considerations

The unavoidable frequency dependence of the open-loop gain of a practical operational amplifier may create instability in an improperly designed circuit using these devices. To illustrate this situation, consider the noninverting amplifier of Fig. 6-3(a) designed using an operational amplifier with an open-loop gain $\mu(s)$ which is a function of the complex frequency variables. If we ignore for simplicity the finite input and output resistances of the operational amplifier, then the closed-loop gain $A_v(s)$ of the noninverting amplifier is given as

$$A_v(s) = \frac{\mu(s)}{1 + \mu(s) \cdot \beta} \tag{6-105}$$

where $\beta = R_2/(R_1 + R_2)$. We recall from our knowledge of circuit theory that a circuit described by a transfer function $T(s)$ is *strictly stable* if and only if all the poles of $T(s)$ are in the left-half s-plane excluding the $j\omega$-axis and it is *marginally stable* if all the poles are in the left-half s-plane including the $j\omega$-axis with the additional condition that poles on the $j\omega$-axis are simple. Now, poles of the closed-loop gain $A_v(s)$ of our noninverting amplifier are given by the roots of the *characteristic equation*:

$$1 + \mu(s) \cdot \beta = 0 \tag{6-106}$$

In most cases of interest to us the feedback network is resistive and hence the feedback factor β is real and positive. Let us now determine the form of the open-loop gain $\mu(s)$ which will ensure stability for any amount of feedback.

Consider first a 3-pole model of the operational amplifier. The expression for open-loop gain $\mu(s)$ is then of the form

$$\mu(s) = \frac{K_0 \omega_0 \omega_1 \omega_2}{(s + \omega_0)(s + \omega_1)(s + \omega_2)} \tag{6-107}$$

where K_0, ω_0, ω_1, and ω_2 are positive. Substituting the above into Eq. (6-106) and rearranging the characteristic equation we arrive at

$$(s + \omega_0)(s + \omega_1)(s + \omega_2) + K_0 \omega_0 \omega_1 \omega_2 \beta = 0 \tag{6-108}$$

The parameters of the feedback amplifier, in particular K_0 and β, vary with temperature, aging, and so on. Consequently a plot of the location of the roots of the characteristic equation as $K_0 \beta$ is varied from zero to infinity displays instantly the stability of the feedback amplifier with respect to variation in $K_0 \beta$. Such a plot, known as a *root locus plot*,[20] is shown in Fig. 6-46 for a typical case. It is seen that when $K_0 \beta$ is equal to zero, the roots are negative real and are at the poles of the open-loop gain $\mu(s)$. As $K_0 \beta$ is increased, one of the roots always remains negative real. However, the other two roots eventually become complex and move from the left-half s-plane to the right-half s-plane making the feedback amplifier unstable. Therefore such a feedback amplifier is potentially unstable. The following numerical example illustrates the possible instability problem due to the pole movement to the right half of the s-plane.

Example 6-16. Determine the roots of the characteristic equation

$$(s + 1)(s + 2)(s + 4) + 8K_0 \beta = 0$$

for values of $K_0 \beta$ in the range 0–15.

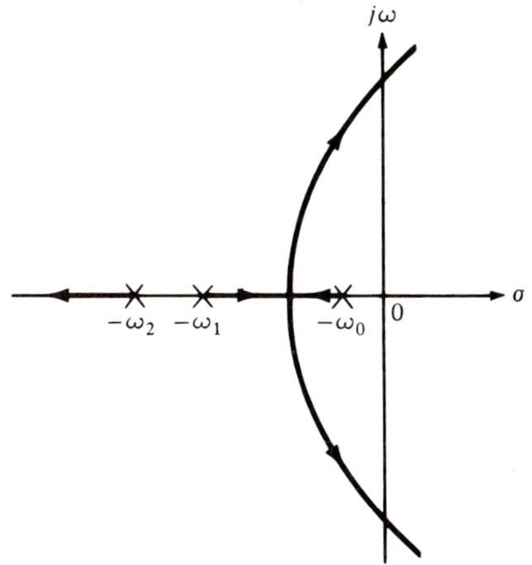

Figure 6-46   Root-locus plot of a 3-pole negative feedback amplifier.

Using a root-finding algorithm, the roots of the above characteristic equation have been calculated for several specific values of $K_0\beta$ and are summarized in Table 6-2. We observe from this table that one of the roots is always on the negative real axis. Two of the roots remain in the left-half s-plane for values of $K_0\beta$ between 0 and 10. However, these two move to the right half-plane for values of $K_0\beta$ greater than or equal to 12.

TABLE 6-2

$K_0\beta$	Roots
0	-1; -2; -4
1.0	$-1.116327 \pm j1.452576$; -4.767345
4.0	$-0.612432 \pm j2.559525$; -5.775136
10.0	$-0.08234313 \pm j3.587134$; -6.835313
12.0	$+0.04664577 \pm j3.828781$; -7.093291
15.0	$+0.2165989 \pm j4.144047$; -7.433197

It is a simple matter to determine the exact value of $K_0\beta$ for which the two roots cross the imaginary axis. To this end we set $s = j\omega$ in the characteristic equation and arrive at

$$-j\omega^3 - 7\omega^2 + j14\omega + 8(1 + K_0\beta) = 0$$

Equating separately the real and imaginary parts of the above equation to zero, we obtain

$$\omega^3 = 14\omega$$
$$7\omega^2 = 8(1 + K_0\beta)$$

which when solved yields

$$K_0\beta = \frac{98}{8} - 1 = 11.25$$

Next we consider the 2-pole model of the amplifier. In this case the expression for the open-loop gain is of the form

$$\mu(s) = \frac{K_0\omega_0\omega_1}{(s + \omega_0)(s + \omega_1)} \tag{6-109}$$

where K_0, ω_0, and ω_1 are nonnegative. The corresponding characteristic equation is now

$$(s + \omega_0)(s + \omega_1) + K_0\omega_0\omega_1\beta = 0 \tag{6-110}$$

The root-locus plot of the above equation is sketched in Fig. 6-47. Note here that for $K_0\beta$ equal to zero the roots are at the poles of the open-loop gain and are negative real. But as $K_0\beta$ increases they eventually become complex and

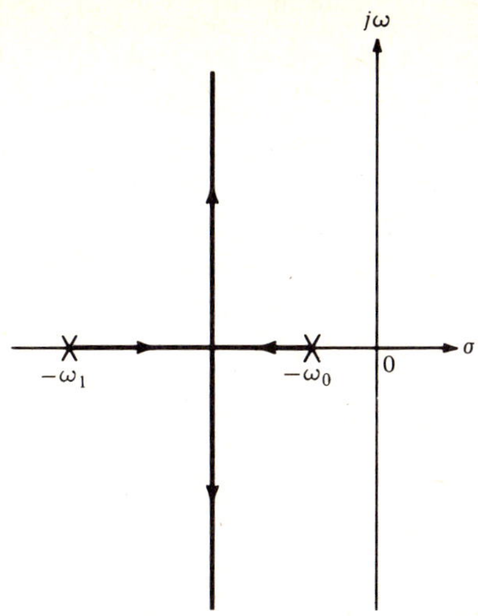

Figure 6-47   Root-locus plot of a 2-pole negative feedback amplifier.

move toward infinity at a constant distance from the $j\omega$-axis. Consequently such a feedback system is strictly stable.

In the case of a 1-pole model, the expression for the open-loop gain is

$$\mu(s) = \frac{K_0 \omega_0}{(s + \omega_0)} \qquad (6\text{-}111)$$

with nonnegative K_0 and ω_0. The characteristic equation for this case is

$$(s + \omega_0) + K_0 \omega_0 \beta = 0 \qquad (6\text{-}112)$$

It is evident that the root of this equation is always negative real for any value of $K_0 \beta$. The corresponding root-locus plot is shown in Fig. 6-48. We conclude that in this case the feedback system is definitely strictly stable.

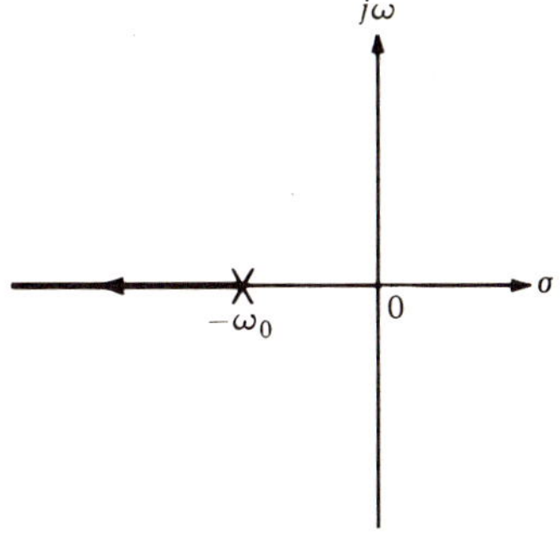

Figure 6-48   Root-locus plot of a single-pole negative feedback amplifier.

A few comments here are in order. We observe that for resistive feedback, both the single-pole model and the 2-pole model lead to strictly stable feedback system. However, this analysis assumed that no other parasitics are present and the open-loop gains are as given by Eqs. (6-109) and (6-111). In any practical feedback amplifier circuit, parasitic and stray capacitances are always present. To illustrate the effect of these capacitances, consider the modified circuit representation of the operational amplifier as shown in Fig. 6-49 where C_o represents the effective output capacitance. The voltage gain of this amplifier is

$$\frac{\mathbf{V}_2}{\mathbf{V}_1} = \frac{\mu(s)}{R_o C_o s + 1} \tag{6-113}$$

It is seen that if the amplifier gain $\mu(s)$ is represented by a 2-pole model, the overall open-loop voltage gain is then a network function with three poles. Consequently this amplifier when used in a negative feedback configuration may become unstable. We can thus conclude that the 2-pole model does not provide any *stability margin* to take care of parasitics. As a rule, therefore, *all operational amplifiers intended for use in a negative resistive feedback configuration should preferably have a single-pole representation so that the feedback amplifier will be stable for any amount of feedback.* Some monolithic operational amplifiers such as the Type 741 amplifier are internally compensated and can be adequately represented by a single-pole model for most applications. Most others require external *RC* compensating networks to achieve a single-pole representation. These ICs have specific terminals where the compensating network are to be connected. It is recommended that the manufacturer's guidelines be followed for frequency response compensation of these latter operational amplifiers.

It should be noted that the sufficiency of a single-pole model in achieving stability holds only for purely resistive feedback. If reactive elements are present in the feedback network, an operational amplifier with negative feedback can still oscillate even though it is adequately modeled by a single-pole open-loop gain.

In the case of noninverting amplifier designed using an operational amplifier of open-loop gain $\mu(s) = K_0 \omega_0/(s + \omega_0)$, the closed-loop gain from Eq. (6-105)

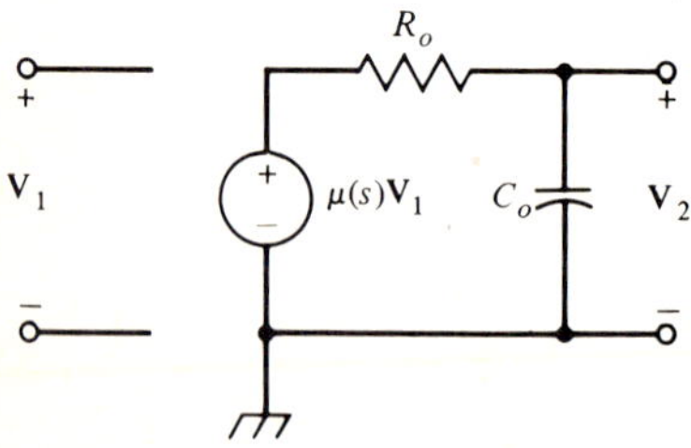

Figure 6-49 Circuit representation of an operational amplifier including the output parasitic capacitance and output resistance.

is given as

$$A_v(s) = \frac{K_0 \omega_0}{s + \omega_0(1 + K_0\beta)}$$

The dc gain $A_v(0)$ is $K_0/(1 + K_0\beta)$ and the 3-dB bandwidth is $\omega_0(1 + K_0\beta)$. The *gain-bandwidth product* is then $K_0\omega_0$ which is the same as that of the operational amplifier itself, and is independent of β. This is an interesting result. Hence, if the closed-loop dc gain* is decreased, the useful bandwidth of the feedback amplifier is increased maintaining the gain-bandwidth product as a constant. On the other hand, if the closed-loop dc gain is increased, the bandwidth of the feedback amplifier is decreased.

6-10 Types of Operational Amplifiers[11]

The characteristics of practical operational amplifiers depend on the specific circuit configuration and the method of fabrication being used to build the modules. With regard to these two considerations, the operational amplifier modules can be grouped into six broad types: (i) bipolar transistor types, (ii) FET-input types, (iii) wide-band types, (iv) chopper-stabilized types, (v) varactor types, and (vi) integrated-circuit types. The operational amplifiers belonging to the first five categories are primarily fabricated using all discrete components and those belonging to the last category are fabricated in monolithic or hybrid integrated circuit forms. Based on the above classification scheme, some general comments can be made concerning the characteristics of practical operational amplifiers. However, it should be kept in mind that as a result of continuing improvements in device and circuit fabrication technology, it is difficult to make precise statements regarding the properties of these or any other modules. Furthermore the classification itself is also not very precise because operational amplifiers belonging to some of the first five categories are also available in IC form.

The general purpose bipolar-type operational amplifier is made with bipolar transistors, discrete resistors, and capacitors. It is almost always internally compensated to provide a frequency response characteristic with a 6 dB/octave roll-off up to the unity-gain crossover frequency. The high offset current and the associated drift with time and temperature exhibited by this type of operational amplifier is much more of a problem than the offset voltage and its drift.

An operational amplifier with field-effect transistor (FET) input stages exhibits very high input resistances (typically in the range of 10^4–10^6 MΩ) and very low offset currents. In the earlier versions, the offset voltage drift used to be worse than that of the bipolar type. However, this is no longer true.

The wide-band type has a very high gain–bandwidth product, typically of the order of 1000 MHz. It usually is designed with a very low dc gain (about

---

* In most cases, the closed-loop dc gain is essentially $1/\beta$ as practical operational amplifiers in general have a very large value of open-loop dc gain K_0.

30–50) which is necessary to make the bandwidth very large. Furthermore this type of operational amplifier is characterized with a low input impedance and high input current. Some of the wide-band modules have a very high slew rate.

The chopper stabilized variety is quite a complex circuit. It employs a dc amplifier, an ac amplifier, a demodulator, *RC* filters, and a mechanical or an electronic switch operating at a very high rate (Fig. 7-67). Because of its special purpose design it is characterized with very low offset voltage and current drifts. It has a very high open-loop gain. However, it is always available in single-ended form with only an inverting input terminal which, along with its higher power requirement and very high cost, limits its use.

The varactor type uses a special type of semiconductor diode, known as a *varactor*, whose internal capacitance varies with varying bias voltages. It is characterized with very low noise, excellent isolation of input from ground, very high common-mode rejection, and extremely low bias currents.

Since its introduction in 1965, the performance of the monolithic operational amplifiers have continued to improve with improvement in IC technology. The high performance IC operational amplifiers have typically a very large dc gain with a reasonably high bandwidth. The offset voltage and current drifts are considerably lower than the discrete version. Some monolithic modules have a very low-power requirement, typically of the order of 100 μW. The monolithic operational amplifiers with FET input stages have very low input offset and bias currents but their voltage offset and its drift are higher than those with bipolar transistor input stages. Some varieties make use of super-beta transistors at the input to obtain reasonably low input currents. The biggest attraction of the IC operational amplifier is its very low cost compared to its discrete counterpart. Many of the attractive features of the monolithic circuit and the discrete version have been made available in the hybrid IC operational amplifier with a moderate increase in the cost. It should be noted that recently monolithic chopper stabilized operational amplifiers have been introduced.

6-11   Summary

The properties of an operational amplifier are described in this chapter along with a number of linear circuit applications. The idealized model of the operational amplifier is defined in Section 6-1 which also points out a simple approach to the analysis of circuits containing these devices. Several simple applications of the operational amplifier are considered in Section 6-2. In particular, a number of useful circuits such as single-ended and differential-input voltage amplifiers, current amplifiers, voltage-to-current and current-to-voltage converters, summing and difference amplifiers, and so on, are described here. The characteristics of a practical operational amplifier are described in Section 6-3. Section 6-4 considers the effects of the finite input and output resistances, and the finite gain of a practical amplifier on the performances of two circuits designed using this device. The effects of input offset voltages and currents are treated in Section 6-5, and that of the finite common-mode rejection is considered in

Section 6-6. A brief discussion on simulation of linear constant coefficient differential equations using operational amplifiers is provided in Section 6-7. Ac applications of the operational amplifier are the subjects of Section 6-8. Here the design of frequency-selective networks with specified frequency-magnitude response characteristic is considered. The subject of stability of circuits containing operational amplifiers is discussed in Section 6-9, where it is shown that an operational amplifier with a single-pole gain model is stable in most (that is, resistive) negative feedback applications. Finally, Section 6-10 provides a brief discussion of some of the characteristics of different types of operational amplifier modules available in the market.

References

1. M. English, "Some Applications of the μA741 Operational Amplifier," Application Note, Fairchild Semiconductor, Mountain View, Calif.
2. M. Kahn, *The Versatile Op Amp*, Holt, Rinehart & Winston, New York, 1970.
3. A. Barna, *Operational Amplifiers*, John Wiley & Sons, New York, 1971.
4. F. C. Fitchen, *Electronic Integrated Circuits and Systems*, Van Nostrand Reinhold Co., New York, 1970.
5. W. Kaelin *et al.*, "Review of Operational Amplifier Principles," Application Bulletin, Fairchild Controls, Mountain View, Calif., December 1968.
6. R. Demrow, "Op Amps as Electrometers or—The World of fA," *Analog Dialogue*, Vol. 5, no. 2, February/March 1971, pp. 6, 7.
7. Application Briefs, Analog Dialogue, Vol. 5, no. 2, February/March 1971, p. 14.
8. S. K. Mitra, *Analysis and Synthesis of Linear Active Networks*, John Wiley & Sons, New York, 1969.
9. S. K. Mitra, Ed., *Active Inductorless Filters*, IEEE Press, New York, 1971.
10. J. Eimbinder, Ed., *Designing with Linear Integrated Circuits*, John Wiley & Sons, New York, 1968.
11. G. J. Deboo and C. N. Burrous, *Integrated Circuits and Semiconductor Devices: Theory and Application*, McGraw-Hill Book Co., New York, 1971, Chap. 4.
12. E. Moustakas and S-P. Chan, *Introduction to the Applications of the Operational Amplifier*, Academic Cultural Company, Santa Clara, Calif., 1974.
13. J. N. Giles, Ed., *Linear Integrated Circuits Applications Handbook*, Fairchild Semiconductor, Mountain View, Calif., 1967.
14. Staff, *The Application of Linear Microcircuits*, SGS-Fairchild, London, 1966.
15. G. E. Tobey, J. G. Graeme, and L. P. Huelsman, Eds., *Operational Amplifiers—Design and Applications*, McGraw-Hill Book Co., New York, 1971.
16. Staff, *Handbook of Operational Amplifier Applications*, Burr-Brown Research Corp., Tucson, Ariz., 1963.
17. J. I. Smith, *Modern Operational Amplifier Circuit Design*, Wiley-Interscience, New York, 1971.
18. H. J. Orchard and A. N. Willson, Jr., "New Active-gyrator Circuit," *Electronics Letters*, Vol. 10, no. 13, June 1974, pp. 261–262.
19. J. G. Graeme, *Operational Amplifiers: Third-generation Techniques*, McGraw-Hill Book Co., New York, 1973.

20. J. G. Truxal, *Automatic Feedback Control System Synthesis*, McGraw-Hill Book Co., New York, 1955.

21. M. E. Van Valkenburg, *Introduction to Modern Network Synthesis*, John Wiley & Sons, New York, 1962.

22. J. V. Wait, L. P. Huelsman, and G. A. Korn, *Introduction to Operational Amplifier Theory and Applications*, McGraw-Hill Book Co., New York, 1975.

23. A. S. Jackson, *Analog Computation*, McGraw-Hill Book Co., New York, 1960.

24. Staff, "A Fast Integrated Voltage Follower with Low Input Current," Application Note AN-5, National Semiconductor Corp., Santa Clara, Calif., May 1968.

Problems

6-1 Design a noninverting voltage amplifier with a gain of 45 dB.

6-2 Design an inverting voltage amplifier with a gain of 30 dB.

6-3 A very high gain inverting amplifier designed using the circuit of Fig. 6-5 would require resistors with very large spread in values. This problem is avoided in the modified inverting amplifier circuit of Fig. 6.50.[17] Determine the expression for the voltage gain v_2/v_1 of this circuit and design an inverting amplifier of 40-dB gain using resistors of moderate values.

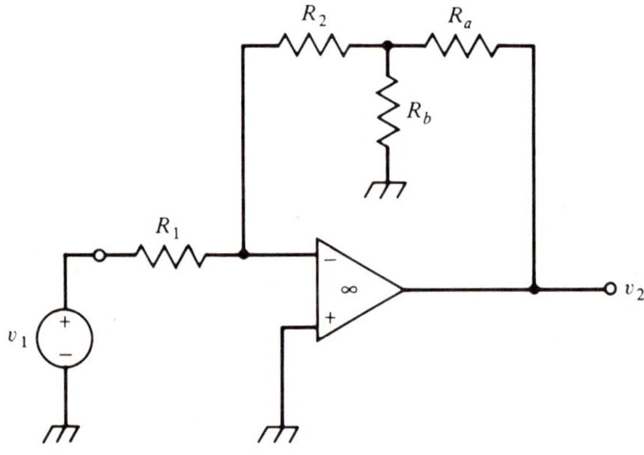

Figure 6-50

6-4 With an additional operational amplifier, the inverting voltage amplifier of Fig. 6-5 can be modified to provide very high input impedance. The pertinent circuit is shown in Fig. 6-51.[19] Determine the expression for the input impedance. Under what condition will it be large?

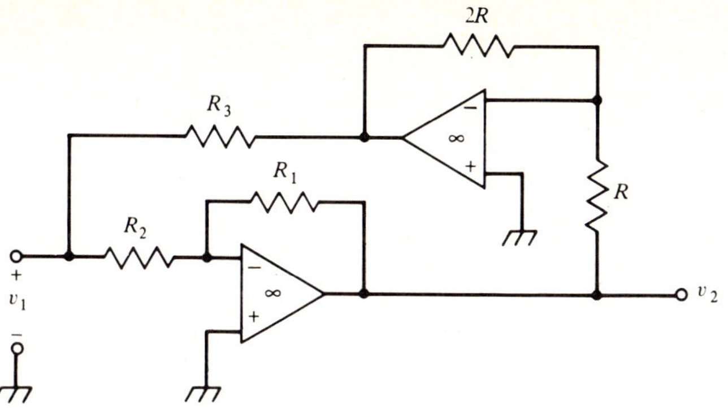

Figure 6-51

6-5 Implement a 6-mA dc current source using an operational amplifier, a 5-V battery, and a resistance.

6-6 An alternate realization[14,15] of a voltage-controlled current source for a floating load is indicated in Fig. 6-52. Show that for this circuit, the load current is related to the signal voltage v_1 as

$$i_2 = -\left(\frac{R_1}{R_2 R_3} + \frac{1}{R_2}\right)v_1$$

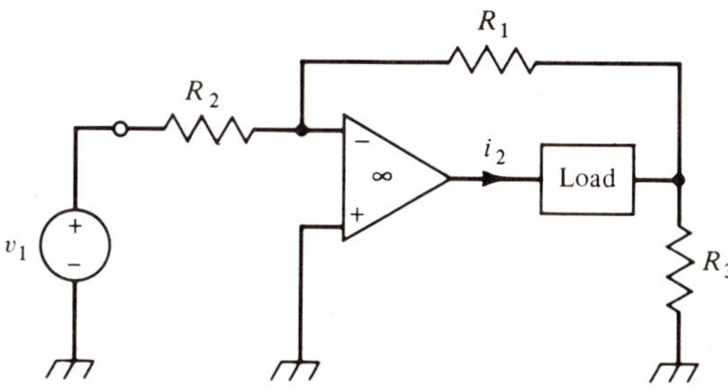

Figure 6-52

6-7 Another implementation of a voltage-to-current converter (transadmittance amplifier) for a grounded load, known as the Howland circuit,[17] is sketched in Fig. 6-53. Determine the expression for the load current i_2 as a function of the input signal voltage v_1.

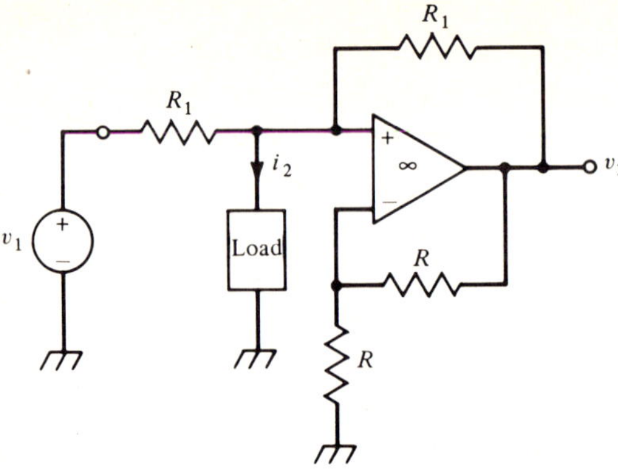

Figure 6-53   Howland circuit.

6-8   A realization of a differential-input, single-ended output voltage-to-current converter, which is a generalization of the Howland circuit of Fig. 6-53, is shown in Fig.
6-54.[17] Derive the expression for the output current i_2 as a function of the input
voltages v_1 and v_2.

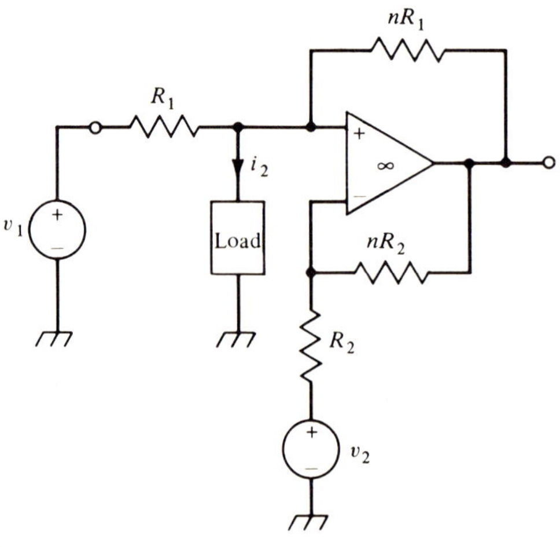

Figure 6-54

6-9   The circuit of Fig. 6-55 is a voltage-to-current converter for a grounded load and
exhibits a very high input impedance.[15] Show that if $R_3 = R_1 - R_2$, then

$$i_2 = -\frac{2}{R_2}\, v_1$$

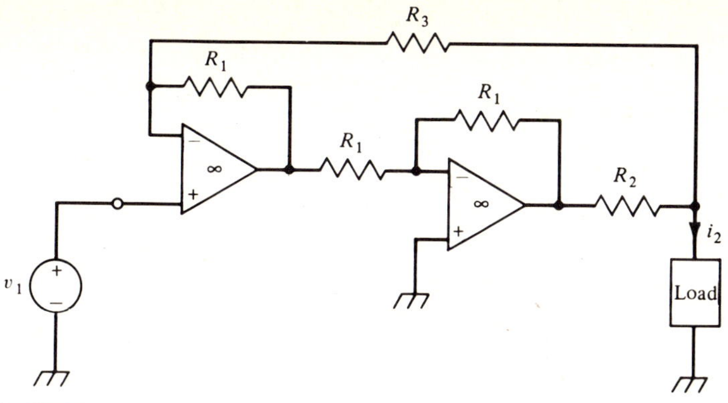

Figure 6-55

6-10 Design a current amplifier by appropriately cascading transresistance and transconductance amplifiers.

6-11 Derive Eq. (6-35).

6-12 Determine the differential-mode gain v_2/v_1 of the differential-input, differential-output amplifier of Fig. 6-56.

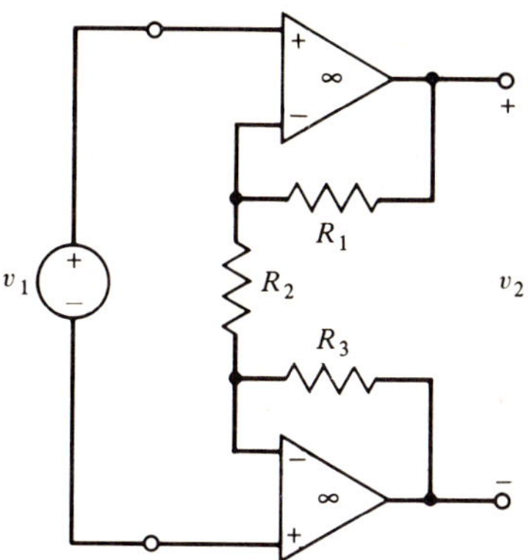

Figure 6-56

6-13 Design a differential amplifier using a negative gain summing amplifier and an inverting voltage amplifier.

6-14 Implement a high input impedance differential amplifier utilizing the difference amplifier of Fig. 6-12 and a noninverting voltage amplifier.

6-15 Implement the following equations using the difference amplifier of Fig. 6-12: **(a)** $v_o = 5v_2 - \frac{3}{2}v_1$, **(b)** $v_o = 2v_2 - \frac{3}{2}v_1$, and **(c)** $v_o = \frac{5}{2}v_2 - 2v_1$.

6-16 The difference amplifier circuit of Fig. 6-12 can also be used for measuring deviation. If we set $R_3 = R_4$, $R_2 = R_1 + \Delta R$, and connect the two input terminals together and excite with a constant voltage source E, show that the output voltage v_o is then linearly proportional to the resistance deviation ΔR.

6-17 The differential-input, single-ended-output voltage-amplifier circuit of Fig. 6-57 is known as the *potentiometric* amplifier[10] and can be designed to exhibit very high common-mode rejection. Show that its common-mode gain is given by

$$A_{vcm} = \left. \frac{v_o}{v_{1A}} \right|_{v_{1A}=v_{1B}} = \frac{K_1 + K_2 - 1}{K_1 K_2}$$

where $K_1 = R_2/(R_1 + R_2)$ and $K_2 = R_4/(R_3 + R_4)$. Determine the expression for the differential-mode gain $A_{vd} = v_o/(v_{1A} - v_{1B})$ under the condition that the common-mode gain has been set to zero.

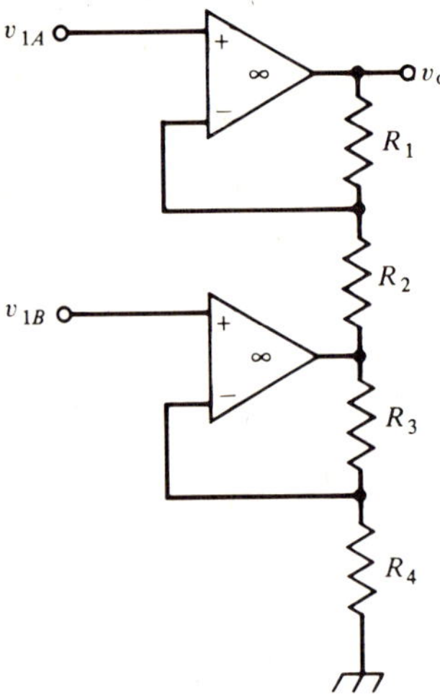

Figure 6-57

6-18 Develop the exact expression for the output voltage v_o of the bridge amplifier of Fig. 6-58 as a function of ε where $R_t = R_0(1 + \varepsilon)$. Show that the relation is linear for $\varepsilon \ll 1$.

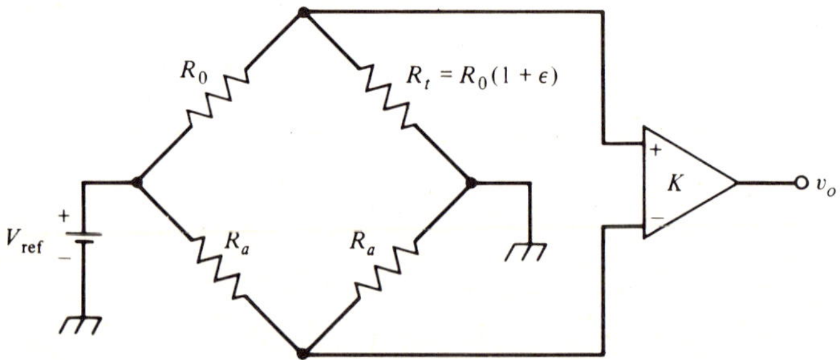

Figure 6-58 Another realization of a bridge amplifier.

6-19 Evaluate the performance of the noninverting amplifier of Problem 6-1 designed using an operational amplifier characterized by the following parameters:

$$\mu = 10{,}000, \qquad R_i = 50\,\text{k}\Omega, \qquad R_o = 250\,\Omega$$

6-20 Analyze the effects of finite gain, and finite input and output resistance on the performance of the inverting amplifier of Problem 6-2. The characteristics of the operational amplifier are as given in Problem 6-19.

6-21 Simulate the differential equation

$$b_2 \frac{d^2 y}{dt^2} + b_1 \frac{dy}{dt} + b_0\, y(t) = a_1 \frac{dx}{dt} + a_0\, x(t)$$

where $y(t)$ is the output voltage and $x(t)$ is the input voltage.

6-22 Design a differential integrator circuit which integrates the difference between two voltage signals.

6-23 A possible implementation of a noninverting integrator is shown in Fig. 6-59.[11] Determine the relation between the input and the output voltages in the time domain.

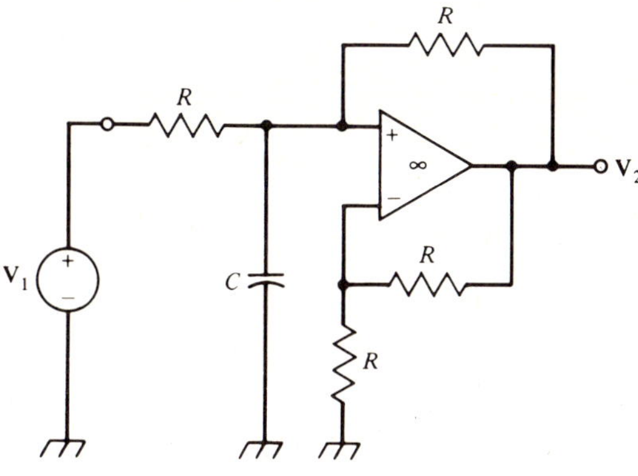

Figure 6-59

6-24 The circuit of Fig. 6-60 was advanced by Antoniou for simulation of an inductance.[9] Determine the value of the equivalent inductance L_{eq} seen between the two input terminals.

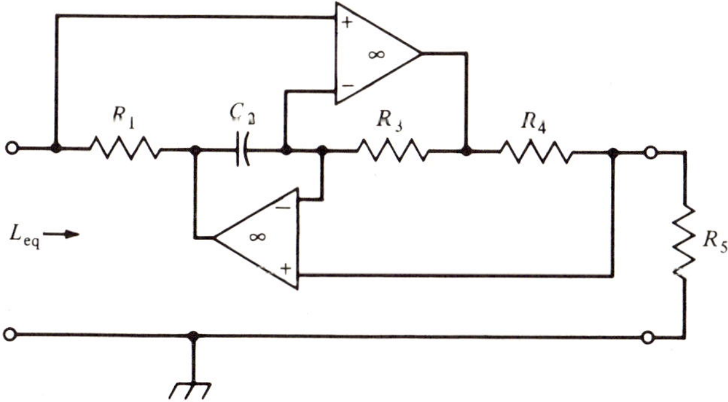

Figure 6-60

6-25 Another circuit for inductor simulation, which uses one operational amplifier, is shown in Fig. 6-61.[18] Show that the circuit simulates an inductance of equivalent inductance $L_{eq} = R_1 R_4{}^2 C (R_2 + R_1)/R_2(R_2 + R_4)$ if the condition $(R_1 + R_2)(R_2 + R_3) = R_1 R_2 R_5/R_6$ is satisfied.

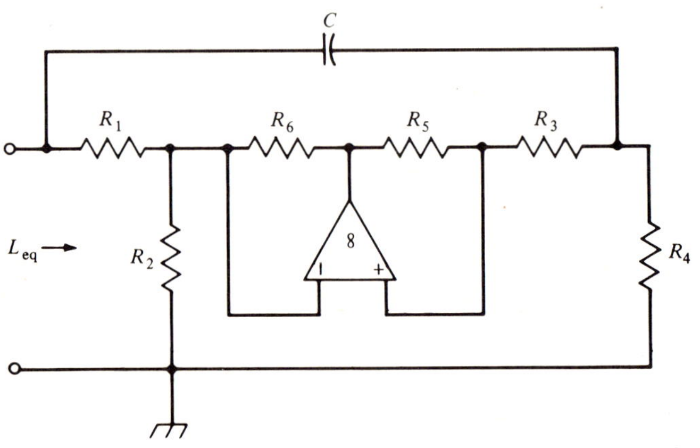

Figure 6-61

6-26 Figure 6-62 shows the circuit configuration of an active RC second-order low-pass filter proposed by Bach.[8] Show that its transfer function is given by Eq. (6-99).

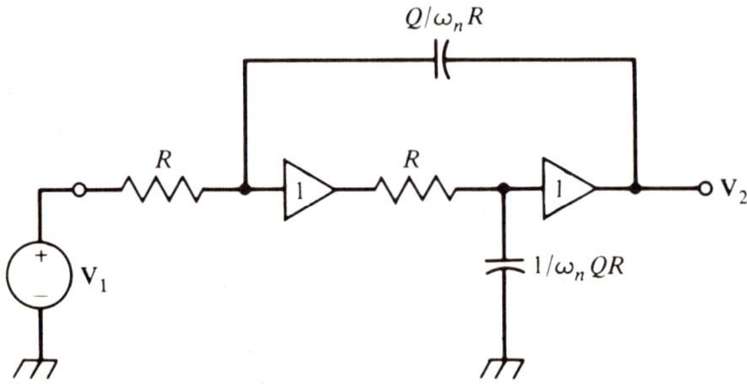

Figure 6-62

6-27 Show Fig. 6-63 (also proposed by Bach[8]) implements the second-order high-pass transfer function of Eq. (6-103).

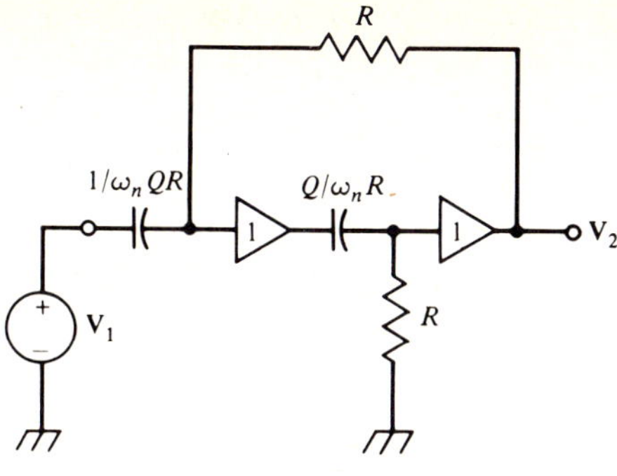

Figure 6-63

6-28 Two additional second-order bandpass filters are shown in Fig. 6-64.[15] Show that the transfer function $\mathbf{V}_2/\mathbf{V}_1$ of each circuit is that given by Eq. (6-94).

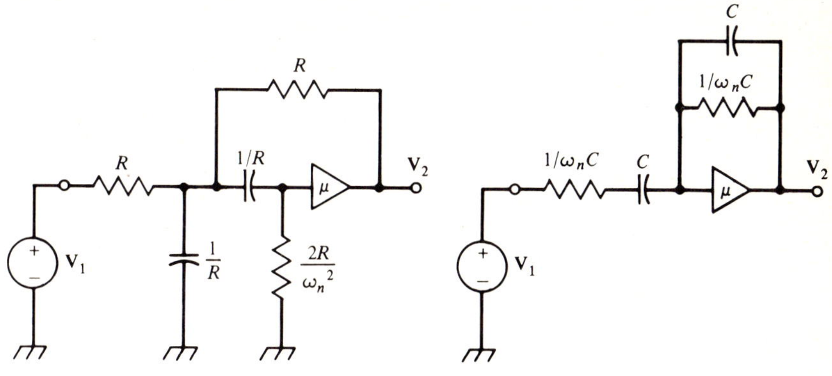

Figure 6-64

6-29 Design a second-order Butterworth low-pass filter with a 3-dB cut-off frequency at 2 kHz using the circuit configuration of Fig. 6-62.

6-30 A second-order Butterworth high-pass transfer function is given by

$$T(s) = \frac{Hs^2}{s^2 + \sqrt{2}\,\omega_n s + \omega_n^2}$$

Show that at $\omega = \omega_n$, the magnitude is 3 dB down from its value at $\omega \to \infty$. Plot the magnitude function.

6-31 Design a second-order Butterworth high-pass filter with a 3-dB cut-off frequency at 500 Hz using the circuit configuration of Fig. 6-43.

6-32 Repeat Problem 6-31 using the circuit configuration of Fig. 6-63.

6-33 Realize the bandpass transfer function

$$T(s) = \frac{Hs}{s^2 + 0.4s + 400}$$

using the circuits of Figs. 6-38 and 6-64. What are the values of the resonant frequency in hertz and the Q value?

6-34 Figure 6-65 shows the circuit schematic of a third-order low-pass active RC filter.[8] Show that its voltage transfer function is given as

$$T(s) = \frac{\mathbf{V}_2}{\mathbf{V}_1} = \frac{a_3}{s^3 + a_1 s^2 + a_2 s + a_3}$$

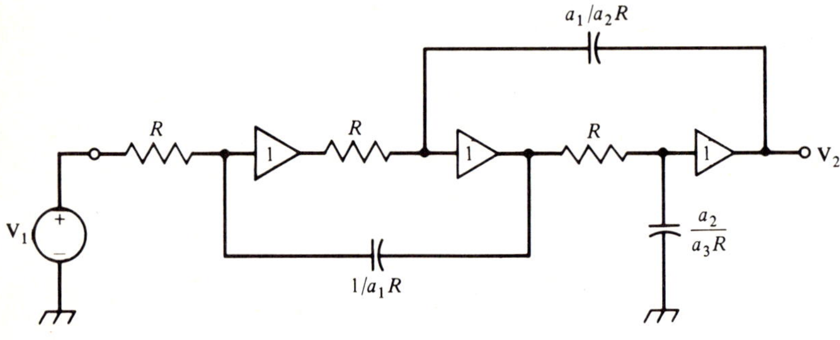

Figure 6-65

6-35 A third-order low-pass Butterworth transfer function is given as

$$T(s) = \frac{\omega_n^{\,3}}{s^3 + 2\omega_n s^2 + 2\omega_n^{\,2} s + \omega_n^{\,3}}$$

(a) Show that at $\omega = \omega_n$, the magnitude of this transfer function is 3 dB less than that at dc.

(b) Design a third-order low-pass Butterworth filter with a cut-off at 1 kHz.

6-36 Analyze the noninverting ac amplifier of Fig. 6-66[16] and determine its voltage transfer function.

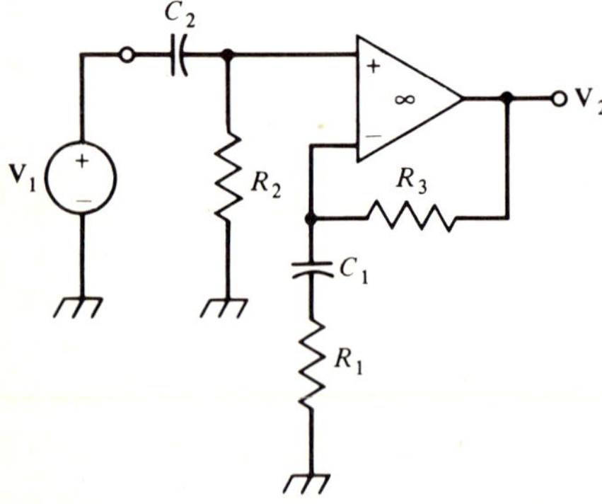

Figure 6-66

CHAPTER 7

Nonlinear Function Modules

The silicon monolithic operational amplifier is probably the single most important contribution in the field of analog integrated circuits. It showed the possibility of fabrication and high volume production of complex analog circuitry in IC form at a per unit cost comparing favorably with its discrete counterparts. A number of other analog monolithic and hybrid IC modules have appeared in recent years. A majority of these are primarily used for performing nonlinear operations such as multiplication, division, integral and fractional powers, logarithms and exponentiations, transcendental and correlation computations, modulation and demodulation, and so on. The nonlinear function modules together with other circuit elements such as operational amplifiers, diodes, and passive elements, are finding increasing new applications in signal processing, telemetering, instrumentation, and measurement systems, often replacing costly discrete circuits, special purpose digital computers, or electromechanical systems. In this chapter we introduce some of the widely used nonlinear function modules and describe a number of useful applications of these devices. In particular we consider here the analog multipliers, the logarithmic amplifier, the trigonometric function generator, the comparator, and the analog switch.

7-1   The Analog Multiplier

To many circuit and system designers, the analog IC multiplier would rank as the second most important advance in the field of analog ICs after the IC operational amplifier. Since its first introduction in 1968, it has gone through several refinements leading to improvements in performance with a corresponding decrease in cost.

The Ideal Model

An ideal single-ended analog multiplier is a 2-input, single-output device such that the output voltage is proportional to the product of the two input analog signal voltages. Figure 7-1 shows the symbolic representation of a single-ended multiplier. With respect to this figure we can describe the operation of an ideal multiplier as

$$v_o = \frac{v_x \cdot v_y}{K} \tag{7-1}$$

where v_o is the output voltage in volts, v_x and v_y are the input voltages in volts, and K is the constant of proportionality in volts. The scale factor constant K in most multiplier modules is fixed at the time of manufacture and is used to establish the full-scale range of the output signal. However, in some commercially available units, it is externally adjustable by means of a voltage or a current. Typically, the range of the input signals is ± 10 V. Hence to restrict the output voltage to ± 10 V, the value of the scale factor K is normally chosen as 10 V.

The two input terminals of a multiplier are often referred to as the *X-channel* and the *Y-channel*, and are usually labeled " *X* " and " *Y* " in an actual module.

The controlled-source model of an ideal multiplier is sketched in Fig. 7-2; it indicates that ideally the input resistances are infinite (i.e., no currents flow through the input terminals) and the output impedance is zero. An ideal multiplier is characterized by zero offset implying $v_o = 0$ if either v_x is zero, v_y is zero, or both are zero.

Figure 7-2 indicates that the multiplier can be considered as an ideal voltage amplifier, having a voltage variable gain which is achieved by applying the input signal to be amplified at one terminal and the control voltage at the other input terminal. This fact will be utilized later in the design of voltage-variable filters. Note that with this interpretation we can consider the ideal multiplier to have an infinite bandwidth.

Multipliers are classified in accordance with the allowable polarities of the two input signals and the output signal. If both inputs v_x and v_y are bipolar, that is, they can assume both positive and negative values, and the multiplier

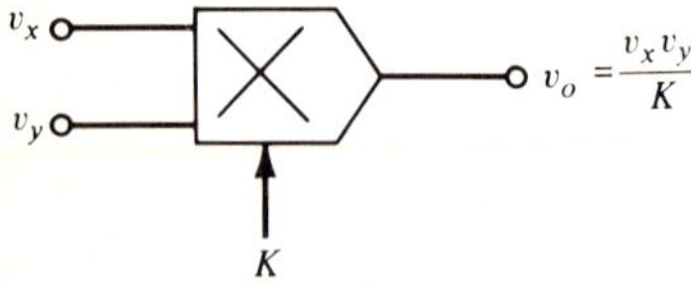

Figure 7-1 Symbolic representation of an analog multiplier.

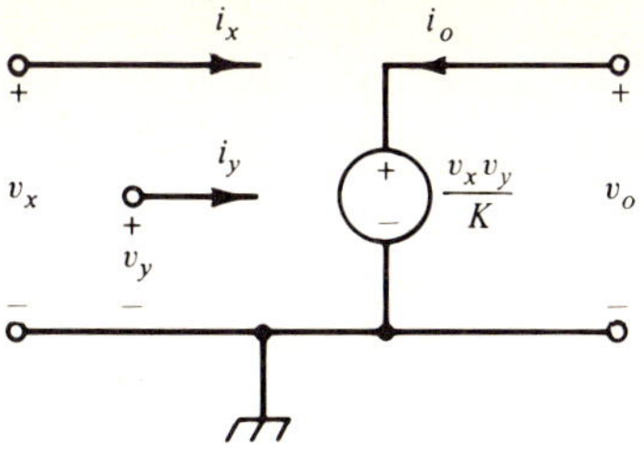

Figure 7-2   Controlled source model of an analog multiplier.

develops a bipolar v_o with the correct sign, then the device is called a *4-quadrant multiplier*. Here the output will be positive if the input signals have the same sign; otherwise it is negative. In a *2-quadrant* multiplier, on the other hand, one of the input terminals accepts only unipolar positive or negative signals with no restriction on the signal at the other input terminal. Here also the output takes the correct sign. For example, if v_x is restricted to negative values only, then v_o will be 180° out of phase with v_y. Finally multipliers accepting only unipolar signals at both inputs are known as *1-quadrant* multipliers. However, such multipliers are rarely used at the present time. It should be noted that the designation "quadrant" comes from the multiplier output–input relationship plotted in a rectangular coordinate system with the output v_o appearing perpendicular to the input v_x–v_y plane (Fig. 7-3).

It is possible to convert a multiplier of one type into another one using additional circuitry. Converting a 4-quadrant multiplier to a 2- or 1-quadrant one is straightforward with the aid of *absolute-value circuits** (Problem 7-1). Likewise, a 2-quadrant unit can be simply modified to work as a 1-quadrant multiplier. The reverse process is slightly more complicated as illustrated in the following example.

Example 7-1.   Design a 4-quadrant multiplier using two 2-quadrant multipliers and operational amplifiers.

Without any loss of generality, let us assume that the 2-quadrant multipliers available accept only positive signals at the X-channel. However, if the input signal v_x can be negative we can add a positive bias E, so that their sum $(E + v_x)$ is always positive. The bias can also be chosen to keep the difference $(E - v_x)$ positive for positive v_x. If we now apply $\frac{1}{2}(E + v_x)$ to the X-channel and v_y to the Y-channel of the one 2-quadrant multiplier, its output will be $(v_x + E)v_y/2K$. Likewise, if we apply $\frac{1}{2}(E - v_x)$ to the X-channel and v_y to the Y-channel of a second 2-quadrant multiplier, its output will be given by $(E - v_x)v_y/2K$.

---

* An *absolute-value circuit* develops an output that is equal to the absolute value of the input.

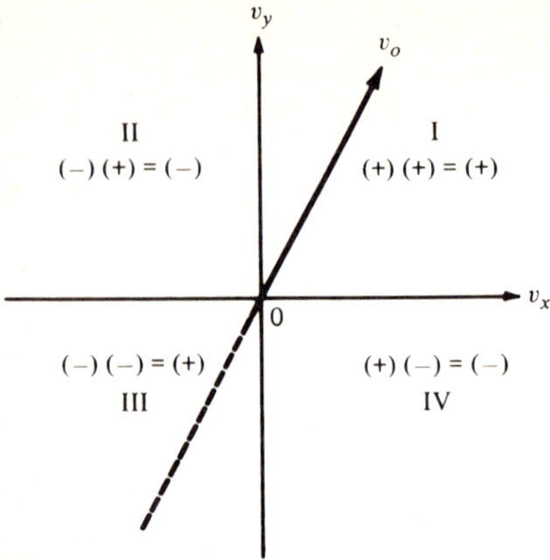

Figure 7-3   Illustration of " quadrant " concept.

By forming the difference of the two multiplier outputs we obtain

$$\frac{(v_x + E)v_y}{2K} - \frac{(E - v_x)v_y}{2K} = \frac{v_x v_y}{K} \tag{7-2}$$

indicating 4-quadrant operation. The complete realization is left as an exercise (Problem 7-2).

We describe several applications next.

Simple Applications

In addition to multiplying two analog signals, multipliers can be used to design circuits to perform other types of nonlinear operations. Some of these are outlined below.

Generation of Integer Powers.   By connecting the two inputs together [Fig. 7-4(a)] a *squarer* is obtained. If the input to the squarer is v_x, then its output is given as v_x^2/K.

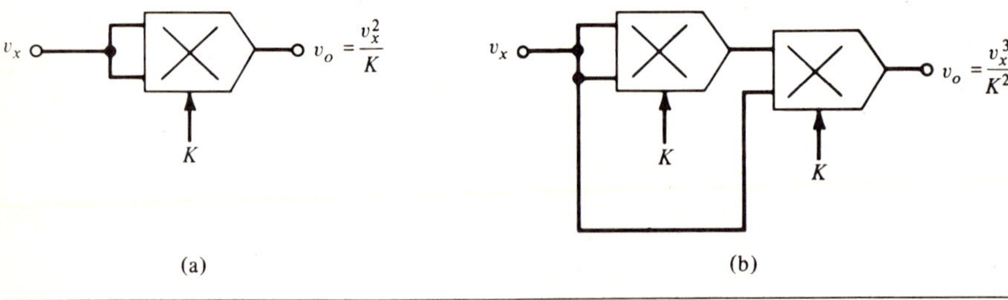

Figure 7-4   (a) A squarer and (b) a cuber.

The squarer designed using a 4-quadrant multiplier offers 2-quadrant operation because the input can be bipolar. A 1-quadrant squarer results if the multiplier is, instead, a 2-quadrant or a 1-quadrant type. An *odd-function* squarer develops an output that is proportional to the square of the input but has the same polarity as that of the input. It can be designed using an *absolute-value circuit* and a multiplier (Problem 7-4).

By feeding in the output of a squarer and the original analog input v_x to another multiplier, as shown in Fig. 7-4(b), the cube of the analog signal can be generated. The same principle can be extended to generate any higher integer powers of analog signals. The symbolic representation of an mth power circuit is sketched in Fig. 7-5.

As will be shown later, by placing these circuits in the feedback path of an operational amplifier, the roots of an analog signal can be developed. Another application of these circuits considered next are in the generation of sinusoidal signals of frequencies which are integer multiples of that of any given sinusoidal signal.

Frequency Multiplier.   A squarer can be used as a *frequency doubler*. Observe that if the input v_x to the squarer is a sinusoidal signal $A \sin \omega t$, then its output is given as

$$v_o = \frac{A^2}{K}(\sin \omega t)^2 = \frac{A^2}{2K} - \frac{A^2}{2K}\cos 2\omega t \qquad (7\text{-}3)$$

Thus the output signal contains a dc component and a cosinusoidal component of frequency 2ω which is twice that of the input signal. The dc term can be removed by passing the signal through a high-pass or a bandpass filter. A bandpass filter of resonant frequency 2ω is preferable[1] because it can then suppress the harmonic distortion due to the inherent nonlinear error present in a practical multiplier.

A *frequency tripler*, on the other hand, can be designed with two multipliers and a summing amplifier without any additional filter. To this end, any one of the following trigonometric identities can be used:

$$\sin 3\omega t = 3 \sin \omega t (\cos \omega t)^2 - (\sin \omega t)^3 \qquad (7\text{-}4a)$$

$$\cos 3\omega t = (\cos \omega t)^3 - 3(\sin \omega t)^2 \cos \omega t \qquad (7\text{-}4b)$$

We consider here the design of the frequency tripler using the identity of Eq. (7-4a). A similar procedure can be followed to implement a tripler with the aid of Eq. (7-4b) (Problem 7-6).

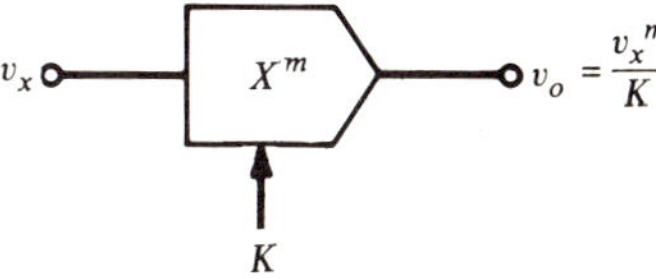

Figure 7-5   Symbolic representation of a circuit generating mth power of an analog signal.

We assume that both the input and output signals have a full-scale range of ± 10 V. In addition, the multiplier scale factor is 10 V. To implement Eq. (7-4a) we rewrite it as

$$\sin 3\omega t = 3 \sin \omega t (1 - \sin^2 \omega t) - \sin^3 \omega t$$
$$= 3 \sin \omega t - 4 \sin^3 \omega t \tag{7-5}$$

It is necessary to rescale Eq. (7-5) appropriately to ensure ± 10 V full-scale input and output voltages. Note that the maximum values of both $\sin \omega t$ and $\sin 3\omega t$ are ± 1. Let us denote the input $10 \sin \omega t$ as v_x and the output $10 \sin 3\omega t$ as v_o. Thus Eq. (7-5) becomes

$$\frac{v_o}{10} = 3\left(\frac{v_x}{10}\right) - 4\left(\frac{v_x}{10}\right)^3$$

To take into account the multiplier scale factor, we rewrite it further as

$$\frac{v_o}{10} = \frac{3}{10} \cdot v_x - \frac{4}{10} \cdot \frac{v_x}{10} \cdot \frac{v_x^2}{10} \tag{7-6}$$

Equation (7-6) is in proper form for implementation. The last term in the above equation can be generated using a cuber [Fig. 7-4(b)], the output of which can then be fed into a difference amplifier along with v_x to implement Eq. (7-6). If we use the circuit of Fig. 6-12 to this end, the complete circuit diagram for the frequency tripler is as shown in Fig. 7-6. To determine the resistor values of the difference amplifier, we compare Eq. (7-6) with Eq. (6-28) which yields the scale factors K_1 and K_2 of the difference amplifier as

$$K_1 = \frac{4}{10}, \qquad K_2 = \frac{3}{10}$$

Hence, from Eq. (6-29a) we arrive at

$$\frac{R_2}{R_1} = K_1 = \frac{4}{10}$$

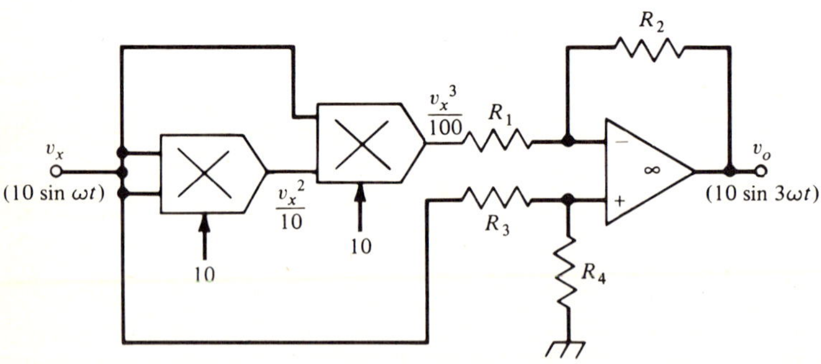

Figure 7-6   A frequency tripler. $R_1 = 10$ kΩ, $R_2 = 4$ kΩ, $R_3 = 11$ kΩ, $R_4 = 3$ kΩ.

and from Eq. (6-30) we obtain

$$\frac{R_3}{R_4} = \frac{K_1 + 1 - K_2}{K_2} = \frac{11}{3}$$

One set of convenient values for the four resistors is

$$R_1 = 10\,\text{k}\Omega, \qquad R_2 = 4\,\text{k}\Omega$$
$$R_3 = 11\,\text{k}\Omega, \qquad R_4 = 3\,\text{k}\Omega$$

Higher-order frequency multipliers can also be designed in a similar manner (Problems 7-7 and 7-8).

Divider.   Two circuit arrangements for dividing an analog signal by another one are sketched in Fig. 7-7. We analyze first Fig. 7-7(a). It follows from the figure that at the output of the multiplier,

$$v_1 = \frac{v_x v_o}{K} \tag{7-7}$$

We observe next that because of a signal path from the output of the operational amplifier to its inverting input terminal, this terminal is at virtual ground. In

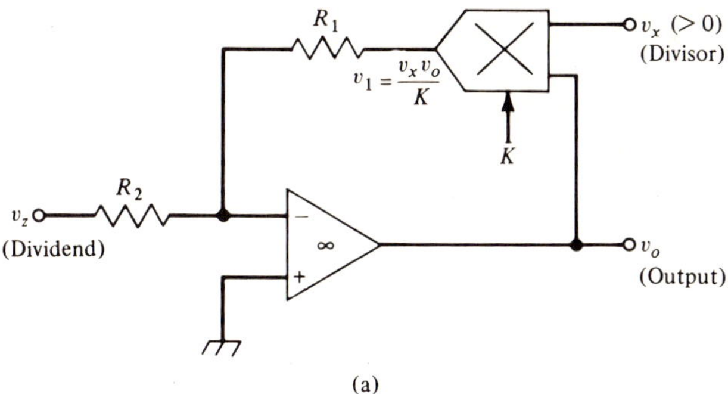

(a)

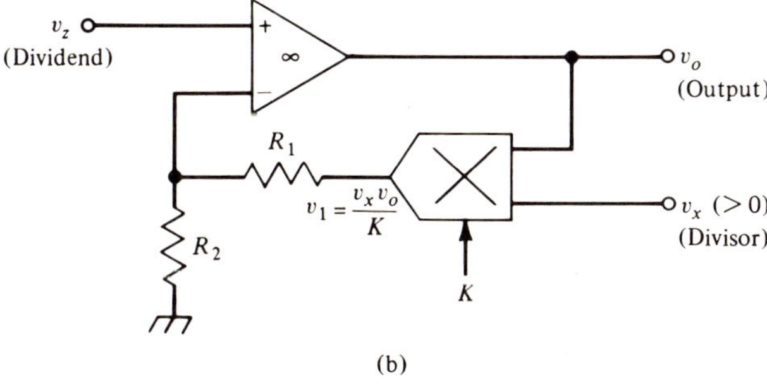

(b)

Figure 7-7   Divider circuits.

addition, ideally no current flows into the inputs of the operational amplifier. Consequently, by summing the currents at the inverting input terminal, we arrive at

$$\frac{v_1}{R_1} + \frac{v_z}{R_2} = 0 \tag{7-8}$$

Eliminating v_1 from Eqs. (7-7) and (7-8) we obtain

$$v_o = -\frac{KR_1}{R_2} \cdot \frac{v_z}{v_x} \tag{7-9}$$

If $R_1 = R_2$, then

$$v_o = -K \cdot \frac{v_z}{v_x} \tag{7-10}$$

Division with a positive scale factor is achieved by the circuit of Fig. 7-7(b). Here also the voltage v_1 at the output of the multiplier is given by Eq. (7-7). Because of negative feedback, the voltage at the inverting input terminal of the operational amplifier is v_z. Applying KCL at this terminal we obtain

$$\frac{1}{R_1}\left(\frac{v_x v_o}{K} - v_z\right) - \frac{v_z}{R_2} = 0$$

which after rearrangement yields

$$v_o = K\left(1 + \frac{R_1}{R_2}\right)\frac{v_z}{v_x} \tag{7-11}$$

Setting $R_1 = 0$ and $R_2 = \infty$, we obtain

$$v_o = K \cdot \frac{v_z}{v_x} \tag{7-12}$$

Note that in both circuits to ensure negative feedback from the output of the operational amplifier to its inverting input terminal, the polarity of v_1 must be the same as that of v_o. This is achieved by ensuring that the input signal v_x remains always positive leading to a 2-quadrant operation for the divider.

In each of the two divider circuits, the value of the scale factor K is determined from desired ranges of the input and output signals. In most cases, the ranges for various voltages are: ± 10 V for v_z, $+0$ to $+10$ V for v_x, and ± 10 V for v_o. The scale factor K is chosen as 10 V. However, to maintain the range for the output, $|v_z/v_x| < 1$ should be ensured.

Equations (7-10) and (7-12) also indicate that the divider circuits of Fig. 7-7 can be considered as amplifiers having a voltage-variable gain by treating v_z as the input signal and v_x as the voltage controlling the closed-loop gain K/v_x. However, the closed-loop gain increases with decreasing values of v_x which results in a decrease of bandwidth and an increase in offset and multiplier errors. This effect usually places a lower limit on v_x. As we show later in this type of dividers, the errors are roughly inversely proportional to the magnitude of the denominator voltage v_x and as a result very large dynamic ranges are difficult to achieve. The logarithmic amplifier-based dividers, discussed in Section 7-4, offer much larger dynamic ranges but are not as accurate as the dividers designed using analog multipliers.

A number of manufacturers sell units containing both the multiplier and the operational amplifier in one package. By appropriate external connections such a unit can be used either as a multiplier or a divider.

Generation of Fractional Powers.   The divider circuits of Fig. 7-7 can be converted to square-rooter circuits by connecting the v_x terminal to v_o. A square-rooter derived from the divider of Fig. 7-7(a) is shown in Fig. 7-8(a). To derive its input–output relation, we set $v_x = v_o$ in Eq. (7-9) which becomes

$$v_o = -K\,\frac{R_1\,v_z}{R_2\,v_o}$$

that is,

$$v_o = \sqrt{-\frac{KR_1}{R_2}\cdot v_z}\qquad (7\text{-}13)$$

It follows from above that v_o would be real if v_z is negative, indicating that the square-rooter is a 1-quadrant device. To prevent "lock-up," it is necessary to insure the input be constrained (with the aid of additional circuitry) not to assume positive or even zero values. The largest errors occur for values of the input signal approaching the zero value because here small changes in the input signal cause the largest percent change in the output values. Modification of the circuit of Fig. 7-7(b) as shown in Fig. 7-8(b) can be used to obtain the square-root of a positive signal. Its analysis is left as an exercise (Problem 7-9). An *odd-function*

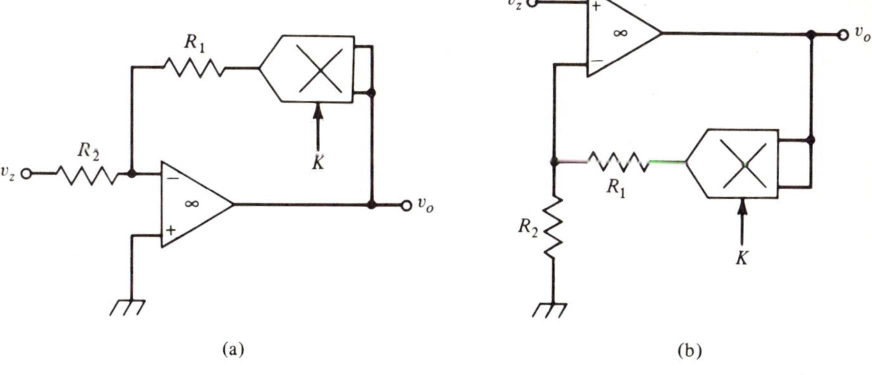

Figure 7-8   Square-rooters.

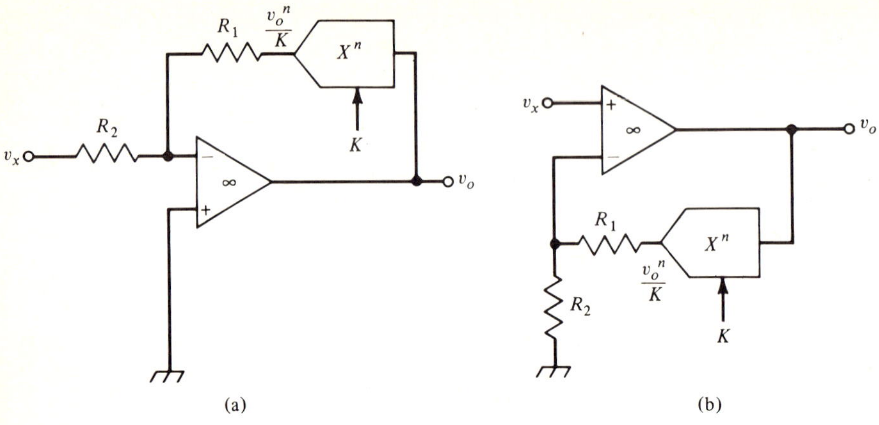

(a)   (b)

Figure 7-9   Schemes for generating nth root of an analog signal.

square-rooter generates an output proportional to the square-root of the input but having the same polarity as that of the input.

Square-rooters find applications in root-mean-square measurements, vector computation, simulation of fluid-flow parameters (Problem 7-35), and so on. For example, a circuit to compute the vector sum $\sqrt{v_x^2 + v_y^2}$ can be designed by summing the outputs of two squarers and then feeding the sum to a square-rooter.

Logarithmic amplifiers can also be used to implement the square-rooting operation (Section 7-4). This latter approach provides a wider dynamic range and good accuracy for very low-valued inputs. On the other hand, the multiplier-based square-rooter is more accurate for large-valued signals and offers faster operation.

It should be noted from Fig. 7-8 that the square-root of an analog signal is obtained by placing the squarer in the feedback path of an inverting or a non-inverting amplifier designed using an operational amplifier. Thus to generate the nth root of an analog signal, we can replace the squarer in the feedback path by a circuit which develops the nth power (Fig. 7-9).

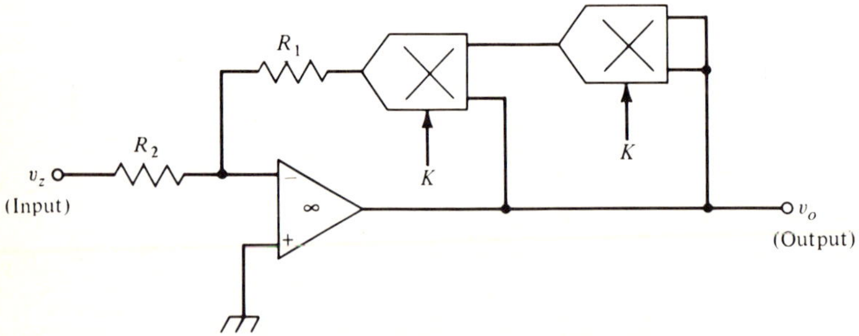

Figure 7-10   A circuit schematic for cube-root extraction.

For example, a possible circuit arrangement for extracting the cube-root of an analog voltage is as shown in Fig. 7-10. Analysis of this circuit is left as an exercise (Problem 7-10).

Cascading an mth power circuit (Fig. 7-5) with an nth root circuit, one can develop (m/n)th power of an analog signal. An alternate way to develop (m/n)th power is with the aid of logarithmic amplifiers (Section 7-4).

7-2 Characteristics of a Practical Multiplier[13, 15]

A practical multiplier not only has finite input and output impedances but also has many additional sources of errors which usually cause some inaccuracy in the multiplication operation. In addition, reasonably accurate multiplication is only valid in practice at low frequencies extending to several kiloHertz. An understanding of the characteristics of a practical multiplier is thus essential to make use of this device to its fullest extent.

Multiplier Accuracy

The effect of almost all nonidealness in a practical multiplier is to produce an output v_o which is not identical to $v_x v_y / K$. The difference between the actual value and the desired value of the output is used to denote the accuracy of the multiplier, a commonly used criterion to estimate the quality of a practical multiplier unit. In analytical form the error term can thus be expressed as

$$v_o = \frac{v_x v_y}{K} + \Delta_v \tag{7-14}$$

A major problem in any practical circuit is that the error Δ_v is caused by a number of factors and consequently it varies with input signal levels, time, temperature, and frequency. Some manufacturers supply two-dimensional equal-error loci plots called *isovers*[1] which are measured with dc input signals and directly show the *static error* for any specified input combination. A more convenient way to describe the error seems to be by specifying worst-case error at full-scale output. Following this approach we rewrite Eq. (7-14) under worst-case as

$$v_o = \frac{v_x v_y}{K} \pm \varepsilon_m \tag{7-15}$$

where ε_m is the worst-case error, that is,

$$\left| v_o - \frac{v_x v_y}{K} \right| \leq \varepsilon_m \tag{7-16}$$

In practice, the worst-case error is specified by the manufacturers as a percentage of full-scale output voltage which is used to describe the multiplier accuracy. Thus,

$$\text{Accuracy} = \frac{\varepsilon_m \times 100}{\text{full-scale voltage}} \text{ percent} \tag{7-17}$$

Invariably the accuracy is measured with dc applied to both inputs and at rated supply voltage and room temperature.

The accuracy of a divider constructed using a multiplier in the feedback path of an operational amplifier can be expressed as a function of the multiplier accuracy as illustrated in the following example.

Example 7-2. Determine the accuracy of the divider.

The accuracy of the divider can be calculated by making use of the key equations for the divider circuit arrangement. Combining Eqs. (7-7) and (7-8) we obtain

$$\frac{v_x v_o}{K} + \frac{R_1}{R_2} v_z = 0$$

Inserting in the above, the multiplier error term ε_m, we get for the worst case

$$\frac{v_x v_o}{K} \pm \varepsilon_m + \frac{R_1}{R_2} v_z = 0$$

Consequently,

$$v_o = -\frac{KR_1}{R_2} \frac{v_z}{v_x} \pm \frac{K\varepsilon_m}{v_x}$$

Thus the worst-case error for the divider is given as

$$\varepsilon_d = \frac{K\varepsilon_m}{v_x} \tag{7-18}$$

assuming of course that the summing resistors R_1 and R_2 are carefully matched such that their ratio R_1/R_2 can be assumed to remain constant. The divider accuracy is again given as a percentage of full-scale voltage.

It is seen from Eq. (7-18) that the divider accuracy is dependent on the value of the divisor signal voltage v_x. For a multiplier with ± 10 V full-scale range, if v_x is ± 10 V, then the divider accuracy is the same as that of the parent multiplier unit. However, the worst-case static error here increases with a decrease in divisor voltage. Figure 7-11 shows the plot of the divider accuracy as a function of the divisor voltage for a dividing circuit constructed using a multiplier having 0.2 percent accuracy with ± 10 V full-scale. R_1 is assumed here to be equal to R_2 as is usually the case. This figure indicates that the divider should not be used for very low-valued divisor voltage. One way to get around this problem would be to rescale the divisor voltage by an amplifier at the input end and then increase the output voltage by an amplifier. This arrangement cannot be used, however, if the divisor must vary through the full-scale range.

Sources of Multiplier Errors

The total error ε_m in Eq. (7-15) is caused by a number of factors. In most units, some of these sources of errors can be externally adjusted to minimize their

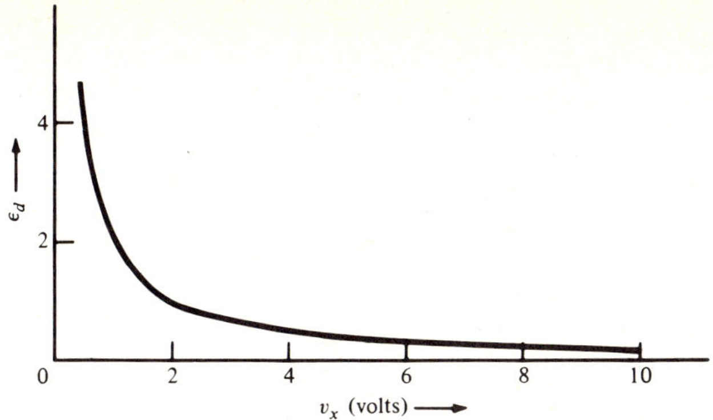

Figure 7-11   A typical plot of the divider accuracy as a function of the divisor voltage.

contributions. It is thus necessary to know what these sources are and how to control them. The most important of these are : (1) output offset, (2) feedthrough, (3) gain variations, (4) nonlinear operation, (5) finite bandwidth, and (6) drift of all errors due to a change in temperature, time, and power supply voltage. The first four sources of errors can be analytically described by rewriting Eq. (7-15) as

$$v_o = \frac{(v_x + \delta_x)(v_y + \delta_y)}{K(1 + \Delta K)} + \varepsilon_o + f(v_x, v_y) \qquad (7\text{-}19)$$

where ΔK indicates the change in proportionality constant from its nominal value of K. δ_x and δ_y are the input offset voltages at the X- and Y-terminals, respectively, and ε_0 is the output offset voltage. $f(v_x, v_y)$ is the error due to unavoidable nonlinear effects of the unit and depends on the technique used to design the multiplier unit. Since δ_x, δ_y, and ΔK are in general very small quantities we can expand the first term on the right-hand side of Eq. (7-19) in a power series and neglect second- and higher-order terms to obtain an expression for the total error as

$$\varepsilon_m(v_x, v_y) \cong -\Delta K\left(\frac{v_x v_y}{K}\right) + \left(\frac{\delta_x}{K} v_y\right) + \left(\frac{\delta_y}{K} v_x\right) + \varepsilon_o + f(v_x, v_y) \qquad (7\text{-}20)$$

Without any loss of generality we assume in the remaining part of this section K is equal to 10 V and the full scale is ± 10 V.

Output Offset.   If we set $v_x = v_y = 0$, then Eq. (7-20) reduces to

$$\varepsilon_m(0, 0) = \varepsilon_o + f(0, 0) \qquad (7\text{-}21)$$

For most multipliers, $f(0, 0)$ can be considered to be equal to zero. Then Eq. (7-21) implies that the voltage v_o measured at the output terminal by grounding the two input terminals is the *output offset voltage* ε_o. This voltage can be trimmed to zero by adjusting the output offset potentiometer.

Feedthrough. Next we set $v_x = 0$. Ideally from Eq. (7-1) if v_x is zero, v_o should be zero irrespective of the value of v_y. However, as Eq. (7-20) points out, the output voltage v_o under this condition is given as

$$v_o|_{v_x = 0} = \frac{\delta_x}{10} v_y + f(0, v_y) \tag{7-22}$$

assuming of course that ε_o has been trimmed to zero as the first step in error minimization procedure. This output voltage is known as *X-channel feedthrough* which is minimized as follows. The procedure is to ground the X-terminal and apply a 20 V p–p sinusoidal signal, for example, $v_y = 10 \sin \omega t$ at the Y-terminal. The test frequency normally is kept very low, typically 50 Hz. This sweeps the voltage at the Y-terminal through its whole range. For this case, then, Eq. (7-22) modifies to

$$v_o|_{v_x = 0, v_y = 10 \sin \omega t} = (\delta_x) \sin \omega t + f(0, 10 \sin \omega t)$$

Note that because of the presence of the nonlinear term, the output in general will consist of a dc term, a fundamental component, and several harmonics. The X-channel offset error is now minimized by adjusting the X-input offset potentiometer for best ac null at the output. Similarly the Y-channel feedthrough error is minimized by grounding the Y input terminal and making $v_x = 10 \sin \omega t$, and then adjusting the Y-input offset potentiometer for best ac null.

Gain Error. The term $\Delta K(v_x v_y/K)$ in Eq. (7-20) is known as the *gain error*. Since there is no way the nonlinearity error $f(v_x, v_y)$ can be trimmed to zero, the total error caused by these two factors can be minimized somewhat by adjusting the gain trim potentiometer. Either dc trimming or ac trimming can be followed. In dc trimming the X and Y inputs are set at various dc values in the range of ± 10 V, and the average gain error is minimized by adjusting the potentiometer for each combination of input values. To understand the procedure we observe that for nonzero input voltages, the actual multiplier output would be

$$v_o = \frac{v_x v_y}{10} + \Delta K\left(\frac{v_x v_y}{10}\right) + f(v_x, v_y) \tag{7-24}$$

assuming all the offset errors have been trimmed to zero. Thus, if $v_x = +10$ V and $v_y = -5$ V, the total output will be the actual output plus 5 V. Hence to measure the error we must add 5 V to the multiplier output.

For ac trimming, one of the inputs is connected to a 10-V dc supply and the other input is connected to a sinusoidal source. As before the ideal output value should be subtracted from the actual multiplier output voltage to obtain the true error.

It should be noted that all of the previous errors are not constant but vary with time, temperature, and power supply voltage. It may be necessary for some applications to trim these errors periodically.

Nonlinearity.   Almost all of the errors discussed earlier can be reduced substantially by external adjustments except the contribution due to nonlinearity which cannot be reduced by any means. If one of the input voltages is dc and the other a sinusoidal signal, the nonlinear term $f(v_x, v_y)$ generates second and higher harmonic components which can be measured using a distortion analyzer.

An alternate procedure is as follows. Note that if either v_x is a dc voltage and v_y a sinusoidal signal or vice versa, the first two terms on the right-hand side of Eq. (7-24) contribute only to the fundamental component in the output signal. These components can be cancelled by subtracting an appropriately scaled ac signal from the multiplier output by means of a summing amplifier. The output of the summer is then an estimate of the contribution due to the nonlinearity, and its rms value gives a measure of the total harmonic distortion. The nonlinear distortion increases at higher frequencies. Normally the specifications supplied by the manufacturers are the measured values at low frequencies.

Frequency Response

The frequency response of a multiplier is normally specified in two different ways. One way is to treat it as a voltage amplifier and obtain its gain characteristic between the output and one input terminal as a function of frequency while the other input terminal is held at a constant dc value. The measurements are made at small signal levels. The *small-signal bandwidth* of the multiplier is then the frequency at which the gain is 3 dB less than its value at dc.

The large-signal response, on the other hand, is obtained by applying full-scale sinusoidal signal to one input and full-scale dc voltage to the other and measuring the multiplying error over a wide frequency range. The *large-signal bandwidth* is defined as the maximum frequency at which the absolute multiplying error is 1 percent (defined again as a percentage of full-scale range). The *absolute error*, also known as the *dynamic error*, is the peak error due to both amplitude and phase-shift error. A typical large-signal frequency response is illustrated in Fig. 7-12.

Input and Output Resistances

The input and output resistances of a practical multiplier have finite values. Typically the input resistances (from terminal to ground) are in the order of

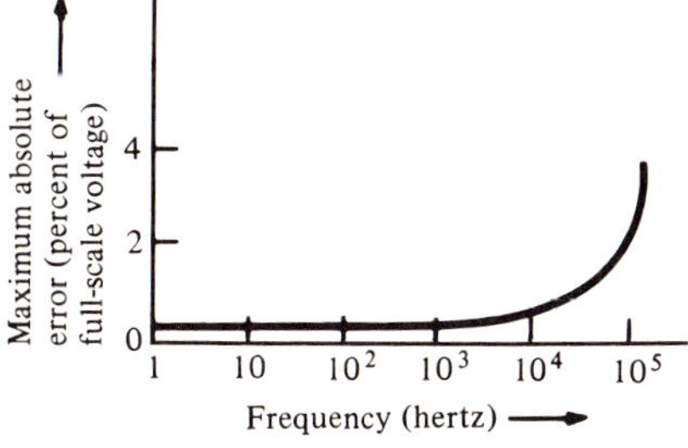

Figure 7-12   Typical large-signal frequency response of a multiplier.

several megohms and the output resistance is several ohms. If driven from a high impedance source, the finite input impedances may cause severe loading problems. In addition, the bias current flowing through the source impedance may cause a finite voltage drop at the input. These effects can be minimized by inserting noninverting voltage amplifiers between the sources and the multiplier input terminals as buffers. The same circuit arrangement can also be used to scale the input analog signals to full-scale value to minimize the errors caused by low-valued signals in some applications.

Improvement of Dynamic Range[1]

A low-pass filter at the output of a multiplier can reduce the feedthrough and noise significantly and eliminate harmonic distortion, thus improving the dynamic range of a multiplier. This approach may be followed when the input signal frequencies range over a portion of the multiplier bandwidth. It may be profitable to use instead a bandpass filter which eliminates drift also by blocking low-frequency components. A second-order active RC filter such as that described in Section 6-8 is quite adequate for this purpose.

Other Characteristics

All of the major contributing terms to multiplier accuracy vary with temperature. Usually the drift of total worst-case error and sometimes its constituent terms with temperature are specified per degree centigrade.

Power supply voltage variation also leads to unwanted variation in the output signal. The error is measured by applying a full-scale dc voltage to one input terminal and a full-scale sinusoidal voltage to the other. The peak multiplication error is then the variation in the output voltage when the power supply voltage is varied by ± 1 V around its rated value. This error divided by the peak change in power supply voltage (2 V) is normally specified by the manufacturers as *power supply rejection* in mV/V units. From this figure the peak offset error due to power supply variation can be easily calculated.

The differential phase shift between X- and Y-channels introduces an additional dynamic error and normally is less significant for most applications. Like the operational amplifier, the multiplier is also characterized by *slewing rate error* which is caused by the inability of the output signal to follow large rapid changes in the input signals.

7-3   Some Additional Applications of the Multiplier

A number of other possible applications of the multiplier unit is considered here. An understanding of the previous applications and those discussed here would undoubtedly trigger many other applications.

Root-Mean-Square Measurement

A frequent use of the multiplier is in the measurement of the root-mean-square (rms) of an analog signal. To this end, the multiplier is used as a squarer and as a square-rooter. In order to understand the basic idea involved, let us first explain how the mean of an analog signal can be obtained by a first-order low-pass filter.

The *mean* or *average* value of a periodic signal $e(t)$ of period T is by definition

$$\overline{e(t)} = \frac{1}{T} \int_0^T e(t)\, dt \tag{7-25}$$

Since the signal is periodic, we can also express the above as

$$\overline{e(t)} = \frac{1}{T} \int_{(n-1)T}^{nT} e(t)\, dt \tag{7-26}$$

where n is any integer.

Consider now a circuit having an impulse response $h(t)$ which is a rectangular pulse of height $1/T$ from $t = 0$ to $t = T$ and zero elsewhere (Fig. 7-13):

$$h(t) = \begin{cases} \dfrac{1}{T}, & \text{for} \quad 0 < t < T \\[2ex] 0, & \text{elsewhere} \end{cases} \tag{7-27}$$

If the signal $e(t)$ is fed into this circuit, the output $e_o(t)$ would be given by the convolution integral as

$$e_o(t) = \int_0^t e(\lambda)h(t - \lambda)\, d\lambda$$

and consequently at $t = T$, the output will be

$$e_o(T) = \int_0^T e(\lambda)h(T - \lambda)\, d\lambda = \frac{1}{T} \int_0^T e(\lambda)\, d\lambda$$

which is nothing but the average value of the input signal.

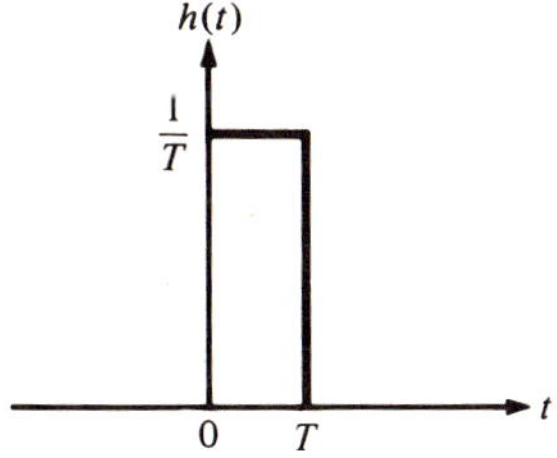

Figure 7-13   Impulse response of an ideal averaging circuit.

It is impossible to design a circuit having an impulse response of the form of Eq. (7-27). However, an alternate practical way of approximately measuring the average value is by means of a circuit having an impulse response given by

$$h_a(t) = \begin{cases} \dfrac{1}{\Delta} e^{-t/\Delta}, & t > 0 \\[2ex] 0, & \text{elsewhere} \end{cases} \tag{7-28}$$

where Δ is the time-constant of the circuit. The plot of $h_a(t)$ is sketched in Fig. 7-14. In order to show that the circuit characterized by Eq. (7-28) can indeed provide us with the average value, we have approximated $h_a(t)$ by an impulse response $h_b(t)$ which is a sum of narrow pulses of width T and height $e^{-nT/\Delta}/\Delta$ as indicated in Fig. 7-14. If the signal is passed through the first-order circuit characterized by Eq. (7-28), the output $e_o(t)$ will be given as

$$e_o(t) = \int_0^t e(\lambda) h_a(t - \lambda)\, d\lambda \cong \int_0^t e(\lambda) h_b(t - \lambda)\, d\lambda$$

For large values of time t we thus obtain

$$e_o(t) = \frac{1}{\Delta} \int_0^T e(\lambda)\, d\lambda + \frac{1}{\Delta} \int_T^{2T} e(\lambda) e^{-T/\Delta}\, d\lambda + \frac{1}{\Delta} \int_{2T}^{3T} e(\lambda) e^{-2T/\Delta}\, d\lambda + \cdots \tag{7-29}$$

It follows from Eq. (7-26) that

$$\int_{(n-1)T}^{nT} e(\lambda)\, d\lambda = T\overline{e(t)} \tag{7-30}$$

Hence Eq. (7-29) reduces to

$$e_o(t) = \frac{T}{\Delta}\left[1 + e^{-T/\Delta} + e^{-2T/\Delta} + e^{-3T/\Delta} + \cdots\right]\overline{e(t)} = \frac{T}{\Delta(1 - e^{-T/\Delta})}\overline{e(t)} \tag{7-31}$$

Now if $T/\Delta \ll 1$, that is, if the time-constant Δ of the circuit is much larger than the period T of the input signal, then

$$e^{-T/\Delta} \cong 1 - \frac{T}{\Delta}$$

and Eq. (7-31) reduces to

$$e_o(t) = \frac{T}{\Delta\left(1 - 1 + \dfrac{T}{\Delta}\right)}\overline{e(t)} = \overline{e(t)} \tag{7-32}$$

This indicates that by measuring the output of the circuit characterized by Eq. (7-28) after a long period of time, we obtain an estimate of the average value of the input signal. Even though the above expression is technically correct as $t \to \infty$, reasonable estimate is obtained after $t = 7\Delta$.

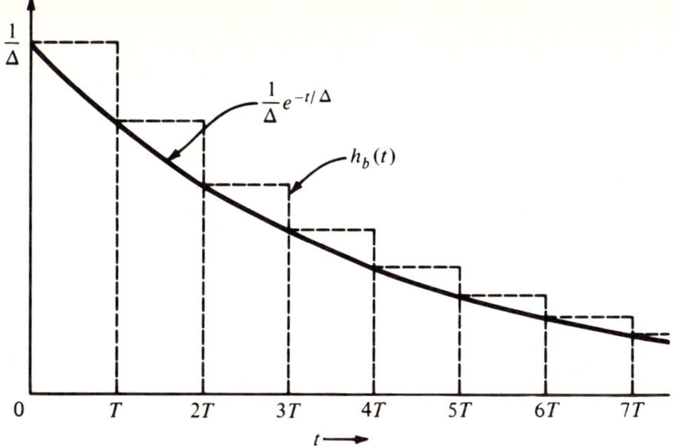

Figure 7-14 Impulse response of a first-order low-pass filter.

The impulse response $h_a(t)$ can be realized by a first-order low-pass filter of the type shown in Fig. 7-15. The transfer function of the filter is obtained by analyzing it in the s-domain. To this end, we apply KCL at the inverting input terminal of the operational amplifier (which is at virtual ground) yielding

$$\frac{V_1}{R} + \left(sC + \frac{1}{R}\right)V_2 = 0$$

Rearranging we arrive at

$$\frac{V_2}{V_1} = \frac{-1/\Delta}{s + (1/\Delta)} \tag{7-33}$$

where we have set the time-constant $RC = \Delta$. Note that the time-constant Δ can be changed by varying the capacitor C. The reciprocal of Δ is the 3-dB bandwidth of the filter.

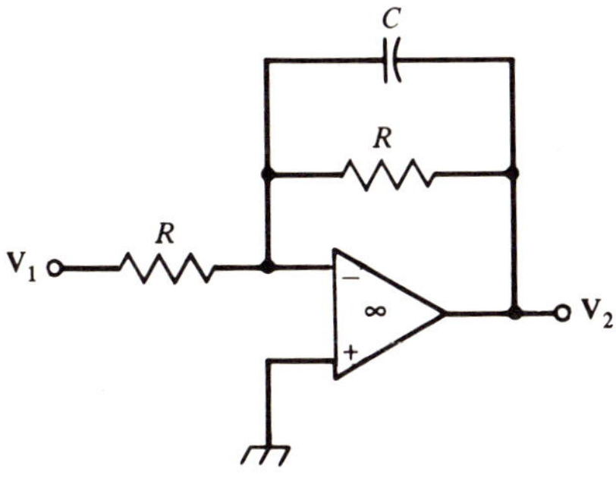

Figure 7-15 A first order low-pass filter.

The mean-square value of a signal is thus obtained by feeding the signal to a squarer and averaging its output by means of the low-pass filter of Fig. 7-15. To obtain the rms value, the output of the low-pass filter is connected to a square-rooter. The rms-generating circuit can be used to construct a rms voltmeter if a dc voltmeter is connected from the output of the square-rooter to ground. This type of rms measurement is more accurate and versatile than most conventional ac voltmeters which provide a true rms value only for sinusoidal voltages.

Voltage-Variable Active Filters[8]

The design of inductorless filters with low-pass, high-pass, and bandpass frequency response characteristics was considered in Section 6-8. These filter configurations have fixed frequency responses and to adjust their responses, the values of a number of components need to be readjusted. Often, it is convenient to have filters whose characteristics are externally adjustable by varying dc voltages. We describe here two such circuits.

First-Order Low-Pass Filter. A single-pole, low-pass filter having voltage-variable 3-dB cut-off frequency can be easily implemented using a multiplier and an operational amplifier as indicated in Fig. 7-16. Note that this circuit is a slight modification of the circuit of Fig. 7-15.

We analyze this circuit next. Let us denote the output voltage of the operational amplifier as $\mathbf{V}_3$. Then the output $\mathbf{V}_2$ of the multiplier is given by

$$\mathbf{V}_2 = \frac{\mathbf{V}_3 E_x}{K} \tag{7-34}$$

The inverting input terminal of the operational amplifier is seen to be at virtual ground. Applying KCL at this terminal we arrive at

$$\frac{\mathbf{V}_1}{R_1} + sC\mathbf{V}_3 + \frac{\mathbf{V}_2}{R_2} = 0 \tag{7-35}$$

Eliminating $\mathbf{V}_3$ from Eqs. (7-34) and (7-35) we obtain

$$\frac{\mathbf{V}_1}{R_1} + \left(s\frac{CK}{E_x} + \frac{1}{R_2} \right)\mathbf{V}_2 = 0$$

which upon rearrangement leads to the voltage transfer function of the filter as

$$\frac{\mathbf{V}_2}{\mathbf{V}_1} = -\frac{E_x/KCR_1}{s + (E_x/KCR_2)} \tag{7-36a}$$

The circuit has a dc gain of $-R_2/R_1$ (obtained by letting $s = 0$), and a 3-db cut-off frequency of E_x/KCR_2 radians/sec which can be varied in practice over a range of 10:1 by varying E_x. Note that the time-constant T of the circuit is the reciprocal of the 3-dB cut-off frequency. It follows from above that for stability E_x must be always positive. Since for most multipliers $K = 10$, maximum value

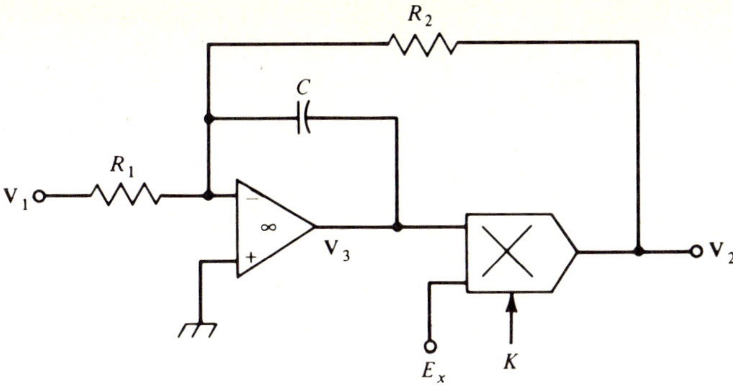

Figure 7-16 A voltage-variable low-pass filter.

of the cut-off frequency is $1/CR_2$. Any value less than this can be obtained by decreasing E_x. Recall that for very low values of E_x, the multiplier error becomes quite significant.

Second-Order Bandpass Filter.   An implementation of a bandpass filter with voltage-controlled resonant frequency is shown in Fig. 7-17. The filter bandwidth or the Q is also voltage controlled. If this second feature is not desired, some of the multipliers can be eliminated.

In analyzing this circuit we assume the multiplier scale factor K to be 10. Observe that the output voltage of the multiplier M_1 is given as

$$V_2 = \frac{E_x V_4}{10} \qquad (7\text{-}36b)$$

and that of the multiplier M_2 is $E_y V_4/10$. The output of the operational amplifier A_2 is $-\alpha E_y V_4/10s$ where we have set $\alpha = 1/RC$. As a result, the output of the multiplier M_3 is

$$V_3 = -\frac{\alpha E_y^2 V_4}{100s} \qquad (7\text{-}37)$$

The output of the operational amplifier A_3 is simply $-V_3$. Finally, applying KCL at the inverting input terminal of A_1 we arrive at the expression for its output voltage as

$$V_4 = -\frac{\alpha}{s}(V_1 + V_2 - V_3) \qquad (7\text{-}38)$$

Eliminating V_2 and V_3 from the above equations we obtain after some algebra

$$\frac{V_4}{V_1} = \frac{-\alpha s}{s^2 + (\alpha E_x/10)s + (\alpha^2 E_y^2/100)} \qquad (7\text{-}39)$$

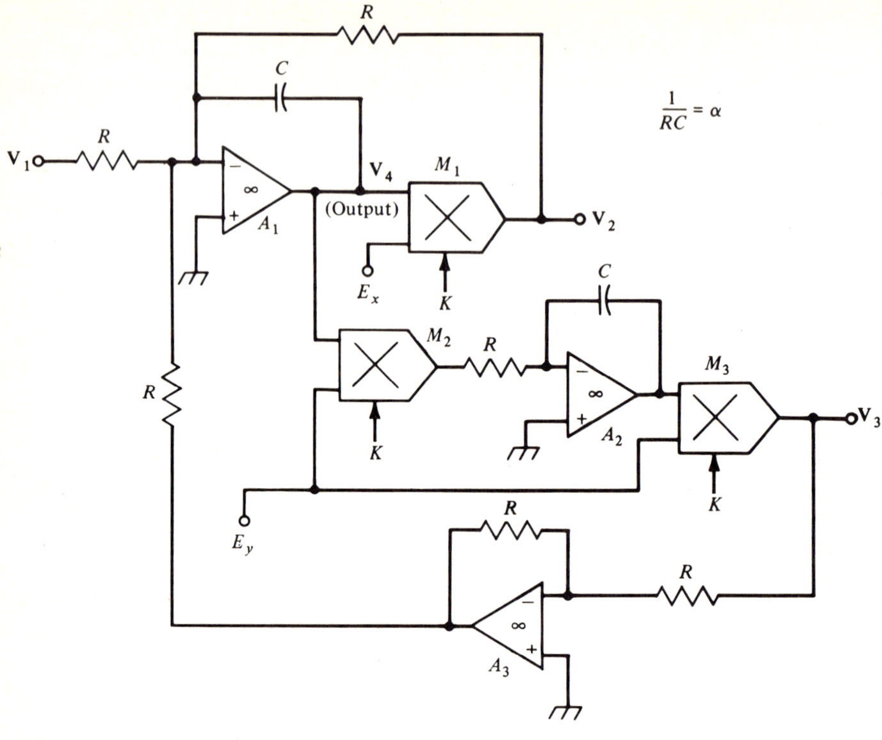

Figure 7-17   A voltage-variable bandpass filter.

which is recognized as a bandpass transfer function. Comparing Eq. (7-39) with Eq. (6-94) we obtain

$$\omega_n = \alpha E_y/10 \tag{7-40}$$

$$Q = E_y/E_x \tag{7-41}$$

The 3-dB bandwidth is thus [see Eq. (6-93)]

$$\text{BW} = \alpha E_x/10 \tag{7-42}$$

Note that the resonant frequency ω_n and Q are directly proportional to the dc voltage E_y and the bandwidth is proportional to the dc voltage E_x. Hence, by keeping E_x constant the bandwidth can be kept constant while ω_n and Q can be varied by adjusting E_y. Also by making E_x a constant fraction of E_y, the resonant frequency can be varied keeping Q constant.

Function Fitting[15]

For many applications it is necessary to implement arbitrary single-valued nonlinear functions for which each set of input values corresponds to a unique output value independent of past input values. Examples of such applications are in the linearization of the outputs of many transducers, bridge amplifiers, and so on, and also in the generation of trigonometric functions. A usual method of implementation of such nonlinear functions is to approximate the input–output relationship by curve fitting. Several approaches are followed for curve fitting. In one approach, the nonlinear function is approximated by a polynomial or a rational function of the independent variable. For realizability, the coefficients of the polynomial of the rational function are chosen to be real coefficients. For implementation of this type of approximation, the multipliers are particularly attractive. The implementation procedure is illustrated with the aid of two examples.

Example 7-3.   Implement

$$y = f(x) = y_0 + a_1 x + a_2 x^2 - a_3 x^3 + a_4 x^4 \tag{7-43}$$

where the coefficients a_i are positive and maximum values of x and y are ± 1. We rewrite Eq. (7-43) as

$$y = y_0 + a_1 x + a_2 x^2 + x^2(a_4 x^2 - a_3 x)$$

$$= y_0 + a_1 x + Ka_2 \frac{x^2}{K} + K^2\left[\frac{x^2}{K^2}\left(Ka_4 \frac{x^2}{K} - a_3 x\right)\right] \tag{7-44}$$

In writing the above expression, we have taken into account the multiplier scale factor K. A possible implementation of Eq. (7-44) is sketched in Fig. 7-18. The resistor values in ohms are

$$\begin{aligned}
R_1 &= R_5 = R_{10} = 1 \\
R_2 &= a_3 \qquad R_3 = a_3 + 1 - Ka_4 \\
R_4 &= Ka_4 \qquad R_6 = 1/K^2 \qquad R_7 = 1/Ka_2 \\
R_8 &= 1/a_1 \qquad R_9 = 1/(a_1 + Ka_2 + K^2)
\end{aligned} \tag{7-45}$$

In order to ensure a positive value of the resistor R_3, the realizability constraint of this configuration is that $a_3 + 1$ must be greater than Ka_4.

Example 7-4.   Implement

$$y = y_0 + \frac{a_1 x + a_2 x^2 + a_3 x^3 + a_4 x^4}{1 - b_2 x^2} \tag{7-46}$$

As before, we assume the coefficients a_1, a_2, a_3, a_4, and b_2 to be positive. Furthermore, maximum values of x and y are also assumed to be ± 1. We rewrite Eq. (7-46) as

$$(y - y_0)(1 - b_2 x^2) = a_1 x + a_2 x^2 + a_3 x^3 + a_4 x^4$$

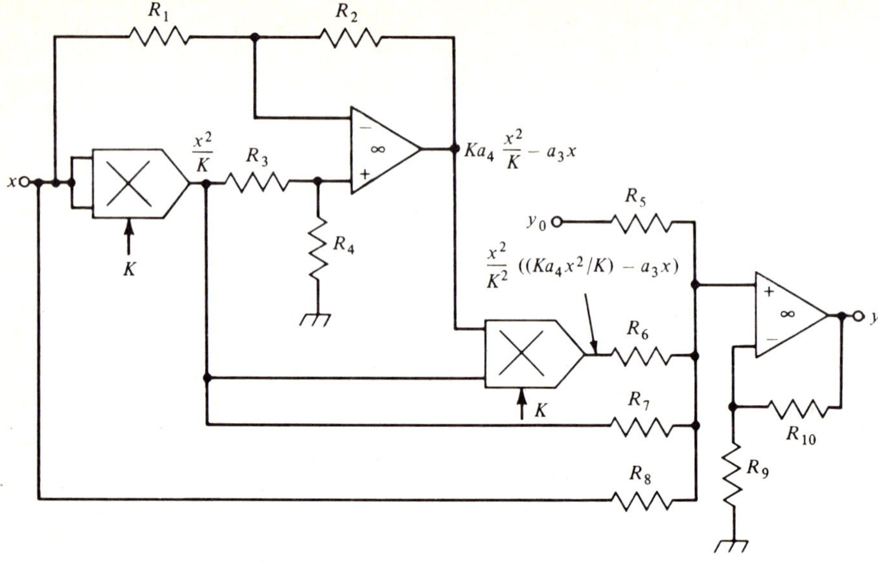

Figure 7-18   Implementation of a fourth-order polynomial.

which is equivalent to

$$y - y_0 = a_1 x + a_2 x^2 + x^2[a_3 x + a_4 x^2 + b_2(y - y_0)]$$

$$= a_1 x + Ka_2 \cdot \frac{x^2}{K} + K^2 \cdot \frac{x^2}{K^2}\left[a_3 x + Ka_4 \frac{x^2}{K} + b_2(y - y_0)\right] \quad (7\text{-}47)$$

A realization of Eq. (7-47) using two multipliers is indicated in Fig. 7-19. The resistor values in ohms are

$$
\begin{aligned}
R_1 &= 1/a_3 & R_2 &= 1/Ka_4 & R_3 &= 1/b_2 \\
R_4 &= 1/(b_2 + a_3 + Ka_2) \\
R_6 &= 1/K^2 & R_7 &= 1/Ka_2 & R_8 &= 1/a_1 \\
R_9 &= 1/(a_1 + Ka_2 + K^2)
\end{aligned}
\qquad (7\text{-}48)
$$

Another approach to curve fitting is by piecewise linear approximation which is implemented using diodes, resistors, and operational amplifiers. This approach is described elsewhere.[16]

Linearization

A number of circuits and devices used for measurement of nonelectrical quantities make use of some type of transducers to convert the parameter being measured into an electrical signal which is then processed and accurately

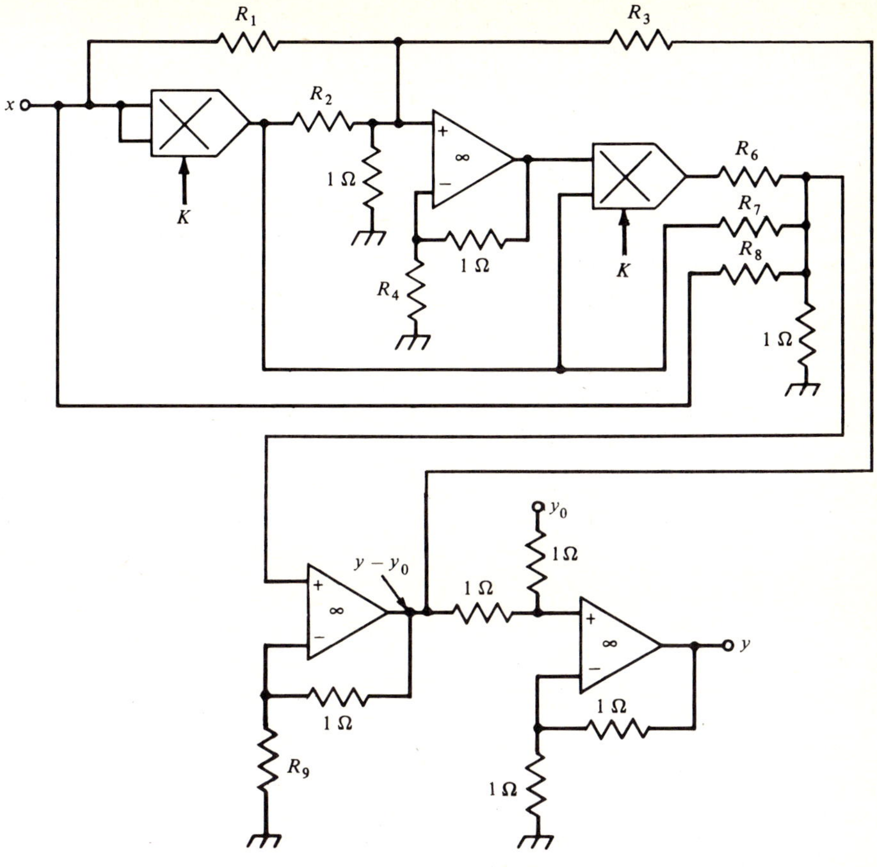

Figure 7-19 An implementation of a fourth-order rational function.

measured using precision electrical instruments. Examples of transducers are: National Semiconductor's hybrid IC pressure transducers (LX14XX, LX16XX, and LX17XX series) and monolithic IC temperature transducers (LX5600 and LX5700). An example of such type of measurements is the bridge amplifier circuit of Fig. 6-58 described in Problem 6-18 for which the output v_o is related to the change ε in the variable resistor $R_t - R_0(1 + \varepsilon)$ as (with $K = 2$)

$$v_o = \frac{\varepsilon/2}{1 + \varepsilon/2} \cdot V_{\text{ref}} \qquad (7\text{-}49)$$

The relation between ε and v_o is seen to be nonlinear. In many other measuring instruments the relation between the electrical equivalent and the original parameter is likewise nonlinear. As a result, these measurements are usually restricted to small values of the parameter being measured for which the relation is approximately linear. One way to expand the region of operation is to "linearize" the relationship with the aid of additional circuitry. To this end, the multiplier modules are quite useful.

As an example of the linearization process,[15] consider the development of a circuit to linearize Eq. (7-49). We rewrite Eq. (7-49) as

$$\frac{\varepsilon}{2} V_{\text{ref}} = v_o + \frac{\varepsilon}{2} v_o = v_o + \frac{K}{V_{\text{ref}}} \cdot \left(\frac{\frac{\varepsilon}{2} V_{\text{ref}} \cdot v_o}{K} \right) \tag{7-50}$$

A realization of Eq. (7-50) is as indicated in Fig. 7-20 with the resistor values as follows:

$$R_1 = R_3 = 1$$
$$R_2 = R_4 = V_{\text{ref}}/K$$

Another bridge linearization circuit using a divider is described in Problem 7-23. In this latter circuit, the output signal is independent of the reference voltage V_{ref}.

Analog Computation[10]

The basic idea behind the analog computation was described earlier in Section 6-7 by considering the simulation of a linear constant coefficient differential equation. In a number of situations, the differential equation characterizing a system are nonlinear, thus requiring the use of nonlinear components for their simulation. To this end, the analog multiplier is a key component. As an example of its application in the simulation of nonlinear differential equations, we consider here the simulation of the rigid-body motion of a spacecraft.[12] The

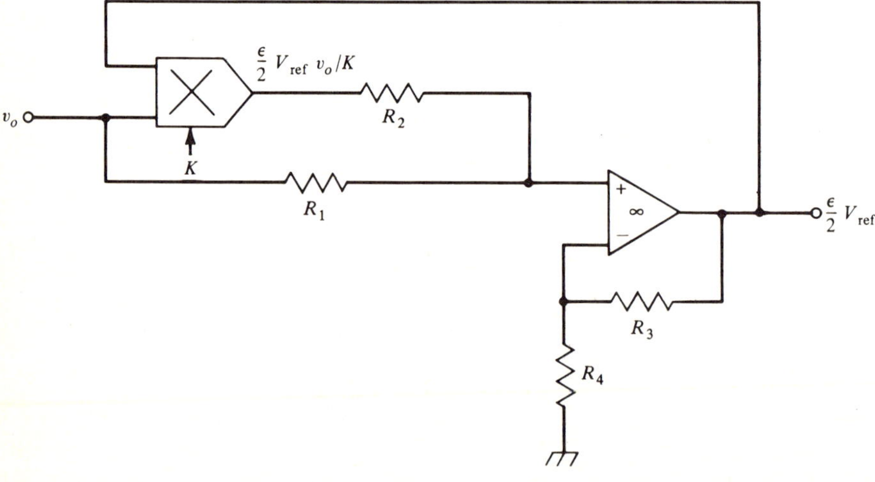

Figure 7-20 A bridge linearization circuit.

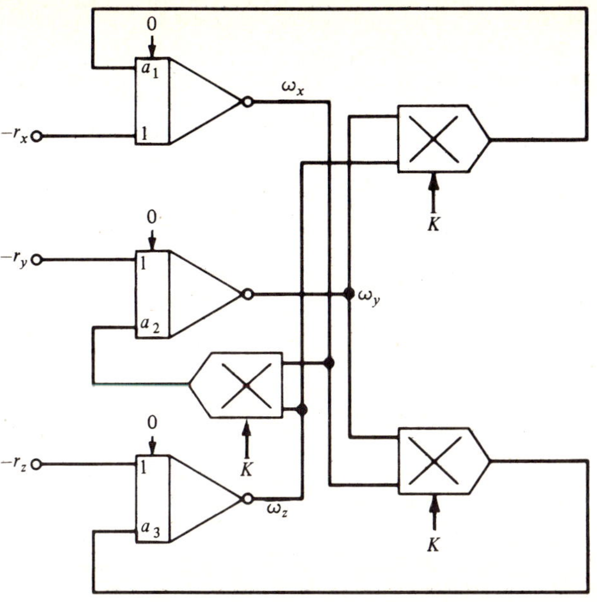

Figure 7-21 Analog simulation of the rigid-body motion of a spacecraft:

$$a_1 = K\left(\frac{J_z - J_y}{J_x}\right),\ a_2 = K\left(\frac{J_x - J_z}{J_y}\right),\ a_3 = K\left(\frac{J_y - J_x}{J_z}\right).$$

pertinent differential equations are

$$\frac{d\omega_x}{dt} = \frac{J_y - J_z}{J_x}\,\omega_y\omega_z + r_x$$

$$\frac{d\omega_y}{dt} = \frac{J_z - J_x}{J_y}\,\omega_z\omega_x + r_y \tag{7-51}$$

$$\frac{d\omega_z}{dt} = \frac{J_x - J_y}{J_z}\,\omega_x\omega_y + r_z$$

where ω_x, ω_y, and ω_z are spacecraft rotation rates; J_x, J_y, and J_z are moments of inertia; and r_x, r_y, and r_z are input torques. Analog simulation of the above equations is shown in Fig. 7-21 in block-diagram form. Additional applications of multipliers in analog simulation of nonlinear differential equations are suggested in Problems 7-27 and 7-28.

7-4 Logarithmic Amplifiers

Another very useful nonlinear function module is the logarithmic amplifier for which the output voltage is the natural logarithm of the input voltage. Because of the nature of the logarithm operation, the logarithmic (or simply *log*)

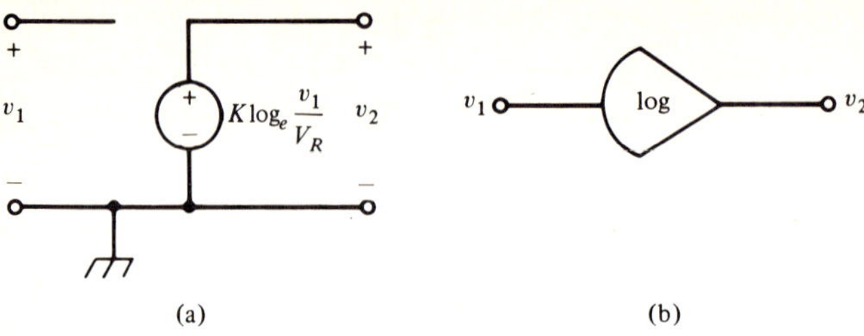

Figure 7-22 Log amplifier: (a) controlled source model, and (b) schematic representation.

amplifier offers only 2-quadrant operation. This type of amplifier, available now in monolithic and hybrid IC forms, is used for signal compression, linearization of certain nonlinear transducers, computation of arbitrary powers, decibel displays, and so on.

The Ideal Model

The controlled source model and the schematic representation of an ideal log amplifier are as shown in Fig. 7-22. It is seen from the figure that the input–output relation of an ideal log amplifier is given as

$$v_2 = K \log_e \left(\frac{v_1}{V_R} \right) \tag{7-52}$$

where V_R is a normalization constant in volts to set the output voltage equal to zero for a prescribed value of the input voltage, and K is the scale factor in volts used to set the output full-scale range. It follows from Eq. (7-52) that for fixed V_R, input v_1 must be a unipolar signal so that v_1/V_R is positive.

The log amplifier in most respects can be considered as a nonlinear voltage amplifier and hence has properties similar to that of a voltage amplifier. Thus for an ideal log amplifier, the input resistance is infinite (i.e., the input current is zero) and the output resistance is zero. In addition to being unilateral, it has an infinite bandwidth.

Simple Applications

With the aid of additional circuit elements, a number of novel nonlinear operations can be implemented by a log amplifier. We illustrate a few of these implementations next.

Antilog Amplifier. A novel way to implement the antilogarithm operation is by placing the log amplifier in the feedback path of an operational amplifier as shown in Fig. 7-23(a). Because of negative feedback, the potential difference between the two input terminals of the operational amplifier is zero. Hence the voltage at the inverting input terminal is v_1. Applying KCL at this terminal we obtain

$$\frac{v_1}{R_2} = \frac{1}{R_1} \left(K \log_e \frac{v_2}{V_R} - v_1 \right)$$

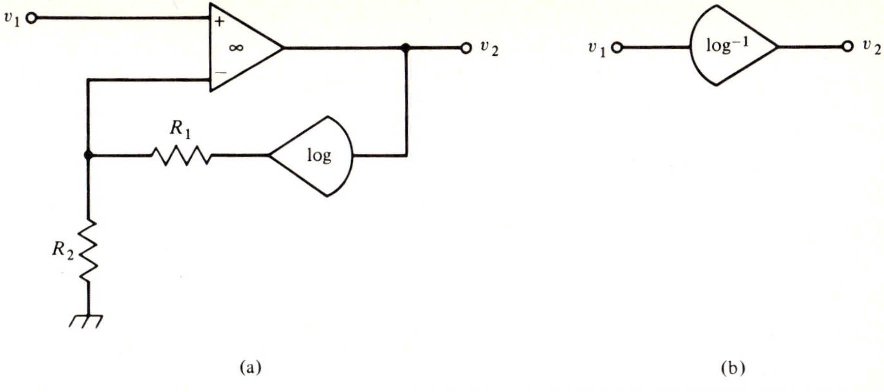

(a)

(b)

Figure 7-23 Antilog amplifier: (a) circuit realization, and (b) schematic representation.

which after rearrangement yields

$$v_2 = V_R \log_e^{-1}\left[\frac{(R_2 + R_1)v_1}{KR_2}\right] \tag{7-53}$$

Usual procedure is to set $R_1 = 0$ and $R_2 = \infty$ which simplifies the above to

$$v_2 = V_R \log_e^{-1}\left(\frac{v_1}{K}\right) \tag{7-54}$$

An alternate realization of an antilog amplifier will be found in Problem 7-29. The circuit schematic of the antilog amplifier is sketched in Fig. 7-23(b). Since an alternate way to write $\log_e^{-1}(v_1/K)$ is $e^{(v_1/K)}$, the antilog amplifier is also known as an *exponentiation amplifier*.

The antilog amplifier is used to linearize the response of devices having logarithmic responses. An example of such application is in the linearization of the output voltage of an oxygen detector used in pollution monitoring.[15] Here the output voltage of the detector is proportional to the log concentration of oxygen. To determine the actual concentration of oxygen, an antilog operation needs to be performed at the output of the detector. Other applications of antilog amplifier are in function fitting, generation of arbitrary powers, and in performing multiplication and division operations.

Log-Ratio Amplifier.   In a log-ratio amplifier, the output voltage is given as the logarithm of the ratio of two analog voltages. A possible implementation of such a device is indicated in Fig. 7-24 where the outputs of two log amplifiers are combined by a difference amplifier (Fig. 6-12). The output voltage v_o is thus given as

$$v_o = \left(K \log_e \frac{v_2}{V_R} - K \log_e \frac{v_1}{V_R}\right) = K \log_e\left(\frac{v_2}{v_1}\right) \tag{7-55}$$

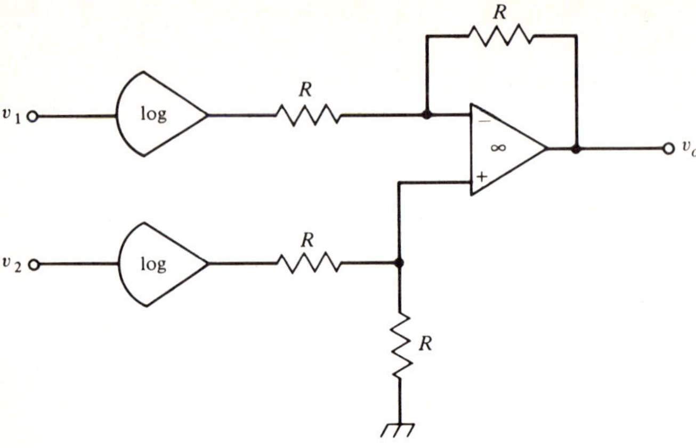

Figure 7-24   A log-ratio amplifier.

An example of an application of the log-ratio amplifier is as a photometer[15] measuring the light absorbance of a medium given as

$$A = -\log_e \frac{I_{sig}}{I_{ref}}$$

where I_{sig} and I_{ref} are currents proportional to the two light intensities generated with the aid of photodiodes. I_{sig} and I_{ref} represent, respectively, the intensities of the light transmitted through space and of the light transmitted through the medium absorbing light.

Generation of the m*th Power.*   The log and antilog amplifiers are particularly suited to raise an analog voltage to any integer or fractional power m. A realization of such a circuit is shown in Fig. 7-25. The finite gain scaling amplifier of voltage gain m can be implemented using either the noninverting voltage amplifier of Fig. 6-3 or the inverting voltage amplifier of Fig. 6-5. If the exponent is less than one, then the scaling amplifier can be replaced by a potentiometer.

Multiplier and Divider.   A straightforward realization of an analog divider is obtained by placing an antilog amplifier at the output of the log-ratio amplifier of Fig. 7-24. This type of divider offers a wider dynamic range and does not exhibit very large errors for values of the denominator voltage approaching zero. An analog multiplier can be realized by adding the outputs of two log amplifiers by a summing amplifier and then taking the antilogarithm of the output of the summing amplifier with the aid of an antilog amplifier.

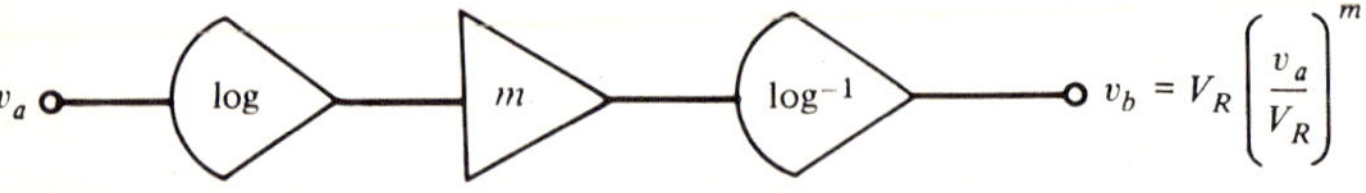

Figure 7-25   A circuit for raising an analog signal to mth power.

Y(Z/X)^m Module.[15] By combining the nonlinear operations described above in various ways, a number of multifunction circuits can be developed. One such realization is the $Y(Z/X)^m$ module. It produces an output voltage that is the product of an analog voltage Y with the mth power of a ratio of two analog voltages Z and X. In this circuit all of the input voltages are required to be positive. In the commercially available unit, the exponent m is externally adjustable and can be any number between $\frac{1}{5}$ and 5. A functional block diagram of this circuit is as shown in Fig. 7-26. This module can be used for all conventional applications of the log amplifier such as taking the logarithm, multiplying, dividing, and raising (or lowering) ratios of analog voltages to an arbitrary power. It can also be used for squaring, square-rooting, rms, and vector computation. It includes a low-drift reference voltage of 9 V which (or its fraction) can be used as one of the input voltages of the module itself for operations involving two analog voltages.

As an application of this module consider the implementation of the vector sum v_o of two analog voltages v_a and v_b; that is, forming

$$v_o = \sqrt{v_a{}^2 + v_b{}^2} \tag{7-56}$$

We rewrite Eq. (7-56) as

$$v_o{}^2 = v_a{}^2 + v_b{}^2$$

or

$$v_o{}^2 - v_o v_b = v_a{}^2 - (v_o v_b - v_b{}^2)$$

or

$$v_o = \frac{v_a{}^2}{v_o - v_b} - v_b \tag{7-57}$$

A realization of Eq. (7-57) using a $Y(Z/X)^m$ module with $m = 1$ and two differential amplifiers is shown in Fig. 7-27. Note that a direct implementation of Eq. (7-56) will require three multipliers and two operational amplifiers (Problem 7-33).

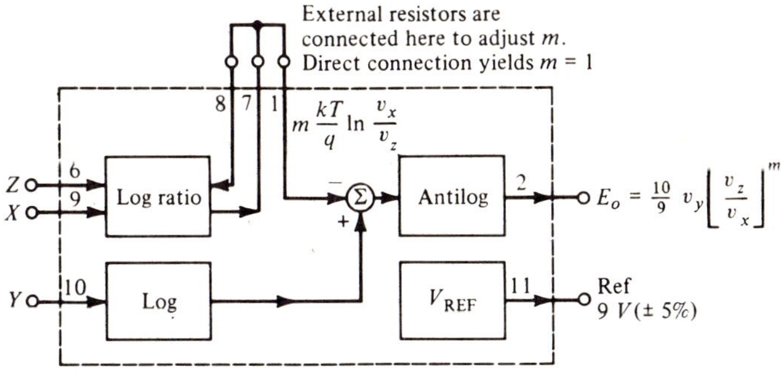

Figure 7-26 Functional block diagram of model 433 $Y(Z/X)^m$ module. (Courtesy Analog Devices, Inc.)

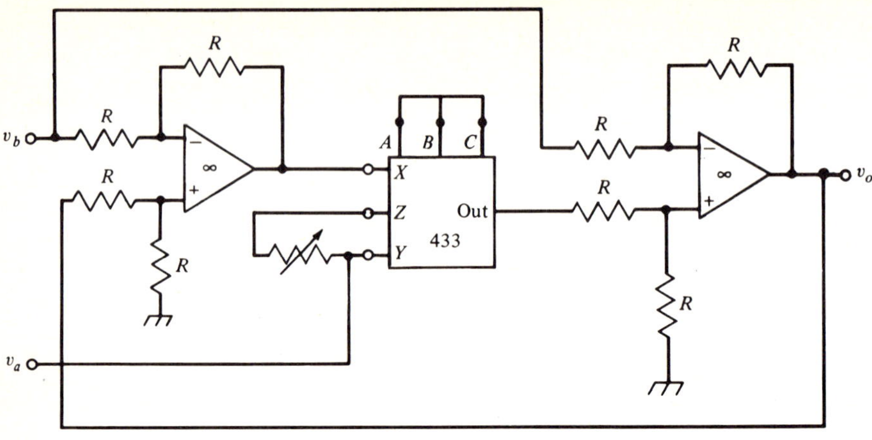

Figure 7-27   Vector-sum circuit using $Y(Z/X)^m$ module.

Log-Amplifier Realization[16]

The circuits for realizing log and antilog amplifiers make use of the nonlinear
i–v characteristic of semiconductor diodes. More specifically, such a diode
(Fig. 7-28) is characterized by the current-voltage relationship

$$i_f = I_0(e^{qv_f/kT} - 1) \tag{7-58}$$

where I_0 is the reverse saturation current, q is a constant equal to the unit charge
1.60219×10^{-19} C, k is the Boltzman's constant (1.38062×10^{-23} J/°K), and
T is the absolute temperature in degrees Kelvin. v_f is known as the forward
voltage across the diode and i_f is the forward current with the polarities as
shown. If the forward voltage is such that $e^{qv_f/kT} \gg 1$, then the current-voltage
relationship of the diode is approximately given by

$$i_f = I_0 e^{qv_f/kT} \tag{7-59}$$

If the temperature effects are ignored, q/kT and I_0 can be considered as constants.

Now consider the circuit of Fig. 7-29 in which the diode has been placed in
the feedback path of an operational amplifier. Because of the feedback connection, the inverting input terminal is at virtual ground. Thus, analyzing this
circuit in a manner similar to that used for the inverting amplifier (Fig. 6-5), we
arrive at

$$i_1 = \frac{v_1}{R} = i_f$$

$$v_2 = -v_f \tag{7-60}$$

Combining Eqs. (7-60) and (7-59) we then obtain

$$v_2 = -\frac{kT}{q} \log_e \left(\frac{v_1}{R}\right) + \frac{kT}{q} \log_e (I_0) \tag{7-61}$$

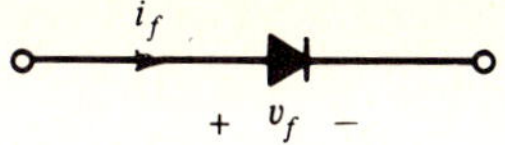

Figure 7-28   A semiconductor diode.

The second term in the above expression generates an offset in the output voltage. The scale factor kT/q and the offset term both vary with temperature. To eliminate the effect of the saturation current I_0, the basic circuit of Fig. 7-29 is modified using a second matched diode (having identical parameters as the one in the feedback path) and a current source in a manner illustrated in Fig. 7-30. The voltage v_3 at the noninverting input terminal of the second operational amplifier is given as

$$v_3 = v_2 + v'_f = v_2 + \frac{kT}{q} \log_e \left(\frac{I_R}{I_0}\right) \tag{7-62}$$

Substituting Eq. (7-61) into (7-62) we arrive at

$$v_3 = -\frac{kT}{q} \log_e \left(\frac{v_1}{RI_R}\right) \tag{7-63}$$

To compensate for the temperature effects on the scale factor, the noninverting amplifier stage at the output is designed to have a temperature-dependent gain by utilizing a temperature-sensitive resistor R_t. The overall output voltage v_o is then given by

$$v_o = -\left(1 + \frac{R_2}{R_1 + R_t}\right) \cdot \frac{kT}{q} \log_e \left(\frac{v_1}{RI_R}\right) \tag{7-64}$$

The diode with logarithmic i–v characteristic used in designing the logarithmic amplifier can be replaced with a grounded base transistor or a diode-connected (base connected to the collector) transistor. A detailed discussion on this type of logarithmic amplifiers including temperature-compensation arrangements will be found elsewhere.[16]

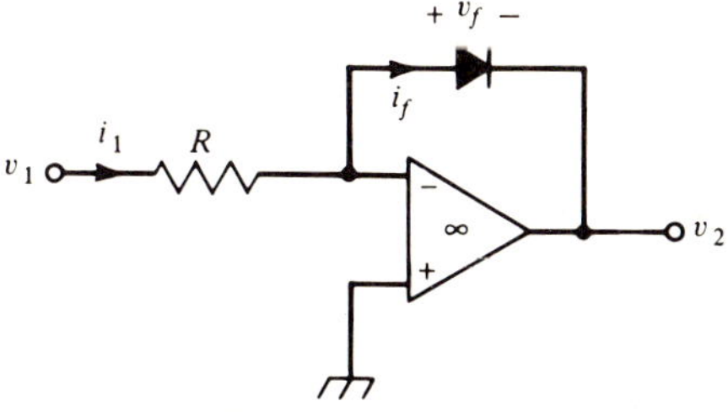

Figure 7-29   Basic log-amplifier circuit.

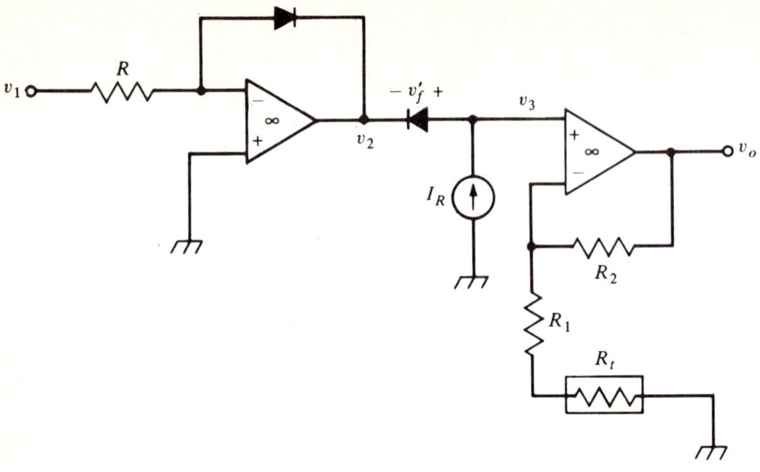

Figure 7-30   Log amplifier with temperature compensation.

The basic circuit for the direct design of an antilog amplifier is shown in Fig. 7-31. An analysis of this circuit yields the relation between the output and input voltages as

$$v_2 = I_0 R e^{-qv_1/kT} \tag{7-65}$$

As before the temperature effects can be compensated for using additional circuitry incorporating a second matched diode and a current source (Problem 7-31).

Characteristics of a Practical Log Amplifier[15]

There are basically two types of error sources in a practical log amplifier. Errors due to the tolerances and changes in the parameters describing the input–output relationship are known as *parametric errors*. The departure from the ideal logarithmic input–output relationship after all parametric errors have been removed by external adjustments is known as the *log-conformity error*. Both types of errors can be referred to either the input or the output. If referred to the

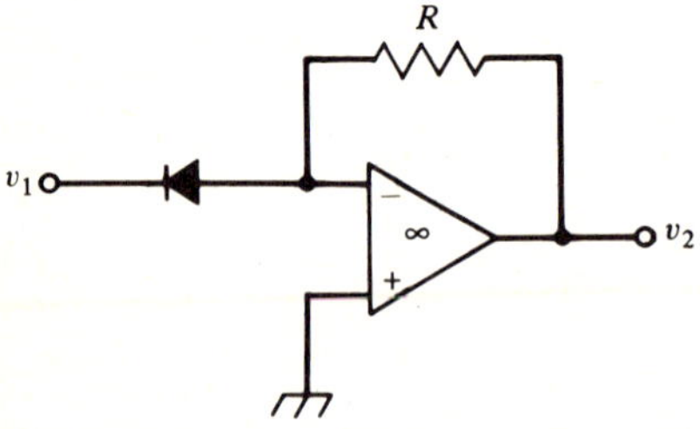

Figure 7-31   The basic antilog amplifier.

input, they are specified as percentage errors. On the other hand, when referred to the output they are usually specified as actual errors in millivolts.

The input–output relation of a nonideal log amplifier takes the form

$$v_2 = -K \log_e \left(\frac{v_1 - E_{os}}{V_R} \right) \tag{7-66}$$

where K is known as the *scale factor*, E_{os} is the *offset voltage*, and V_R is the *reference voltage*.

The offset voltage is a function of the operational amplifier being used to design the log amplifier. It essentially appears as a voltage source in series with the input resistor R and can be trimmed to zero by feeding an adjustable dc voltage in series with the input voltage source. Typically it is of the order of several hundred microvolts. The scale factor is given as the change in the output voltage for a decade change in the input voltage after E_{os} has been trimmed to zero. The error in scale factor is expressed as a percent of its nominal value. Finally, as can be seen from Eq. (7-63), the reference voltage V_R is equal to $R \cdot I_R$. The input resistance R is more stable than the current source I_R. As a result the variations in V_R are essentially due to variations in I_R. Error in V_R is specified as a percent of the nominal value. All of the above three parametric errors drift with temperature.

After all parametric errors have been compensated for, the log-conformity error should be considered. It is usually specified as the deviation from a straight-line of the plot of the output as a function of input drawn on semilog paper over the useful range.

7-5   Trigonometric Function Generators

The trigonometric function generators in which the output voltage is a specified trigonometric function of the input voltage are another useful class of nonlinear function modules. Three of the most commonly used such modules are the sine, the cosine, and the inverse tangent generators which are available as off-the-shelf packages. These modules find applications in wave shaping, function generations, vector resolutions, and coordinate transformations. Invariably these are designed by approximating the pertinent trigonometric functions in suitable forms and then implemented using other nonlinear elements.

Strictly speaking, the actual input–output relationship of a practical module is given in scaled form. For example, if v_2 and v_1 are the output and the input voltages of a sine module, they are related as

$$v_2 = V_F \sin \left(\frac{\pi}{2} \cdot \frac{v_1}{V_R} \right) \tag{7-67}$$

where V_R is voltage corresponding to $90°$ and V_F is full-scale output voltage when $v_1 = V_R$. The trigonometric function modules are classified in accordance with the allowable range of the input voltages. In a 1-quadrant sine module, the input is restricted to be in the range $0 \leq v_1 \leq V_R$. For 2-quadrant operation, the input range is $-V_R \leq v_1 \leq V_R$. The schematic representations of the sine, the cosine, and the $\tan^{-1}$ modules are depicted in Fig. 7-32.

Multipliers are often used in implementing the trigonometric function modules. We illustrate the design procedure by considering next the implementation of the sine module.

The Sine Module[15]

Consider first a polynomial approximation of the sine function using a second-degree polynomial. Thus we write

$$\sin x \cong y(x) = y_0 + a_1 x + a_2 x^2 \tag{7-68}$$

where x is the input and $y(x)$ is the output. For simplicity, we restrict x to be in the first quadrant range, that is, $0 \leq x \leq \pi/2$.

The three constants, y_0, a_1, and a_2 can be determined by fitting the approximation to the curve of $\sin x$ in accordance with some criteria. For example, we might choose $y(x) = \sin x$ for three prescribed values of x in the given range, say, $x = 0$, $x = \pi/2$, and $x = 1$ radian. For these values, we obtain from Eq. (7-68)

$$y_0 + a_1(0) + a_2(0)^2 = \sin 0 = 0$$

$$y_0 + a_1 \left(\frac{\pi}{2}\right) + a_2 \left(\frac{\pi}{2}\right)^2 = \sin \frac{\pi}{2} = 1 \tag{7-69}$$

$$y_0 + a_1(1) + a_2(1)^2 = \sin (1) = 0.8415$$

Solution of the above equations yield

$$y_0 = 0 \qquad a_1 = 1.2004 \qquad a_2 = -0.3589$$

Hence, an approximation to $\sin x$ is given as

$$\sin x \cong y_1(x) = 1.2004x - 0.3589x^2 \tag{7-70}$$

The approximation error $\varepsilon_1(x) = y_1(x) - \sin x$ is usually given as a percent of the full-scale (F.S.) value. The maximum error is at $x = 0.37705$ radians and is of the value 3.341 percent of F.S. The plot of the error as a function of x is sketched in Fig. 7-33.

The maximum error can be minimized while ensuring that error be zero at the endpoints of the prescribed range by choosing a different approximation

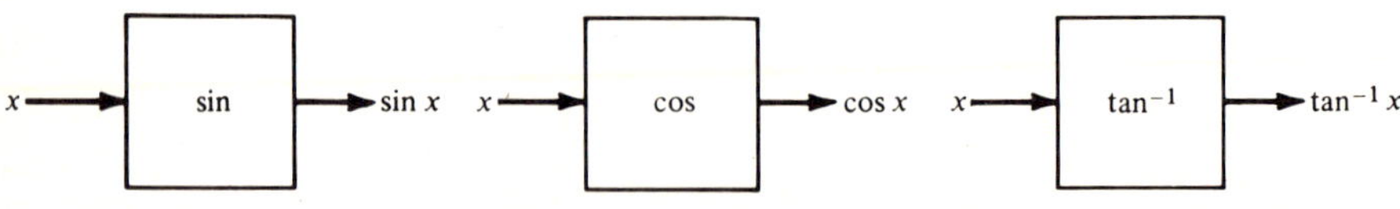

Figure 7-32   Schematic representations of the sine, cosine, and $\tan^{-1}$ modules.

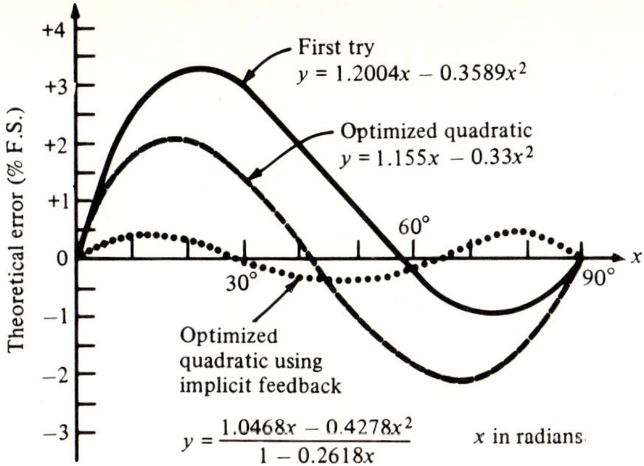

Figure 7-33 Error plots of various approximations to sine functions (Reference 15). (Courtesy Analog Devices, Inc.)

procedure. The lowest error is obtained for the following approximation:

$$\sin x \cong y_2(x) = 1.155x - 0.33x^2 \tag{7-71}$$

The corresponding error $\varepsilon_2(x) = y_2(x) - \sin x$ is also plotted in Fig. 7-33 which is seen to be symmetrical with a maximum error of ± 2.1 percent of F.S.

Still further improvement in the approximation is obtained by choosing a rational function to approximate $\sin x$. For example, we can express

$$\sin x \cong y_3(x) = y_0 + \frac{a_1 x + a_2 x^2}{1 + b_1 x} \tag{7-72}$$

Since there are four constants, we can ensure $y_3(x) = \sin x$ at four values of x. The best approximation is then given by

$$y_3(x) = \frac{1.0468x - 0.4278x^2}{1 - 0.2618x} \tag{7-73}$$

The error $\varepsilon_3(x) = y_3(x) - \sin x$ is again plotted in Fig. 7-33. Now the maximum error is ± 0.44 percent of F.S.

In all plots of Fig. 7-33 the errors are expressed in percentage of F.S. voltage. To express these errors in percentage of actual value of $\sin x$, they should be divided by $\sin x$. All of the above three approximations given in Eqs. (7-70), (7-71), and (7-73) can be implemented using a single multiplier.

It should be noted that the maximum error between the ideal and the approximation can be reduced further by choosing an approximation with nonintegral exponents. For example, the maximum error in the approximation

$$y_4(x) = \frac{0.999642x - 0.1073254x^{3.02}}{1 + 0.0604426x^{2.02}} \tag{7-74}$$

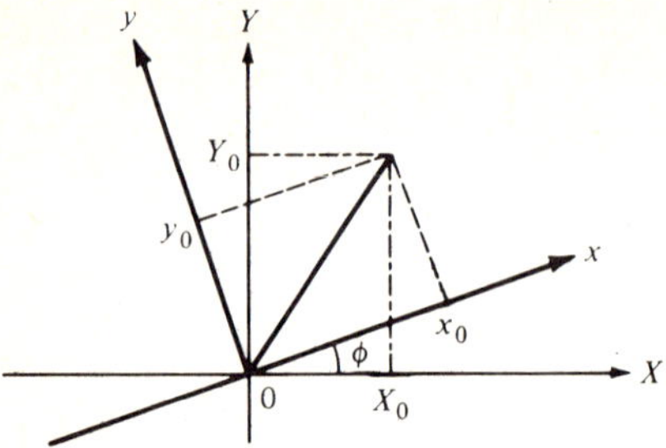

Figure 7-34   Illustration of coordinate rotation.

is about ± 0.004 percent of F.S. Realization of the above requires the use of log amplifiers or the $Y(Z/X)^m$-type module.

Sine generators are also designed using piecewise linear approximation to the sine curve. This second type of curve-fitting technique is described elsewhere.[16] It should be noted that both sine and cosine function generators are also available in a single package.

Applications of the Sine-Cosine Module[8]

Several straightforward applications of the sine and the cosine modules are outlined below.

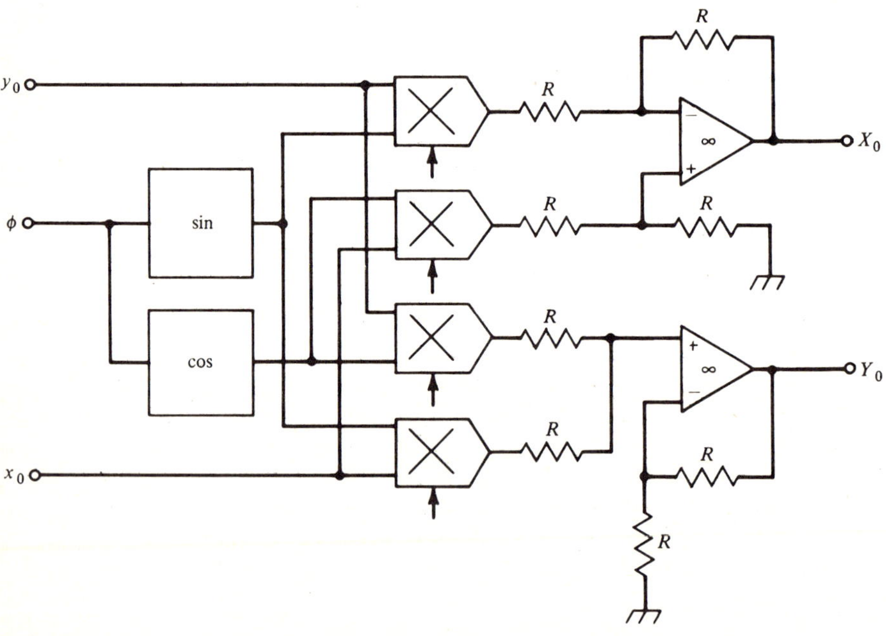

Figure 7-35   Coordinate rotation circuit.

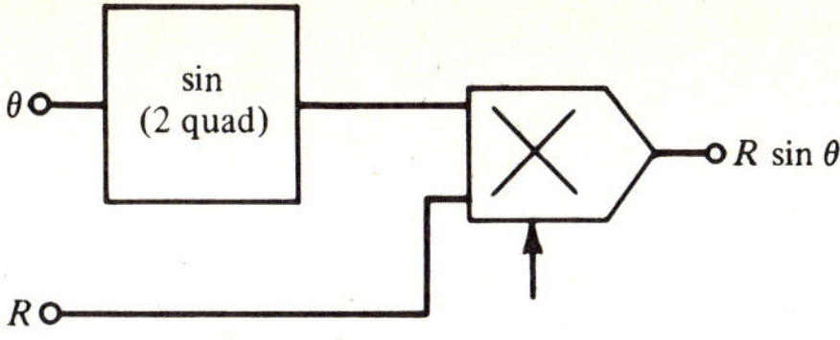

Figure 7-36   Circuit for computing vertical height.

Coordinate Rotation.   Consider a vector whose components are x_0 and y_0 in a given rectangular coordinate system. Let X_0 and Y_0 be the components of the same vector in another rectangular coordinate system which is obtained by rotating the original coordinates by an angle ϕ (Fig. 7-34). The components of the vector in the two coordinate systems are related as

$$X_0 = x_0 \cos \phi - y_0 \sin \phi$$
$$Y_0 = x_0 \sin \phi + y_0 \cos \phi \tag{7-75}$$

An implementation of above is shown in Fig. 7-35.

Conversion of Polar Coordinates to Rectangular Coordinates.   The circuits for converting polar coordinates to rectangular coordinates are usually known as *resolvers* which find applications in some fire control systems, navigational computers, and so on. A typical 2-quadrant resolver is shown in Fig. 7-36. It can be used, for example, to compute the vertical height of a target from the range R and angle θ of the target obtained in an aircraft radar.

A second type of resolver is needed for a radar plan-position-indicator (PPI) display with sector-scanning capabilities. Figure 7-37 shows a graphical representation of the PPI scan of an airborne radar, where the angle Θ is the aircraft

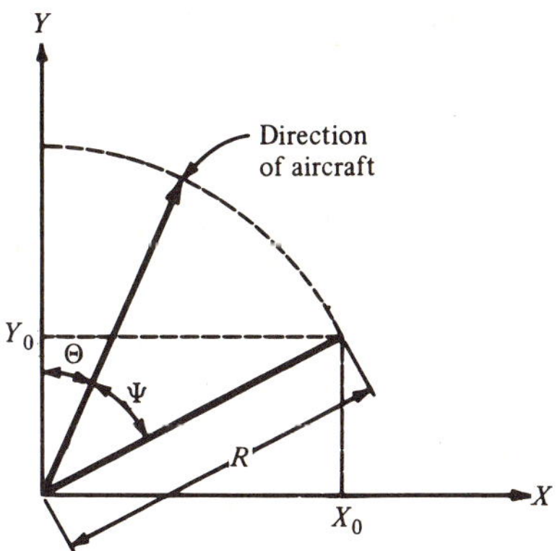

Figure 7-37   Radar scan computation.

heading, angle Ψ is the angle of the antenna boresight with respect to the direction of flight of the aircraft, and R is the radar range. It follows from the figure then that the corresponding components in the rectangular coordinates are given as

$$Y_0 = R \sin\left(\frac{\pi}{2} - \Theta - \Psi\right) = R\,(\cos\Theta \cdot \cos\Psi - \sin\Theta \cdot \sin\Psi)$$

$$X_0 = R \cos\left(\frac{\pi}{2} - \Theta - \Psi\right) = R\,(\sin\Theta \cdot \cos\Psi + \cos\Theta \cdot \sin\Psi)$$

(7-76)

A circuit realization of above equations is straightforward and is left as an exercise.

7-6   The Analog Comparator

The analog comparator is a unique type of circuit. Its input signals are analog and its output is essentially digital in nature. It is a major component in the design of analog-to-digital and digital-to-analog converters, which are considered in Chapter 8. In this section we define the ideal comparator, describe the properties of the practical comparator, and outline several simple applications.

The Ideal Model

The basic purpose of an analog comparator is to compare two analog voltages and indicate which one is larger. Thus it is a 2-input, single-output device as shown symbolically in Fig. 7-38. With reference to this figure, the operation of a comparator can be mathematically represented as

$$
\begin{aligned}
V_o &= V^+, \quad \text{if} \quad V_1 > V_2 \\
&= V^-, \quad \text{if} \quad V_1 < V_2
\end{aligned}
$$

(7-77)

where $V^+ > V^-$. Equation (7-77) states that if the analog voltage level V_1 is greater than the analog voltage V_2, then the output voltage level is V^+, and if V_1 is less than V_2, then V_o is equal to V^-. Note that the output voltage can ideally take one of two possible values, that is, it is binary variable. Usually these devices are designed such that their output voltage levels are compatible with digital IC logic families. For example, for the Fairchild μA710 comparator $V^+ = 3.1$ V and $V^- = -0.5$ V and its output is thus compatible with DTL, RTL, and TTL logic circuits.

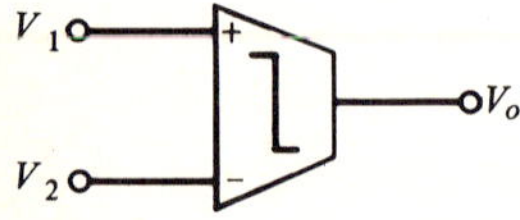

Figure 7-38   Symbolic representation of an analog comparator.

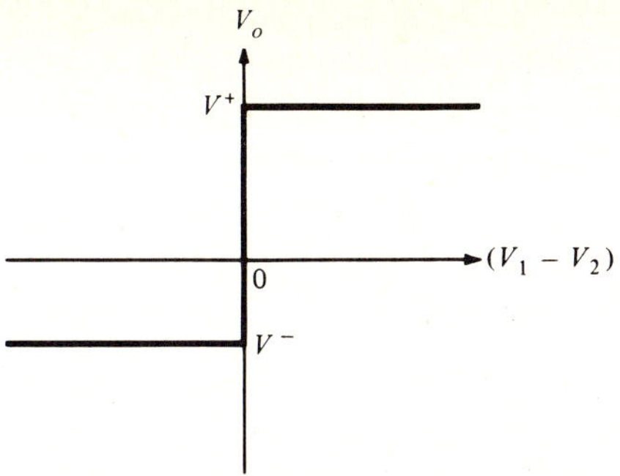

Figure 7-39 Transfer characteristic of an ideal analog comparator.

An alternate way to interpret Eq. (7-77) is to note that if the differential input voltage $(V_1 - V_2)$ is positive, then $V_o = V^+$; and if $(V_1 - V_2)$ is negative, then $V_o = V^-$. Thus the output responds to the difference of the two input signals. Following the convention of the differential-input operational amplifier, the upper terminal in Fig. 7-38 is designated as the *noninverting* input terminal (marked " $+$ ") and the lower terminal is designated the *inverting* terminal (marked " $-$ ").

A convenient way to characterize an ideal comparator is by its transfer characteristic as shown in Fig. 7-39. Another way to draw the transfer characteristic is to treat one of the input voltages as a fixed reference voltage V_{ref} (as is normally the case in most applications) and plot the transfer characteristic as a function of the other input voltage. This is illustrated in Fig. 7-40 where we have considered V_1 as the reference voltage having a positive value V_{ref}.

Properties of a Practical Comparator

A comparator is essentially a very high-gain difference voltage amplifier with output limiting. The basic difference between a comparator and an operational

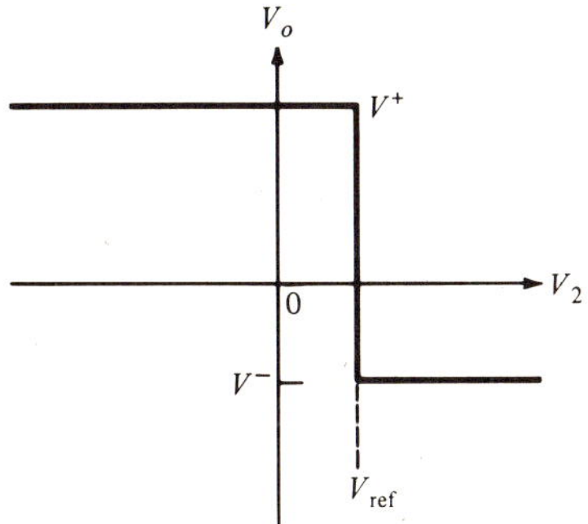

Figure 7-40 Alternate representation of the transfer characteristic of an analog comparator.

amplifier is that in the case of the former the output is designed to "saturate" very fast to one of two possible voltage levels. In addition, the comparator is able to switch quickly from one level to the other. The input stage of a comparator is also a differential amplifier like that of an operational amplifier. Thus one naturally would expect some similarity between the characteristics of these two devices. Some of the important properties of a practical comparator are reviewed next.

Resolution.   The typical transfer characteristic of a practical comparator is as shown in Fig. 7-41. This plot indicates that the difference voltage $(V_1 - V_2)$ must exceed $+V'_{min}$ in order to drive the output voltage to its positive saturation level V^+. Similarly the difference voltage must be more negative than $-V''_{min}$ to drive the output to its negative saturation level. Thus there exists a minimum voltage difference between the two inputs which results in a distinguishable output. This minimum difference voltage is called the *resolution* or *sensitivity* of the comparator and typically it is of the order of several millivolts. It follows from Fig. 7-41 that the resolution is a function of the *gain μ* of the comparator (given by the slope of the transfer characteristic before saturation) and the output voltage swing $(V^+ - V^-)$. For fixed resolution an increase in $V^+ - V^-$ would require an increase in μ (Problem 7-41).

Offset Voltage.   Ideally when the two input voltages are equal, that is, $V_1 = V_2$, the output V_o is just at the threshold. Then slight changes in the input voltage levels should move the output level to one of its normal operating states, V^+ or V^-. In practice, however, because of the mismatch of components and a host of other factors, even if $V_1 = V_2$, V_o is not at the threshold because of offset error. The *offset error voltage V_{os}* can thus be defined as the voltage that should be applied at the input terminals to bring the output voltage to the threshold level, indicating $V_1 = V_2$ (see Fig. 7-42). Note from the transfer characteristic of Fig. 7-41, for $V_1 = V_2$, $V_o = 0$. However, the threshold voltage is defined as the approximate voltage level at the output of the comparator at which the digital

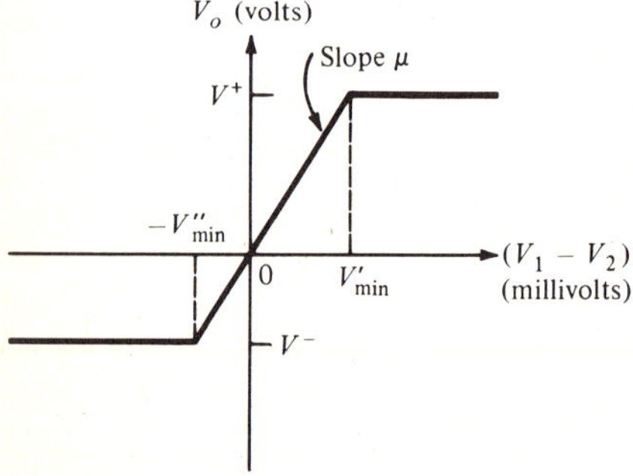

Figure 7-41   Transfer characteristic of a practical comparator.

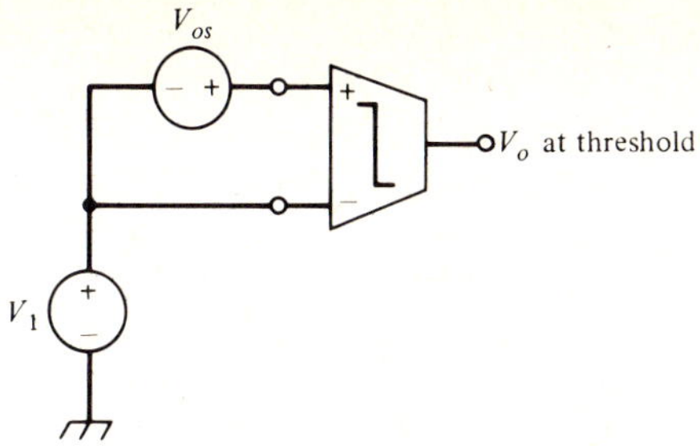

Figure 7-42   Model illustrating offset voltage.

logic circuit connected to the output of the comparator changes its digital state. The threshold voltage level thus depends on the voltage swing of the comparator and digital circuits driven by it, and may or may not have a zero value.

Example 7-5.   A comparator is excited by the same signal at both of its inputs. The measured output voltage is 1.6 V above the threshold level. Determine the input offset error voltage if the gain of the comparator is 800.

The input offset error voltage is obtained by dividing the negative of the output error by the comparator gain, that is,

$$V_{os} = -\frac{1.6}{800} = -2 \text{ mV}$$

Offset and Bias Currents.   Some fixed amount of current always flows through the input terminals irrespective of the input voltage levels. These error currents can be defined as the currents that should be fed into the input terminals to bring the output to the logic threshold level. Their effects can be calculated by setting the input signal sources to zero as shown in Fig. 7-43 where R_1 and R_2 are the effective resistances seen by the two input terminals. The difference between the two input currents I_{s1} and I_{s2} is defined as the *input offset current* I_{os} and the average of them is the *input bias current* I_B. Thus,

$$I_B \cong (I_{s1} + I_{s2})/2$$
$$I_{os} = I_{s1} - I_{s2}$$

$$(7\text{-}78)$$

It is seen from Fig. 7-43 that the input error currents create an effective input offset voltage

$$I_{s1}R_1 - I_{s2}R_2$$

which is equivalent to

$$I_B(R_1 - R_2) + \frac{I_{os}}{2}(R_1 + R_2)$$

$$(7\text{-}79)$$

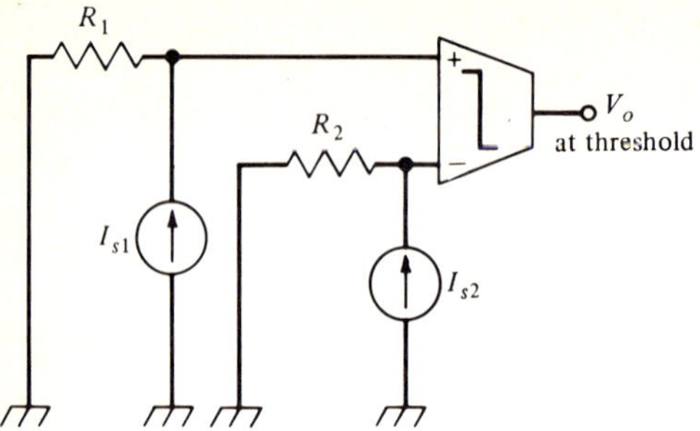

Figure 7-43   Calculation of effects of offset and bias currents.

Like the operational amplifier, the input stage of a monolithic IC comparator is a well-matched differential amplifier. Consequently the two input currents are approximately equal. Hence, as Eq. (7-79) points out, their contribution to the input offset voltage can be minimized by making the two input resistances equal. To minimize the effect of the input offset current, the input resistances should be small. Typical values of the input offset current for μA710 is less than 2 μA and that of the input bias current is 27 μA.

Drift.   The input offset voltage and currents also drift with temperature, time, and so on, which may be detrimental for some applications. Normally the average temperature coefficients of the input offset voltage and input offset current are specified in μV/$^\circ$C and nA/$^\circ$C, respectively.

Response Time.   An important factor determining the speed of operation of a circuit containing the comparator is the *response time* of the comparator. It is defined as the time taken by the output to cross the logic threshold voltage after the application of a step input to one terminal while the other input terminal is at a fixed reference voltage.[24] Before the step function is applied, the difference voltage across the two input terminals is equal to the fixed reference voltage which causes the output of the comparator to saturate to either V^+ or V^- level depending on the sign of the reference voltage. The input step should be chosen to drive the output in the opposite direction. The height of the input step should be in excess of the reference voltage plus the total input offset voltage. The excess amount is called the voltage overdrive. The response time is then the time taken to bring the output level from saturation to logic threshold voltage. The response time for Fairchild μA710 for a 100-mV step input with 5-mV overdrive is about 40 nsec.

Input Voltage Range.   For a differential-input comparator, two different voltage ranges at the input must be observed. One is the voltage range of each input analog signal and the other is the range of the difference voltage across the two inputs. Exceeding either one of these two voltage limits may cause degradation in the input offset and bias currents.

Example 7-6.   One of the input terminals of a μA710 comparator is at $+4$ V. Determine the range of safe voltage level at the other input terminal.

The input voltage range of the μA710 is ± 5 V and the differential input voltage range is also ± 5 V. Hence, if one of the input voltage V_1 is at $+4$ V, the range of the other input voltage V_2 should be

$$-1 \leq V_2 \leq 5 \text{ V}$$

Common-Mode Rejection Ratio.   An ideal comparator responds to the difference voltage across its input terminals and should be insensitive to the voltage common to both input terminals. However, the actual input voltage levels do affect the performance of a practical comparator, a measure of the effect of the input voltages is given by its *common-mode rejection ratio* (CMRR) defined as

$$\text{CMRR} = \frac{\text{Common-mode input voltage range}}{\text{Maximum change in input offset voltage}} \tag{7-80}$$

where the maximum change in input offset voltage is measured over the input voltage range. The CMRR is usually given in decibels and the typical value for μA710 is about 100 dB.

Logic Compatibility.   If the comparator is used to drive digital logic circuits, it is necessary to ensure that the comparator output levels and logic threshold voltage are compatible with those of the logic circuits connected to its output terminals. Compatibility implies that the comparator is capable of driving the loading logic circuits to saturation under a worst-case condition.[24] Comparators having very low output impedances are also used as buffers for going from high-level pulse circuits to low-level IC logic circuits.* It should be noted that the large output current from the comparator can damage the logic circuits and the use of a series resistor at the output is recommended to limit the *output current*. Excessive output current may also cause damage to the comparator and most manufacturers specify the maximum output current limit as peak output current.

Some Applications

In addition to its application in the design of analog-to-digital and digital-to-analog converters, analog comparator also finds applications in the design of pulse forming and shaping circuits, oscillators, pulse-height discriminator, digital line receivers, core-memory sense amplifiers, and so on. A few of these applications are described next.

Voltage Level Detector.   The comparator easily lends itself for voltage level detection. The circuit arrangement is indicated in Fig. 7-44. The corresponding transfer characteristic is sketched in Fig. 7-40. Thus, if the output voltage level

---

* The maximum number of logic circuits that can be driven by the comparator is known as its *fan-out.*

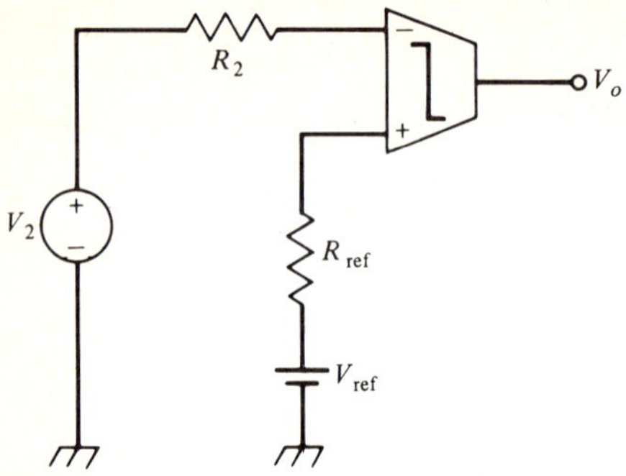

Figure 7-44   Voltage level detector.

is V^+, then it means that the input analog signal V_2 is less than V_{ref}. On the other hand, if V_o is V^-, then V_2 is greater than V_{ref}. To minimize the effect of input bias currents, the input resistances R_2 and R_{ref} should be made equal.

One simple application of the above arrangement is in detecting pulses in the presence of noise. If V_2 is a pulse train contaminated with noise and V_{ref} is the threshold level, then only the signal peaks in V_2 exceeding V_{ref} would result in negative output pulses. The threshold level should be chosen to exceed the noise level.

The circuit of Fig. 7-44 can also be used as a square-wave generator. If V_{ref} is set to 0 V and V_2 is a sinusoidal signal of frequency ω, then the output V_o will be a square wave of period $2\pi/\omega$ sec, having a peak-to-peak amplitude of $(V^+ - V^-)$ V.

Two comparators can be used to design a window discriminator as shown in Fig. 7-45(a).[22,24] The circuit arrangement is used to detect whether a voltage or pulse amplitude $|V_s|$ is within two limits E_1 and E_2 where $E_1 < E_2$. The transfer characteristic of the complete circuit is as shown in Fig. 7-45(b) which indicates that if V_o is V^-, then the input analog signal V_s satisfies the condition

$$E_1 < V_s < E_2 \tag{7-81}$$

To minimize variations of the discrimination levels with change in temperature, the input impedances should be equal, that is, $R_s = R_1 = R_2$.

The basic circuit of Fig. 7-45(a) can be readily adapted for use as a core-memory sense amplifier, whose purpose is to amplify the output of core memory used in digital systems and eliminate the large common-mode signals appearing during read-out. The complete circuit arrangement using two μA710 comparators is shown in Fig. 7-46.[22] The threshold levels here are provided from a positive supply V_{adj} through the resistors R_5 and R_6. For the values shown, the currents through R_5 and R_6 are each approximately 1 mA. These currents flow through the resistors R_3 and R_4 producing a voltage drop of 20 mV across each resistor. Thus the threshold sense levels are -20 and $+20$ mV. The input resistors R_1 and R_2 provide the termination of the sense line. The sense levels can

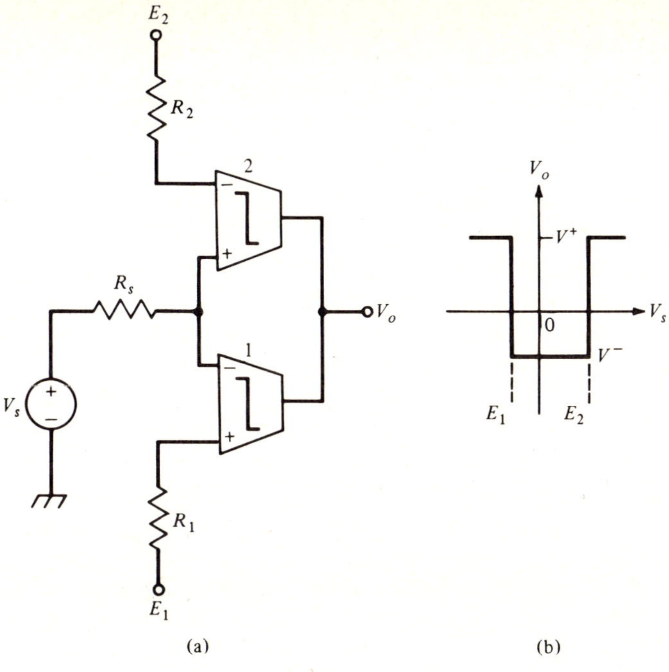

Figure 7-45   Window discriminator: (a) circuit diagram, and (b) transfer characteristic.

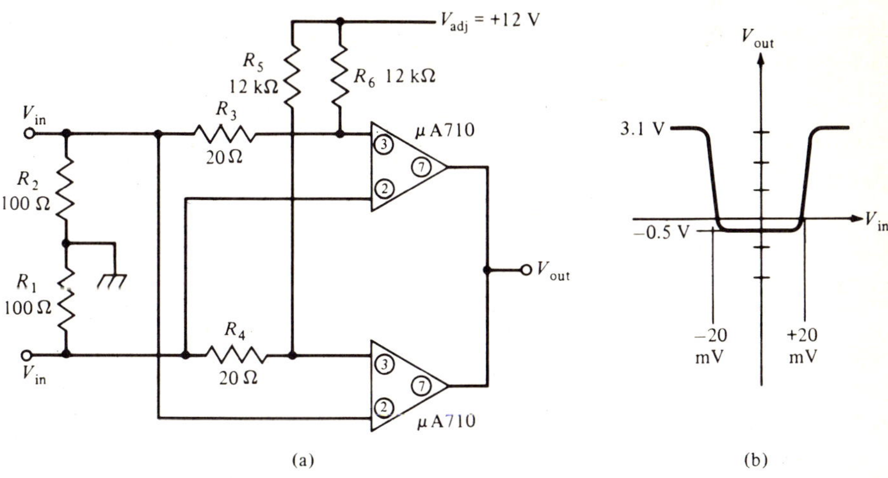

Figure 7-46   Core memory sense amplifier: (a) circuit diagram, and (b) transfer characteristic. (Courtesy Fairchild Camera and Instruments Corp.)

be set by adjusting the supply voltage V_{adj}. Independent adjustment of sense level of each comparator to take into account the offset voltages can be achieved by trimming R_5 and R_6.

Dual comparators in one IC package are also available for applications of the type described above. One such package is the Fairchild μA711 dual comparator.

Multivibrators. The monostable and astable multivibrators can easily be constructed using the differential-input comparator. The basic circuit arrangement for the monostable multivibrator is sketched in Fig. 7-47(a).[18,22] To explain the operation of the circuit let us assume that the input pulse $v_2 = -K\delta(t)$ is applied at $t = 0$. Note that in order to operate properly the amplitude of the input pulse K should be greater than V_{ref}. It follows from the figure that for $t < 0$, $v_2 = 0$ and $v_1 = -V_{ref}$, implying $v_1 - v_2 = -V_{ref}$ which is negative. Hence, according to the transfer characteristic of the comparator (see Fig. 7-40), the output v_o equals V^-. At $t = 0$, $v_2 = -K$ and $v_1 = -V_{ref}$, implying $v_1 - v_2 = K - V_{ref}$ which is positive. The positive difference voltage then drives the output of the comparator to V^+ level. At $t = 0+$, v_2 goes back to ground potential and the difference voltage is equal to v_1 whose value varies with time. To emphasize this fact, let us denote the voltage level at the noninverting terminal as $v_1(t)$. As we show a little later, $v_1(t)$ decreases exponentially from its positive value at $t = 0+$ approaching the value $-V_{ref}$ as $t \to \infty$. However, as soon as $v_1(t)$ crosses the zero value, the difference voltage across the comparator input terminals becomes negative which drives the output of the comparator back to its V^- level. Thus the output waveform of the circuit of Fig. 7-47(a) is a positive square wave as shown in Fig. 7-47(b). To determine $v_1(t)$ and hence the duration of the pulse, we analyze the equivalent circuit of Fig. 7-48 which represents the external circuit.

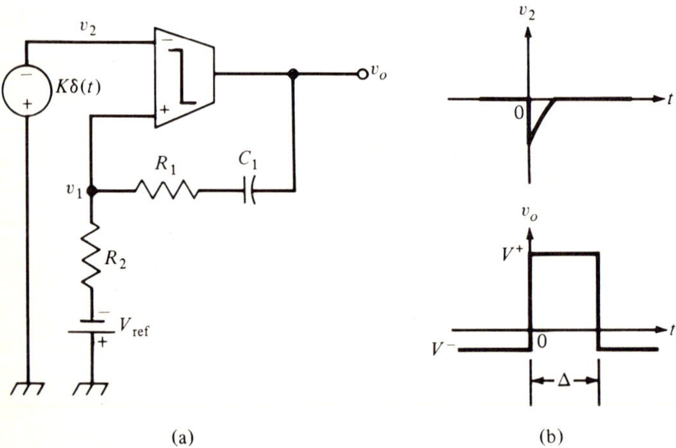

(a)   (b)

Figure 7-47 Monostable multivibrator: (a) circuit diagram, and (b) time-domain response.

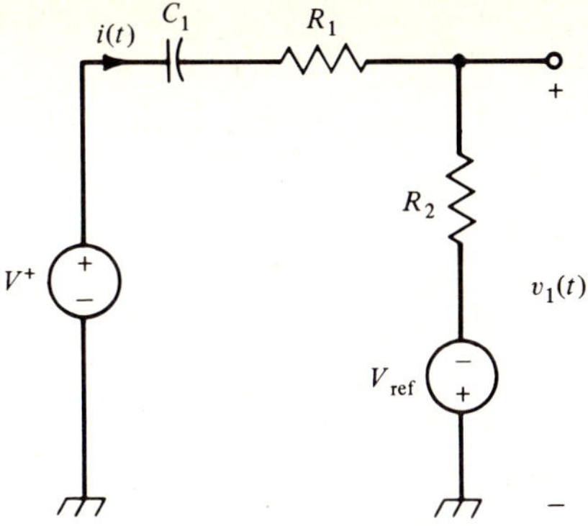

Figure 7-48 Equivalent circuit for analysis.

Kirchhoff's voltage law leads to the first-order differential equation describing the circuit:

$$(R_1 + R_2)i(t) + \frac{1}{C_1}\int_0^t i\,dt = V^+ + V_{\text{ref}} \tag{7-82}$$

To solve the above equation we differentiate both sides to get

$$\frac{di}{dt} + \frac{i(t)}{(R_1 + R_2)C_1} = 0 \tag{7-83}$$

the solution of which is recognized as[23]

$$i(t) = ke^{-\frac{t}{(R_1 + R_2)C_1}} \tag{7-84}$$

It follows from the circuit of Fig. 7-48 [or from Eq. (7-82)], that at $t = 0$

$$i(0) = \frac{V^+ + V_{\text{ref}}}{R_1 + R_2} \tag{7-85}$$

which is equal to the constant k in Eq. (7-84). Thus,

$$i(t) = \frac{V^+ + V_{\text{ref}}}{R_1 + R_2}e^{-\frac{t}{(R_1 + R_2)C_1}} \tag{7-86}$$

Hence,

$$v_1(t) = R_2 i(t) - V_{\text{ref}} = \frac{R_2}{(R_1 + R_2)}(V^+ + V_{\text{ref}})e^{-\frac{t}{(R_1 + R_2)C_1}} - V_{\text{ref}} \tag{7-87}$$

At $t = 0$

$$v_1(0) = \frac{R_2}{R_1 + R_2}(V^+ + V_{\text{ref}}) - V_{\text{ref}} = \frac{R_2 V^+ - R_1 V_{\text{ref}}}{R_1 + R_2} \tag{7-88}$$

For proper operation $v_1(0)$ must be positive, that is, the condition

$$V_{\text{ref}} < \frac{R_2}{R_1} V^+ \tag{7-89}$$

must be satisfied. Normally V_{ref} is a few millivolts and $R_2 V^+/R_1$ is several volts.

To determine the time Δ at which $v_1(\Delta)$ becomes zero we set

$$\frac{R_2(V^+ + V_{\text{ref}})}{(R_1 + R_2)} e^{-\frac{\Delta}{(R_1 + R_2)C_1}} = V_{\text{ref}} \tag{7-90}$$

which can be solved to yield

$$\Delta = (R_1 + R_2)C_1 \log_e \frac{R_2(V^+ + V_{\text{ref}})}{(R_1 + R_2)V_{\text{ref}}} \tag{7-91}$$

The resistor R_1 is used to reduce the recovery time. Satisfactory operation can still be obtained with R_1 set to zero.

In practice it is preferable to use an RC differentiating circuit (Fig. 7-49) at the inverting input terminal of the comparator to obtain a sharp trigger pulse.

By making $V_{\text{ref}} < 0$, it is possible to obtain a negative output pulse with a positive trigger pulse (Problem 7-48).

A free-running multivibrator[18,22,24] realization is sketched in Fig. 7-50. The explanation of the operation of this arrangement is left as an exercise (Problem 7-49).

Rectification.   The process of producing a dc output from an ac input is known as *rectification*. Dc power supplies are needed for biasing IC modules and discrete active circuits. On the other hand, most power sources produce 60 Hz sinusoidal voltages. The concept of full-wave rectification can be simply explained with the aid of a comparator and a multiplier as illustrated next.

The proposed circuit arrangement is sketched in Fig. 7-51.[4] Here the output levels of the comparator must be equal and opposite ($V^+ = -V^-$). A suitable comparator for this purpose is the Burr-Brown Model 9892/25 comparator which has a full-scale swing of ± 6 V. Let $A \sin \omega t$ be the ac voltage. Since the noninverting input terminal of the comparator is grounded, the positive half-cycle of the ac signal will drive the output of the comparator to the V^- level and

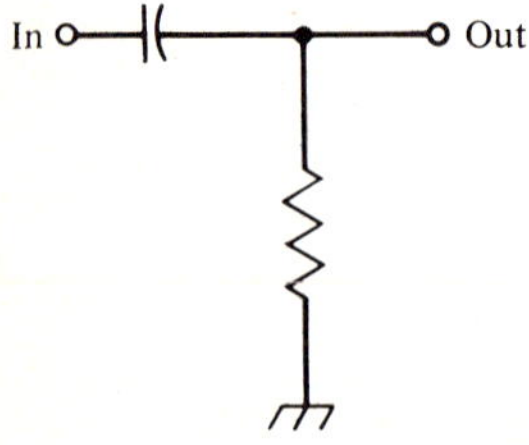

Figure 7-49   A differentiating circuit.

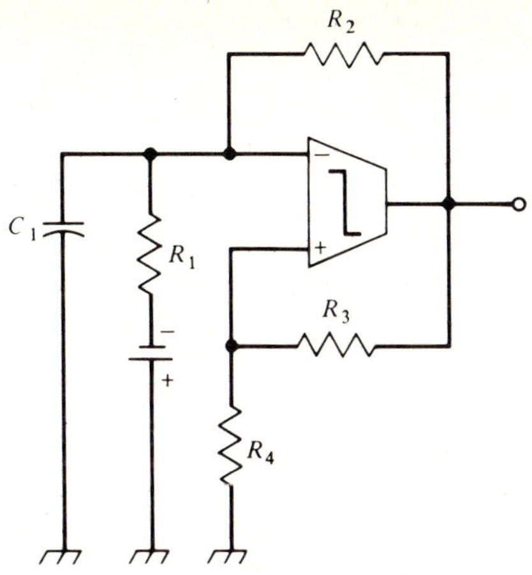

Figure 7-50   A free running multivibrator.

the negative half-cycle will drive the output of the comparator to the V^+ level. Thus the output of the comparator is a continuous train of square wave which is $180°$ out of phase with the ac input.

The output of the multiplier will thus be

$$v_o(t) = \frac{AV^+}{K}\,|\sin \omega t| = B|\sin \omega t| \tag{7-92}$$

which is plotted in Fig. 7-52 for $A = V^+ = K = 10$ V along with the original ac voltage. This rectification process is known as full-wave rectification in contrast to half-wave rectification where the output voltage in alternate half-cycles is zero. The dc component V_{dc} in the output voltage $v_o(t)$ is given as

$$V_{dc} \triangleq \frac{1}{2\pi}\int_0^{2\pi} v_o(t)\,d(\omega t) = \frac{1}{\pi}\int_0^{\pi} B\sin(\omega t)\,d(\omega t) = \frac{2B}{\pi} \tag{7-93}$$

Note from Fig. 7-52 that the rectified output is not a constant; rather it is a periodic wave composed of a dc part and an ac component. In an ideal rectifier

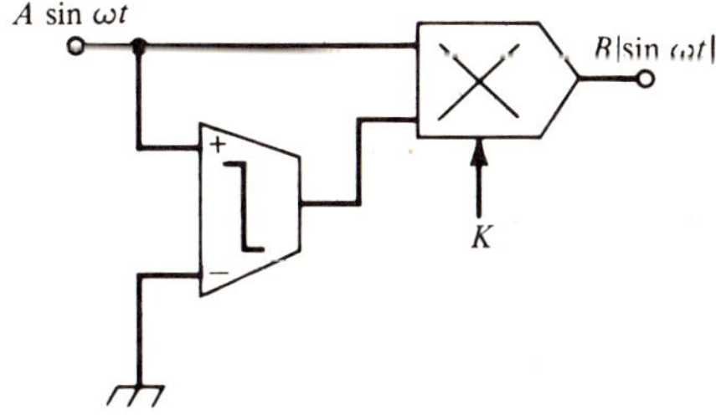

Figure 7-51   A full-wave rectifier.

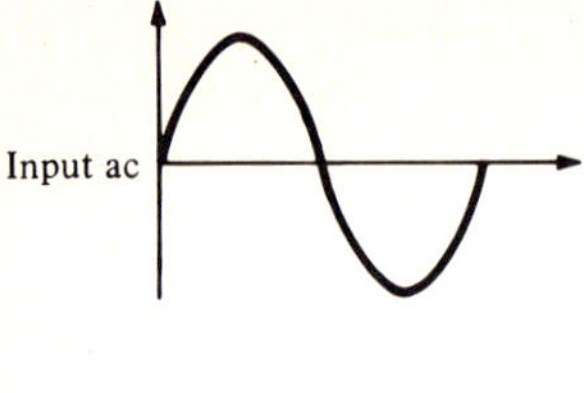

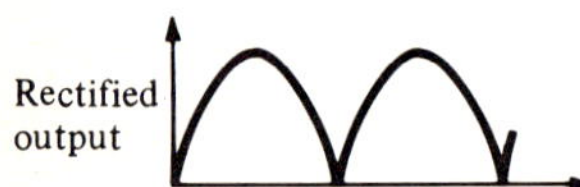

Figure 7-52   Input and output waveform of a full-wave rectifier.

the ac component should be zero. The ac component can be filtered out to obtain a better approximation to the dc. A measure of the rectification achieved is given by the *ripple factor r* defined as[21]

$$r \triangleq \frac{V'_{\text{rms}}}{V_{\text{dc}}} \tag{7-94}$$

where V'_{rms} is the rms value of the ac components in the output of the rectifier. Now

$$V'_{\text{rms}} = \left[\frac{1}{2\pi} \int_0^{2\pi} (v_o(t) - V_{\text{dc}})^2 \, d(\omega t) \right]^{1/2} \tag{7-95}$$

which after some algebra (Problem 7-50) leads to

$$r = \sqrt{\left(\frac{V_{\text{rms}}}{V_{\text{dc}}} \right)^2 - 1} \tag{7-96}$$

where V_{rms} is the rms value of the output of the rectifier. Observe for the full-wave rectifier

$$V_{\text{rms}} \triangleq \left[\frac{1}{2\pi} \int_0^{2\pi} v_o{}^2(t) \, d(\omega t) \right]^{1/2} = \left[\frac{1}{\pi} \int_0^{\pi} B^2 \sin \omega t \, d(\omega t) \right]^{1/2} = \frac{B}{\sqrt{2}} \tag{7-97}$$

As a result

$$\frac{V_{\text{rms}}}{V_{\text{dc}}} = \frac{B/\sqrt{2}}{2B/\pi} = 1.11 \tag{7-98}$$

Substituting above in Eq. (7-96) we thus obtain

$$r = [(1.11)^2 - 1]^{1/2} = 0.482 \tag{7-99}$$

It should be noted that the ripple factor for half-wave rectification is 1.21 (Problem 7-51) which is considerably higher than that of the full-wave rectification, as is to be expected.

A power supply should supply power at constant dc voltage from an ac line. The full-wave rectifier described earlier produces an output with ripples, that is, it contains ac components in addition to the dc component. In order to obtain a ripple-free voltage, a filter is used to eliminate the ac components as much as possible.[21]

7-7 Modulation and Demodulation

Signals whose frequency components are in the kiloHertz range or over can be efficiently transmitted over large distances by converting them into electromagnetic waves. On the other hand, low-frequency signals such as speech cannot be easily transmitted directly over a long distance. To circumvent this problem, a low-frequency signal is usually used to vary the amplitude or frequency or phase of a very high-frequency sinusoidal signal. This process is known as *modulation*. The modulated high-frequency signal is then transmitted. At the receiving end, the modulated high-frequency signal is *demodulated* to recover the low-frequency signal. In this section, we describe the basic principles of modulation and discuss in details a commonly used modulation process and its associated demodulation technique.

Basic Concepts[25]

Let $v_c(t) = E \cos(\omega_c t + \phi)$ be the high-frequency signal to be modulated. $v_c(t)$ is commonly known as the *carrier*, $f_c = \omega_c/2\pi$ is the carrier frequency in Hertz. In its unmodulated form the amplitude E, phase ϕ, and the carrier frequency ω_c are constants. Let $v_m(t)$ represent the low-frequency modulating signal. As indicated earlier, there are three basic ways $v_m(t)$ can be used to modulate the carrier $v_c(t)$.

If we vary the amplitude of the carrier as

$$E = E_c + \alpha v_m(t)$$

then the modulated carrier is given as

$$v_o(t) = \left[E_c + \alpha v_m(t)\right] \cos(\omega_c t + \phi) \tag{7-100}$$

and the corresponding process is known as *amplitude modulation*. For *phase modulation*, the phase ϕ of the carrier is varied as

$$\phi = \beta v_m(t)$$

resulting in a modulated carrier of the form

$$v_o(t) = E \cos\{\omega_c t + \beta v_m(t)\} \tag{7-101}$$

In the case of *frequency modulation*, the modulated carrier is of the form

$$v_o(t) = E \cos\left\{\omega_c t + \gamma \int v_m(t)\,dt\right\} \tag{7-102}$$

If we denote the argument of the cosine function above as $\psi(t)$, then we can define the instantaneous frequency of the modulated carrier as

$$\omega_i = \frac{d\psi}{dt} = \omega_c + \gamma v_m(t)$$

which clearly depicts how the carrier frequency is modulated by the modulating signal $v_m(t)$.

In the remaining part of this section we concentrate only on amplitude modulation and demodulation process. For a discussion of the other two types of modulation mentioned above, the interested reader is referred to References 28 and 29 listed at the end of this chapter. An analog signal can also be transmitted as pulses by sampling it at uniform intervals and coding the samples in digital form. This type of modulation is called the *pulse-code-modulation* (PCM) and is discussed elsewhere.[28,29]

Amplitude Modulation

To understand the effects of amplitude modulation (AM), let us assume first that the modulating signal is also sinusoidal in form [Fig. 7-53(a)]:

$$v_m(t) = E_m \cos \omega_m t \tag{7-103}$$

Substituting Eq. (7-103) in Eq. (7-100) and letting $\phi = 0$ we arrive at

$$v_o(t) = (E_c + \alpha E_m \cos \omega_m t) \cos \omega_c t = E_c(1 + m \cos \omega_m t) \cos \omega_c t \tag{7-104}$$

where we have set $m = \alpha E_m/E_c$. The parameter m is known as the *modulation index*. A typical plot of the modulated carrier given by Eq. (7-104) is shown in

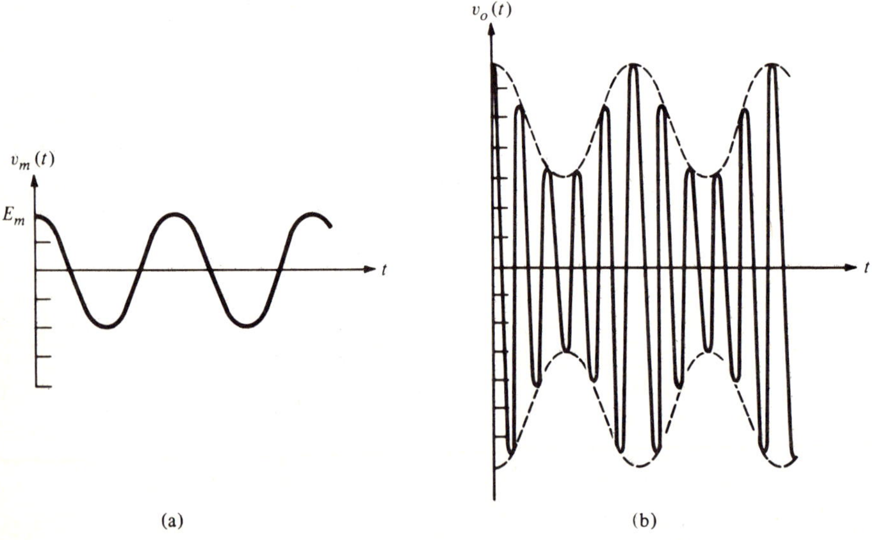

Figure 7-53   Illustration of the amplitude modulation process.

Fig. 7-53(b) from which it is seen that the envelope of the modulating carrier is a replica of the modulating signal. Note that to avoid distortion and possible loss of faithful recovery of the modulating signal at the receiving end, m is usually restricted to be a positive number between 0 and 1.

Using trigonometric identities, we can rewrite Eq. (7-104) as

$$v_o(t) = E_c \cos \omega_c t + \frac{mE_c}{2} \cos(\omega_c + \omega_m)t + \frac{mE_c}{2} \cos(\omega_c - \omega_m)t \quad (7\text{-}105)$$

It is seen from above that the modulated carrier is composed of the original carrier and two additional sinusoidal components of frequencies above and below the carrier frequency. These additional components are called *sidebands*. A plot of the amplitudes of the carrier and the sideband components is known as the *frequency spectrum* of the modulated carrier. Such a plot for Eq. (7-105) is sketched in Fig. 7-54.

In the case of a modulating signal containing sinusoidal components of various frequencies, the result is quite analogous. Each frequency component will generate a pair of sideband components of frequencies symmetrically situated on both sides of the carrier frequency and differing from the carrier frequency by an amount equal to the original frequency. For example, if the modulating signal contains n components of frequencies $\omega_1, \omega_2, \ldots, \omega_n$ with ω_n being the largest, then it can be shown that the *upper sideband* of the modulated carrier will contain n components of frequencies $\omega_c + \omega_1, \omega_c + \omega_2, \ldots, \omega_c + \omega_n$. Likewise, the *lower sideband* will also contain n components of frequencies $\omega_c - \omega_1, \omega_c - \omega_2, \ldots, \omega_c - \omega_n$. Thus the modulated carrier will have frequency components in the band $\omega_c - \omega_n$ to $\omega_c + \omega_n$. Or, in other words, the modulated carrier is of bandwidth $2\omega_n$ centered at ω_c and is of twice the bandwidth of the modulating signal (Fig. 7-55). In the case of a typical broadcast program, the audio signal is usually in the frequency band 100 Hz to 5 kHz. Thus the modulated carrier in AM radio transmission requires a bandwidth of 10 kHz. As a result each carrier frequency used by radio stations must be separated by at least 10 kHz.

A straightforward implementation of an amplitude modulator is with the aid of a multiplier and a summing amplifier. One such possible implementation

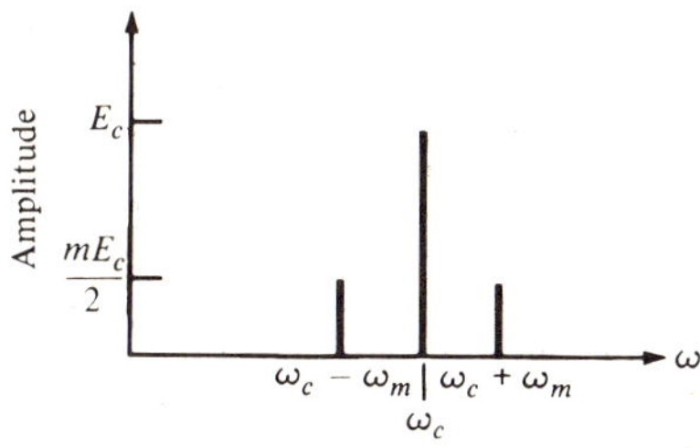

Figure 7-54 Frequency spectrum for amplitude modulation.

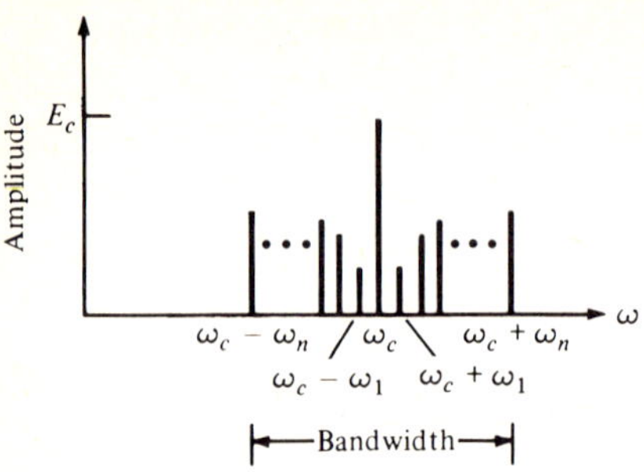

Figure 7-55   Frequency spectrum for frequency modulation.

is sketched in Fig. 7-56. Monolithic circuits specially designed for modulation and demodulation are also available. One such unit is the Motorola MC1596/MC1496 package. Figure 7-57(a) shows its use as an AM modulator circuit.

Let us now examine the energy contained in the modulated carrier. It is given by the sum of energies of the unmodulated carrier and the individual frequency components in the two sidebands. If the modulating signal contains only one sinusoidal frequency component, then from Eq. (7-105) we observe that the power of each of the sideband components is proportional to $m^2 E_c^2/8$ and that of the carrier is proportional to $E_c^2/2$. When $m = 1$ (i.e., 100 percent modulation), the power of each of the sideband components is one-fourth of that of the carrier. Thus of the total energy transmitted, two-thirds is in the carrier and only one-third is in the sidebands. If the modulation is less than 100 percent, the energy contained in the sidebands is proportionally less. Since the sidebands contain the information regarding the modulating signal, it is preferable to make the modulation index as close to one as possible so that the sidebands will contain the maximum amount of energy possible.

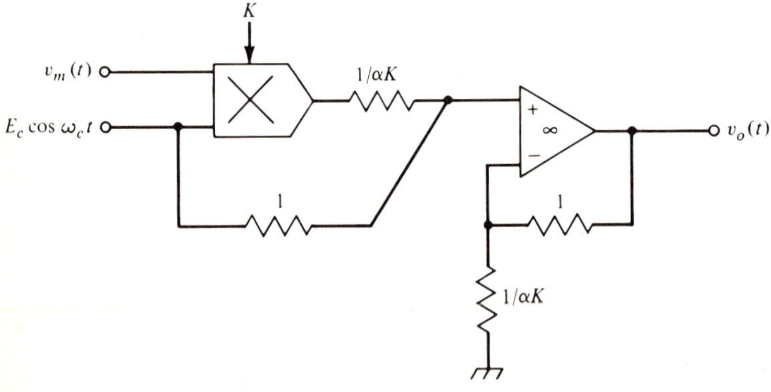

Figure 7-56   Amplitude modulation circuit.

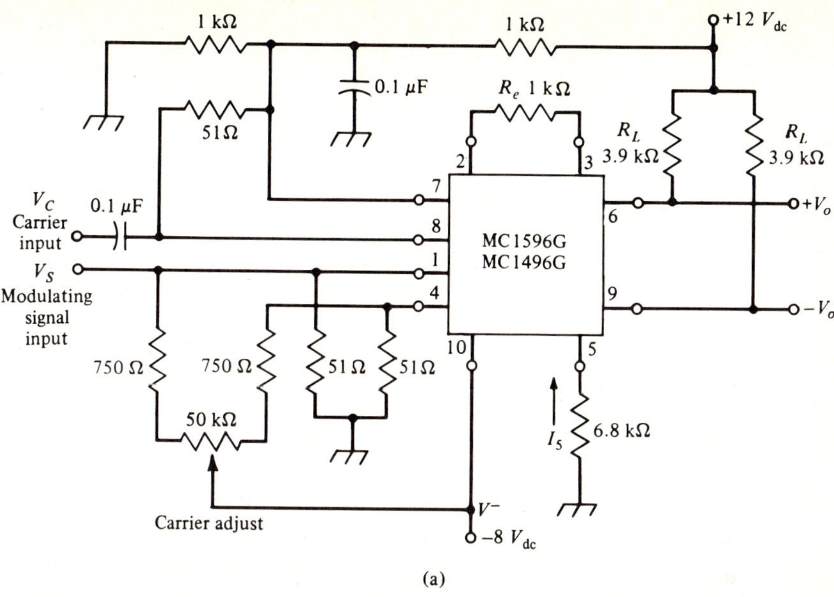

(a)

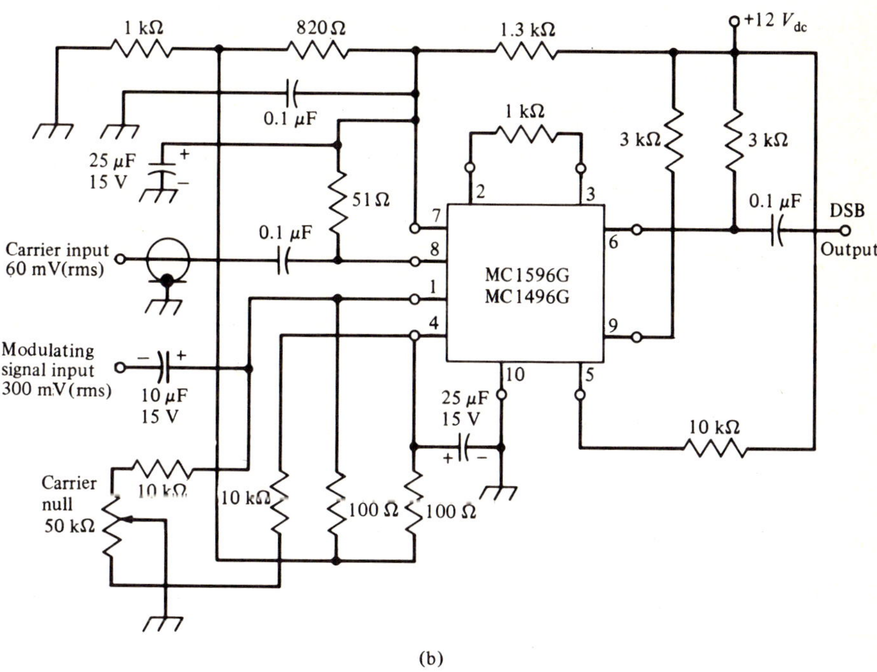

(b)

Figure 7-57 The use of a MC1596/MC1496 as (a) an amplitude modulator, and (b) a balanced modulator. (Courtesy Motorola Semiconductor Products Inc.)

Balanced Modulation

The carrier frequency component in the amplitude-modulated signal does not contain any information regarding the modulating signal but a large amount of the power transmitted belongs to the carrier component. Hence the suppression of the carrier is quite attractive for efficient transmission. This is achieved by removing the dc part of the amplitude from the previous modulation. The result of the modified modulation process, known as *balanced modulation*, is given as

$$v_o(t) = (E_m \cos \omega_m t)(E_c \cos \omega_c t)$$

$$= \frac{E_c E_m}{2} \cos (\omega_c + \omega_m)t + \frac{E_c E_m}{2} \cos (\omega_c - \omega_m)t \qquad (7\text{-}106)$$

The modulated signal now contains only the upper and the lower sideband components and does not contain the carrier. As a result, this process is also called a *suppressed-carrier modulation*. Normally, one of the sidebands is removed by filtering and the other sideband is transmitted, thus requiring considerably less power and less bandwidth than that needed in the transmission of an AM signal.

A multiplier by itself can be used as a balanced modulator by feeding in the carrier to one input terminal and the modulating signal to the other. The MC1596/MC1496 can also be used as a balanced modulator. The pertinent circuit connection is shown in Fig. 7-57(b).

Demodulation

The technique of recovering the modulating signal from the modulated carrier depends on the type of modulation being used. If the modulated carrier has been obtained by balanced modulation, then the demodulation is achieved simply by multiplying the modulated signal by a locally generated sinusoidal signal whose frequency is identical to that of the carrier frequency. If $v_o(t) = E_o \cos (\omega_c + \omega_m)t$ is the upper sideband component of the modulated signal and $v_y(t) = E_y \cos \omega_c t$ is the output of the local oscillator, then their product is given as

$$E_o E_y \cos (\omega_c + \omega_m)t \cdot \cos \omega_c t = \frac{E_o E_y}{2} \cos \omega_m t + \frac{E_o E_y}{2} \cos (2\omega_c + \omega_m)t \quad (7\text{-}107)$$

The product signal contains two components, one that is proportional to the original modulating signal and the other that is a very high-frequency component. The low-frequency signal can be recovered by passing the product signal through a low-pass filter of cut-off frequency $2\omega_c$ radians/sec. This type of demodulation is achieved readily by an analog multiplier or by an IC modulator of the type described above.

7-8 The Analog Switch[26]

The switch is conceptually the simplest of all electrical circuit elements. Its basic purpose is to either provide a conducting path between two parts of a circuit or to isolate the two parts. The connection or the disconnection is instituted by means of a control input. In an ordinary household ON–OFF switch, the control is provided mechanically. In an electronic analog switch, on the other hand, the control input is usually provided by a digital signal. With the recent availability of hybrid and monolithic IC units, analog switches are finding increasing applications in many system designs. In this section we review the properties of the ideal and the practical switches and outline several applications.

The Ideal Switch

The model of an ideal analog switch is as shown in Fig. 7-58. When the control signal is at one logic level, the switch is "closed" or ON providing a direct connection between the two analog signal terminals ① and ②. When the control signal is at the other logic level, the switch is "open" or OFF, disconnecting the two analog signal terminals. The resistance of the switch as measured between terminals ① and ② assumes two distinct values depending on whether the switch is ON or OFF. For an ideal switch, the ON resistance is zero and the OFF resistance is infinite.

The ideal switch is also assumed to connect or "turn-on" and disconnect or "turn-off" instantaneously. Or, in other words, the *turn-on* and *turn-off* times are zero, respectively. The circuit that interprets the digital control signal and causes the switch to turn on or off accordingly is called the *driver*. In a monolithic or hybrid IC switch the switching element and the driver are available in a single package. In an ideal switch there is also no coupling of the control signal logic level transitions to the analog signal path.

Switch Configurations

IC analog switches are available in a variety of configurations similar to those provided by the more conventional electromechanical relays. The switch configuration shown in Fig. 7-58 is of the *single-pole single-throw* (SPST) type. Other configurations available are the *single-pole double-throw* (SPDT), *double-pole single-throw* (DPST), and the *double-pole double-throw* (DPDT) types as shown in Fig. 7-59. Each of these configurations employs a single driver circuit. Both *break-before-make* and *make-before-break* switch actions are available in the IC

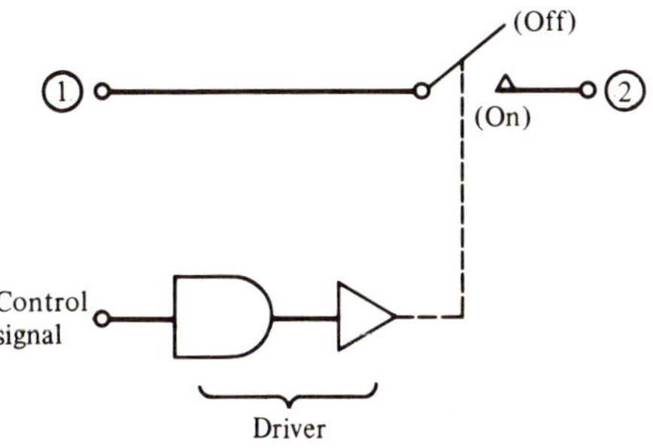

Figure 7-58   Model of an ideal single-pole single-throw (SPST) switch.

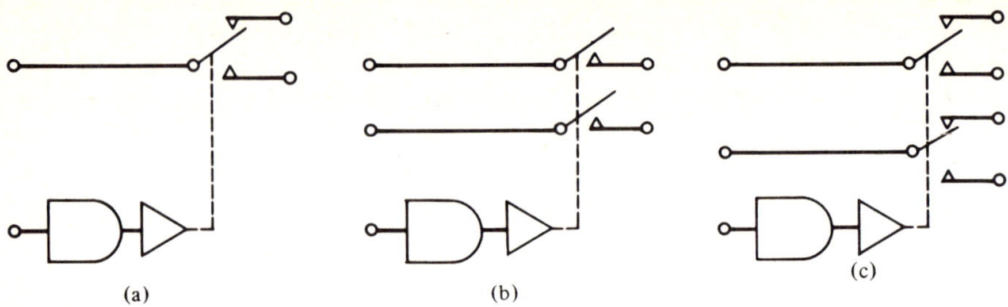

Figure 7-59　Other switch configurations: (a) SPDT, (b) DPST, and (c) DPDT.

packages. The switch packages usually contain one to four individual switch
and driver circuits. Switches connected to operate as a multiplexer (see Section
8-6) contain four to sixteen switch and driver circuits in a single package.

Two basic arrangements for inserting a switch in a circuit are the series and
the shunt arrangements shown in Fig. 7-60. In the series-connection scheme,
the analog signal appears across the load when the switch is ON; however, in
the shunt-connection case, the analog signal is transmitted to the load when the
switch is OFF. For an ideal switch there is no difference between the two
arrangements. However, as is shown later, in practice due to the nonideal properties of the switch, one arrangement may be more suitable than the other.

Characteristics of a Practical Switch

A practical switch is a nonideal device characterized by a number of parameters
that can be grouped essentially into three classes: (i) analog channel parameters,
(ii) digital control line parameters, and (iii) switching parameters. These parameters cause output voltage errors which if large enough may affect the performance of a circuit containing switches. An understanding of these parameters
and their effects on the system performance is necessary in order that a suitable
switch package may be selected for a particular application.

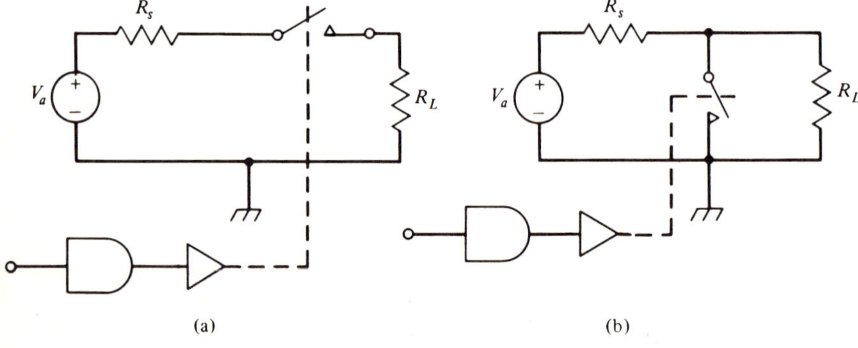

Figure 7-60　Basic switch arrangements: (a) series connection, and (b) shunt connection.

Analog Channel Parameters.   The resistances of a practical switch in the ON
and OFF positions are finite. Typically the ON resistance is in the range 10–100
Ω, and the OFF resistance is in the range 10^{10}–10^{12} Ω. These resistances along
with the resistance of the analog source and the load resistance seen by the
switch should be taken into account in analyzing the performance of a circuit
designed using analog switches. The ON resistance causes the output voltage
with the switch in the ON state to assume a value lower than the analog input
voltage. The OFF resistance, on the other hand, causes a small amount of output voltage to appear across the load instead of perfect isolation for the switch
in the OFF state. The following example illustrates the determination of output
voltage errors in both cases.

Example 7-7.   An analog switch with an ON resistance of 75 Ω and an OFF
resistance of 10^{10} Ω is connected in the series form. If the analog source
resistance is 40 Ω, determine the errors in the switch output voltage in the
ON and OFF conditions for two different load resistance values: (a) 2 kΩ
and (b) 100 kΩ.

The circuit model of the series connection for output voltage error analysis
is as shown in Fig. 7-61, where R_s is the analog source resistance, R_{sw} is the
switch resistance, and R_L is the load resistance seen by the switch. If the
switch is in ON state, $R_{sw} = R_{ON}$, the ON resistance of the switch, and V_o
$= V_{ON}$. When the switch is in the OFF state, $R_{sw} = R_{OFF}$, the OFF resistance
of the switch, and $V_o = V_{OFF}$. The output voltage in either case is obtained
using the voltage divider relation

$$V_o = \frac{R_L}{R_L + R_{sw} + R_s} \cdot V_a \qquad (7\text{-}108)$$

V_{ON} and V_{OFF} are determined by substituting R_{ON} and R_{OFF} for R_{sw}, respectively, in the above equation. Note that ideally V_{ON} should be equal to V_a and
V_{OFF} should be zero.

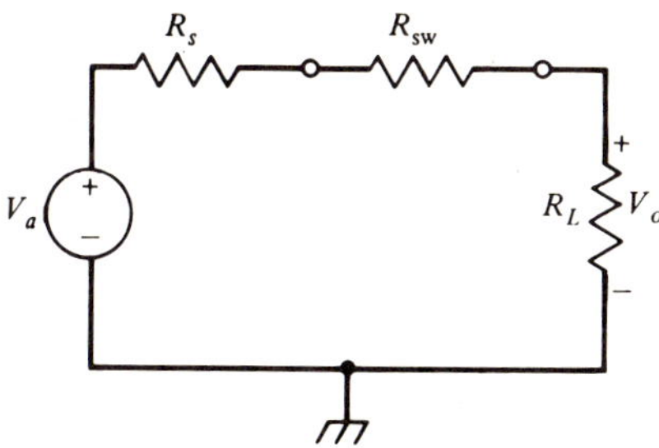

Figure 7-61   Circuit model for voltage error analysis.

The output voltage error in the ON state is given as

$$V_a - \frac{R_L}{R_s + R_{ON} + R_L} \cdot V_a = \frac{R_s + R_{ON}}{R_s + R_{ON} + R_L} \cdot V_a$$

and the relative error is therefore

$$\frac{100(R_s + R_{ON})}{R_s + R_{ON} + R_L} \text{ percent} \tag{7-109}$$

which, for our example, reduces to $100 \times 115/(115 + R_L)$ percent. For $R_L = 2\ \text{k}\Omega$, the relative error is 5.44 percent, and for $R_L = 100\ \text{k}\Omega$, the relative error is 0.115 percent.

Similarly the output voltage in the OFF state is

$$V_{OFF} = \frac{R_L}{R_s + R_{OFF} + R_L} \cdot V_a = \frac{R_L}{R_{OFF}} \cdot V_a \tag{7-110}$$

since $R_{OFF} \gg (R_s + R_L)$ for most practical switches. The *OFF isolation* is given by V_a/V_{OFF}, and is usually specified in decibels. Thus,

$$\text{OFF isolation in decibels} = 20 \log_{10} \frac{V_a}{V_{OFF}} = 20 \log_{10} \frac{R_{OFF}}{R_L} \tag{7-111}$$

For our example, $R_{OFF} = 10^{10}\ \Omega$. Thus, for $R_L = 100\ \text{k}\Omega$, the OFF-isolation is 100 dB and for $R_L = 2\ \text{k}\Omega$, OFF isolation is 134 dB.

As pointed out by the above example, both the ON voltage error and the OFF isolation decrease with increasing load resistance. We later describe alternate switch connections that can improve OFF isolation without increasing the ON voltage errors.

Another source of voltage error is caused by the leakage currents of the switching transistor. In most IC switches, the leakage currents are very small, in the range 1–5 nA. Their effects are negligible when the switch is ON. However, error caused by leakage in the OFF state may be appreciable. For example, a 5-nA leakage current flowing into a load of 10 kΩ will cause a 50 μV to appear across the load.

The voltage errors at dc or low frequencies are caused primarily by the switch resistances and the leakage currents. On the other hand, the voltage errors at high frequencies are caused mainly by the analog channel capacitance, capacitances from analog channel input and output terminals to ground and to the driver output, driver output impedance, and so on. A typical high-frequency model of a switch is as shown in Fig. 7-62. These capacitances, which are typically 5–20 pF, increase the ON voltage error and reduce considerably the OFF isolation at higher frequencies, thus limiting the frequency range of application of the switch. Effects of the capacitances can be determined using s-domain analysis (Problem 7-57).

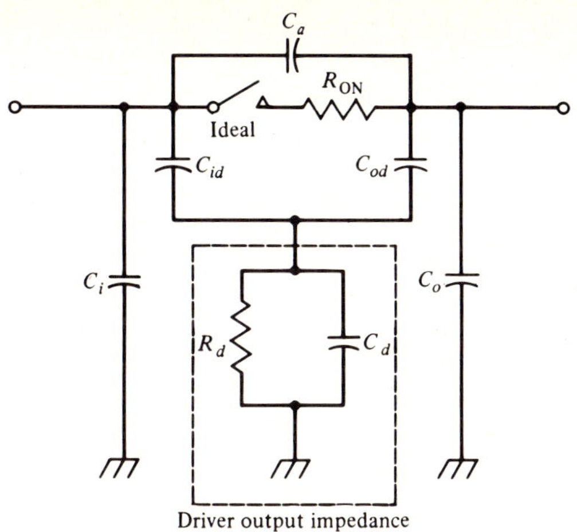

Figure 7-62 A typical high-frequency model of an analog switch.

Digital Control Line Parameters.   The parameters belonging to this class are the driver input thresholds and the driver input leakage currents. The two input thresholds are the maximum allowable voltage V_L and the minimum voltage V_H that can be recognized by the driver as LOW and HIGH voltage levels, respectively. These voltage levels determine the *logic compatibility* of the IC switch. Drivers designed to interface with TTL, DTL, or RTL logic circuits have typically $V_L = 0.8$ V and $V_H = 4.6$ V. The two input leakage currents of interest are the currents I_L and I_H measured at the driver input with the input at LOW and HIGH voltage levels, respectively.

Switching Parameters. These parameters affect the dynamic performance of the switch. Two important parameters in this category are the switch turn-on and turn-off times which are determined with the analog input voltage at a constant value and an *RC* load connected at the output. The *turn-on time t_{ON}* is given by the delay between the time the activating edge of the control input logic level crosses the threshold value (usually 50 percent) and the time the output is at 90 percent of its steady-state value with the switch ON. Likewise, the *turn-off time t_{OFF}* is defined as the delay between the time when the activating edge of the control input logic level crosses the threshold value and the time the output is at 10 percent of its steady-state value with the switch ON. Typical value of t_{ON} is 0.3 μsec and that of t_{OFF} is 1 μsec.

In practice, the digital switch control signal may also be coupled into the analog signal path inductively and capacitively at each instant the switch is turned ON or OFF causing a voltage spike to appear on the output. The initial amplitude of the spike and its exponential decay time depends on the value of the line resistances and the capacitances.

Improving Isolation

As indicated earlier, because of the nonideal properties of a practical switch, for a given switch either the series or the shunt connection arrangement of Fig. 7-60 may be more suitable, depending on the source and load. To illustrate this, consider the switch of Example 7-7 with ON and OFF resistances of 75 and 10^{10} Ω, respectively. Let the analog source resistance be 40 Ω and the load resistance be 100 kΩ. The OFF isolation for the series connection, as worked out in Example 7-7, is 100 dB. To determine the OFF isolation in the shunt case, we first observe that the load is " disconnected " from the source when the switch is ON. Then the load R_L is shunted by R_{ON} and the effective load resistance is given by the parallel combination of R_L and R_{ON}. Since $R_{ON} \ll R_L$, the effective load resistance is essentially R_{ON}. As a result, in the shunt arrangement

$$\text{OFF isolation in decibels} = 20 \log_{10} \frac{V_a}{V_{OFF}} = 20 \log_{10} \frac{R_s + R_{ON}}{R_{ON}} \quad (7\text{-}112)$$

which for our example reduces to 3.7 dB. Obviously, with regard to the OFF isolation the series arrangement is far superior to the shunt connection in this particular case.

Additional improvement in OFF isolation can be obtained by using more than one switch. Two widely used multiswitch arrangements are shown in Fig. 7-63. Here the switches S_1 and S_2 are driven in antiphase. Thus, when S_1 is ON, S_2 is OFF and vice versa.

In the series-shunt arrangement, the effective ON characteristic of the overall combination is essentially same as that of the basic series arrangement, whereas, in the OFF mode, the effective load resistance is almost equal to the ON resistance of the switch S_2, thus improving the OFF isolation considerably [see

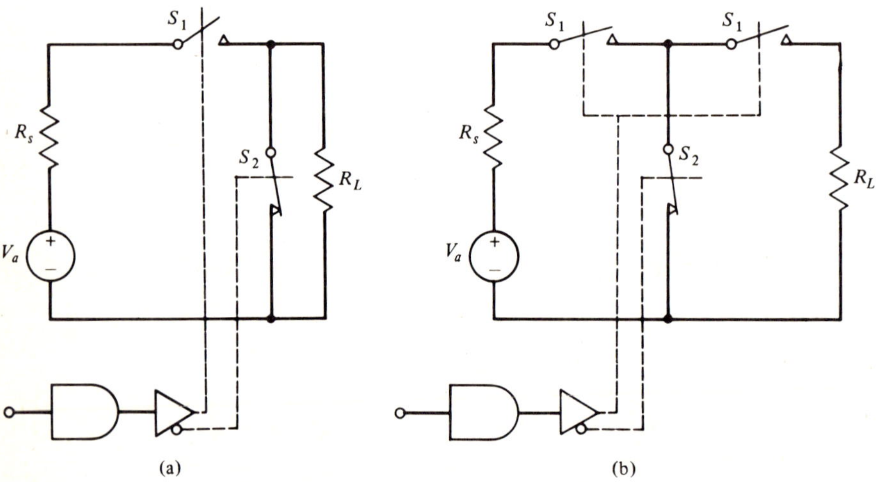

(a)   (b)

Figure 7-63   (a) Series-shunt arrangement, and (b) TEE arrangement.

Eq. (7-111)]. This arrangement can be used provided the shunting of the load in the OFF mode can be tolerated.

The TEE configuration improves the OFF isolation further without shunting of the load. However, in the ON mode, two switches are in series doubling the effective ON resistance which in turn increases the ON voltage error.

Types of Switches

IC analog switches invariably use field-effect-transistor (FET) as the switching element because of their very high OFF-to-ON resistance ratio. Three different technologies are usually used to fabricate the switch: n-channel junction FET (JFET), p-channel metal-oxide-semiconductor FET (PMOS), and complementary metal-oxide-semiconductor FET (CMOS).

A JFET switch is characterized with a constant and low ON resistance independent of variations of the input analog or the supply voltage. However, it

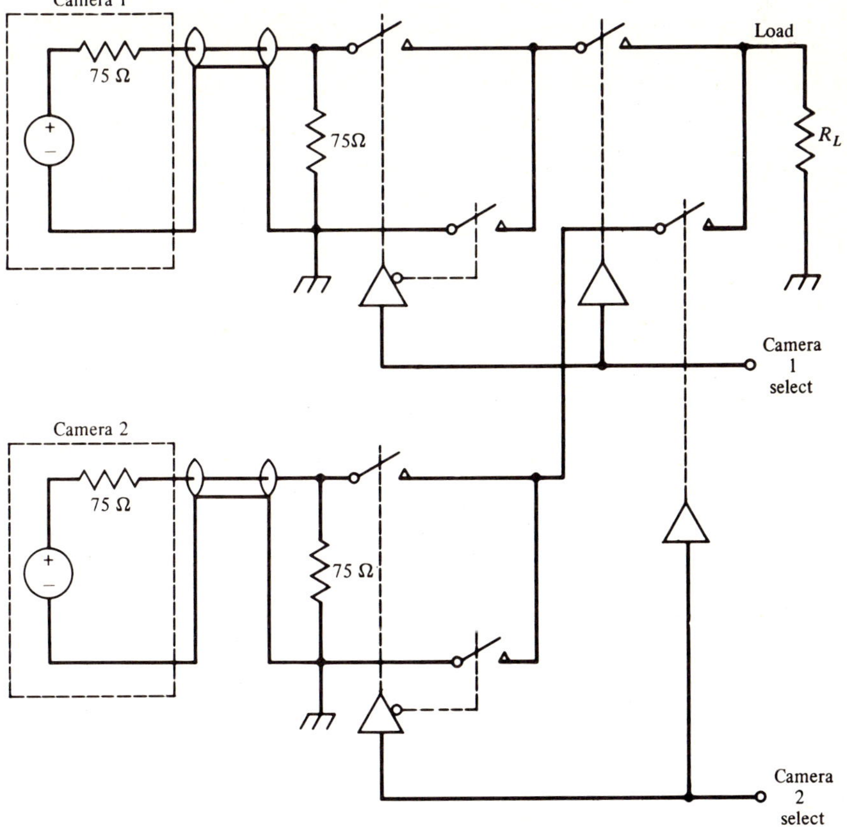

Figure 7-64 A video switch.

requires a fairly complex driver circuit. The input analog signal here must be unipolar in the range 0 to about -7.5 V.

The PMOS switch is easier and cheaper to fabricate. It also restricts the input analog signal to be negative in the range 0 to about -10 V. A serious drawback of a PMOS switch is the high dependence of the value of the ON resistance on the analog and supply voltages. As a result, it can be used in applications where variation of the ON resistance is not very critical to the performance of the system.

A parallel combination of PMOS and NMOS transistors are used in designing the CMOS switch which allows the handling of bipolar analog signals. Here the ON resistance is fairly independent of the analog voltage but varies slightly with the supply voltage. Overall power dissipation is considerably lower compared to the other two types of switches.

Some Applications

Analog switches are used in a variety of applications. An important application of the switch is in routing signals in a telephone exchange. It is also the key

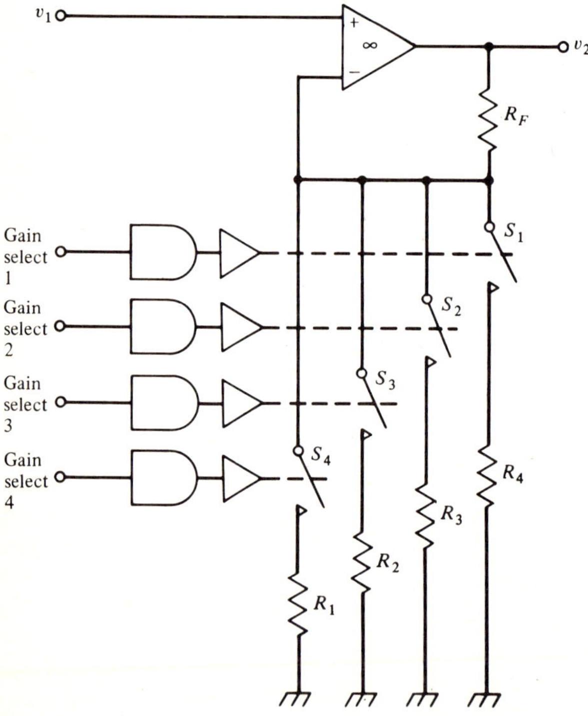

Figure 7-65 A noninverting amplifier with programmable gain and $R_1 = 1\text{ k}\Omega$, $R_2 = 2\text{ k}\Omega$, $R_3 = 4\text{ k}\Omega$, $R_4 = 8\text{ k}\Omega$, and $R_F = 8\text{ k}\Omega$.

element in the interface circuit between analog and digital systems. Such applications are considered in the next chapter. Here we outline several other circuits employing the switch as a component.

Video Switch.   Analog switches are often used for switching a video monitor between several cameras. A possible implementation of a video switch with very high OFF isolation is sketched in Fig. 7-64 whose operation should be self-evident.

Programmable Gain Amplifier.   Amplifiers with digitally controllable gain can be easily constructed by switching in different gain setting resistors of a non-inverting or inverting type amplifier. For example, Fig. 7-65 shows a non-inverting amplifier with sixteen different gain options that are digitally programmable. The gain values available are 1 through 16 at increments of unity by closing and opening switches S_1, S_2, S_3, and S_4 in various combinations.

Active Filters with Programmable Characteristics.   By switching in different resistors and/or capacitors, filters with adjustable characteristics can be implemented. An illustration of such a design is provided in Fig. 7-66 which shows a first-order low-pass filter with digitally programmable 3-dB cut-off frequencies. For the element values shown, the four different cut-off frequency options available are as shown in Table 7-1.

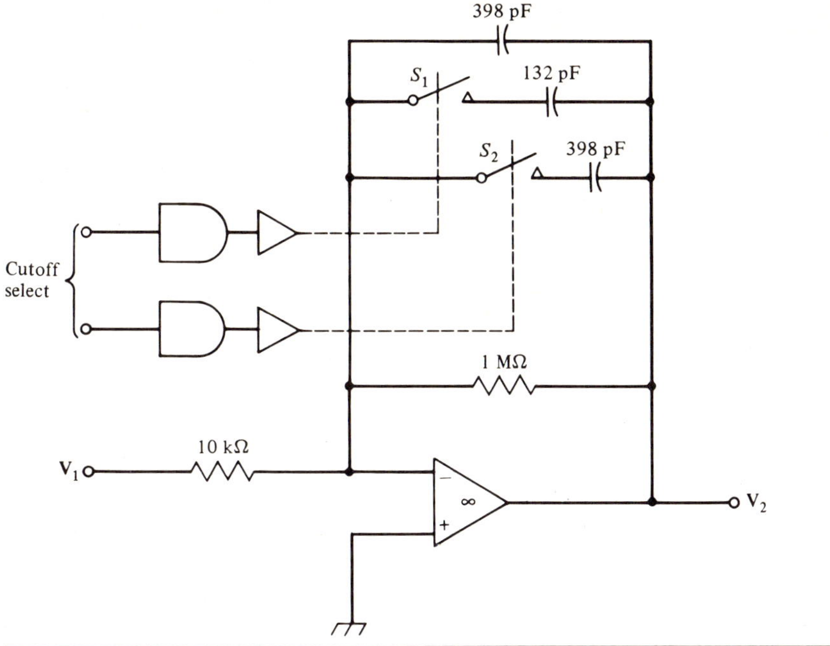

Figure 7-66   Low-pass filter with programmable cut-off.

TABLE 7-1

S_1	S_2	Cut-off Frequency (hertz)
ON	ON	171.5
OFF	ON	200
ON	OFF	300
OFF	OFF	400

Chopper Stabilized Amplifier.   Amplification of analog signals with very slowly varying amplitudes is difficult with conventional ac amplifiers because they would require extremely large coupling capacitances. As a result, it is preferable to use dc amplifiers with very small temperature and long-term drifts. One approach to the design of such a low-drift dc amplifier is to use an analog switch to sample or " chop " periodically the slowly varying analog signal into a train of pulses whose amplitude is equal to that of the analog signal at sampling instants. The high-frequency pulse train is then amplified by a low-drift ac amplifier. A second analog switch at the output of the ac amplifier operating synchronously with the switch at the input demodulates the amplified pulse train. A low-pass filter at the output of the second switch develops an amplified version of the analog input. A block-diagram representation of the complete arrangement is shown in Fig. 7-67. Here the analog switches determine the over-all chopper stabilized amplifier system offset and drift. Monolithic chopper stabilized amplifiers are now commercially available.

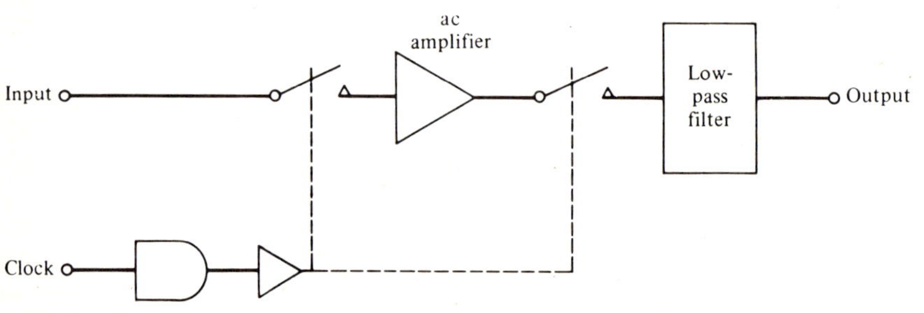

Figure 7-67   Block-diagram representation of a chopper-stabilized amplifier.

7-9   Summary

Several frequently used nonlinear function modules are introduced in this chapter and a number of applications of these devices are described. A major part of this chapter has been devoted to the analog multiplier which is finding increasing applications in system design. The ideal analog multiplier is defined in Section 7-1. The multiplier is used to perform arithmetic operations such as division and generations of integer and fractional powers which are outlined in this section along with schemes to develop sinusoidal signals of frequencies which are integer multiples of that of the input sinusoidal signal. The properties of a practical multiplier module are described in Section 7-2. Several additional applications of the multiplier unit such as rms measurement, function fitting, and so on, are considered in Section 7-3. Section 7-4 is concerned with the log and antilog amplifiers and their applications. Generation of trigonometric functions such as sine and cosine functions and their applications are the subject of Section 7-5. Section 7-6 deals with the analog comparator and its applications. The basic concepts of modulation and demodulation are described in Section 7-7. Emphasis has been placed here on amplitude modulation process which can be simply implemented using an analog multiplier. Finally Section 7-8 describes the operation of the analog switch and outlines several of its applications.

References

1. "Evaluating, Selecting, and Using Multiplier Circuit Modules for Signal Manipulation and Function Generation," Analog Devices, Norwood, Mass., 1970.
2. A. Bilotti, "Applications of a Monolithic Analog Multiplier," *IEEE J. Solid-State Circuits*, Vol. SC-3, December 1968, pp. 373–380.
3. R. S. Burwen, "A Complete Multiplier/Divider on a Single Chip," *Analog Dialogue*, Vol. 5, no. 1, January 1967, pp. 3–5, 15.
4. R. Burwen, "Save Money with Analog Multipliers," *Electronic Design*, Vol. 19, no. 7, April 1, 1971, pp. 44–47.
5. R. S. Burwen, "Linearize Almost Anything with Multipliers," *Electronic Design*, Vol. 19, no. 8, April 15, 1971, pp. 74–75.
6. T. Cate, "Separate the Signals from the Noise," *Electronic Design*, Vol. 18, no. 25, December 6, 1970, pp. 60–68.
7. T. Cate, "Extraction of Square Roots—A Useful Analog Instrumentation Technique," *Electronic Instrument Digest*, January 1971, pp. 7–11.
8. T. Cate and H. Handler, "Designing with Packaged Analog Mulipliers," *EEE Magazine*, May 1969.
9. B. Gilbert, "A Precise Four-Quadrant Multiplier with Subnanosecond Response," *IEEE J. Solid-State Circuits*, Vol. SC-3, December 1968, pp. 365–373.
10. A. S. Jackson, *Analog Computation*, McGraw-Hill Book Co., New York, 1960.
11. G. A. Korn and T. M. Korn, *Electronic Analog and Hybrid Computers*, McGraw-Hill Book Co., New York, 1964, pp. 254–306.
12. W. R. Perkins and J. B. Cruz, Jr., *Engineering of Dynamic Systems*, John Wiley & Sons, New York, 1969, Chap. 4.
13. R. Strata, "Don't Be Fooled by Multiplier Specs," *Electronic Design*, Vol. 19, no. 6, March 15, 1971, pp. 66–70.
14. B. Welling and L. Kinsey, "Using the MC1595 Multiplier in Arithmetic Operations," Application Note AN-490, Motorola Semiconductor Products Inc., Phoenix, Ariz.
15. Engineering Staff (D. H. Sheingold, Ed.), *Nonlinear Circuits Handbook*, Analog Devices, Inc., Norwood, Mass., 1974.

16. J. G. Graeme, G. E. Tobey, and L. P. Huelsman, Eds., *Operational Amplifiers: Design and Applications*, McGraw-Hill Book Co., New York, 1971, Chap. 7.

17. "The Digital Logic Handbook," Digital Equipment Corporation, Maynard, Mass., 1967 Ed., part V.

18. J. N. Giles, Ed., "Fairchild Semiconductor Linear Integrated Circuits Applications Handbook," Fairchild Semiconductor, Mountain View, Calif., 1967.

19. D. F. Hoeschele, Jr., *Analog-to-Digital/Digital-to-Analog Conversion Techniques*, John Wiley & Sons, New York, 1968, pp. 222–268.

20. H. V. Malmstadt and C. G. Enke, *Digital Electronics for Scientists*, W. A. Benjamin, New York, 1969.

21. J. Millman, *Vacuum-tube and Semiconductor Electronics*, McGraw-Hill Book Co., New York, 1958, pp. 337–346, 499–526.

22. "The Application of Linear Microcircuits," SGS-Fairchild, London, 1966.

23. M. E. Van Valkenburg, *Network Analysis*, 2nd Ed., Prentice-Hall, Englewood Cliffs, N.J., pp. 401–416.

24. R. J. Widlar, "The Operation and Use of a Fast Integrated Circuit Comparator," Application Bulletin APP-116, Fairchild Semiconductor, Mountain View, Calif., 1966.

25. P. M. Chirlian, *Electronic Circuits: Physical Principles, Analysis, and Design*, McGraw-Hill Book Co., New York, 1971.

26. Staff, *Analog Switches and Their Applications*, Siliconix Inc., Santa Clara, Calif., 1976.

27. H. Taub and D. Schilling, *Digital Integrated Circuits*, McGraw-Hill Book Co., New York, 1977, Chap. 13.

28. D. J. Sakrison, *Communication Theory: Transmission of Waveforms and Digital Information*, John Wiley & Sons, New York, 1968.

29. A. B. Carlson, *Communication Systems: An Introduction to Signals and Noise in Electrical Communication*, 2nd Ed., McGraw-Hill Book Co., New York, 1975.

Problems

7-1   Convert a 4-quadrant multiplier to a 2-quadrant device using an absolute-value circuit.

7-2   Implement a 4-quadrant multiplier using two 2-quadrant multipliers in accordance with Eq. (7-2).

7-3   Design a circuit to implement $v_o = (v_{x1} - v_{x2})(v_{y1} - v_{y2})/K$ where v_{x1}, v_{x2}, v_{y1}, and v_{y2} are analog voltages.

7-4   Design an odd-function squarer using a multiplier and an absolute-value circuit.

7-5   Design an odd-function squarer using two 1-quadrant squarers and operational amplifiers.

7-6   Implement a frequency tripler using the identity of Eq. (7-4b).

7-7   Design a frequency quadrupler using the trigonometric identity

$$\cos 4\omega t = 8\,(\cos \omega t)^4 - 8\,(\cos \omega t)^2 + 1 \tag{7-113}$$

7-8   Design a harmonic generator whose output signal is the fifth harmonic of the input signal. Use the trigonometric identity

$$\cos 5\omega t = 16\,(\cos \omega t)^5 - 20\,(\cos \omega t)^3 + 5\,(\cos \omega t) \tag{7-114}$$

7-9 Analyze the operation of the square-rooter circuit of Fig. 7-8(b).

7-10 Analyze the circuit of Fig. 7-10 and determine the expression for the output voltage v_o as a function of the input voltage v_z. Is there any restriction on the polarity of v_z for this circuit?

7-11 Design a suitable circuit to detect the phase difference θ between two sinusoidal signals $V_x \sin \omega t$ and $V_y \sin (\omega t + \theta)$.

7-12 Realize the circuits shown in Fig. 7-68 using a multiplier and operational amplifiers.

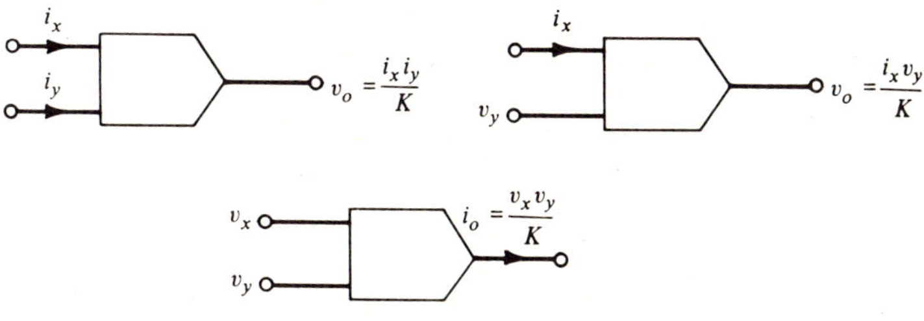

Figure 7-68

7-13 Determine the accuracy of the square-rooter circuit based on the design of Fig. 7-8 (a) as a function of the accuracy of the multiplier. Plot the accuracy of the squarer as a function of the input voltage assuming that the constituent multiplier has a 0.25 *percent* accuracy with ± 10 V full-scale range. Assume the two resistors, R_1 and R_2, to be equal.

7-14 Determine the worst-case outputs of a 0.5 percent accuracy multiplier having a 10 V full-scale range for $v_x = 10$ V and v_y taking the following values: 2 V, 4 V, 6 V, 8 V, and 10 V.

7-15 The worst-case error measured for a multiplier having a 15 V full-scale range is ± 5 mV. What is the accuracy of the multiplier?

7-16 A divider is designed using a 0.5 percent accuracy multiplier and equal matched summing resistors. The full-scale voltage of the multiplier is 10 V. Determine the worst-case outputs for a numerator signal $v_z = 10$ V and the denominator signal v_x taking the values: 2 V, 4 V, 6 V, 8 V, and 10 V.

7-17 Determine the worst-case outputs of a square-rooter for the following input signals: 2 V, 4 V, 6 V, 8 V, and 10 V. The multiplier used is of 0.5 percent accuracy and has a 10 V full-scale range. Summing resistors are assumed to be equal.

7-18 Calculate the peak offset error of a multiplier with ± 50 mV/V power supply rejection. The ± 15 V power supply supplied has a 60-Hz ripple of rms magnitude of 0.04 V on each supply.

7-19 Determine the time t_0 at which the impulse response defined in Eq. (7-28) is down to 0.1 percent of its value at $t = 0$.

7-20 Obtain a simplified realization of a bandpass filter with only voltage-controlled resonant frequency from the circuit of Fig. 7-17.

7-21 Show that the circuit of Fig. 7-17 can be used as a voltage-variable low-pass filter if the output is chosen as V_3.

7-22 The circuit of Fig. 7-69 can be used either as a first-order low-pass or high-pass filter with voltage variable characteristic. Analyze the circuit and determine the transfer functions V_2/V_1 and V_3/V_1.

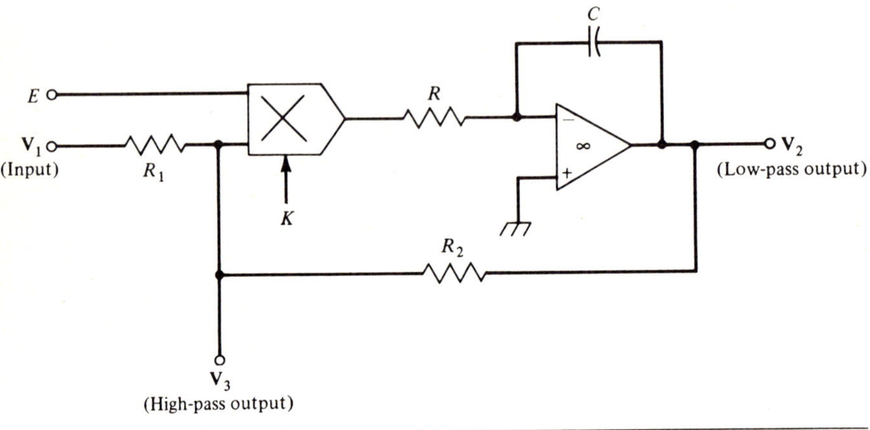

Figure 7-69

7-23 An alternate bridge linearization circuit can be implemented by rewriting Eq. (7-49) as[15]

$$\frac{\varepsilon}{2} = \frac{v_o}{v_o - V_{\text{ref}}} \qquad (7\text{-}115)$$

Implement Eq. (7-115).

7-24 Implement the equation

$$y = y_0 + a_1 x - a_2 x^2 + a_3 x^3 - a_4 x^4$$

where x and y are the input and output voltages and all coefficients (y_0, a_1, a_2, a_3, and a_4) are positive.

7-25 Implement

$$y = y_o + \frac{a_1 x + a_2 x^2}{1 - b_1 x}$$

using a single multiplier and at most three operational amplifiers.

7-26 Develop a circuit to generate $\tan x$ where x is an analog voltage.

7-27 Develop a suitable diagram for solving the Van der Pol's equation[10]

$$\frac{d^2 x}{dt^2} - \varepsilon(1 - x^2)\frac{dx}{dt} + x = 0$$

by analog simulation.

7-28 Develop a suitable diagram for solving the Rayleigh's equation[10]

$$\frac{d^2x}{dt^2} - \varepsilon\left[1 - \left(\frac{dx}{dt}\right)^2\right]\left(\frac{dx}{dt}\right) + x = 0$$

by analog simulation.

7-29 Analyze the antilog amplifier circuit of Fig. 7-70.

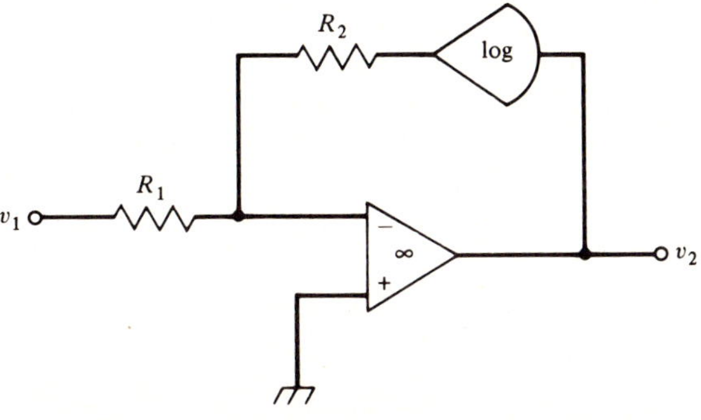

Figure 7-70

7-30 Implement a circuit to develop the logarithm of the product of two analog signals.

7-31 Analyze the operation of the temperature compensated antilog amplifier of Fig. 7-71.[16]

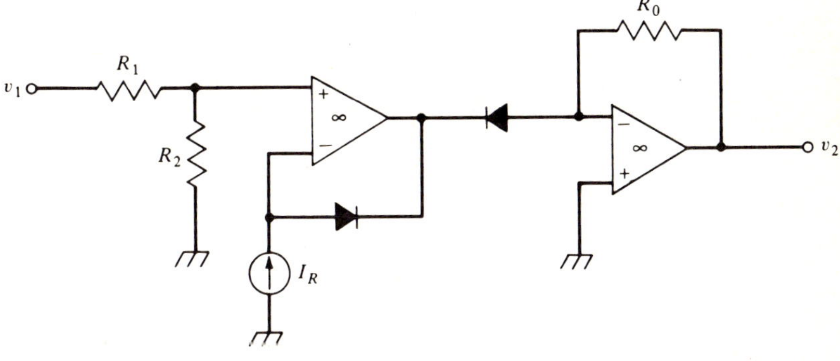

Figure 7-71   Antilog amplifier with temperature compensation.

7-32 The generation of X^Y can be achieved by first forming the product of Y and $\log X$ and then taking the inverse logarithm of the product. Design a circuit to generate X^Y based on this approach.

7-33 Design a vector-sum circuit implementing Eq. (7-56) using multipliers and operational amplifiers.

7-34 Design a vector-difference circuit to implement the equation

$$v_o = \sqrt{v_a{}^2 - v_b{}^2}$$

using a $Y(Z/X)^m$ module.

7-35 The gas flow F through an orifice or a nozzle can be measured by knowing the absolute temperature T, the absolute pressure P, and the drop in pressure ΔP. The pertinent equation is given as[15]

$$F = \alpha\left(1 - \beta\frac{\Delta P}{P}\right)\sqrt{\frac{P \cdot \Delta P}{T}} \tag{7-116}$$

Obtain a circuit realization of Eq. (7-116).

7-36 Design a circuit to measure the true output power P of a three-phase generator. The output voltages and currents of the generator are given as

$$v_\kappa(t) = V_m \sin\left(\omega t + \frac{2\kappa\pi}{3}\right)$$

$$\kappa = 1, 2, 3$$

$$i_\kappa(t) = I_m \sin\left(\omega t + \frac{2\kappa\pi}{3} + \phi\right)$$

The output power P is given as

$$P = \sum_{\kappa=1}^{3} v_\kappa(t) \cdot i_\kappa(t)$$

7-37 A proposed approximation to the tangent of an angle x is given as[15]

$$\tan x \cong \frac{2.81065\left(\dfrac{x}{\pi/2}\right) - \left(\dfrac{x}{\pi/2}\right)^2}{1.81065 - 0.81065\left(\dfrac{x}{\pi/2}\right) - \left(\dfrac{x}{\pi/2}\right)^2} \tag{7-117}$$

Show that for the above approximation the relative error is less than 0.1 percent for x less than 72°, less than 0.5 percent for x less than 81°, and less than 1.2 percent for all values of x up to 90°. Implement the approximation using multipliers and operational amplifiers.

7-38 Implement a sine module using each of the three approximations of the sine function given by Eqs. (7-71), (7-73), and (7-74). Assume the multipliers to be used in the circuit realizations have a ± 10 V full-scale range.

7-39 Design the resolver described by Eq. (7-76).

7-40 The positive and negative output levels of a comparator are $+4.3$ and -2.1 V, respectively. Draw the transfer characteristics of the comparator for the following input conditions: **(a)** Inverting input held at 3 V, **(b)** inverting input held at -4 V, **(c)** noninverting input held at 3 V, and **(d)** noninverting input held at -4 V.

7-41 Show graphically that to keep the resolution constant, an increase in output voltage swing must be accompanied by an increase in gain of the comparator.

7-42 The input offset voltage of a comparator is 3 mV. If the gain of the comparator is 950, by what amount should the output differ from the threshold level if the two inputs are at the same level? Assume that the input bias and offset currents are negligible.

7-43 The input bias and offset currents of a comparator are 30 and 1.5 μA, respectively, at room temperature. If the input resistors are 100 and 200 Ω, what is the effective input offset error voltage due to the input error currents?

7-44 If the input error currents are 29 and 30 μA, calculate the approximate values of the input bias current and input offset current.

7-45 The input offset current of μA710 is a decreasing function of temperature. If the offset current at 20°C is 0.8 μA and the average temperature coefficient of the offset current is 5 nA/°C, what would be the value of the input offset current at 100°C?

7-46 The input voltage range of a comparator is ± 10 V and the differential input voltage range is ± 7 V. If V_1 is the voltage at the noninverting input terminal and V_2 is that at the inverting input terminal, which of the following input level pairs are admissible:

(a) $V_1 = 10$, $V_2 = 4$ (b) $V_1 = 10$, $V_2 = -4$
(c) $V_1 = 7$, $V_2 = 1$ (d) $V_1 = 7$, $V_2 = -1$
(e) $V_1 = 4$, $V_2 = 8$ (f) $V_1 = -4$, $V_2 = 8$
(g) $V_1 = 0.006$, $V_2 = 0.004$ (h) $V_1 = 0.006$, $V_2 = -0.004$
All values in volts.

7-47 If the output levels of the comparator of Problem 7-46 are $+6$ and -3 V, and the resolution is 3 mV, determine the output voltage for each of the admissible pairs of input voltages in Problem 7-46.

7-48 Show that by making $V_{\text{ref}} < 0$ in the circuit of Fig. 7-47 a positive trigger pulse will produce a negative output pulse.

7-49 Show that the circuit of Fig. 7-50 is a free-running multivibrator.

7-50 Derive Eq. (7-96).

7-51 Compute the dc value and the rms value of the output of a half-wave rectifier and show that the ripple factor for this case is 1.21.

7-52 Rectification can also be achieved by feeding the ac input to a squarer. Calculate the ripple factor for this case and compare it with that of the full-wave rectifier of Fig. 7-51.

7-53 The Fourier series representation of a periodic signal $x(t)$ is given as

$$x(t) = \beta_0 + \sum_{m=1}^{\infty} \beta_m \cos m(\omega t) + \sum_{m=1}^{\infty} \alpha_m \sin m(\omega t) \tag{7-118}$$

where the coefficients α_m, β_m are given by

$$\beta_0 = \frac{1}{2\pi} \int_0^{2\pi} x(t)\, d(\omega t) \tag{7-119}$$

$$\beta_m = \frac{1}{\pi} \int_0^{2\pi} x(t) \cos m(\omega t)\, d(\omega t) \tag{7-120}$$

$$\alpha_m = \frac{1}{\pi} \int_0^{2\pi} x(t) \sin m(\omega t)\, d(\omega t) \tag{7-121}$$

Show $\alpha_m = 0$ for an even periodic function for which $x(t) = x(-t)$. Likewise, show that for an odd periodic function, $\beta_0 = 0$ and $\beta_m = 0$.

7-54 Using the results of Problem 7-53, show that in the Fourier series representation of the output $v_o(t)$ of a full-wave rectifier is given as

$$v_o(t) = \frac{2B}{\pi} - \frac{4B}{\pi}\left[\tfrac{1}{3}\cos 2\omega t + \tfrac{1}{15}\cos 4\omega t + \tfrac{1}{35}\cos 6\omega t + \cdots\right]$$

7-55 A low-frequency signal $V_x(t) = V_1 \cos \omega_1 t + V_2 \cos \omega_2 t + V_3 \cos \omega_3 t$ is used to modulate a carrier $V_c \cos \omega_c t$ by means of a balanced modulator. Determine the lower and upper sideband components in the output of the modulator.

7-56 An analog switch is to be inserted between an analog source of internal resistance $100\ \Omega$ and a load resistance of $50\ \Omega$. Compare the performances of the series, shunt, series-shunt, and TEE-arrangements if the analog switch to be used has an ON resistance of $50\ \Omega$ and an OFF resistance of $5\ M\Omega$.

7-57 A simplified ON state equivalent circuit of any of the four switch arrangements is as shown in Fig. 7-72. Determine the ON transfer function $T_{ON}(s) = V_{ON}/V_i$. Show that the 3-dB cut-off frequency ω_c is given as

$$\omega_c = \frac{R_L + R_{ON}}{R_L R_{ON}(C_{ON} + C_L)}$$

Determine the cut-off frequency in hertz for the following element values:

$$R_{ON} = 25\ \Omega, \qquad C_{ON} = 30\ pF, \qquad R_L = 75\ \Omega, \qquad C_L = 10\ pF$$

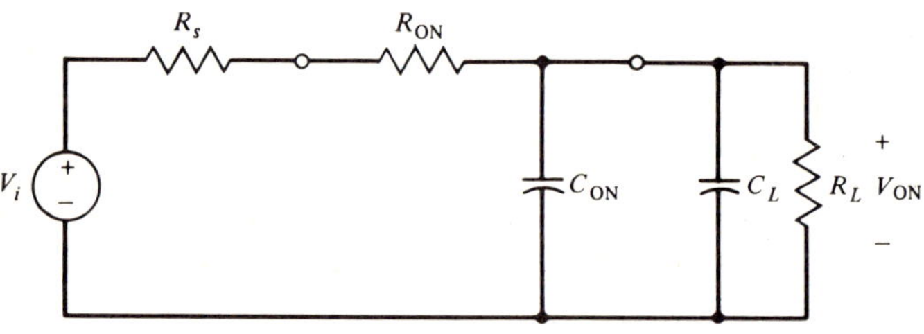

Figure 7-72 Simplified ON equivalent circuit.

7-58 A simplified dc equivalent circuit of the series switch arrangement incorporating the leakage currents is sketched in Fig. 7-73. Determine the ON voltage error and the OFF isolation for the following component values:

$$R_{ON} = 50\ \Omega, \quad R_s = 10\ \Omega, \quad R_L = 50\ k\Omega, \quad I_{li} = I_{lo} = 1\ nA$$

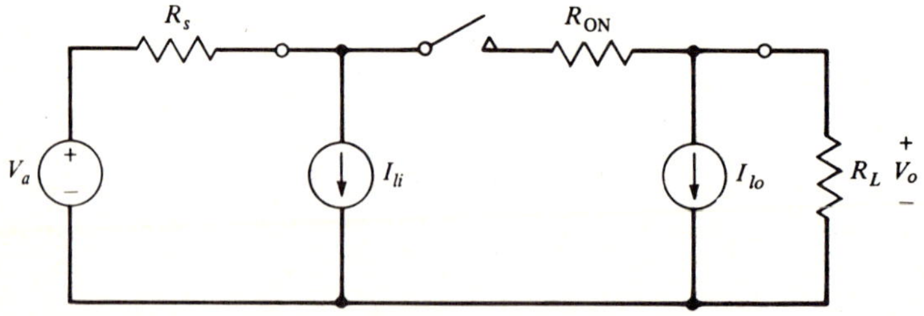

Figure 7-73 Simplified dc equivalent circuit of a series switch.

7-59 Design an inverting voltage amplifier with programmable gain values of -10, -30, -60, and -100.

7-60 Design a polarity reversing voltage amplifier by modifying the difference amplifier of Fig. 6-12 with the aid of two SPDT switches.

7-61[27] The circuit of Fig. 7-74(a) can be used to select periodically the largest of the three input voltages V_a, V_b, and V_c. The switches S_1, S_2, S_3, and S_4 close periodically once every T sec driven by their respective clock signals CP_1, CP_2, CP_3, and CP_4 whose timing diagram is shown in Fig. 7-74(b). Analyze the operation of this circuit.

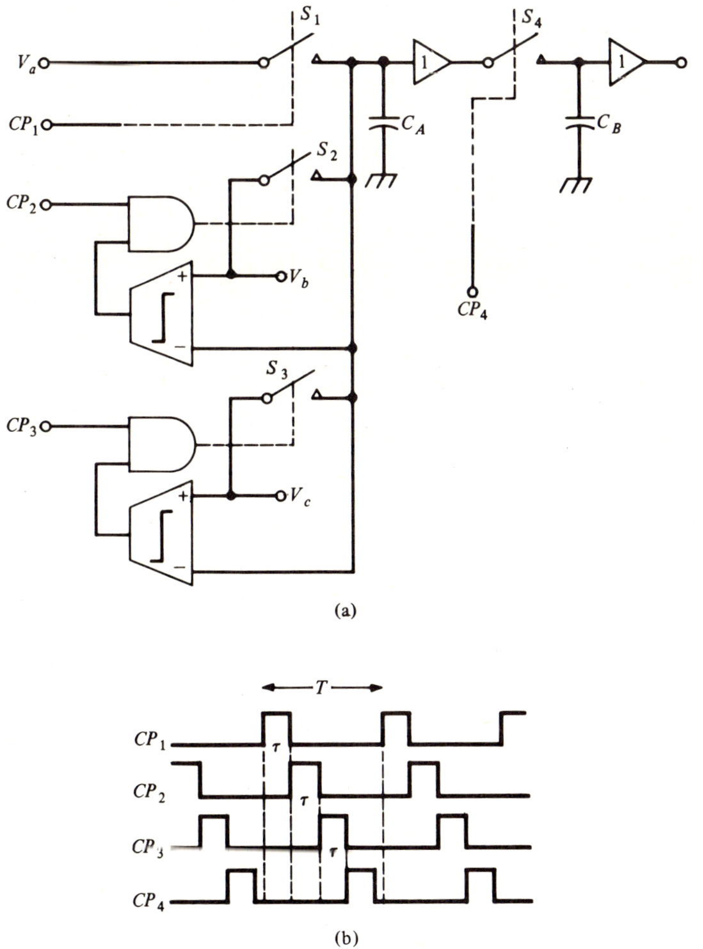

Figure 7-74 (a) Maximum voltage decision maker, and (b) clock timing diagrams.

CHAPTER 8

Analog-to-Digital and Digital-to-Analog Converters

In Chapters 2 through 4 we were concerned with the analysis and design of digital circuits for which all pertinent signal variables are discrete functions of time taking one of two possible values. In the next three chapters we discussed the analysis and design of analog circuits where the signals are instead continuous functions of time and can take any real values between some specified ranges. As indicated earlier, many functions performed almost exclusively by analog circuits are increasingly being performed by digital circuits. In many situations, the signal to be processed is naturally in analog form. In order to process this signal digitally, it is first necessary to convert it into an equivalent digital form. The conversion of an analog signal to a digital signal is performed by an *analog-to-digital converter* (often abbreviated as an *A/D converter* or simply as *ADC*). Similarly, in some cases the processed digital signal has to be converted into an equivalent analog form for further processing by an analog system. This reverse process is accomplished by the *digital-to-analog converter* (abbreviated as *D/A converter* or simply as *DAC*). A digital voltmeter, for example, measures and displays an analog voltage digitally. Thus such an instrument just makes use of only an A/D converter. On the other hand, in the pulse-code-modulation (PCM) telephone system, the analog speech signal is converted into a digital form for transmission and at the receiving end the digital signal is converted back into an analog form for hearing. Such a system then makes use of both A/D converters and D/A converters.

These converters are the subject of discussion of this chapter. There are many different types of such converters and it is not possible to discuss in detail all of them in one chapter. Rather we plan to concentrate on a few conceptually simple types of such units to provide the reader with some of the fundamental concepts of the area.

8-1   Basic Considerations

Even though a time-varying analog signal can be directly fed into an A/D converter, for most applications it is necessary to use a sample-and-hold circuit preceding the converter to minimize the conversion error. This is illustrated next.

Necessity of Sampling

Let us calculate the effect of A/D conversion of a time-varying signal.[1] For simplicity assume that the analog signal is a sinusoidal signal given by

$$x(t) = A \sin \omega t \tag{8-1}$$

Let the A/D conversion time be τ sec. If we are interested in converting $x(t_0)$, the output of the converter at the end of conversion process may actually represent any value of $x(t)$ between $x(t_0)$ and $x(t_0 + \tau)$. The maximum error between the input analog signal and the output of the A/D converter is thus.

$$\Delta x = |x(t_0) - x(t_0 + \tau)| \tag{8-2}$$

The maximum value of Δx takes place when the rate of change in amplitude of $x(t)$, that is, the slope dx/dt, is greatest. It is evident for sinusoids that dx/dt is largest at crossover points, that is, at points where $x(t)$ changes signs. Now from Eq. (8-1)

$$\frac{dx}{dt} = A\omega \cos \omega t \tag{8-3}$$

which is maximum when $\cos \omega t = \pm 1$. Thus we obtain from Eq. (8-3)

$$(\Delta x)_{\text{max}} = A\omega(\Delta t) = A\omega\tau \tag{8-4}$$

$(\Delta x)_{\text{max}}$ is then the uncertainty error voltage that is due to the uncertainty in knowing exactly the time of occurrence of the signal at the input of the converter. Uncertainty error can be reduced by decreasing the time τ. One way to reduce this would be to use a sampler and produce a pulse train whose amplitude is modulated by the analog signal $x(t)$. However, the width of the pulses are invariably much smaller than the A/D conversion time. Consequently it is necessary to hold the output of the sampler for a longer period to allow complete conversion. Let t_s denote the sampling time of the sample-and-hold circuit. We show later (Section 8-3) that during this time, there also will be some uncertainty introduced. If it is assumed that no further uncertainty is introduced during hold time, then the uncertainty error voltage is essentially

$$(\Delta x)'_{\text{max}} = A\omega t_s \tag{8-5}$$

which in practice would be much smaller than that given by Eq. (8-4).

Sampling Process

As indicated above, sampling is a key operation performed to enable digital processing of a time-varying analog signal. Sampling is done by a device called a *sampler* which is essentially an analog switch operating periodically at a constant rate. The switch closes for a very short time t_s sec once every T sec. The time interval t_s is known as the *sampling time*. The time interval T between two adjacent samples is the *sampling period* and its reciprocal is the *sampling rate*. Thus, for example, if the sampling period is 125 μsec, the sampling rate is 8000/sec.

If we denote the analog input signal by $x(t)$, the output $x^*(t)$ of the sampler is an amplitude-modulated pulse train as shown in Fig. 8-1. Each pulse is of constant amplitude and is of width t_s sec. The amplitude of the pulse at $t = nT$ is equal to that of the analog signal at the same time, that is,

$$x^*(nT) = x(nT) \qquad n = 0, 1, 2, \ldots \qquad (8\text{-}6)$$

The question that naturally arises at this point is as follows: Given an analog signal, how fast should it be sampled? If the analog signal is sampled faster, the sampling period decreases. Consequently the speed of operation of the digital circuits processing the sampled signals (actually their digitally coded versions) should be increased with an associated increase in cost. Thus the question should be rephrased as: Given an analog signal, how slow can it be sampled? The answer to this question lies in the sampling theorem stated below without the proof.*

The sampling theory states that a band-limited signal having no frequency components above a frequency f_0 Hz can be uniquely determined by its values sampled uniformly at intervals less than or equal to $1/2f_0$ sec apart.

The maximum allowable sampling interval, also known as the *Nyquist interval*, is thus $1/2f_0$ sec.

> ***Example 8-1.***   An analog signal $x(t)$ contains the following frequency components: 10, 50, and 100 Hz. Determine the minimum allowable sampling rate.
>
> Note that the highest-frequency component present in $x(t)$ is of frequency 100 Hz. Hence the samples should be at least $1/200 = 5$ msec apart. Thus the minimum sampling rate is 200 sec, that is, the switch should close once every 5 msec.

Even though the sampling theorem states that a sampling rate of 2 cycle of the highest-frequency component present is adequate, normally the sampling is done much faster at a rate two to five times the minimum rate.

If the sampling rate is less than the *Nyquist rate*, the sampling is improper resulting in frequency *aliasing* errors. Thus, if the sampling interval $T > 1/2f_0$,

---

* The proof is found in a number of texts. See, for example, Lathi, Reference 3.

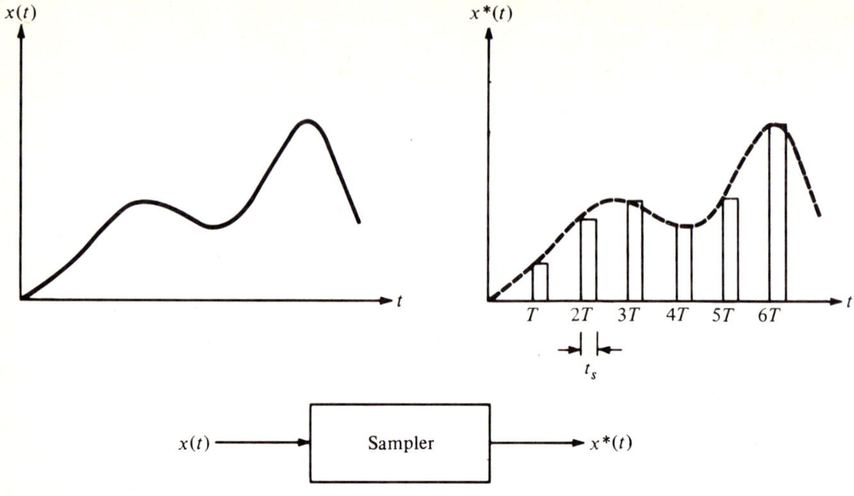

Figure 8-1   A typical input and output waveform of a sampler.

then frequency components in $x(t)$ that are above $1/2T$ Hz get inseparably mixed up with the frequencies below $1/2T$ Hz making it impossible to reconstruct $x(t)$ from its sampled form.

It is instructive at this point to examine the characteristics of a hybrid electronic system, a block-diagram representation of which is shown in Fig. 8-2. An analog signal $x(t)$ generated by the analog system on the left is fed into a sample-and-hold module which samples the input signal periodically once every T sec. The sampled value is then held constant at the input of the A/D converter which produces a binary digital signal. This digital signal is processed by a digital system producing an output signal that is also in digital form. Finally the output of the digital processor is converted into an analog signal by a D/A

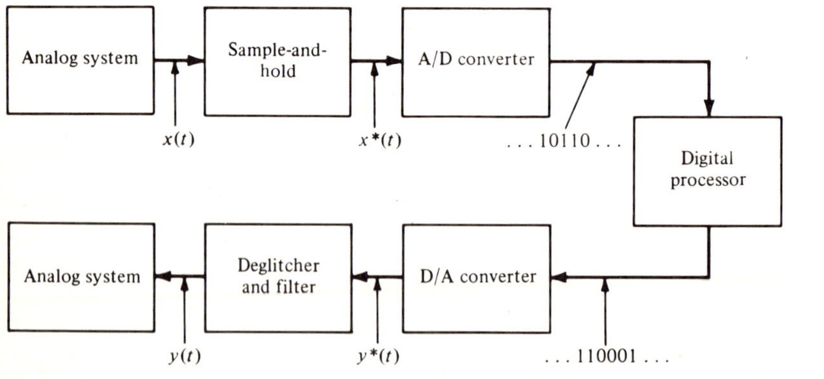

Figure 8-2   Block-diagram representation of a typical hybrid electronic system.

converter followed by a smoothing filter. This digitally processed analog signal can then be fed into another analog system.

Let us now examine this hybrid system in a little more detail. To understand some of the special features of such a system, we have plotted in Fig. 8-3 several signal variables appearing at various points in the system. The waveform in the solid line is a typical analog signal $x(t)$ appearing at the input of the sample-and-hold module. The output $x^*(t)$ of the sample-and-hold is shown with a dashed line. Finally at the bottom is the digital version of the analog input as appearing at the output of the A/D converter.

An examination of Fig. 8-3 reveals a number of interesting points. We first note that $x(t)$ takes all possible values between 5 and 8.6 V in the interval $t = T$ to $t = 2T$. Thus, if $x(t)$ is fed directly to the A/D converter, accurate conversion would not have been possible. As indicated earlier, it is preferable to sample $x(t)$ at uniform intervals at $t = 0$, T, $2T$, and so on, and hold each sampled value at the input of the A/D converter until the next sample arrives. The time interval T between two consecutive samplings is the *sampling period*. Note from Fig. 8-3 that the sampled value $x(T)$ is 5 V and the output $x^*(t)$ is held at this value until $t = 2T$. It is evident that the A/D conversion must be complete before $x(t)$ is sampled next at $t = 2T$. $x(2T)$ is 8.6 V which is held until $t = 3T$, and so on.

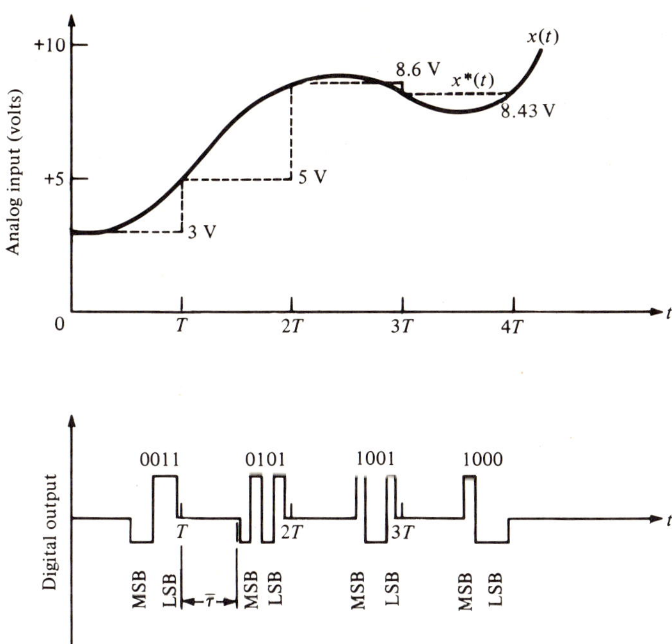

Figure 8-3   A typical input and output waveform of a sample-and-hold and an A/D converter.

We next observe that the output of the A/D converter does not appear immediately after each sampling. Instead, it is delayed by τ sec which is the *conversion time* taken by the converter. In Fig. 8-3 the digital output appears serially with MSB first. In some applications the digital output may appear in parallel immediately after the conversion is complete.

Finally we observe that while $x(T)$ has been accurately coded as 0101, the next two signals have not been converted accurately. $x(2T)$ which is 8.6 V, appears as 1001. The decimal equivalent of 1001 is 9; thus there is an error of -0.4 V in the conversion process. On the other hand, $x(3T)$ has been coded as 1000. The difference between the actual and the coded representation is now $+0.43$ V. This type of inaccuracy in digital representation is known as *quantization error* which is discussed in detail later in Section 8-3.

8-2 Sample-and-Hold Circuit

For rapidly varying analog signal, it is preferable to sample periodically the signal and hold the amplitude of each sampled signal at the input of the A/D converter for a sufficient time to allow accurate conversion. The sampling and holding operation is performed by a sample-and-hold circuit also available as modules usually in hybrid IC form. In addition to their use in A/D conversion, these circuits are often used in pulse-amplitude detection, pulse height to pulse width conversion, and time interval measurements.

There are a number of variations of this circuit. In principle, they operate as follows: The *sampling* is performed by an analog switch which when closed charges a capacitor to a voltage equal to the value of the input analog signal and the capacitor then *holds* the voltage across it during the time the switch is open. The switch is controlled by a digital control signal.

Basic Circuit

The basic sample-and-hold circuit is shown in Fig. 8-4. The voltage follower at the output acts as a buffer between the shunt capacitor C and the input stage of the A/D converter to prevent loading. The analog switch S is a single-pole single-throw type electronic switch described earlier in Section 7-8. The resistance R_s shown is the generator resistance.

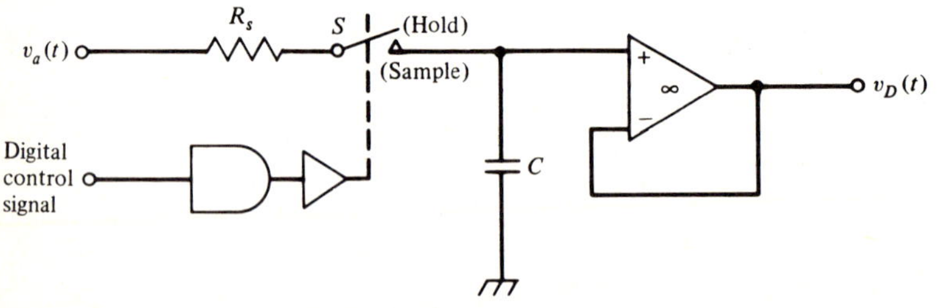

Figure 8-4 The basic sample-and-hold circuit.

To understand the operation of this circuit we have drawn in Fig. 8-5 an equivalent representation of the circuit during the time the switch is closed. In the figure, R_{ON} represents the impedance of the switch in its ON position. The sampling time t_s is in general very small, of the order of a few microseconds. Hence we can assume for analysis purpose that during the sampling time, the analog voltage $v_a(t)$ is practically constant. To simplify the notational problem, consider the switch closure at $t = 0$. By assumption,

$$v_a(t) = v_a(0) = V_A \qquad \text{for} \qquad 0 \le t \le t_s \tag{8-7}$$

It is a simple exercise to show that the voltage $v_D(t)$ across the capacitor is given by

$$v_D(t) = V_A[1 - e^{-t/RC}] \qquad \text{for} \qquad 0 \le t \le t_s \tag{8-8}$$

where we have set $R = R_{ON} + R_s$. In deriving Eq. (8-8), we assume that the initial voltage across the capacitor at $t = 0$ is zero.

Ideally $v_D(t_s)$ should be equal to V_A when the switch opens. But as Eq. (8-8) points out, this can happen only as $t \to \infty$ requiring an infinite acquisition time— not a very practical solution. This implies, then, that we would have to accept some error in the value of analog voltage across the capacitor in representing V_A. The difference between $v_D(t_s)$ and V_A is defined as the *allowable error* due to sample-and-hold operation, which is usually referred to the full-scale analog voltage. If we want $v_D(t_s)$ to be 90 percent of V_A corresponding to a maximum of 10 percent allowable error, then from Eq. (8-8),

$$v_D(t_s) = V_A[1 - e^{-t_s/RC}] = 0.9\, V_A \tag{8-9}$$

which can be solved for t_s to yield

$$t_s = 2.3RC \ \text{sec} \tag{8-10}$$

On the other hand, if we require $v_D(t_s)$ to be 99.9 percent of V_A, then the required sampling time is

$$t_s = 6.9RC \ \text{sec} \tag{8-11}$$

Then the maximum allowable error is 0.1 percent.

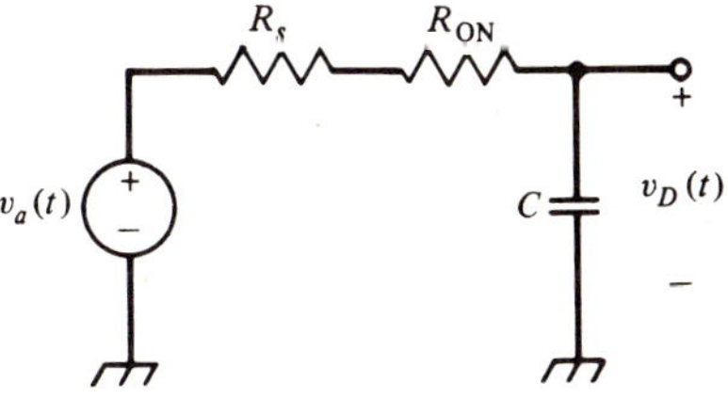

Figure 8-5 An equivalent representation of the basic sample-and-hold circuit during the sample mode.

If the capacitor has zero loss and the leakages through the switch during its OFF period and the amplifier input impedance are zero, then the voltage $v_D(t_s)$ across the capacitor will be maintained at the same level during the time the switch is OFF (open). Thus it follows during each successive switch closure, the capacitor will have some initial voltage across it instead of zero voltage as assumed previously. Moreover, the change in amplitude of the analog signal from sample to sample is, in practice, a fraction of the full scale. Consequently less sampling time would be required to achieve a given allowable error as the following example illustrates.

Example 8-2.   Determine the sampling time for a ± 0.1 percent allowable error assuming the change in amplitudes of successive samples is less than ± 1 percent.

Let V_A^+ be the amplitude of the present sample and V_A^- that of the previous sample. For simplicity, assume

$$V_A^- = 0.99\, V_A^+ \tag{8-12}$$

The initial voltage across the capacitor before the arrival of the sample V_A^+ is approximately V_A^-. It can be shown, then, that due to the presence of the initial voltage, the capacitor voltage $v_D(t)$ due to an application of a voltage V_A^+ is now given as

$$v_D(t) = V_A^+ + (V_A^- - V_A^+)e^{-t/RC} \tag{8-13}$$

If V_A^+ is assumed to be full-scale voltage, then 0.1 percent allowable error implies

$$v_D(t_s) = 0.999\, V_A^+ \tag{8-14}$$

Substituting Eqs. (8-12) and (8-14) in Eq. (8-13), we readily obtain

$$t_s = 2.3\, RC \text{ sec} \tag{8-15}$$

which is one-third of the time needed to obtain the same allowable error if the initial voltage across the capacitor was zero [see Eq. (8-11)].

A basic drawback of the simple sample-and-hold circuit of Fig. 8-4 is that the capacitor loads the input analog source $v_a(t)$ which may be rapidly varying or may not have sufficient current supply to charge the capacitor fast enough. This problem can be avoided by isolating the source and the capacitor with a voltage follower.

Practical Considerations

Due to the nonideal properties of the switch and the buffer amplifier at the output end, a practical sample-and-hold does not operate exactly the same way as described earlier. It is thus appropriate to discuss the characteristics of a practical unit. This will enable us either to improve the performance of an existing unit or to select an appropriate module for a specific application.

Mode Control Signal.   The operation of the switch is controlled by an external mode control signal usually generated by a TTL compatible digital circuit. The control signal is a periodic wave of period T sec. A typical control waveform is sketched in Fig. 8-6. Here we have assumed that the sampling is done when the control signal is HIGH and the holding takes place when the control signal is LOW. The sampling mode is initiated when the rising edge of the mode control signal crosses the threshold value (for example, point A in Fig. 8-6). Likewise, the hold mode is initiated when the falling edge of the mode control signal crosses the threshold value (as shown by point B in Fig. 8-6).

Acquisition Time.   This parameter characterizing the performance of a practical sample-and-hold module is defined as the total time needed to switch from hold to sample mode and acquire the input signal within specified accuracy. By acquiring we mean that the output signal is tracking the input signal, that is, its value is "equal" to that of the signal during the sample mode. The acquisition time is thus the time interval between the time when the rising edge of the mode control signal exceeds the threshold value and the time when the output of the amplifier has settled within a prescribed error as illustrated in Fig. 8-7. The acquisition time t_{ac} thus depends on the operation of the switch, the time constant of the RC circuit, and the dynamic performance of the output operational amplifier. The total time t_{ac} is then given as

$$t_{ac} = t_{delay} + t_{slew} + t_{settling} \tag{8-16}$$

where t_{delay} is the switching delay time (usually a very small quantity), t_{slew} is the time needed by the output voltage of the amplifier to slew and converge to the desired value, and $t_{settling}$ is the settling time of the amplifier. It is evident from our earlier discussion that the minimum sampling time should be greater than the acquisition time needed to acquire the input signal within the allowable error. The actual value will thus depend on the difference between the initial voltage across the capacitor (i.e., the output) and the input signal at the instant the switch is closed. Normally the acquisition time is specified under worst-case condition of a full-scale difference between the input and output. This can be measured by applying a step input of full-scale amplitude as shown in Fig. 8-7 for a typical full-scale range of -10 to $+10$ V.

   Acquisition time can be decreased by decreasing the time constant RC. However, the circuit of Fig. 8-4 also acts as a low-pass filter of bandwidth $(1/RC)$

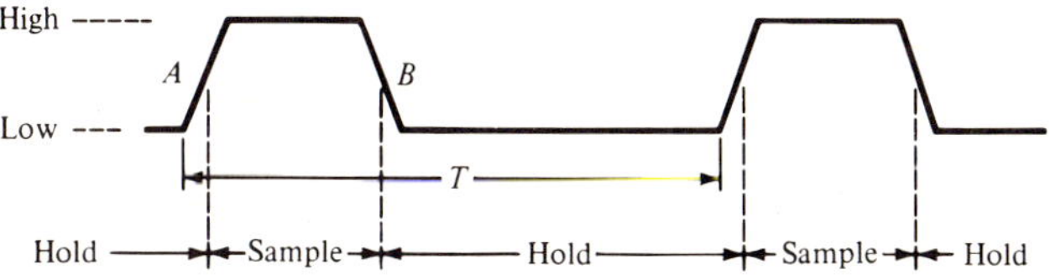

Figure 8-6 A typical mode-control signal waveform.

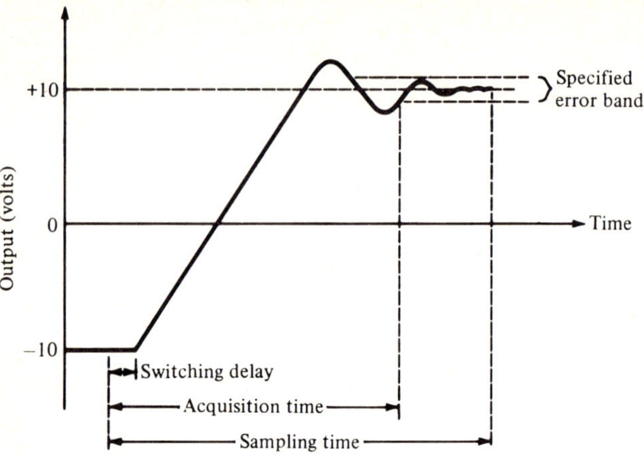

Figure 8-7 A typical worst-case waveform at the output of a sample-and-hold circuit.

radians/sec. Hence, from the point of eliminating any superimposed high-frequency noise, the bandwidth should be small requiring as large a time constant as possible. Thus a compromise between the two extremes has to be chosen and the actual value chosen depends on the particular application. A decrease in the time constant can be achieved by decreasing both the total resistance R in the path and the holding capacitor value. However, as we show later, in practice it is preferable to use as large a capacitor as possible to minimize the drift in the hold voltage due to leakage during the holding phase. A decrease in the value of R can be achieved by minimizing R_{ON} and R_s. Minimization of R_{ON} is obtained by choosing an analog switch with the smallest ON resistance. To minimize R_s a voltage follower can be inserted between the analog source and the input of the sample-and-hold circuit.

Aperture Time.   Another factor determining the performance of a sample-and-hold circuit is the *aperture time* defined as the time required to switch from the sample mode to the hold mode. More precisely, it is the time interval between the time when the falling edge of the mode control signal crosses the threshold value and the time when the output has stopped following the input. Aperture time is typically of the order of 100 nsec. It is usual practice to specify along with the aperture time a settling time to take into account the voltage transients after the switch is turned OFF.

Droop (Hold Voltage Drift).   To minimize the conversion error, it is necessary to insure that the variation in the output voltage $v_D(t)$ be as small as possible during the hold mode. The variation or drift is primarily due to the charge leakage out of the holding capacitor through the amplifier input terminals and the switch. If the capacitor is discharged at a constant rate, we can express the drift in the capacitor voltage by the equation[1]

$$\frac{\Delta V_d}{\Delta t} = \frac{I}{C} \tag{8-17}$$

where ΔV_d is the output voltage decay in Δt sec, and I is the constant current in amperes flowing out of the capacitor. Total drift during the hold period is then

$$(\Delta V_d)_{\text{total}} = \frac{I}{C}(T - t_s) \cong \frac{I}{C} T \tag{8-18}$$

This drift or droop rate is specified usually in mV/sec.

To minimize the drift, then, we should minimize the leakage current I, maximize the value of capacitor, and decrease the hold time $(T - t_s)$. Note that some amount of leakage current will always be present due to the unavoidable input current present in all operational amplifiers. Hence to keep I as small as possible, the OFF resistance of the switch should be very high and the operational amplifier used at the output end should have very small input bias and offset currents. The hold time can be decreased by increasing the sampling rate except the minimum value of hold time necessary should exceed the maximum conversion time taken by the A/D converter. The best approach to decrease the drift is then to increase the capacitor value. Again, this increases the time constant of the circuit which should be kept small for reasons indicated earlier.

The output voltage drift is also a function of temperature and power supply variation. These data are provided by the manufacturers and should be considered in selecting a module for a particular application.

Other Factors.   Since operational amplifiers are usually used in the design of sample-and-hold circuits, their nonideal properties such as offset voltage and currents, power supply rejection, and so on invariably affect the performance of the complete circuit. Input and output ratings, and logic compatibility of the mode control input are also important and should be carefully checked before application.

Practical Circuits

It is possible to minimize the effect of hold voltage decay at the output by modifying the circuit as indicated in Fig. 8-8.[4] During the sampling phase, both the switches are closed and thus the operation of this new circuit is identical to that discussed earlier. Just prior to the beginning of the holding mode, the voltages at the two input terminals and the output of the operational amplifier would be the same having a value equal to that of the sampled amplitude. During the holding mode, since the currents through the input terminals I_1 and I_2 are almost equal, the voltage drifts across the two capacitors would be equal resulting in practically no change in the output voltage V_D (Problem 8-6). Hold times can be increased significantly by this modification using an operational amplifier with moderate input currents without increasing the capacitor value.

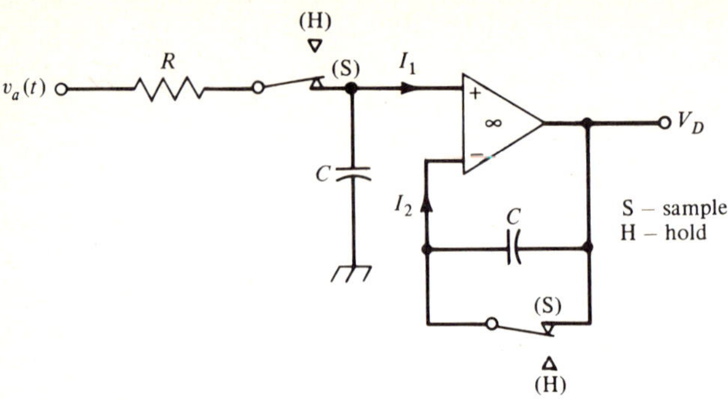

Figure 8-8   Input current compensated sample-and-hold circuit.

Considerably better tracking of the input analog signal is obtained by the feedback arrangement of Fig. 8-9 where the operational amplifier at the input also provides the necessary isolation. When the switch is in the sample position, a path is provided from the output of the leftmost operational amplifier to its inverting input terminal forcing both the input terminals to be at the same potential. As a result, the output of the voltage follower, which is identical to the voltage across the capacitor C, is forced to track the input analog voltage faithfully. However, the open-loop arrangements provide faster acquisition and settling in comparison to the above circuit with feedback.

Since the operational amplifier in both circuits of Figs. 8-4 and 8-9 is used in the noninverting configuration, the full-scale analog voltage cannot exceed the common-mode voltage of the amplifier. Consequently, in some applications, it may be necessary to attenuate the signal at the input end. This problem can be avoided in sample-and-hold circuits which make use of inverting amplifiers as illustrated in Figs. 8-48 and 8-49.

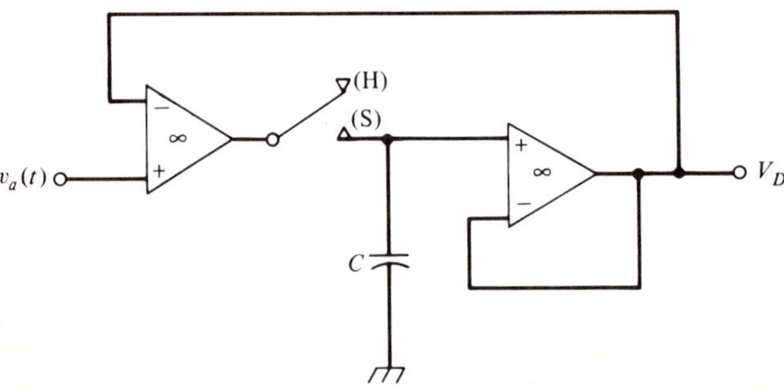

Figure 8-9   A sample-and hold circuit using feedback in noninverting form with unity gain.

A Typical Sample-and-Hold Module

A number of manufacturers market hybrid and monolithic integrated circuit sample-and-hold modules. Figure 8-10 shows the block diagram of the Datel Systems Model SHM-IC-1. It is an all monolithic IC available in a 14-pin DIP and requires an external hold capacitor that can be chosen to achieve desired speed and accuracy requirements. It is used in a closed-loop arrangement in either noninverting or inverting mode. The circuit connection for the non-inverting unity gain operation is sketched in Fig. 8-11. During the sample mode, the switch is closed allowing the input operational amplifier to charge the hold capacitor. Because of outside feedback loop, the output follows the input signal during this mode. During the hold mode, the switch is gated OFF keeping the voltage across C_H constant. Note that during this mode, the operational amplifier A_1 has no feedback but the gating circuit prevents its saturation.

The input operational amplifier is used in a noninverting configuration providing a very high input impedance of the order of 100 MΩ. This sample-and-hold unit has a 0.01 percent maximum gain error in the sample mode. The acquisition time for the output to settle within 0.1 percent for a 10-V step input is 4 μsec with a 0.001-μF external holding capacitor. The aperture time is 50 nsec. The maximum hold voltage drift within the operating temperature range is 50 mV/sec. Output voltage offset during sample mode is a few millivolts and can be nulled externally. The required power supply voltage is ± 15 V.

The sample mode is initiated when the mode control signal is LOW (0 to $+0.8$ V) and the hold mode is effected when the mode control signal is HIGH

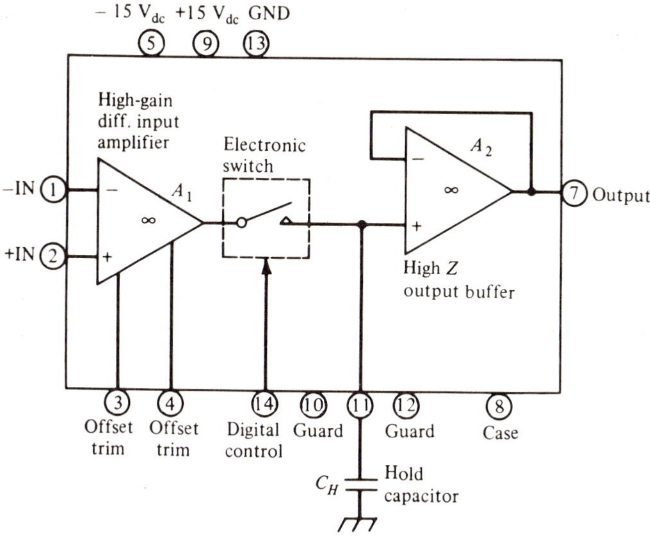

Figure 8-10 Schematic representation of the Datel Systems sample-and-hold module SHM-IC-1. (Courtesy Datel Systems, Inc.)

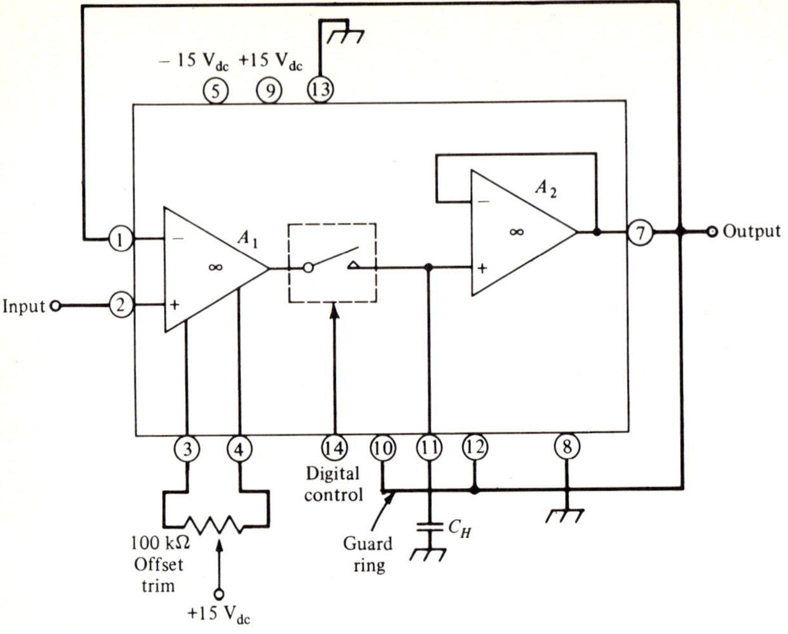

Figure 8-11   Circuit connection for noninverting unity gain operation of the module SHM-IC-1. (Courtesy Datel Systems, Inc.)

($+2$ to 5.5 V). The mode control signal can thus be provided by DTL or TTL logic circuits.

Additional details on the module will be found in the manufacturer's brochure along with the circuit connections for inverting and noninverting operations with gain.

8-3   Quantization Error

We indicated before in Section 8-1 that there will almost always be an error in the digital representation of an analog signal. Let us examine this in little more detail.

Consider, for example, an analog voltage $x(t)$ whose full-scale value is ± 7 V. Then at any time t_1, $x(t_1) = X$ can take any value between -7 and $+7$ V. If X is digitally represented using say 3 bits plus a sign bit, then the coded version can assume one of $2^4 - 1 = 15$ possible levels as illustrated in Fig. 8-12. Thus any value of X which does not coincide with one of 15 discrete values has to be rounded off to the nearest discrete level. This rounding-off procedure is known as *quantization* and correspondingly the difference between two adjacent quantized levels is the *quantization step*. In Fig. 8-12, the quantization step is 1 V. Note

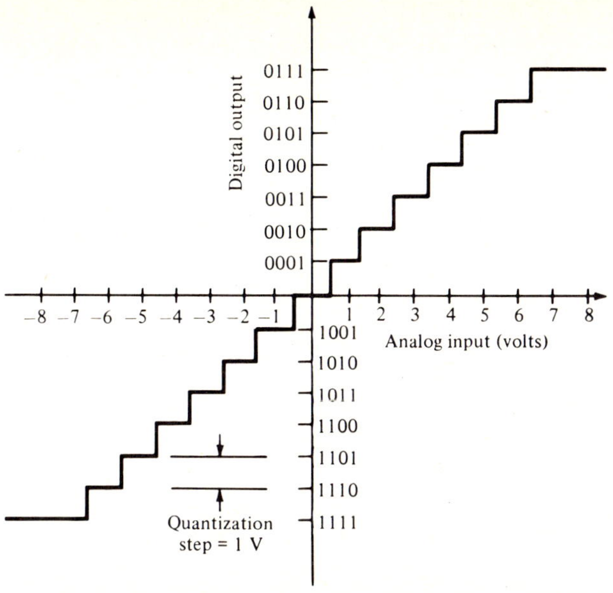

Figure 8-12   Input–output relations of an A/D converter.

that if natural binary coding is used, the quantization step is also equal to the weight of the least significant bit (LSB) of the digital representation.

Example 8-3.   Determine the digital representation using a natural binary code of 3 bits plus a sign bit of the following analog signals: (a) 4 V, (b) 4.20136 V, (c) 4.9 V, (d) 4.97 V, (e) 5 V, (f) 3 V, (g) -4.9 V, (h) -3.1 V. The quantization step is 1 V. Determine the error in representation in each case.

With the aid of Fig. 8-12, the digital representations are readily established as shown in Table 8-1 which also lists the error.

TABLE 8-1

Analog Signal	Digital Representation[a]	Error
4.0	0100	0
4.20136	0100	0.20136
4.9	0101	-0.1
4.97	0101	-0.03
5.0	0101	0
3.0	0011	0
-4.9	1101	0.1
-3.1	1011	-0.1

[a] The leftmost bit is the sign bit where 0 implies positive and 1 implies negative.

As the previous example points out, a given input signal is always represented by its nearest quantized value. For example, any analog signal X which lies in the range

$$3.5 < X \leq 4.5$$

is quantized to a value of 4. The quantization error ε then lies in the range

$$-\tfrac{1}{2} < \varepsilon \leq \tfrac{1}{2}$$

It follows from above that if q represents the magnitude of the quantization step, then the quantization error at any instant can take any value in the range

$$-\frac{q}{2} \leq \varepsilon \leq \frac{q}{2} \tag{8-19}$$

To minimize the quantization error ε, the quantization step has to be decreased; this is achieved by increasing the number of quantization levels for a fixed full-scale voltage. The accuracy of conversion of an ideal A/D converter thus depends on the number of quantization levels. This is expressed in terms of *resolution* of the converter. If M is the number of discrete levels at the output, then the resolution is said to be 1 part in M or $(100/M)$ percent. However, the number of discrete levels depend on the number of bits and the type of coding being used to represent the magnitude of the input analog signal. Thus, if the output is coded in natural binary, n bits would give rise to 2^n levels. The accuracy is then 1 part in 2^n or $100/2^n$ percent. On the other hand, if output is coded in binary coded decimal (BCD) using k decades ($4k$ bits), the number of discrete levels would be 10^k. In this case now, the resolution is 1 part in 10^k or 10^{2-k} percent.

Example 8-4. Determine the resolution of an A/D converter having an output of 12 bits for both natural binary and BCD coding.

For binary coding, resolution will be $100 \times 2^{-12} = 0.0244$ percent. On the other hand, 12 bits imply 3 decades. Thus the resolution in BCD coding will be $10^{-1} = 0.1$ percent.

This example illustrates the deterioration of resolution in BCD coding. However, in a number of applications like digital voltmeters, and so on, BCD coding is used because of other conveniences.

Let us now plot the quantization error as a function of the analog amplitude, illustrated in Fig. 8-13. Since the exact value of the quantization error at any instant is a function of the input analog signal amplitude at that instant which is not known in most applications, it is more meaningful to talk about its expected or mean-square value $\overline{\varepsilon^2}$, an approximate estimation of which is derived next.[6]

If the input analog signal has a constant slope, then the plot of the quantization error as a function of time would be similar to that shown in Fig. 8-13. In the general case (Problem 8-10), if the quantization step is very small, the portion

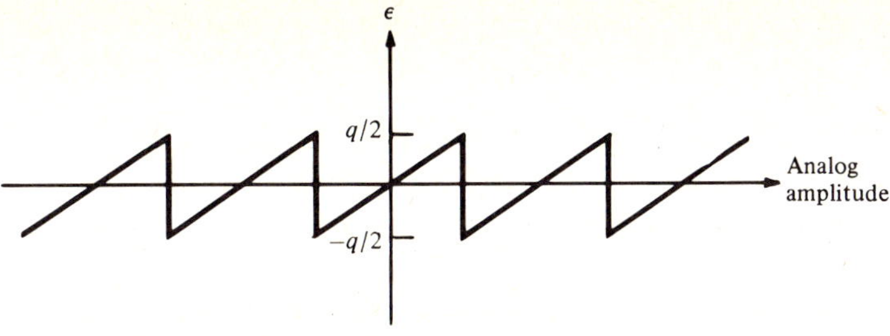

Figure 8-13   Quantization error as a function of analog amplitude.

between two adjacent discontinuities in the quantization error plot with respect to time can be approximated by a straight line with slope m. For convenience, assume the center of this straight line is at $t = 0$ as shown in Fig. 8-14. Then the error takes the maximum positive value $+q/2$ at $t = q/2m$ and the maximum negative value $-q/2$ at $t = -q/2m$. Thus we can represent the error in the time interval $-q/2m \leq t \leq q/2m$ as

$$\varepsilon(t) = mt \tag{8-20}$$

The mean-square error is then given by

$$\overline{\varepsilon^2} = \frac{m}{q} \int_{-q/2m}^{q/2m} \varepsilon^2(t)\, dt = \frac{m}{q} \int_{-q/2m}^{q/2m} m^2 t^2\, dt = \frac{q^2}{12} \tag{8-21}$$

As expected, the mean-square error is also decreased by decreasing the quantization step. Decreasing the quantization step by one-half amounts to a 75 percent decrease in the error—a significant decrease.

Now we are in a position to discuss several specific circuits for the D/A and the A/D conversion process.

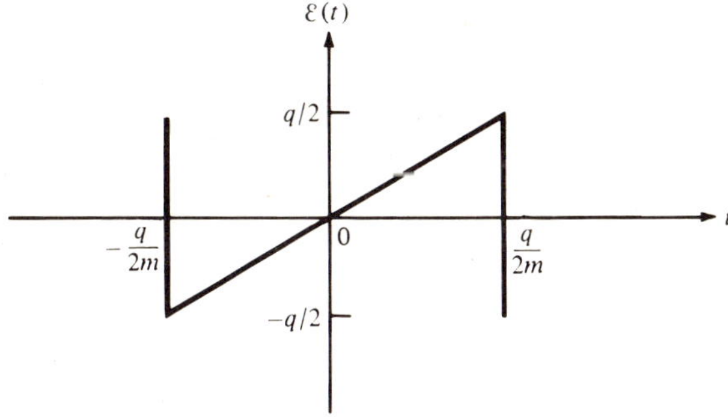

Figure 8-14   Quantization error as a function of time.

8-4 Digital-to-Analog Converters

The conversion of a digital signal to an analog signal is easier to implement than the reverse process. Moreover, some A/D converters make use of a D/A converter in an iterative fashion to arrive at the digital equivalent. Thus we consider first the analysis and design of D/A converters and review their properties.

The basic idea behind most commonly used D/A converters can be explained by means of the simplified block-diagram representation shown in Fig. 8-15 for a positive digital word of 6-bit length. Here V_R is a fixed reference voltage. The digital signal is contained in the register on the left. Each flip-flop in the register individually controls an analog switch as shown by the dotted lines. For example, if the state D_5 of the flip-flop containing the most significant bit (MSB) is a logical ONE, then the corresponding analog switch S_5 is in its ON position—that is, closed. Likewise, if the state D_5 is a logical ZERO, then S_5 is in its OFF position—that is, open, and so on. The output V_o of the D/A converter can thus be expressed as

$$V_o = (a_5 D_5 + a_4 D_4 + a_3 D_3 + a_2 D_2 + a_1 D_1 + a_0 D_0)V_R \qquad (8\text{-}22)$$

where D_k represents the state of the kth bit, and hence the kth flip-flop. If kth bit is ZERO, then $D_k = 0$ and if the kth bit is ONE, then $D_k = 1$. The constant a_k determines the weight of the kth bit. The reference voltage V_R is used to obtain the desired full-scale voltage at the output.

Table 8-2 lists the appropriate values of the constants $\{a_k\}$ for a digital signal in natural binary and 8-4-2-1 BCD code.

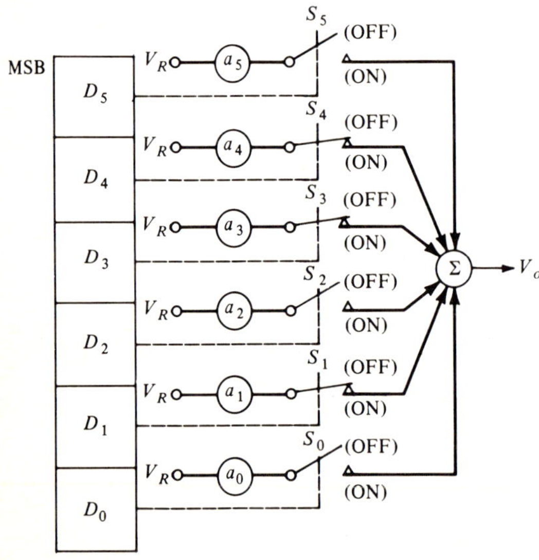

Figure 8-15   Schematic representation of a 6-bit D/A converter.

TABLE 8-2

Constant	Binary	8-4-2-1
a_0	1	1
a_1	2	2
a_2	4	4
a_3	8	8
a_4	16	10
a_5	32	20

Example 8-5. The digital signal to be converted is 101001. Determine the output V_o of the D/A converter of Fig. 8-15 if the digital signal is coded in 8-4-2-1 BCD code.

Substituting the values of the coefficients from Table 8-2 in Eq. (8-22) we obtain

$$V_o = [(20 \times 1) + (10 \times 0) + (8 \times 1) + (4 \times 0) + (2 \times 0) + (1 \times 1)]V_R$$
$$= 29V_R$$

We now discuss two different implementations of the D/A converter. Other realizations will be found elsewhere.[1, 10, 11]

Weighted-Resistor D/A Converter

The weighted-resistor–type D/A converter for a 6-bit positive digital signal is shown in Fig. 8-16. Let us analyze the operation of this circuit. Let Y_o be the output admittance of the D/A converter. If the reference source is ideal having

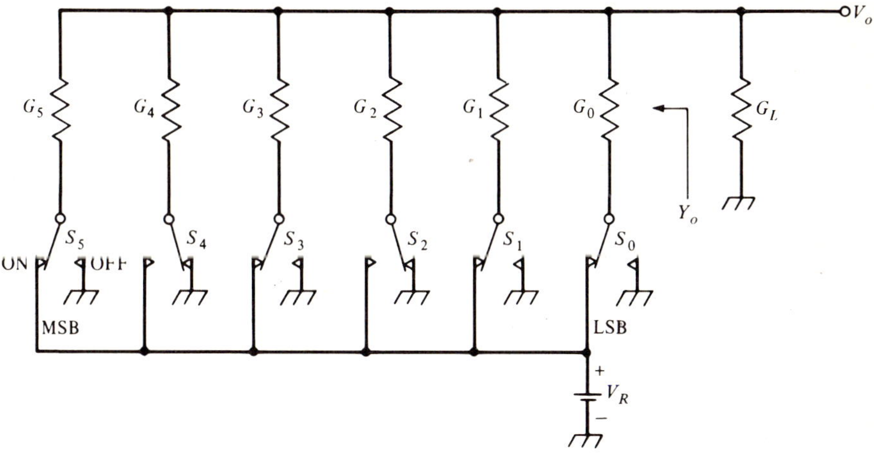

Figure 8-16 A 6-bit weighted-resistor D/A converter.

a zero impedance, then it follows from the figure that

$$Y_o = G_0 + G_1 + G_2 + G_3 + G_4 + G_5 \tag{8-23}$$

For our analysis we consider the effect of each bit separately and then determine the output voltage by the superposition theorem. Without any loss of generality, let us assume that the digital signal is positive. Now, if only the MSB is ONE and the rest are ZERO (the digital signal is 100000), the corresponding equivalent representation is as shown in Fig. 8-17(a). Note that the total admittance of the five conductances in parallel is $(Y_o - G_5)$, which is in parallel with the load conductance G_L. This simplifies the equivalent representation further as shown in Fig. 8-17(b). The output voltage due to MSB only is, by voltage-divider relation, obtained as

$$V_{o5} = \frac{G_5}{G_L + Y_o - G_5 + G_5} V_R = \frac{G_5}{G_L + Y_o} V_R \tag{8-24}$$

Following the same procedure we can show that the output voltage V_{ok} due to the kth bit being ONE and the remaining bits as ZERO's is

$$V_{ok} = \frac{G_k}{G_L + Y_o} V_R, \qquad k = 4, 3, 2, 1, 0 \tag{8-25}$$

In general, then, we can write by the superposition theorem

$$V_o = (G_5 D_5 + G_4 D_4 + G_3 D_3 + G_2 D_2 + G_1 D_1 + G_0 D_0) \frac{V_R}{G_L + Y_o} \tag{8-26}$$

For a prescribed code, the design of the weighted-resistor D/A converter is thus straightforward. A specific example is considered next.

Example 8-6. Design a weighted-resistor D/A converter for a 6-bit positive digital signal in 8-4-2-1 BCD code.

Comparing Eq. (8-26) with Eq. (8-22) we get

$$\frac{G_k}{G_L + Y_o} = a_k, \qquad k = 0, 1, 2, \ldots, 5 \tag{8-27}$$

This indicates that the conductance value G_k is directly proportional to the weight a_k of the corresponding bit. The weights $\{a_k\}$ for 8-4-2-1 BCD code are given in Table 8-2 from which we then arrive at

$$\begin{aligned} G_5 = 20G \qquad G_4 = 10G \qquad G_3 = 8G \\ G_2 = 4G \qquad G_1 = 2G \qquad G_0 = G \end{aligned} \tag{8-28}$$

In Eq. (8-28), G determines the admittance level.

The reference voltage, if not specified, can be chosen to obtain any desired full-scale voltage.

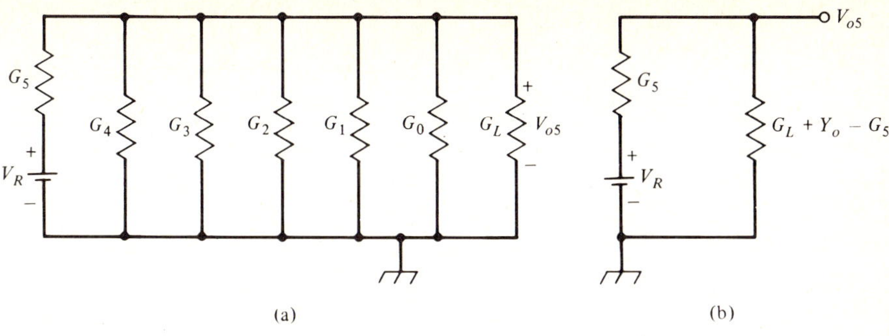

Figure 8-17 Equivalent circuits of the 6-bit weighted-resistor D/A converter with the MSB switch ON and others OFF.

Example 8-7. For the converter of Example 8-6 determine the value of the reference voltage to obtain a full-scale voltage of 15 V.

The full-scale output voltage V_{FSO} can be obtained when all the bits are ONE. Then from Eq. (8-26)

$$V_{FSO} = (G_5 + G_4 + G_3 + G_2 + G_1 + G_0)\frac{V_R}{G_L + Y_o}$$

$$= \frac{Y_o}{G_L + Y_o}\, V_R \tag{8-29}$$

Usually $G_L \ll Y_o$. Then $V_{FSO} \cong V_R$, implying the reference voltage should be +15 V.

Often an operational amplifier is placed at the output of the D/A converter of Fig. 8-16 to provide gain. It also prevents loading of the weighted-resistor network which otherwise may cause accuracy problems. The operational amplifier may be connected in an inverting or a noninverting mode. Figure 8-18 shows the former connection. This circuit can be recognized as basically a negative-gain summing amplifier [Fig. 6-11(a)] and achieves the weighted sum of voltages by summing the currents generated by each source at the inverting input of the operational amplifier. It follows from this figure that the total current i is given as

$$i = (G_5 D_5 + G_4 D_4 + G_3 D_3 + G_2 D_2 + G_1 D_1 + G_0 D_0)V_R \tag{8-30}$$

which is converted to a voltage output $V_o = i/G_f$ by the current-to-voltage converter.

A basic problem with this conceptually simple D/A converter network is that the spread of the resistance values becomes quite large for a D/A converter with moderate resolution. For example, in the 6-bit D/A converter for BCD code designed in Example 8-6, the spread is 20:1. In a 12-bit D/A converter for BCD code, the spread will be 800:1, and for a 12-bit D/A converter for natural binary

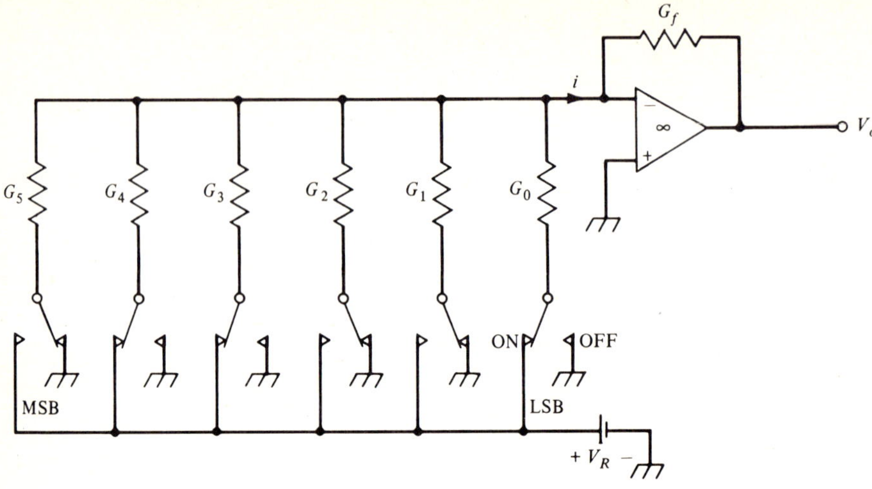

Figure 8-18　A 6-bit weighted-resistor D/A converter with negative gain.

code the spread will be 4096:1. This type of resistor value spread is quite difficult to achieve in a thin-film or thick-film fabrication process. This type of manu-facturing, on the other hand, is preferable for lower cost, smaller size, and better tracking of resistor values. The above problem can be circumvented by employ-ing a resistor-ladder configuration described next.

Resistor-Ladder D/A Converter

Probably the most widely used D/A converter is the resistor-ladder–type D/A converter shown in Fig. 8-19, where G_L again represents the load admittance.

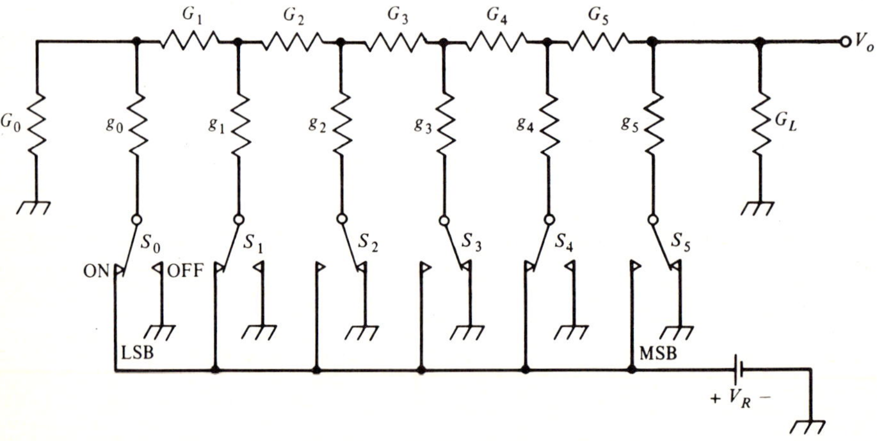

Figure 8-19　A 6-bit resistor-ladder D/A converter.

Here the analog switch S_k is in the ON position if the kth bit is ONE and is in the OFF position if it is ZERO.

The design procedure for this type of D/A converters for any specified code with positive weights has been developed by Aaron and Mitra.[8] An explanation of the procedure is beyond the scope of this book. To understand the operation of the converter consider the case of natural binary coding. One set of conductance values for a 6-bit positive digital word is as follows:

$$g_0 = g_1 = g_2 = g_3 = g_4 = g_5 = G_0 = G$$
$$G_1 = G_2 = G_3 = G_4 = G_5 = 2G \tag{8-31}$$

As before we consider the effect of each bit separately. If the signal is positive with the MSB being ONE and the rest as ZERO's, that is, the digital signal is 100000, then the circuit reduces to that shown in Fig. 8-20(a). Note that the admittances Y_0 through Y_4 seen at various points in the circuit as shown can be readily determined by inspection. Thus,

$$Y_0 = 2G$$

$$Y_1 = G + \frac{2GY_0}{2G + Y_0} = G + \frac{4G^2}{4G} = 2G$$

$$Y_2 = G + \frac{2GY_1}{2G + Y_1} = 2G$$

$$Y_3 = G + \frac{2GY_2}{2G + Y_2} = 2G \tag{8-32}$$

$$Y_4 = G + \frac{2GY_3}{2G + Y_3} = 2G$$

This leads to the simplified equivalent circuit of Fig. 8-20(b) which can be further simplified as indicated in Fig. 8-20(c). Using the voltage-divider relation, we then obtain

$$V_{o5} = \frac{G}{2G + G_L} V_R \tag{8-33}$$

Next assume the digital signal is 010000, that is, the second most significant bit is ONE and the rest are ZERO's. The corresponding equivalent circuit is shown in Fig. 8-21(a). Following the procedure outlined earlier we obtain the simplified equivalent circuit of Fig. 8-21(b).

Finally, by replacing the circuit to the left of the conductance $(G + G_L)$ by its Thevenin equivalent form, the circuit of Fig. 8-21(c) results. By voltage-divider relation, it follows that

$$V_{o4} = \frac{G}{2G + G_L} \cdot \frac{V_R}{2} = \frac{1}{2} \frac{G}{2G + G_L} V_R \tag{8-34}$$

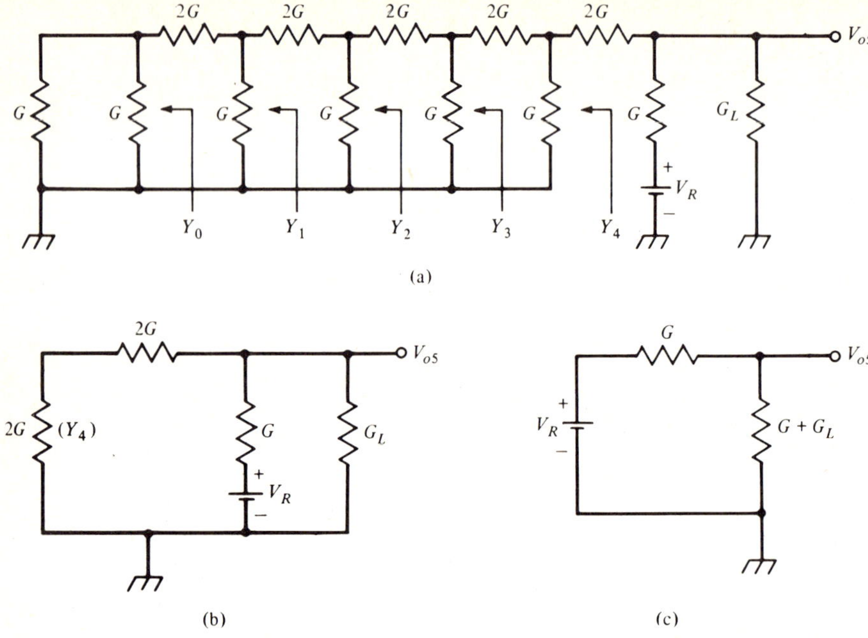

Figure 8-20 Equivalent circuits of a 6-bit resistor-ladder D/A converter for natural binary code with the MSB switch ON and other switches OFF.

This procedure can be repeated to yield the output voltage V_{ok} due to the kth bit being ONE and the rest as ZERO's:

$$V_{ok} = \frac{1}{2^{5-k}} \frac{G}{2G + G_L} V_R, \qquad k = 0, 1, \ldots, 5 \tag{8-35}$$

In general, then, the output voltage is given as

$$V_o = (D_5 + \tfrac{1}{2}D_4 + \tfrac{1}{4}D_3 + \tfrac{1}{8}D_2 + \tfrac{1}{16}D_1 + \tfrac{1}{32}D_0)\frac{G}{2G + G_L} V_R$$

$$= (32D_5 + 16D_4 + 8D_3 + 4D_2 + 2D_1 + D_0)\frac{G}{2G + G_L} \cdot \frac{V_R}{32} \tag{8-36}$$

Example 8-8. Determine the quantum level for the binary resistor-ladder D/A converter.

The quantum level is the output voltage due to LSB only which from Eq. (8-36) is seen to be

$$\frac{G}{2G + G_L} \cdot \frac{V_R}{32}$$

In most applications $G_L \ll G$ which makes the quantum level $V_R/64$ V. Again the reference voltage V_R can be chosen to produce the desired full-scale voltage if necessary.

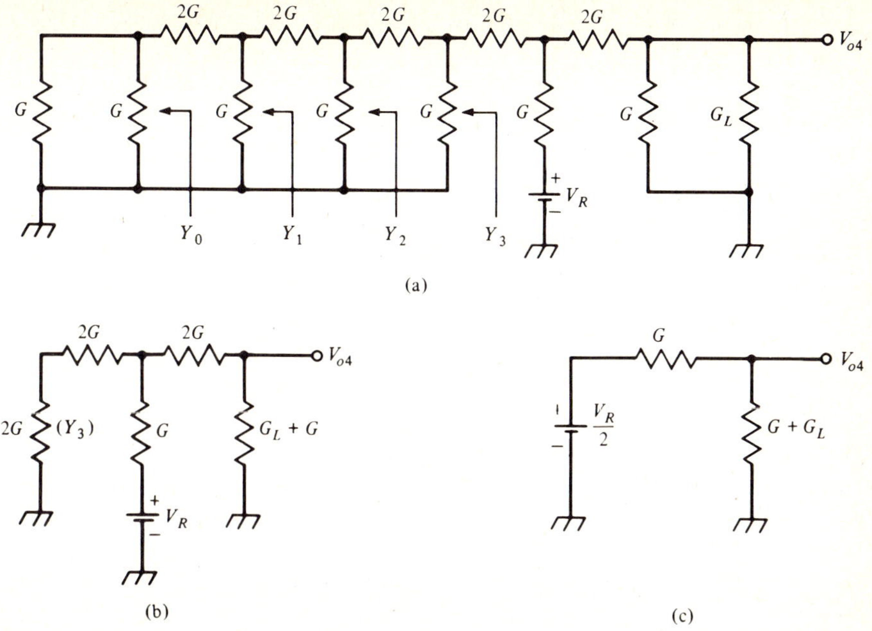

Figure 8-21   Equivalent circuits for the 6-bit R-$2R$ resistor-ladder D/A converter with the second-most significant bit switch ON and other switches OFF.

Generalization of the 6-bit binary R–$2R$ ladder-type D/A converter to the n-bit case is straightforward and is left as an exercise. In this case the expression for the output voltage is given as

$$V_o = (2^n D_n + 2^{n-1} D_{n-1} + \cdots + 2D_1 + D_0)\frac{G}{2G + G_L} \cdot \frac{V_R}{2^n} \qquad (8\text{-}37)$$

The full-scale output voltage V_{FSO} is obtained when all the bits are ONE. Then Eq. (8-37) becomes

$$V_{\text{FSO}} = (2^n + 2^{n-1} + \cdots + 2 + 1)\frac{G}{2G + G_L} \cdot \frac{V_R}{2^n}$$

$$= \frac{2^{n+1} - 1}{2^n}\frac{G}{2G + G_L} \cdot V_R \qquad (8\text{-}38)$$

For $G_L \ll G$,

$$V_{\text{FSO}} = \frac{2^{n+1} - 1}{2^{n+1}} V_R = \left(1 - \frac{1}{2^{n+1}}\right) V_R \qquad (8\text{-}39)$$

Unlike the weighted-resistor D/A ladder, the full-scale voltage here depends on the number of bits n.

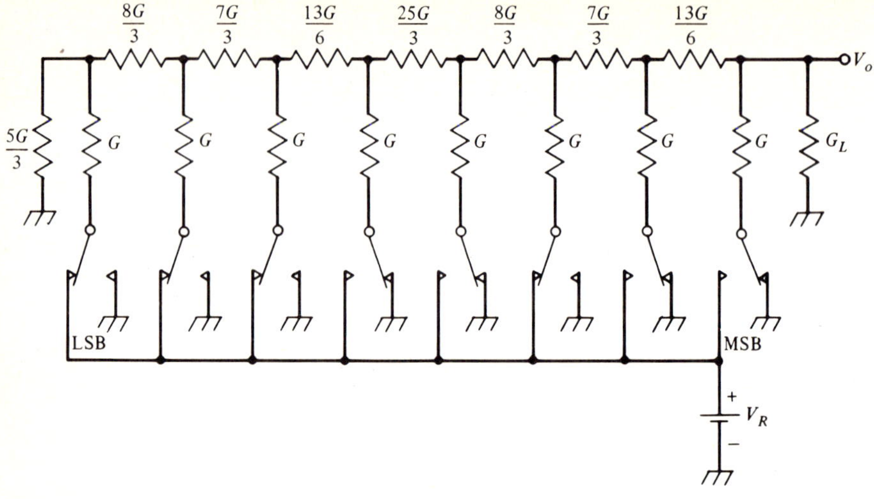

Figure 8-22   An 8-bit resistor-ladder D/A converter for BCD code.

The resistor-ladder D/A converter for an 8-bit digital word in 8-4-2-1 BCD code is illustrated in Fig. 8-22. Realizations for other BCD codes will be found in References 8 and 9.

As in the case of the weighted-resistor D/A converter, an operational amplifier is also usually used at the output of the resistor-ladder D/A converter to provide isolation and gain. Here, also, the weighted sum of the voltages is obtained by summing the currents at the input of this operational amplifier which is connected to provide either a negative or a positive gain.

Bipolar Converters

For conversion of bipolar digital signals, a number of schemes are used to modify the basic D/A converters described earlier. If the digital word is in sign-magnitude form, the sign bit can be used to switch in one of two voltage reference sources of identical magnitudes but of opposite polarities, $+V_R$ and $-V_R$. An alternate scheme is to use a single positive reference voltage but switch the gain of the output amplifier network to positive if the sign bit is a logical 0, and to negative if the sign bit is a logical 1. A possible implementation of such a scheme is shown in Fig. 8-23.[11] To analyze the output circuit, note that the input terminals of the two operating amplifiers on the left side are at ground potential. Hence summing the currents at these input terminals we arrive at

$$i + \frac{v_1}{R_1} = 0$$

$$\frac{v_1}{R_2} + \frac{v_2}{R_3} = 0$$

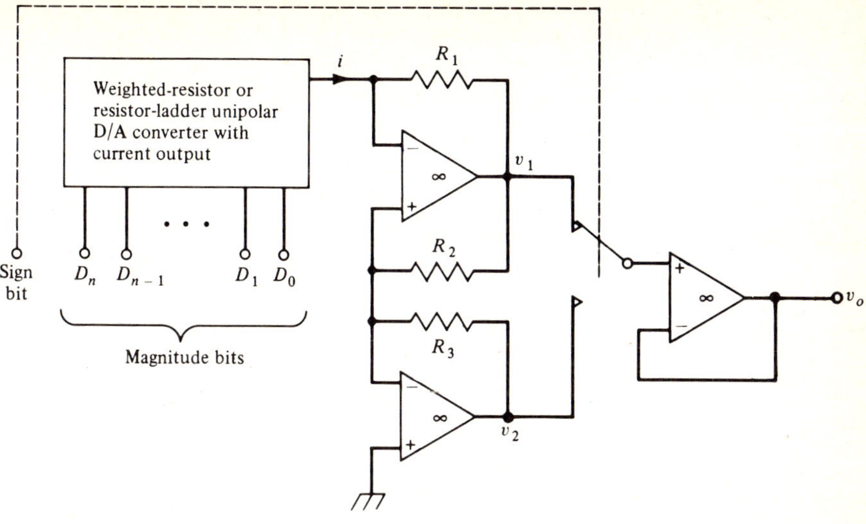

Figure 8-23 A bipolar D/A converter for digital words in sign-magnitude form.

which when solved yields

$$v_1 = -R_1 i, \qquad v_2 = -\frac{R_3}{R_2} v_1$$

If we choose $R_2 = R_3$, then

$$v_2 = -v_1 = R_1 i$$

Conversion of digital words in offset binary code is achieved by subtracting a voltage, equal to that generated by the MSB only, from the output normally developed by the unipolar D/A converters described above. Figure 8-24 shows a possible implementation of a 4-bit bipolar weighted-resistor D/A converter for digital words in offset binary code. Instead of using a separate offset reference voltage as shown here, it is usual practice to obtain the offset voltage from the converter's main reference which minimizes the drift of the zero output with temperature.

Excellent discussions on the operation and design of various types of D/A converters are found in References 1, 10, and 11.

Practical Considerations[11]

The D/A converter modules are available in either monolithic or hybrid IC forms. Their designs are based either on the two approaches discussed earlier or on variations thereof. A number of factors affect the performance of a practical D/A converter. An understanding of these sources of errors is essential to ensure satisfactory operation of these modules in a complex system. It should be noted that a D/A converter is almost invariably used with a buffer amplifier at the

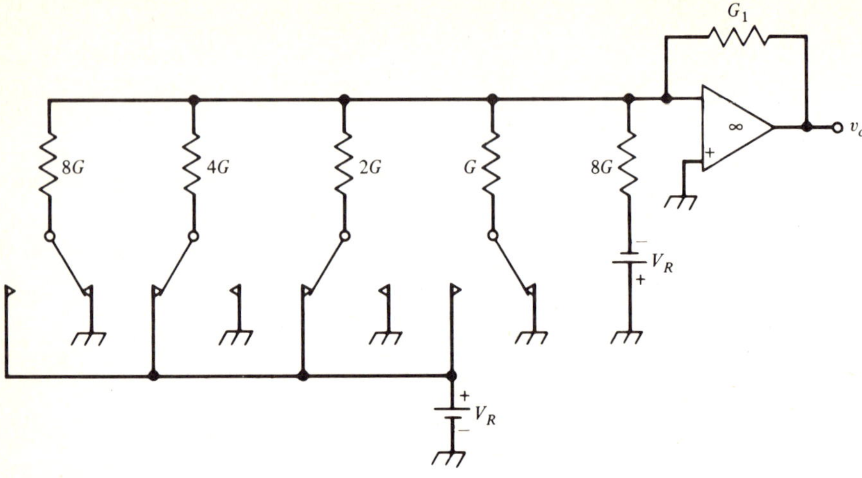

Figure 8-24   A 4-bit bipolar D/A converter for digital words in offset binary code.

output to provide a low-impedance output. As a result, the effect of this amplifier should be considered in evaluating the performance of a practical D/A converter.

To understand the definitions of various parameters characterizing a nonideal D/A converter, it is convenient to indicate the output–input relationship by means of a graph. Such a graph is shown in Fig. 8-25 for an ideal 3-bit unipolar binary D/A converter. Here the analog outputs for each possible digital inputs are shown as vertical "bars." Let us now discuss some of the most important characteristics of a practical D/A converter.

Resolution.   The *resolution* of a D/A converter is defined in a manner identical to that of an A/D converter. Specifically, the resolution of a D/A converter is determined from the number of bits of the input digital word and hence it is

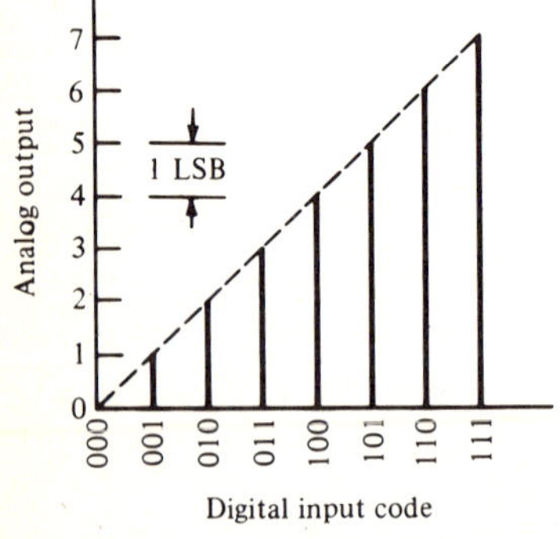

Figure 8-25   Input–output relationship of an ideal 3-bit unipolar binary D/A converter.

directly related to the number of discrete levels at the input. If the number of discrete steps is m, then the resolution is 1 part in m or $100/m$ percent. If, for example, the input is coded in natural binary, then for an n-bit word, the resolution is 1 part in 2^n.

Linearity and Montonicity.   For an ideal D/A converter, the output as a function of the discrete input levels will be on a straight line as shown in Fig. 8-25. The difference between output signals for two consecutive input digital signals is equal to 1 LSB.

In a practical D/A converter circuit, because of unavoidable deviations in resistor values, nonideal switches, and sources, the actual output when evaluated for each digital input combination in general may not be on a straight line but rather unevenly distributed. Figure 8-26(a) shows a typical case. The *linearity error* or *nonlinearity* is defined as the maximum distance of the top of the output bars from a straight line drawn connecting the tops of the end bars (zero and full scale). It is tacitly assumed here that the user has adjusted the zero and the full-scale outputs to their nominal values by calibration before measuring the nonlinearity. The linearity error should be less than $\pm\frac{1}{2}$ LSB. A measure of the variation in the difference of the analog outputs corresponding to two adjacent digital words is given by its *differential nonlinearity*.

If the analog output for each digital input is always greater than or equal to that of the adjacent smaller digital word, the response of the D/A converter is said to be *monotonic* [see Fig. 8-26(a)]. If the analog output for a digital input is smaller than that obtained for the adjacent smaller digital input, the D/A converter is nonmonotonic [see Fig. 8-26(b)]. As before, it is assumed that the endpoints have been calibrated by the user before monotonicity is checked.

Accuracy.   This parameter determines the maximum deviation of measured output of the D/A converter from that of an ideal D/A converter. Note that an

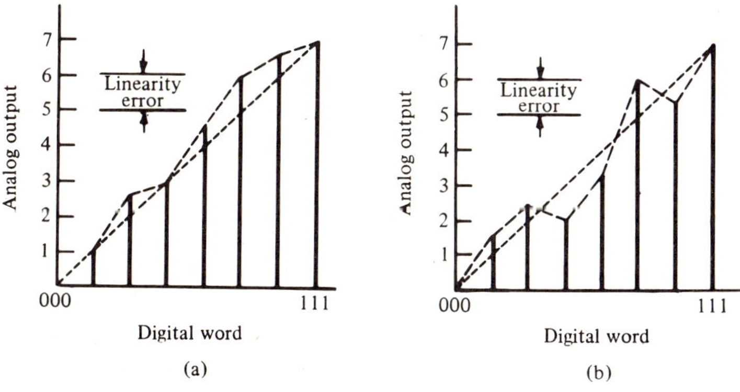

Figure 8-26   Linearity errors in a practical D/A converter: (a) monotonic, and (b) nonmonotonic input–output relationships.

incremental change in the reference voltage does not affect the resolution and the linearity but does indeed affect the accuracy if the absolute magnitude of the output is critical. The accuracy parameter cannot be specified for D/A converters not having full-scale output trimming arrangement or internal reference supplies.

Glitch.　　The analog switches in a D/A converter are controlled by the digital input signal. As can be seen in Figs. 8-16 and 8-19, the analog switch is in the OFF position if the corresponding digital input bit is ZERO and it is in the ON position if the digital input bit is ONE. In a practical analog switch, the *turn-on time* for a ZERO-to-ONE transition is not equal to the *turn-off time* for a ONE-to-ZERO transition. Consider the case of a 6-bit binary D/A converter for which the turn-on times of the switches are greater than the turn-off times. Suppose that at $t = t_n$ the input digital word is 100000 whose decimal equivalent is 32. Then at $t = t_{n+1}$ the input digital word changes to 011111 whose decimal equivalent is 31. The switch corresponding to the MSB should turn-off and all the other switches are required to turn-on. As the turn-on time is longer than the turn-off time, the MSB switch is turned off first and momentarily all the switches are in the OFF positions before the five other switches are turned ON. Thus, momentarily, the input digital word appears to be 000000 leading to a false output of 0 V before it settles to the correct output due to the digital input of 011111. The temporary state of all ZERO's will thus cause a narrow pulse or spike of height half of full scale to appear at the output. These types of pulses appearing at the output during the transition periods are known as *glitches.*

　　If glitches are undesirable then they can be eliminated with the aid of a sample-and-hold circuit at the output. This circuit holds the previous value of the D/A converter until the glitch disappears and then acquires and holds the new output. For most applications the glitches can be ignored.

Word-Conversion Rate.　　This parameter is defined as the maximum number of complete digital words that can be decoded per second. The reciprocal of the conversion rate is the *word-conversion time* which is then the time in seconds taken to decode a digital word. The theoretical minimum value of the conversion time is the minimum time between two readings of the converter register which can be as low as 0.1 μsec. However, there is always an effective capacitance across the load seen by the decoder due to the input capacitance of the amplifier and other stray capacitances. This capacitance along with the frequency response of the amplifier should be taken into account in determining the conversion time. The effect of the unavoidable capacitance across the load can be seen by re-examining one of the D/A converter discussed earlier. For example, in the case of the 6-bit binary resistor ladder, the equivalent representation for digital input 100000 as shown in Fig. 8-20(c) will be modified as in Fig. 8-27. The output voltage will now be a function of time and is given by

$$v_{o5}(t) = \frac{GV_R}{2G + G_L}\left[1 - \exp\left(-\frac{2G + G_L}{C}t\right)\right] \qquad (8\text{-}40)$$

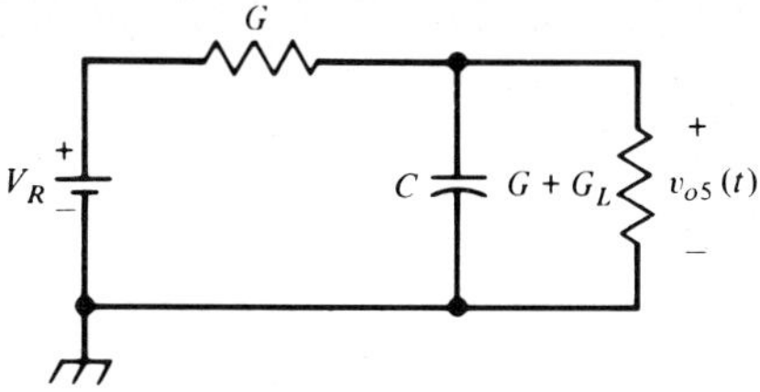

Figure 8-27   Model to determine the effect of capacitance across the load of a D/A converter.

Note that if the capacitor is neglected by letting $C = 0$, then v_{o5} takes the desired value of $GV_R/(2G + G_L)$ to achieve exact conversion. This checks with the value derived in Eq. (8-33) with no loading capacitance. However, for finite value of the capacitance C, exact conversion is achieved only as $t \rightarrow \infty$ which would require an infinite conversion time. The output will reach within ± 0.1 percent of the idealized value in $6.9C/(2G + G_L)$ sec. If $G = 10^{-4} \mho$, $G_L = 10^{-6} \mho$, and $C = 100$ pF, then the time required to achieve ± 0.1 percent accuracy is 3.433 μsec.

In addition, because of the frequency-dependent properties of the output amplifier, the output of the D/A converter will have high-frequency transients causing ripples in the waveform. Total conversion time must also include the settling time which is primarily determined by the slew rate of the amplifier. In a practical D/A converter, the conversion time is thus specified as the time taken by the output to settle within $\pm\frac{1}{2}$ LSB of desired value from the time the conversion is initiated. For example, the Burr-Brown DAC 20 Series D/A converter has a conversion time of 3 μsec for decoding a 10-bit bipolar digital word.

Other Sources of Error.   The accuracy of the conversion is also dependent on the *output offset error* caused primarily by the amplifier and the *gain error* due to the tolerances of the resistors. In many units, these errors can be externally adjusted to zero. The output voltage also drifts due to the changes in temperature and this factor should be taken into account. Error may be caused by power supply variation which is specified as *power supply rejection factor.*

Most practical D/A converters are resistor-ladder type because smaller spread in resistor values are required. Consequently it is easier to fabricate the complete circuit in thin-film or thick-film form resulting in resistors with matched temperature coefficients (TCR). On the other hand, the weighted-resistor type requires a large spread in resistor values making it difficult to obtain all resistors with equal TCR. However, weighted-resistor–type D/A converters having fewer resistors would have less power dissipation and may be preferable in some applications.

Multiplying D/A Converters

For the weighted-resistor D/A converter and the resistor-ladder D/A converter, the output voltage in each case is directly proportional to the reference voltage

V_R as well as the digital input as can be seen from Eqs. (8-26) and (8-36). It is thus evident that by making the reference voltage directly proportional to an external analog signal, it is possible to develop an output that is the product of a digital input signal and an analog input signal. These types of D/A converters that are designed to accept varying reference voltage are known as *multiplying D/A converters* (MDAC) or sometimes as *hybrid multipliers*. The analog reference voltage can also be generated by another D/A converter leading to an analog output which is a product of two digital numbers.

A simple application of an MDAC is in designing an oscillator with digital-controlled amplitude. This is achieved by feeding the output of an analog oscillator into the analog input terminals of the MDAC; the digital input signal is generated by an appropriate digital system. The MDACs can be used to construct digital-to-synchro and resolver converters. Current-producing MDACs are used in graphic display units.

A Typical D/A Converter Module

Monolithic and hybrid IC D/A converters are available. One such unit is the Precision Monolithics' 6-bit binary D/A converter monoDAC-01 which is available in 14-pin DIP and 14-lead F/P package. The converter uses the current-steering approach to switch weighted currents generated by R–$2R$ ladder networks to the summing node of an operational amplifier. The 70×106 mil all monolithic chip contains 85 transistors, 6 capacitors, and a large number of resistors of total value 210.6 kΩ. The package includes an internal voltage source, precision current source, current-steering logic switches, the resistor network, and an internally compensated operational amplifier.

Either bipolar or unipolar operation is possible with this module. A terminal for external adjustment of the full-scale output voltage by a potentiometer is provided. Possible full-scale output voltage ranges are 0 to $+10$ V, -5 to $+5$ V, and -10 to $+10$ V. Settling time to $\pm\frac{1}{2}$ LSB is 3 μsec. Voltage levels for logic input are ZERO < 0.5 V and ONE > 2.1 V. Thus the unit is compatible with DTL and TTL logic circuits. The linearity at $25°$C is 0.40 percent of full-scale. The required power supply voltages are ± 18 V.

8-5 Analog-to-Digital Converters

The A/D conversion methods are not as straightforward as the D/A conversion methods discussed in the previous section. A number of different schemes for the design of A/D converters have been proposed;[1,10,11] however, only a few of these have been used to implement off-the-shelf, small-size converter modules of reasonable cost. The most commonly available and widely used modules are based on one of four different schemes that we describe in this section. In all of these designs, the analog comparator plays a key role.

Direct Conversion Method

In this approach the analog signal amplitude is compared simultaneously with a number of reference signals by means of comparators. The outputs of the

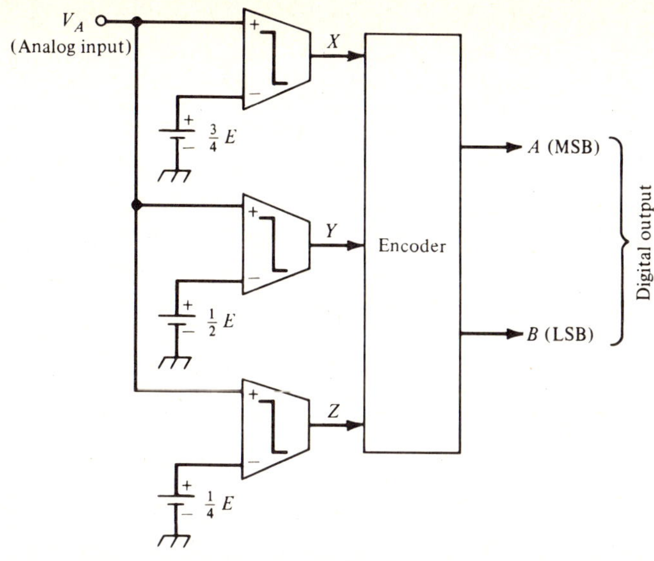

Figure 8-28   A 2-bit A/D converter based on direct conversion method.

comparators immediately indicate the digital equivalent of the analog signal.

Figure 8-28 shows the basic circuit for the conversion of a positive analog signal in the range 0 to E V to 2-bit binary form. Let V^+ and V^- denote the HIGH and LOW levels at the output of the comparator. In practice, the three reference voltages are derived from a single reference source by a resistor potential-divider chain. The states of the comparator output levels X, Y, and Z are determined by the actual analog amplitude V_A. Thus $Z = V^-$ if $V_A < E/4$ and $Z = V^+$ if $V_A \geq E/4$. If $V_A < E/2$, then Y is V^-; otherwise it is V^+. Finally $X = V^-$ indicates $V_A < 3E/4$ and $X = V^+$ means $V_A \geq 3E/4$. If we use the positive logic assignment to denote the output of the comparator, then V^+ represents logical ONE and V^- represents logical ZERO. The relation of the comparator outputs to the input analog voltage is given by Table 8-3. Here three comparators are needed to divide the input voltage range into four parts and the quantum level is $E/4$. We need two bits to code the four parts. The design of the combinational circuit to convert the outputs of the comparators into a binary number is next considered.

Let A denote the MSB and B the LSB. The truth table relating A and B to the switching variables X, Y, and Z is as indicated in Table 8-4. From the table we obtain the expressions for the switching functions A and B as

$$A = \bar{X}YZ + XYZ = YZ$$

$$B = \bar{X}\bar{Y}Z + XYZ \tag{8-41}$$

TABLE 8-3

Range of V_A	X	Y	Z
$\frac{3}{4}E$ to E	1	1	1
$\frac{1}{2}E$ to $\frac{3}{4}E$	0	1	1
$\frac{1}{4}E$ to $\frac{1}{2}E$	0	0	1
0 to $\frac{1}{4}E$	0	0	0

TABLE 8-4

X	Y	Z	A	B
0	0	0	0	0
0	0	1	0	1
0	1	1	1	0
1	1	1	1	1

One possible realization of the encoder is as shown in Fig. 8-29. Note that the encoder can be designed to provide the digital output in any other code.

>   ***Example 8-9.***   Determine the number of comparators needed to obtain an
> n-bit binary representation.
> n bits would lead to 2^n levels. Then the number of comparators needed to
>   obtain 2^n levels is $2^n - 1$. For example, if $n = 10$, then 1023 comparators will
>   be needed.

In the direct method of A/D conversion, all the output bits are developed simultaneously. The conversion time is essentially given by the comparator switching time plus three times the propagation delay of the gates, which is considerably smaller than that obtainable with any other type of A/D converters. A basic drawback of this approach is that the hardware requirements increase very rapidly with an increase in resolution. Such a converter is thus used where speed is important but a low resolution is acceptable. Monolithic A/D converters using the direct conversion method are currently available.

Double-Ramp Method[11, 12]

In this approach the voltage-to-time conversion process is used twice by first converting the input analog voltage to a function of time which is then compared with that generated by a precision reference voltage. A schematic diagram of such a converter is sketched in Fig. 8-30.

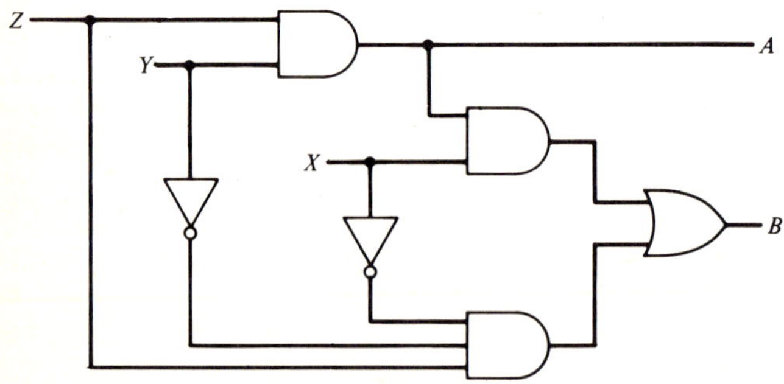

Figure 8-29   A realization of the encoder.

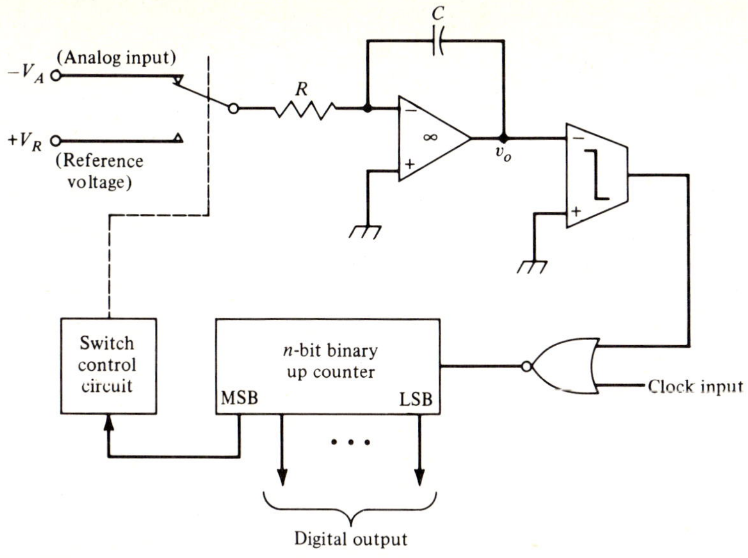

Figure 8-30 A double-ramp A/D converter.

Before the conversion begins the capacitor C of the integrator is discharged and the counter is reset to a zero count. The conversion process is initiated by connecting the input switch to the analog input which is then integrated by the integrator. If the input to the converter has been obtained from a sample-and-hold circuit, it can be assumed to be a constant. In addition, if it is insured to be negative, the output of the integrator will be a linearly rising ramp (see Fig. 8-31). That is,

$$v_o(t) = -\frac{1}{RC}\int_0^t (-V_A)\, dt = \frac{V_A}{RC}\, t \qquad (8\text{-}42)$$

Thus, immediately after the conversion process starts, the comparator output switches to a LOW state, which in turn starts the up-counter. When the MSB of

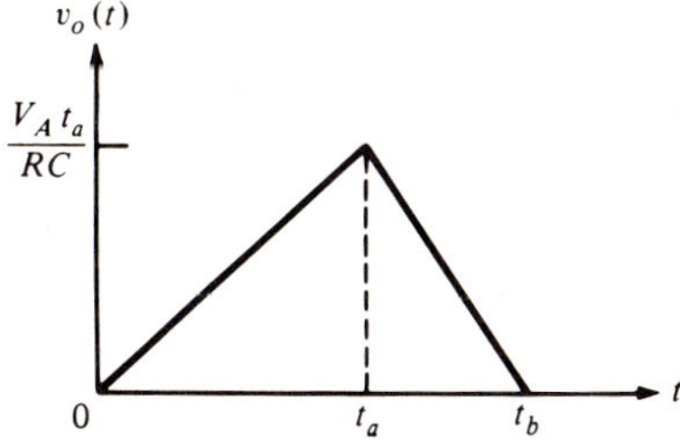

Figure 8-31 The double-ramp output of the integrator of Fig. 8-30.

the counter becomes a logical ONE (at time ($t = t_a$), it switches the input of the integrator to a fixed positive reference voltage $+V_R$. The output of the integrator is then given as

$$v_o(t) = \frac{V_A t_a}{RC} - \frac{1}{RC}\int_{t_a}^{t} V_R\,dt = \frac{V_A t_a}{RC} - \frac{V_R(t - t_a)}{RC} \qquad (8\text{-}43)$$

Thus the output of the integrator starts decreasing linearly (see Fig. 8-31) and at time $t = t_b$ it becomes zero which changes the output of the comparator to a HIGH state. This stops the counter. From Eq. (8-43)

$$V_A = \frac{t_b - t_a}{t_a} \cdot V_R \qquad (8\text{-}44)$$

Since the frequency f of the clock pulses is fixed, the time t_a required for the counter to set the MSB to a ONE is given as

$$t_a = 2^{n-1}/f \qquad (8\text{-}45)$$

If M is the number of pulses counted in the time period $t_b - t_a$, then

$$t_b - t_a = M/f \qquad (8\text{-}46)$$

Substituting Eqs. (8-46) and (8-45) in Eq. (8-44) we obtain

$$V_A = MV_R/2^{n-1} \qquad (8\text{-}47)$$

It follows from above that the contents of the counter (ignoring the MSB) M is directly proportional to the input analog voltage V_A and can be read out in parallel. Furthermore, Eq. (8-47) indicates that the conversion accuracy does not depend on the resistor and capacitor values, and the clock frequency. Since the counter is incremented for each clock pulse, the conversion is monotonic as there are no missing codes. Differential linearity of the dual ramp converter is also very excellent. An additional attractive feature of this converter is that the integrator averages out changes of the analog level during the sampling. It also acts as a low-pass filter and eliminates high-frequency noise. The double-ramp A/D converter can provide quite good accuracy but is somewhat slow in comparison to other converters.

An offset binary representation of a bipolar input can be obtained by attenuating the input and adding a bias equal to half of the reference voltage.

This type of converter finds application in digital voltmeters (DVM) where accuracy is more important than speed. It can also be used in converting the outputs of thermocouples and other transducers to digital form.

Counter-Comparison Method[11, 12]

In this A/D conversion scheme, the analog equivalent of a counter, which begins counting clock pulses at the initiation of the conversion process, is compared to the analog input. The analog equivalent is obtained via a D/A converter. As soon as the output of the D/A converter exceeds the analog input level, the conversion process is stopped by the end of conversion (stop) signal generated by the comparator. A direct read-out of the counter then provides the digital equivalent of the input analog signal. A schematic diagram of the counter-comparison–type A/D converter is shown in Fig. 8-32 together with the input–output waveform. The counter is reset to zero at the start of the conversion process.

It should be noted that the conversion time varies with the value of the analog input. The conversion time is maximum when the counter is set to its maximum value. For example, if the unit is an n-bit binary counter, then the maximum conversion time is equal to $(2^n - 1)T$ where T is the clock period. This limits the speed of such a converter. For proper operation the clock period must exceed the maximum value of the sum of the propagation delay of the counter, propagation delay and slew rate of the comparator, and the D/A converter response time.

Successive-Approximation Method

This method essentially uses a trial-and-error approach successively to obtain the correct digital word. The basic idea behind this method can be explained

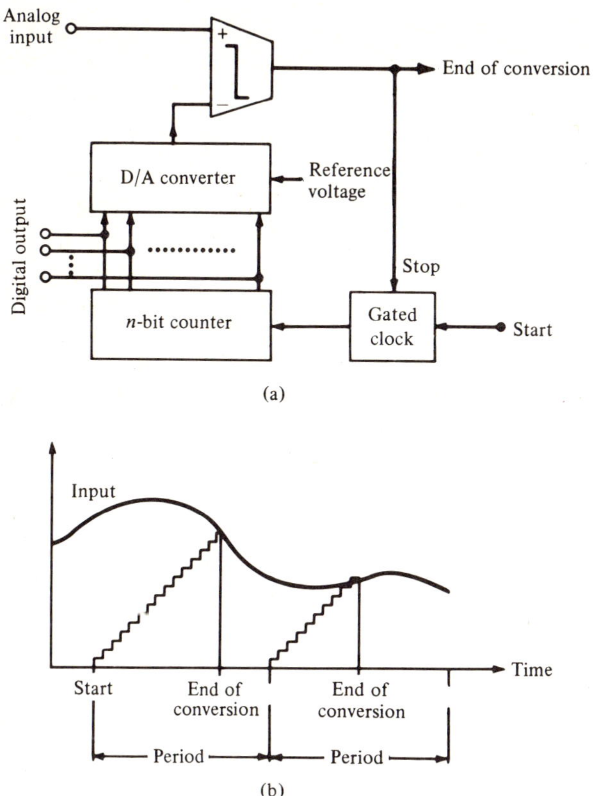

Figure 8-32 (a) Block diagram of a counter-comparison-type A/D converter, and (b) a typical input–output waveform.

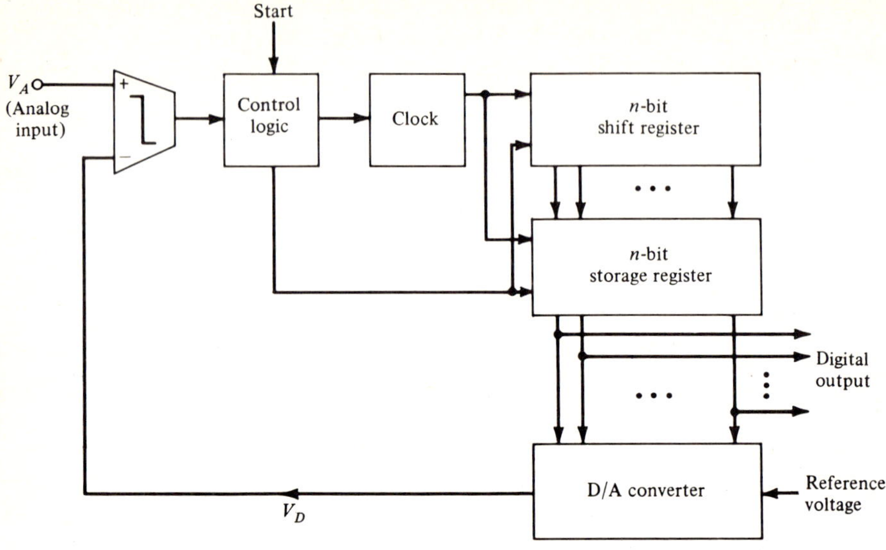

Figure 8-33   Block diagram of a successive approximation type A/D converter.

with the aid of the block diagram of Fig. 8-33. At each step of the conversion procedure, an approximate digital word is converted to an analog signal by a D/A converter, the output V_D of which is compared with the input analog signal V_A. If V_D is less than V_A, then the digital number is increased, and if V_D is greater than V_A, then the digital number is decreased. This trial-and-error procedure is followed systematically as follows. First, at the beginning of the conversion process, the MSB in the register is set to ONE and all other bits are set to ZERO. If $V_D < V_A$, then the ONE in MSB is not disturbed, the bit next to MSB is set to ONE, and the second step begins. If $V_D > V_A$, then the MSB is set to ZERO, the bit next to MSB is set to ONE, and the second step begins. At the second step, the decision to keep ONE in next MSB depends on the sign of difference $(V_A - V_D)$. If it is positive, ONE is kept and if it is negative, it is replaced by a ZERO. This process is followed successively with respect to each successive bits Finally when the last bit has been examined, the conversion process is complete. Then the contents of the register are the digital equivalent of the analog signal. The following example illustrates the approach.

Example 8-10.   Obtain a 4-bit binary representation of an analog signal V_A of value 10.6 V using the successive-approximation technique.

The first approximation is 1000 whose analog equivalent V_D is 8 V. $V_A - V_D = 2.6$ V which is positive. Hence we keep the ONE in the MSB position. Next we set a ONE in the bit next to the MSB. The second approximation is thus 1100. Now $V_D = 12$ V making the difference $V_A - V_D = -1.4$ V. Since the difference is negative, the second MSB is reset to ZERO and the third MSB is set to ONE. The third approximation is now 1010. This implies $V_D = 10$ V. The difference $V_A - V_D$ is now positive implying we keep the ONE in the third MSB. Next the LSB is set to ONE yielding the fourth approximation as 1011. V_D is now 11 V, making the difference $V_A - V_D$ negative. Consequently the LSB is reset to ZERO. Since the LSB has been examined, the conversion process is complete. Hence the digital approximation to 10.6 V is 1010.

In practice, to round off the approximation to the nearest discrete level, $\frac{1}{2}$ LSB is added to the analog signal before it is converted. The number in Example 8-10 will thus be changed to $10.6 + 0.5 = 11.1$ V whose digital representation would be 1011. The discrete level nearest to 10.6 is 11 for the example.

Complete realization of the successive-approximation D/A converter is given by Hoeschele.[1]

The successive-approximation–type converters can be designed with high resolution and reasonably high speed at a moderate cost, and as a result, are used quite frequently. Here the performance of the A/D converter module is primarily determined by the characteristics of the constituent D/A converter and the comparator. For example, if the D/A converter is not monotonic, the resultant A/D converter will also be nonmonotonic. For conversion of bipolar analog signals, a bipolar D/A converter should be used.

Characteristics of a Practical A/D Converter[11]

The A/D converter modules available in the market are mostly fabricated in hybrid IC form, although recently a number of monolithic IC versions have been introduced. Just as in the case of the practical D/A converter circuit, there are a number of sources of errors in the operation of a practical A/D converter unit. As indicated earlier in Section 8-3, an A/D converter is always associated with a quantization uncertainty error of $\pm \frac{1}{2}$ LSB, which can be reduced by increasing the number of discrete levels available at the output, that is, by increasing the *resolution*.*

Linearity and Monotonicity. These characteristics are defined in a manner similar to that of the D/A converter. The errors of a practical A/D converter are usually defined (and measured) at the input analog values at which the transitions in the digital output occur because these transitions can be more accurately determined than the midrange values. The A/D converter exhibits *linearity error* if the difference between two consecutive transition values of the input is not equal for the complete range of the input [Fig. 8-34(a)]. The variation in this difference value over the full range is known as its *differential nonlinearity error*.

If, for continuously increasing input analog signal, all the digital output codes are produced in an increasing order, the A/D converter response is said to be

---

* The resolution of an A/D converter has been defined earlier in Section 8-3.

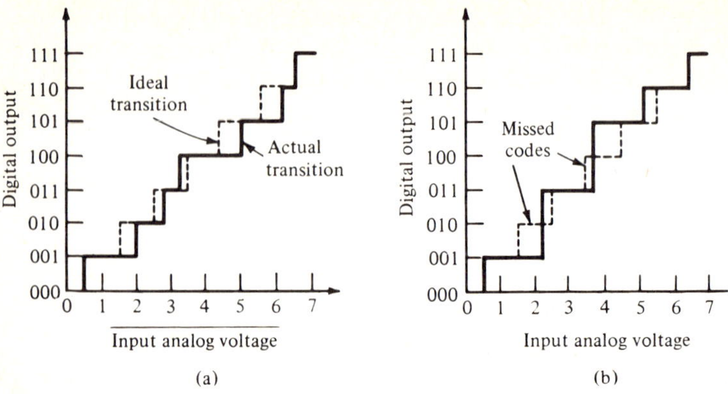

Figure 8-34   Linearity errors in A/D conversion: (a) monotonic, and (b) nonmonotonic.

monotonic. On the other hand, in a *nonmonotonic* A/D converter, one or more codes are skipped in the output as the input is increased [Figure 8-34(b)]. The latter behavior may occur if the differential nonlinearity error is greater than or equal to 1 LSB.

Gain and Offset Errors.   The A/D converter exhibits gain or scale-factor error, if the difference between the values of the last and the first transitions is not equal to full-scale value minus $\frac{1}{2}$ LSB. The *offset error* is said to exist if the first transition does not occur at the prescribed value which for a unipolar converter is $\pm\frac{1}{2}$ LSB.

Absolute and Relative Accuracy.   In practice, the full-scale point of a converter is set with respect to a reference voltage derived from an absolute voltage standard. The tolerance of the full-scale set point in terms of the absolute standard is known as the *absolute accuracy error*. The deviation of the value of the analog input at which a transition in the digital output occurs from its nominal value is also a measure of the accuracy of the A/D converter. The maximum of these deviations over the complete input range relative to the full scale of the analog value is specified as its *relative accuracy error*.

Input Impedance.   The input impedance of the A/D converter is an important factor in determining the performance of the converter. Finite input impedance will load the source causing error in conversion. If the input impedance of the A/D converter is R_i and the analog source impedance is R_a, then the analog signal seen by the converter will be $R_i/(R_a + R_i)$ times the actual analog signal introducing an error of $100\, R_a/(R_a + R_i)$ percent. For example, if $R_i = 2$ MΩ and $R_a = 1$ kΩ, then the error is 0.05 percent.

Word-Conversion Time.   It is the time needed by the converter to generate the digital equivalent of the input analog signal. In an A/D converter controlled by an external-mode control signal, the conversion time is the time interval between the time the control signal initiates the conversion and the time the complete digital word is available at the output. The *bit-conversion time* is the time required to obtain one bit and is thus equal to the word-conversion time divided by the total number of bits in the digital word. *The conversion rate* is the rate at which conversion can be achieved and is the reciprocal of the sum of the word-conversion time and the recovery time. Note that the sampling rate must be equal to or less than the conversion rate.

In the direct (simultaneous) conversion approach, the word-conversion time is essentially equal to the bit-conversion time and it is approximately given as

$$\text{Conversion time} = t_c + m \cdot t_g \text{ sec}$$

where t_c is the response time of the comparator, t_g is the delay caused by a gate, and m is the maximum number of levels in the realization of the encoder. This method is consequently very fast except the design gets more complicated as the resolution increases with a rapid increase in cost.

In the case of the successive-approximation–type A/D converter, since each bit is generated successively, the word-conversion time in this case is equal to the product of bit-conversion time and the number of bits. This type of converter can generate 6 to 12 bits at a bit rate from about 100 kHz to 1 MHz.

Other Factors. Additional characteristics specified for a practical converter are the *temperature coefficients* of the differential linearity, gain, and zero level stability. Power supply sensitivity, input trigger pulse height and width, output voltage levels, and so on, are also of interest.

A Typical A/D Converter Module[11]

Most packaged A/D converters available in the market are of the successive-approximation type. Analog devices ADC-8H series general purpose A/D converters fall in this category. These converters have 8-bit resolution and are available in a package of size 2 in. × 4 in. × 0.4 in. The conversion is monotonic with a relative accuracy of ±0.2 percent and a differential linearity of ±$\frac{1}{2}$ LSB. The maximum conversion time is 12 μsec. Available full-scale ranges for analog input are 0 to +10 V, 0 to +5 V, −10 to +10 V, −5 to +5 V, and 0 to −10 V. Typical input impedance is about 5 kΩ. It is available for unipolar or bipolar parallel output and serial output. The convert command is initiated by an external positive pulse of at least 50-nsec width. The leading edge of the command pulse resets previous count and initiates conversion. Additional information on these modules will be found in the manufacturer's brochure.

8-6   Analog Multiplexers

In many applications it may be necessary to process digitally a number of analog signals. A common technique often used for such applications to minimize the overall system cost and size is to timeshare the digital processor among all input signals. To this end two different approaches are usually followed. In the first approach, as shown in Fig. 8-35(a), all analog inputs are time-division multiplexed to a single A/D converter whose output is then fed into the digital processor. In the second approach, as depicted in Fig. 8-35(b), each analog signal is first converted to its digital form by its own A/D converter and then the outputs of the converters are multiplexed digitally and fed into the processor.

A number of factors need to be considered before one approach of multiplexing is selected over the other. If the analog sources are far apart, the second

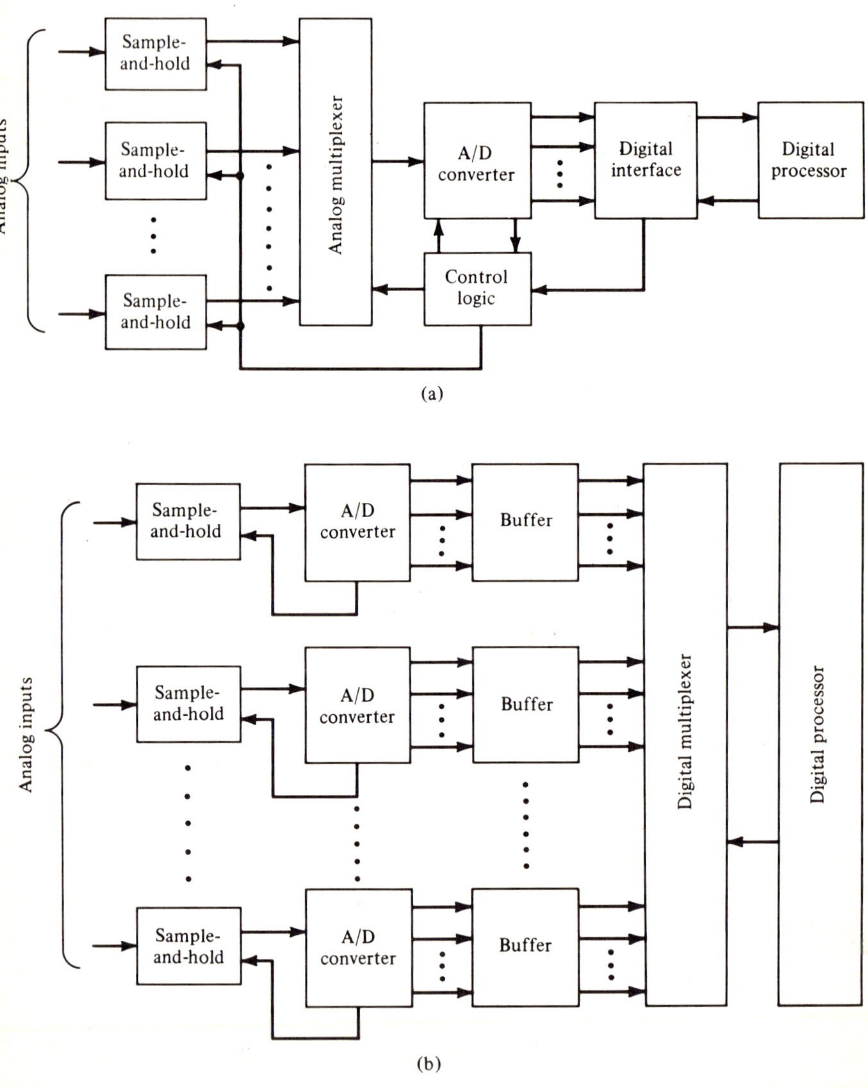

Figure 8-35   Two basic approaches to multichannel multiplexing.

approach is preferable as here the conversion of each analog signal to its digital form can be performed right near the source and the transmission of digital signals over great distances provide better noise immunity. If necessary some preprocessing of the digital signals can be made at the source end before transmission. The second approach allows the use of a slower speed (and hence cheaper) A/D converter in comparison to the first for a prescribed digital data rate to the common processor. Since, in general, a high-speed A/D converter is needed in implementing the first approach, the individual sample-and-hold units should have a reasonably small droop rate as they have to wait longer in the hold mode. If higher resolution analog-to-digital conversion is desired, then the first approach may be more economical as the cost of the converter increases rapidly with an increase in resolution.

Digital multiplexers have been described in Section 3-14. We review briefly below the characteristics of some of the commonly used analog multiplexers.

Multiplexer Circuits[11]

There are basically two types of analog multiplexers: high-level and low-level multiplexers. The former is used for multiplexing analog signals with voltage levels greater than 1 V (often going to 100 V) whereas the latter type is suitable for analog voltages less than 1 V (usually several millivolts). In all analog multiplexers, the analog inputs are sequentially connected to the common output amplifier with the aid of digitally controlled analog switches (Section 7-8).

One form of high-level multiplexer is sketched in Fig. 8-36. The input voltage range here depends on the type of analog switches being used. With solid-state switches, the voltage range is typically ± 20 V. For higher input voltages (up to ± 100 V), the inverting current switching multiplexer configuration of Fig. 8-37 is used. In this circuit, usually, a diode-limiting circuit is placed on the

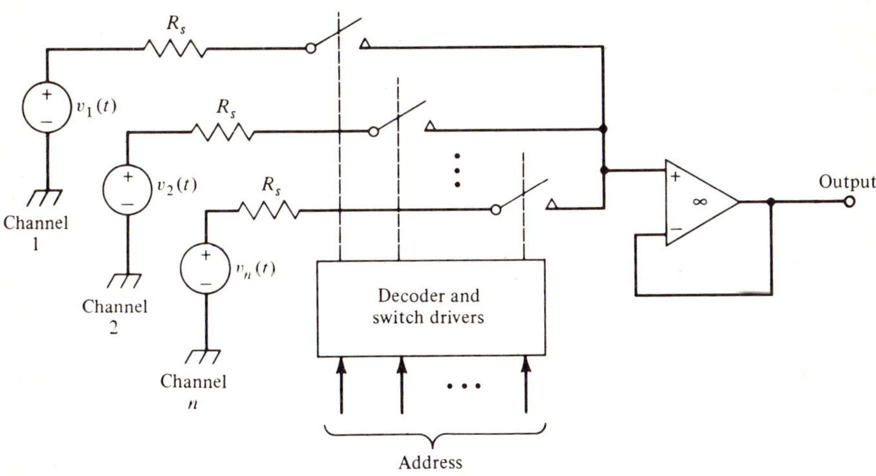

Figure 8-36   Voltage switching high-level multiplexer.

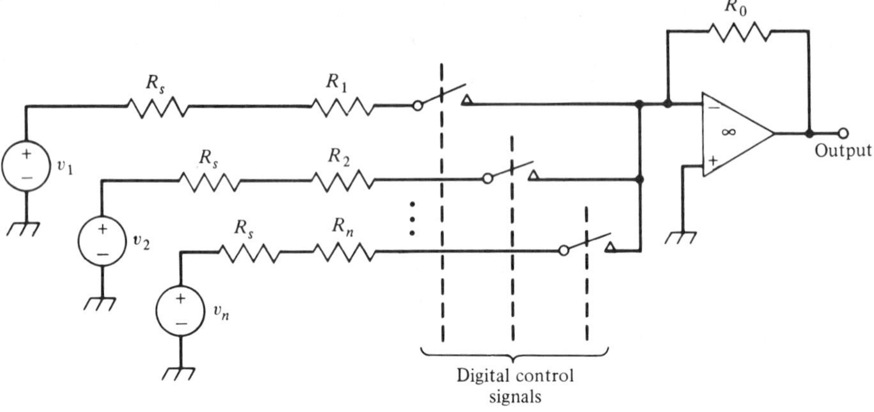

Figure 8-37 Current switching high-level multiplexer.

analog signal end of the switches. If the channel resistance R_k is chosen large enough to swamp the ON resistance of the switch, then the kth channel gain is $-R_o/(R_k + R_s)$ where R_s is the corresponding source resistance. The voltage developed across the switches are not very large as the output ends of the switches are at virtual ground and the other ends are protected by diode-limiting circuits.

Low-level analog inputs are in general multiplexed in differential modes to reduce the effect of any common-mode noise. A possible implementation scheme is sketched in Fig. 8-38. The differential amplifier at the output should have a very high differential mode gain and a very large common-mode rejection. Because of the low levels of the input analog signals, this type of circuit requires careful design which includes matching of the characteristics of the channels and switches for each input channel pair.

Characteristics of Practical Multiplexers[11, 14]

Since the multiplexer is formed by connecting together a number of analog switches, the parameters characterizing a practical multiplexer are essentially the same as those of an individual switch and those caused by the interconnection.

The analog channel parameters of the switch that are of interest are various unavoidable resistances affecting the dc and low-frequency accuracy of the ON channel gain, leakage currents associated with the analog switches causing voltage errors, and various switch capacitances causing high-frequency errors.

The resistances causing gain error are the *ON resistance* R_{ON} of the switch, the *input resistance* R_i of the output buffer amplifier, the *internal resistance* R_s of the analog source, and the *leakage resistance* R_l of the channel between the source and the switch. A model of the high-level voltage-switching multiplexer incorporating these resistances is shown in Fig. 8-39.

With careful design, the leakage resistance R_l, which is distributed along the whole input channel, can be made very large in comparison to source resistance

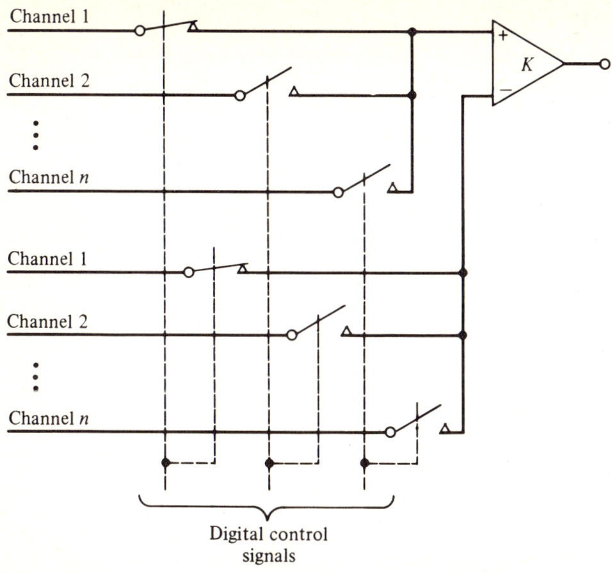

Figure 8-38 Differential-type low-level multiplexer.

R_s, and thus can be neglected for gain error analysis purposes. The dc gain of the ON channel is then given as

$$\frac{v_{o,k}}{v_k} = \frac{\mu_c R_i}{R_i + R_{\mathrm{ON}} + R_s} \tag{8-48}$$

where μ_c is the gain of the output amplifier. Usually a noninverting voltage amplifier (Fig. 6-3) designed using an operational amplifier with open-loop gain μ_o is used as the output amplifier. Then the input resistance R_i depends on the differential-input and common-mode resistances of the operational amplifier (Problem 8-20). Input resistance of the order of 100 MΩ can be obtained with most IC operational amplifiers connected as a voltage follower. Higher input

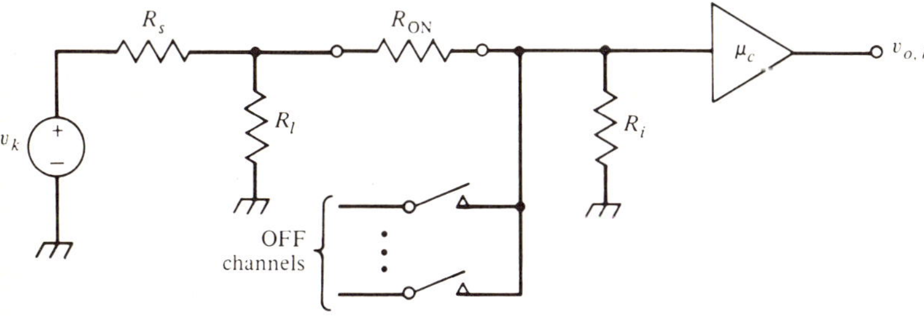

Figure 8-39 A model of a multiplexer incorporating additional resistances affecting gain.

resistances can be achieved with FET-input operational amplifiers. In most cases $R_i \gg R_s \gg R_{ON}$. Then it is seen from Eq. (8-48) that the effect of these resistances are negligible.

The unavoidable leakage currents present in most solid-state switches generate a voltage error as they also flow through the resistances indicated above. A model of the high-level voltage switching multiplexer incorporating the ON-channel leakage current $I_{ON}^{(1)}$ and the OFF-channel leakage current $I_{OFF}^{(j)}$ is depicted in Fig. 8-40 where we have assumed for simplicity that the analog channel 1 is ON and the remaining $n - 1$ channels are OFF. The error voltage v_{e1} developed at the input of the buffer amplifier is then given as

$$v_{e1} = \frac{R_i(R_s + R_{ON})}{R_i + R_s + R_{ON}} \cdot \left(I_{ON}^{(1)} + \sum_{j=2}^{n} I_{OFF}^{(j)} \right) \tag{8-49}$$

The expression for v_{e1} can be approximated as

$$v_{e1} \cong R_s[(n - 1) \cdot I_{OFF}^{(j)} + I_{ON}^{(1)}] \tag{8-50}$$

since in practice $R_i \gg R_s \gg R_{ON}$ and the OFF-channel leakage currents are approximately equal to each other.

The voltage and current offsets (and their drifts) of the output operational amplifier along with its common-mode properties also introduce additional error in the overall performance of the multiplexer unit and must be taken into account.

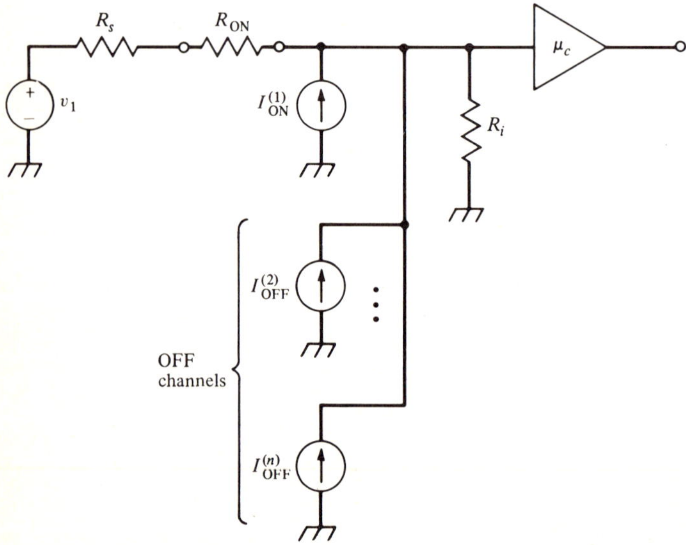

Figure 8-40   Model of a typical multiplexer incorporating the leakage current sources.

The switch capacitances and the finite switch turn-on and turn-off times affect the dynamic performance of the multiplexer circuit. For example, the wire connecting the switches to the input of the buffer amplifier (the output bus) sees a number of extraneous capacitances to ground. These are the channel output capacitances $C_{o,k}$, the switch capacitances $C_{a,k}$ seen across the OFF switches, and the amplifier input capacitance C_i (Fig. 8-41). As a result, when a new channel is switched in, the output voltage cannot immediately become equal to the new input source voltage. If C_{total} denotes the total capacitance seen to ground at the input of the amplifier and if the source resistance is neglected, then to settle within 0.1 percent of the input value, the time required is given by 6.9 $R_{\text{ON}} C_{\text{total}}$ sec. If the source resistance R_s is not negligible, then the effect of the channel capacitance to ground may become appreciable and lead up to a longer settling time than that obtained with $R_s = 0$. Furthermore the settling time of the output amplifier should also be taken into account in determining the overall settling time of the complete multiplexer unit.

The switch *turn-on time* here is defined as the delay between the time the digital address is applied to the decoder input to the time 90 percent value of the full-scale analog input appears at the output. Likewise, the switch *turn-off time* is the delay between the time when the digital address is turned off to the time at which the output is at 10 percent of the full-scale analog input. Note that the delay between the applications of digital address and that of the conversion command signal should be greater than the sum of the total settling time and the switching time.

The digital control line parameters of the switches such as the input decoder thresholds and the input decoder leakage currents should also be taken into account in evaluating the performance of a multiplexer package.

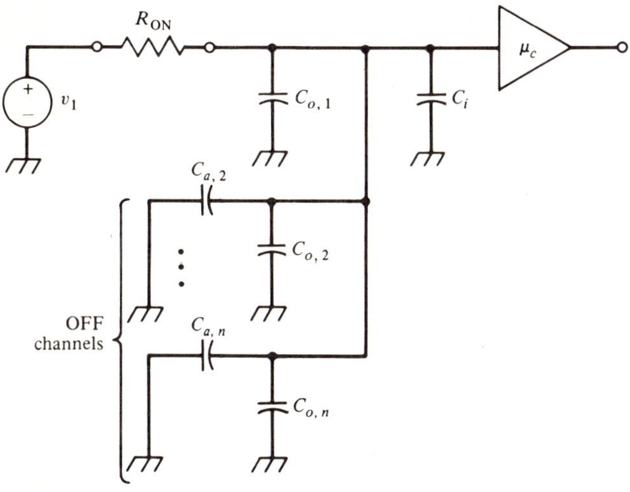

Figure 8-41   Parameters affecting settling time of a multiplexer.

Another switching parameter of interest is the *crosstalk*, which indicates the amount of coupling of the signals connected to the OFF channels to the ON channel. This appears due to the close proximity of the channels and the non-ideal properties of the switches. The crosstalk is given in decibels.

Submultiplexing[11]

A common way to multiplex a very large number of channels is by submultiplexing the input channels in groups. An implementation of a 16-channel multiplexing system using 4-input multiplexers by submultiplexing is sketched in Fig. 8-42. There are several advantages to submultiplexing. First, the leakage error is reduced considerably. For example, if the number of stages in implementing the submultiplexing arrangement using identical multiplexers is M, then the leakage error is given as (Problem 8-22)

$$v_e = R_s M[(\hat{n} - 1)I_{\text{OFF}}^{(j)} + I_{\text{ON}}^{(j)}] \tag{8-51}$$

where $\hat{n}$ is total number of input channels of any multiplexer. For example, if a 64-channel multiplexer system is implemented using 8-channel multiplexers in a two-stage realization, the leakage error is approximately about $\frac{1}{4}$ of that produced in a one-stage direct implementation.

Submultiplexing also reduces the output bus capacitance which in turn leads to a smaller overall settling time.

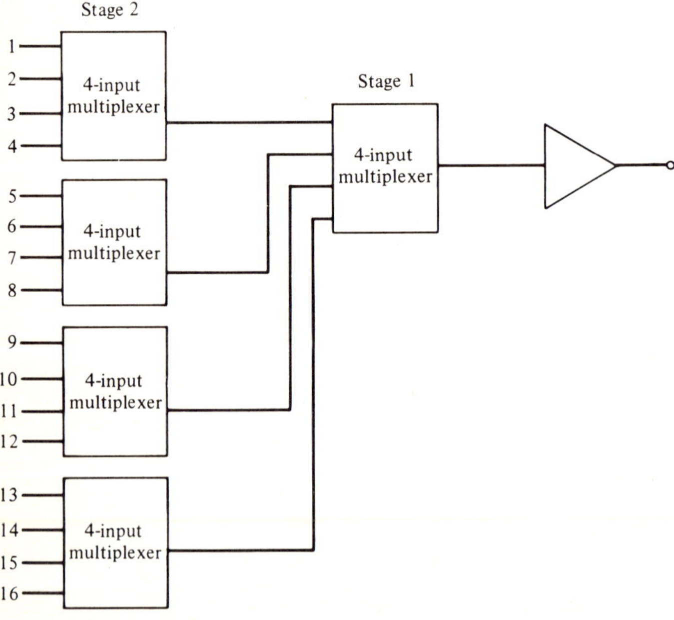

Figure 8-42   Illustration of submultiplexing.

8-7   Additional Applications

The most commonly used applications of the A/D converter, D/A converter, and the sample-and-hold module are in the digital processing of analog signals. In this section we outline several other applications of these circuits.

Maximum Amplitude Detector

The sample-and-hold module with the aid of an analog comparator can be used to determine the maximum amplitude of an analog waveform. The pertinent circuit arrangement is shown in Fig. 8-43. The sample-and-hold module is set to sample mode with the digital control signal at logical ONE and is in hold mode when the digital control signal is at logical ZERO. When the input analog voltage v_i becomes greater than the output v_o of the sample-and-hold module, the output of the comparator is at logical ONE causing the sample-and-hold to track the input. On the other hand, when the input v_i becomes less than the output v_o, the output of the comparator goes to the logical-ZERO state which puts the sample-and-hold circuit in the hold mode.

Digitally Controlled Scale Factor

A multiplying digital-to-analog converter (MDAC) can be considered as an amplifier with a digitally controlled gain. The analog input voltage v_{in} is connected to the reference input terminal of the MDAC and the output of the converter is then equal to $N \cdot v_o$ where N is the analog equivalent of the digital input signal. To obtain a scale factor which is the reciprocal of N, an MDAC can be connected in the feedback path of an operational amplifier as shown in Fig. 8-44 for a noninverting amplifier arrangement. The corresponding inverting configuration is given in Problem 8-23.

Function Generation

Any type of single-valued functional relationship between two analog variables v_1 and v_2 given as $v_2 = f(v_1)$ can be readily implemented using an A/D converter, a ROM, and a D/A converter. A straightforward implementation of the

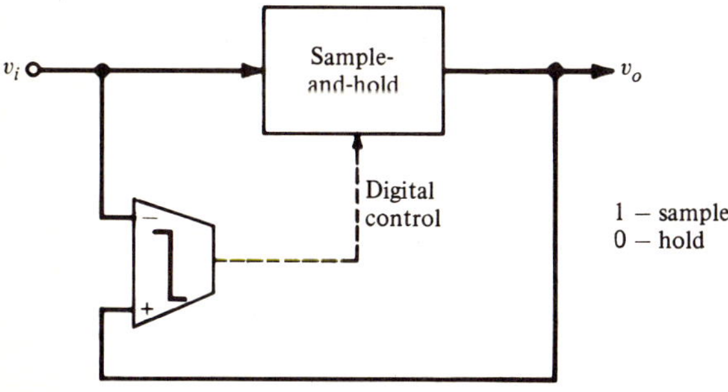

Figure 8-43   Maximum amplitude detector.

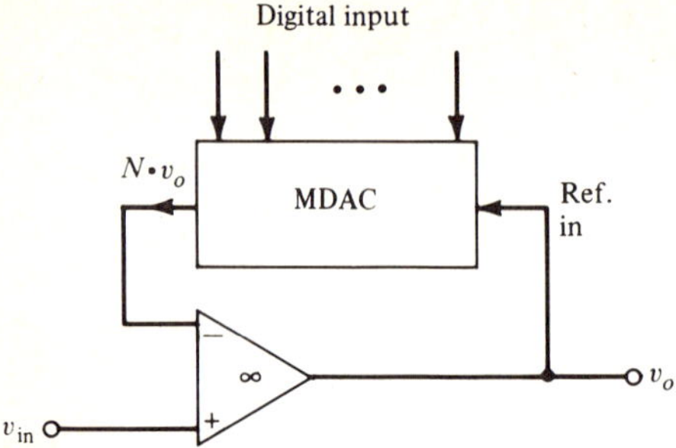

Figure 8-44   Circuit arrangement to obtain a scale-factor inversely proportional to a digital number.

functional relationship is as indicated in Fig. 8-45. It should be noted that such an approach can be used both for linear and nonlinear function generation. A number of preprogrammed ROMs realizing certain types of standard functions, such as trigonometric relationships, are available as off-the-shelf items. Other types of relationships can be programmed by the user with the aid of PROMs.

Waveform Generation

The D/A converter also finds applications in converting digitally generated waveforms into analog functions of time. The basic components in the digital generation of periodic waveforms are a variable frequency clock generator, a counter, and a D/A converter. To generate periodic waveforms with arbitrary shape, the output of the counter can be used to address a ROM whose output is then fed into a D/A converter. The output of the D/A converter is a staircase-type waveform and may also contain glitches. A smooth continuous waveform may be obtained, if necessary, by deglitching and filtering.

To generate a sawtooth waveform, the output of an up counter can be directly fed into a D/A converter as shown in Fig. 8-46. A typical output waveform is also in this figure. Generation of a triangular wave is accomplished with an up/down counter and additional logic circuits which make the counter to count down after it is full and to count up after it is empty. One possible realization of such a triangular waveform generator using the Fairchild 9366/74193 up/down counter is shown in Fig. 8-47 along with its output waveform. By feeding the

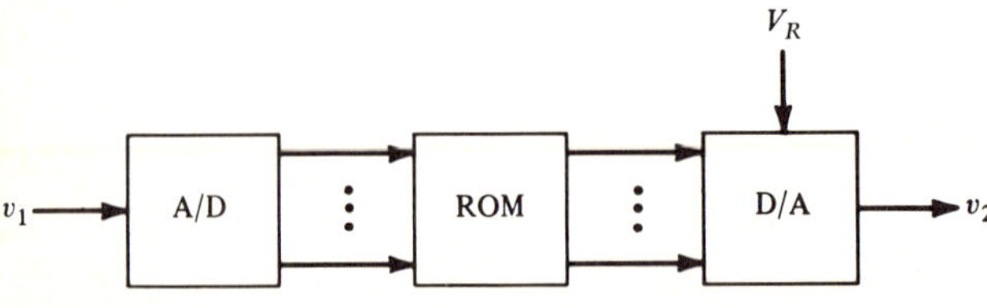

Figure 8-45   A scheme for arbitrary function generation.

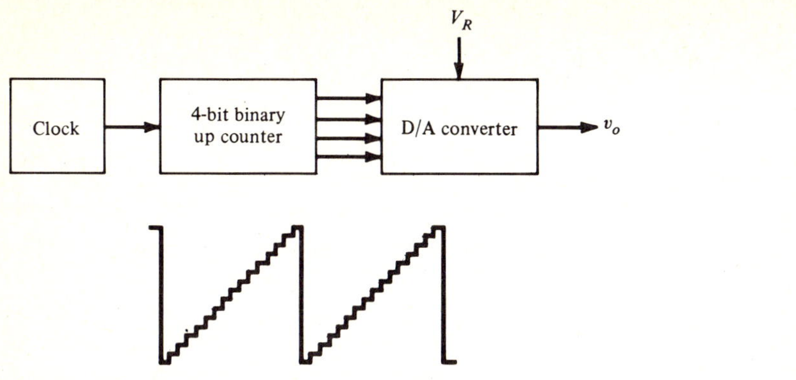

Figure 8-46   A saw-tooth wave generator.

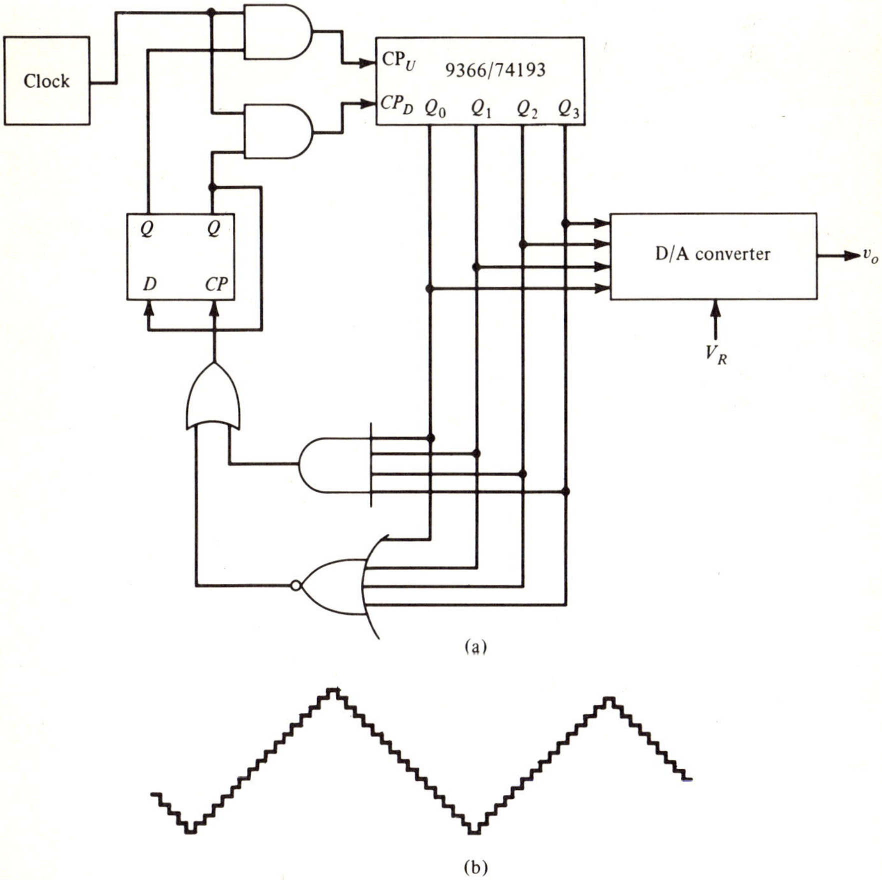

Figure 8-47   A triangular wave generator (a) block diagram, (b) output waveform.

output of the counter in the above circuit to a 1-quadrant sinusoidal ROM and then feeding the output of the ROM to a sign-magnitude-coded D/A converter along with the polarity information, a 4-bit quantized sine wave is generated at the output of the converter.

In all of the waveform generators discussed above, the number of bits of the counter determines the quantization levels in the output waveform. This accuracy can be increased by increasing the counter size. The period of the output waveform can be varied by varying the frequency of the clock.

8-8   Summary

Basic concepts of the analog-to-digital (A/D) and digital-to-analog (D/A) conversion process are introduced along with several conceptually simple types of converters. In order to minimize the conversion error, a time-varying analog signal should be sampled periodically and each sampled value held constant at the input of an A/D converter (Section 8-1). An analog signal containing a number of frequency components should be sampled at a rate no less than two samples per cycle of the highest-frequency component to avoid frequency aliasing errors (Section 8-1). The sampling-and-holding operation is performed by a sample-and-hold circuit. These circuits are described in Section 8-2 which also outlines the practical characteristics of these devices. Because of a finite number of bits in the digital representation of an analog signal, A/D conversion introduces quantization errors which are discussed in Section 8-3. Section 8-4 describes a number of specific D/A converters along with the characteristics of practical converters. In Section 8-5 the basic ideas behind several A/D converter design methods are outlined. The properties of practical A/D converters are also described here. Section 8-6 is concerned with analog multiplexer and their properties. Finally, several simple applications of the converters and sample-and-hold modules are described in Section 8-7.

References

1. D. F. Hoeschele, Jr., *Analog-to-Digital/Digital-to-Analog Conversion Techniques*, John Wiley & Sons, New York, 1968.
2. Staff, *Analog ↔ Digital Conversion Handbook*, Digital Equipment Corporation, Maynard, Mass., 1964.
3. B. P. Lathi, *Signals, Systems, and Communication*, John Wiley & Sons, New York, 1965, pp. 435–442.
4. H. V. Malmstadt and C. G. Enke, *Digital Electronics for Scientists*, W. A. Benjamin, New York, 1969, pp. 309–344.
5. Application Engineering Department, *Digital-to-Analog Converter Handbook*, 1st Ed., Hybrid Systems Corporation, Burlington, Mass., 1970.
6. J. T. Tou, *Digital and Sampled-data Control Systems*, McGraw-Hill Book Co., New York, 1959, pp. 69–92.
7. H. B. Aasnaes and T. J. Harrison, "Triple Play Speeds A–D Conversion," *Electronics*, Vol. 41, April 29, 1968, pp. 69–72.

8. M. R. Aaron and S. K. Mitra, "Synthesis of Resistive Digital-to-Analog Conversion Ladders with Fixed Positive Weights," *IEEE Trans. on Computers*, Vol. EC-16, June 1967, pp. 277–281.

9. M. R. Aaron and S. K. Mitra, "A Note on the Design of Digital-to-Analog Converters," *IEEE Trans. on Computers*, Vol. EC-16, October 1967, pp. 685–686.

10. H. Schmid, *Electronic Analog/Digital Converters*, Van Nostrand Reinhold, New York, 1970.

11. D. H. Sheingold, Ed., *Analog-Digital Conversion Handbook*, Analog Devices Inc., Norwood, Mass., 1972.

12. A. Barna and D. I. Porat, *Integrated Circuits in Digital Electronics*, John Wiley & Sons, New York, 1975.

13. "Digital-to-Analog and Analog-to-Digital Converters," Catalog No. PDS-233, April 1970, Burr-Brown Research Corp., Tucson, Ariz.

14. J. A. Connelly, Ed., *Analog Integrated Circuits*, John Wiley & Sons, New York, 1975.

15. E. Renschler, "Analog-to-Digital Conversion Techniques," Application Note AN-471, Motorola Semiconductor Products, Inc., Phoenix, Ariz.

16. "Mono DAC-01-6 Bit Monolithic D/A Voltage Converter," Advanced Data, Precision Monolithics Inc., Santa Clara, Calif., January 1971.

17. "Single Chip D/A Converter," *Electronic Products*, June 1970.

18. J. O. Bowers, "Modern Monolithic Modules for D-to-A Systems Applications," *Electronic Instrumentation Digest*, September 1970, pp. 8–12.

Problems

8-1 An audio signal is to be sampled for digital transmission. If the bandwidth of signal is 4 kHz, which of the following sampling periods are allowable: **(a)** 10 μsec, **(b)** 50 μsec, **(c)** 0.1 msec, **(d)** 125 μsec, **(e)** 0.2 msec.

8-2 An analog signal containing the following frequency components: 10 Hz, 100 Hz, 330 Hz, 1.2 kHz, and 2.3 kHz is passed through an ideal bandpass filter whose bandwidth is 200 Hz to 2 kHz. If the output of the bandpass filter is to be sampled for A/D conversion, what is the minimum sampling rate?

8-3 Derive Eq. (8-8).

8-4 Derive Eq. (8-10).

8-5 Derive Eq. (8-13).

8-6 Show analytically that if the two input currents of the amplifier in Fig. 8-8 are equal during the hold mode, the output voltage V_D will remain constant.

8-7 Modify the sample-and-hold circuit of Fig. 8-9 to have a controllable gain dependent on resistor values.

8-8 Analyze the operation of the sample-and-hold circuit of Fig. 8-48 and compare it with that of Fig. 8-9.

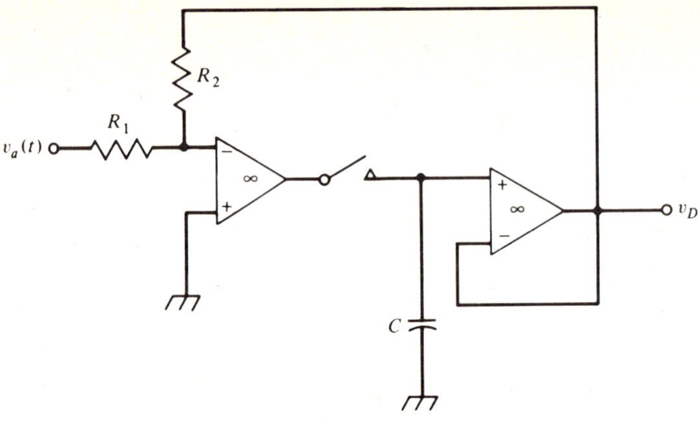

Figure 8-48

8-9 Analyze the operation of the inverting sample-and-hold circuit[4] of Fig. 8-49.

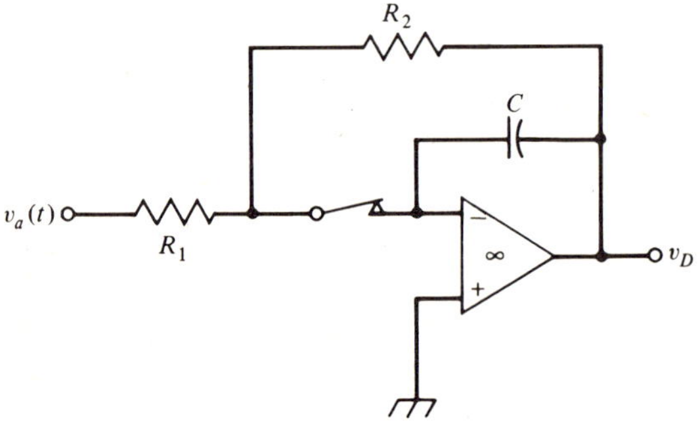

Figure 8-49

8-10 A sinusoidal analog voltage $7.5 \sin (200\pi t)$ is to be converted into a digital signal in natural binary code. The number of bits to be used is 4 with the leftmost bit as the sign bit. Plot the quantization error as a function of time for one period of the input. Assume the sampling rate is 2 kHz.

8-11 Show the circuit schematic of a 6-bit weighted-resistor D/A converter with positive gain.

8-12 Determine the spread in the resistor values of an n-bit weighted-resistor binary D/A converter.

8-13 Determine the spread in the resistor values of weighted-resistor D/A converter for a $4k$-bit digital input in 8-4-2-1 code.

8-14 Design a weighted-resistor D/A converter for a 7-bit positive digital input in **(a)** 5-2-1-1 code and **(b)** 3-3-2-1 code.

8-15 Analyze the 8-bit resistor-ladder BCD D/A converter of Fig. 8-22.

8-16 A modified form of a weighted-resistor BCD D/A converter is shown in Fig. 8-50 for an 8-bit positive BCD number in 8-4-2-1 code. Such a modification decreases the spread in resistor values with modest increase in the total number of resistors. Determine the values of the conductances G_a and G_b.

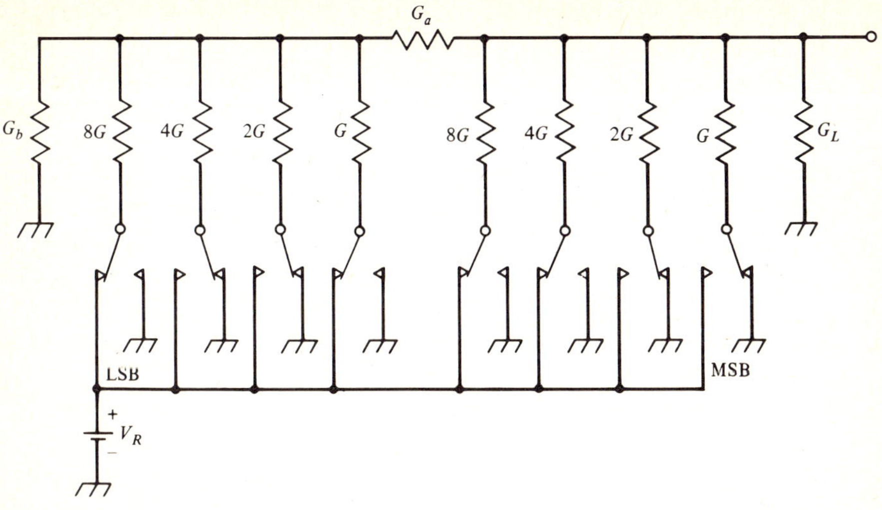

Figure 8-50

8-17 The D/A converter of Fig. 8-50 can be further modified to a resistor-ladder-type converter in a manner shown in Fig. 8-51. Analyze the operation of this circuit and determine the values of the conductances G_a and G_b.

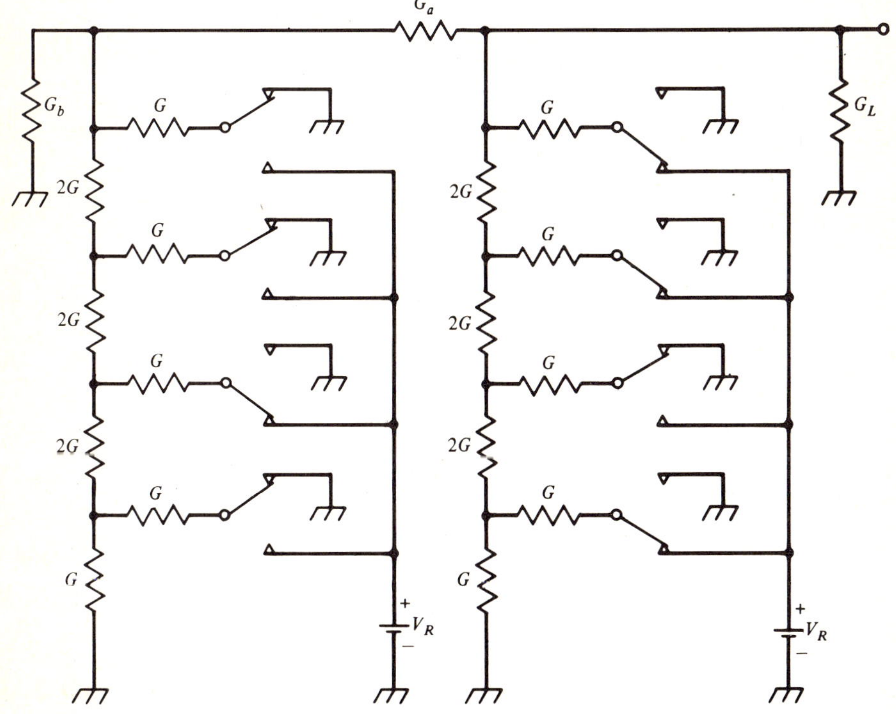

Figure 8-51

8-18 Design a 3-bit A/D converter for the conversion of positive analog inputs in the range 0–10 V using the direct conversion scheme.

8-19 Modify the A/D converter circuit designed in Problem 8-18 to convert a bipolar analog input in the range -5 to $+5$ V. The output of the A/D converter is required to be in the offset binary code.

8-20 Show that the input resistance R_i of the noninverting voltage amplifier of Fig. 6-3 is given as

$$ R_i = \left(\frac{\mu_c}{\mu_o R_d} + \frac{1}{R_{cm}} \right)^{-1} $$

where μ_c is the closed-loop gain of the amplifier circuit, μ_o is the open-loop gain of the operational amplifier, R_d and R_{cm} are the differential-input and the common-mode input resistances of the operational amplifier.

8-21 Derive Eq. (8-49).

8-22 Derive Eq. (8-51).

8-23 Analyze the operation of the circuit of Fig. 8-52.

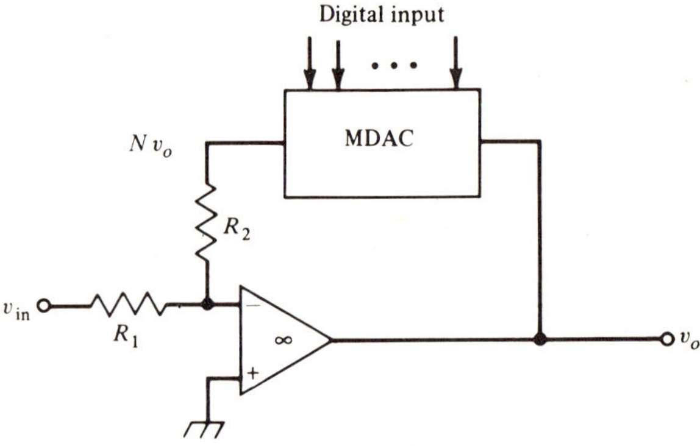

Figure 8-52

INDEX

Absolute-value circuit, 343, 345, 410
Accumulator, 199
Acquisition time, 427
ADC. *See* Analog-to-digital converter
Adder, 7, 111–115, 137, 141, 150–151
   carry-ahead, 114
   full-, 111–114, 139, 150
   half-, 111–113, 150
   modulo-2, 112
   parallel, 199
   ripple-carry, 112–113
   ROM-based, 141
   serial, 196–198
Aliasing error, 421
All-NAND realization, 149, 150
All-NOR realization, 149
Amplifier, 5. *See also* Operational
     amplifier; Voltage amplifier
   ac, 255, 311–313, 340, 408
   audio-frequency, 255–257
   bandwidth, 245, 254, 255, 258, 259, 266
   bridge, 281, 284–285, 336, 365
   broadband, 255
   buffer, 236, 324
   cascaded, 231, 235, 265, 266
   charge, 276–277
   chopper stabilized, 408
   core-memory sense, 387
   current, 229, 230, 239, 261, 335
   cut-off frequency, 254, 266
   dc, 255
   delay time, 246
   difference, 5, 232, 233, 282–285,
     335–336, 346
   differential. *See* difference
   differential input, 232, 233, 259, 336
   differential output, 232, 233
   error, 281

   feedback, 251, 328
   frequency limitation, 251
   frequency response, 242, 251–254
   high-frequency model, 253
   IC (monolithic), 230, 236, 241, 250,
     256, 257, 258
   ideal, 226, 228, 240, 260
   input limitations, 240–241
   input resistance, 241, 249, 263, 265
   instrumentation, 281
   intermediate frequency (i-f), 256
   narrow-band, 255
   negative resistance, 261
   nonideal characteristics, 240–247
   nonlinear distortion in, 243, 267
   output resistance, 241, 249, 264
   packages, 256–259
   potentio-metric, 336
   power, 230, 256
   practical, 240, 259, 260
   pre-, 236, 256
   pulse, 258, 259
   radio frequency (r-f), 255
   rise-time, 246
   settling-time, 246
   single-pole model, 252, 327–328
   slew rate, 246, 259, 266
   stability, 251, 324–329
   steady-state response, 243
   step response, 266, 245–246
   stereo, 257
   summing, 5, 263, 280–281, 284, 439
   terminated, 247–251
   time-domain behavior, 251, 260
   transconductance, 260–261, 264,
     277–279
   transfer characteristic, 235, 267
   transient response of, 245–247